Gas-Particle and Granular Flow Systems

Coupled Numerical Methods and Applications

Gas-Particle and Granular Flow Systems

Coupled Numerical Methods and Applications

Nan Gui
Institute of Nuclear and New Energy Technology
Collaborative Innovation Center of Advanced Nuclear Energy Technology
Key Laboratory of Advanced Reactor Engineering and Safety
Ministry of Education, Tsinghua University
Beijing, China

Shengyao Jiang
Institute of Nuclear and New Energy Technology
Collaborative Innovation Center of Advanced Nuclear Energy Technology
Key Laboratory of Advanced Reactor Engineering and Safety
Ministry of Education, Tsinghua University
Beijing, China

Jiyuan Tu
Institute of Nuclear and New Energy Technology
Collaborative Innovation Center of Advanced Nuclear Energy Technology
Key Laboratory of Advanced Reactor Engineering and Safety
Ministry of Education, Tsinghua University
Beijing, China

School of Engineering, RMIT University
Melbourne, VIC, Australia

Xingtuan Yang
Institute of Nuclear and New Energy Technology
Collaborative Innovation Center of Advanced Nuclear Energy Technology
Key Laboratory of Advanced Reactor Engineering and Safety
Ministry of Education, Tsinghua University
Beijing, China

ELSEVIER

Elsevier
Radarweg 29, PO Box 211, 1000 AE Amsterdam, Netherlands
The Boulevard, Langford Lane, Kidlington, Oxford OX5 1GB, United Kingdom
50 Hampshire Street, 5th Floor, Cambridge, MA 02139, United States

Copyright © 2020 Elsevier Inc. All rights reserved.

No part of this publication may be reproduced or transmitted in any form or by any means, electronic or mechanical, including photocopying, recording, or any information storage and retrieval system, without permission in writing from the publisher. Details on how to seek permission, further information about the Publisher's permissions policies and our arrangements with organizations such as the Copyright Clearance Center and the Copyright Licensing Agency, can be found at our website: www.elsevier.com/permissions.

This book and the individual contributions contained in it are protected under copyright by the Publisher (other than as may be noted herein).

Notices

Knowledge and best practice in this field are constantly changing. As new research and experience broaden our understanding, changes in research methods, professional practices, or medical treatment may become necessary.

Practitioners and researchers must always rely on their own experience and knowledge in evaluating and using any information, methods, compounds, or experiments described herein. In using such information or methods they should be mindful of their own safety and the safety of others, including parties for whom they have a professional responsibility.

To the fullest extent of the law, neither the Publisher nor the authors, contributors, or editors, assume any liability for any injury and/or damage to persons or property as a matter of products liability, negligence or otherwise, or from any use or operation of any methods, products, instructions, or ideas contained in the material herein.

Library of Congress Cataloging-in-Publication Data
A catalog record for this book is available from the Library of Congress

British Library Cataloguing-in-Publication Data
A catalogue record for this book is available from the British Library

ISBN: 978-0-12-816398-6

For information on all Elsevier publications
visit our website at https://www.elsevier.com/books-and-journals

Publisher: Susan Dennis
Acquisition Editor: Anneka Hess
Editorial Project Manager: Michael Lutz
Production Project Manager: Vignesh Tamil
Designer: Matthew Limbert

Typeset by VTeX

Contents

PART 1 THEORIES & MODELS

CHAPTER 1 Introduction to two-phase flow ... 3
- 1.1 Flow classifications ... 3
  - 1.1.1 Single-phase flow ... 3
  - 1.1.2 Gas-particle flow ... 6
  - 1.1.3 Granular flow ... 10
  - 1.1.4 Pebble flow ... 11
- 1.2 Flow regimes ... 14
  - 1.2.1 Dilute and dense flows ... 14
  - 1.2.2 Inertial and elastic flows ... 14
- 1.3 Numerical methods ... 15
  - 1.3.1 Methods for single-phase flow ... 15
  - 1.3.2 Methods for two-phase flow ... 17
- 1.4 Summary ... 20

CHAPTER 2 Discrete particle model ... 21
- 2.1 Spherical particle model ... 21
- 2.2 Generalized hard particle model (GHPM) ... 21
  - 2.2.1 Governing equations ... 21
  - 2.2.2 Consistency with the hard sphere model ... 29
  - 2.2.3 Numerical procedures for implementation ... 30
  - 2.2.4 GHPM model validation ... 32
  - 2.2.5 Application in a lifting hopper ... 36
  - 2.2.6 Application for the particle-wall collision ... 39
  - 2.2.7 Numerical procedure for multiple contacts ... 44
  - 2.2.8 Simulation test and experimental validation ... 45
  - 2.2.9 Application in the particle-wall collision ... 50
- 2.3 SIPHPM model ... 53
  - 2.3.1 Description of the model basis ... 53
  - 2.3.2 Governing equation ... 58
  - 2.3.3 Equation of motion ... 58
  - 2.3.4 Numerical procedures and techniques ... 59
  - 2.3.5 SIPHPM model validation ... 60
- 2.4 EHPM-DEM model ... 64
  - 2.4.1 Extended general hard particle model (EHPM) ... 64
  - 2.4.2 Extended discrete element method ... 67
  - 2.4.3 EHPM-DEM coupling strategy ... 69
  - 2.4.4 Collision detection strategy ... 70

2.4.5 Governing equations of motion 71
2.4.6 Demonstration and validation of EHPM-DEM 72
2.4.7 Simulation efficiency 76
2.5 Heat transfer extensions 80
2.5.1 Particle-particle conduction 80
2.6 Summary 82

CHAPTER 3 Coupled methods **85**
3.1 LES-DEM coupled methods 85
3.1.1 Governing equations 85
3.1.2 Discrete element method (DEM) 86
3.1.3 Approach A: conventional method 87
3.1.4 Approach B: smoothed void fraction method 88
3.2 DNS-DEM coupled methods 92
3.2.1 Immersed boundary method 92
3.2.2 Point-force method 96
3.3 LBM-DEM coupled methods 102
3.3.1 Recovery of governing equation 102
3.3.2 Coupled with heat transfer 111
3.3.3 Multiple schemes LBM-IBM-DEM 114
3.4 Summary 118

PART 2 APPLICATIONS

CHAPTER 4 Application in gas-particle flows **123**
4.1 Homogeneous turbulence 123
4.1.1 Governing equations 123
4.1.2 Collision rates and statistics 127
4.2 Planar jets 134
4.2.1 2D case with the heat transfer 134
4.2.2 3D case with the two-way coupling 144
4.3 Swirling jets 148
4.3.1 Vortex breakdown 148
4.3.2 Coherent oscillation 151
4.3.3 Particle-vortex interaction 158
4.3.4 Four-way coupling 168
4.4 Bubbling fluidized bed 173
4.4.1 3D bubbling fluidized testing bed 173
4.4.2 Pulsed fluidization 177
4.5 Spouted bed 186
4.5.1 CFD-DEM vs. SVFM-based fine LES-DEM 190
4.5.2 Particle phase behavior 192
4.5.3 Gas phase behavior 195
4.5.4 Additional remark 203
4.6 Summary 204

CHAPTER 5 Application in granular flows 207
5.1 Some functions 207
5.1.1 Evaluation functions of mixing degree 207
5.1.2 Evaluation functions of heat transfer degree 214
5.2 Circular drum mixers 217
5.2.1 Flow pattern and mixing evolution 217
5.2.2 Dimension analysis 218
5.2.3 Information entropy analysis 221
5.2.4 Heat conduction features 225
5.3 Wavy drum mixers 243
5.3.1 Wavy wall configuration 243
5.3.2 Analysis and prediction 245
5.3.3 Effects of phase velocity, wave number and amplitude . 247
5.3.4 Driven force analysis 256
5.3.5 Heat conduction features 261
5.4 Mixing of nonspherical particles 269
5.4.1 Polygonal particle mixing in 2D drum 270
5.4.2 Cubic particle mixing in one layer 287
5.4.3 Tetrahedral particle mixing in one layer tumbler 298
5.4.4 Cubic particle mixing in 3D cylinder 308
5.5 Hopper discharge 320
5.5.1 Geometric features on quasi-static discharge 320
5.5.2 Shaken discharge 329
5.5.3 Discharge of 2D polygonal particles 347
5.6 Summary 357

Bibliography 359
Index 373

PART 1

Theories & models

CHAPTER

Introduction to two-phase flow

1

1.1 Flow classifications

In physics and engineering, a flow system is usually characterized by the movements of the material medium, which can be continuous or discrete, e.g. solids, liquids, gases, and plasmas. For example, particle flow or granular flow is mainly composed of solid materials, such as sands, grain, gravels or rocks. A fluid flow usually refers to the flow motion of liquids, gases, or melted metals, etc.

To characterize the common and essential features of different materials in a flow system, the concept of 'phase' is frequently utilized. A phase is a physically distinctive form of a substance, such as the solid, liquid, and gaseous states of ordinary matter – also referred to as a macroscopic state [1]. Therefore, a flow composed of one state of matter is called a single-phase flow, and a flow system with two or more states of matter is called a two-phase or multiphase flow. It is possible for a system of a single material to be considered two-phase or multiphase flow, if at least two distinct dynamical properties of the same material are displayed. For example, a flow consisting of particles of a single material may be regarded as a two-phase system if there are fine light particles and heavy inertia particles as two components with clearly different dynamical behavior. In conclusion, a phase is a continuous or discretized system characterized by the same physical properties, mechanical states, and similar dynamical behaviors.

1.1.1 Single-phase flow

In this section, the challenges in modeling of single-phase flows are discussed. One of the well-known fundamental challenges, which have puzzled the scientists and engineers over hundreds of years, is turbulence. Turbulence is frequently described as 'eddies' of varying scales defined in the Kolmogorov's theory in 1941. Typical features include irregularity, diffusivity, rotationality and dissipation. Conventional investigation of turbulence was mainly done in the Eulerian framework [2], where the fluid motion is depicted by the nonlinear three dimensional Navier–Stokes equation. It cannot be solved analytically, and current computational algorithms are not supported by computational technology to solve problems of practical scale and conditions, in particular at high Reynolds numbers. Recently, Liu et al. [3] proposed to reconsider the definition of vortex and vorticity to characterize the wall turbulence. They have tried to clarify the onset mechanism and structure of the turbulence and stated that the vor-

Gas-Particle and Granular Flow Systems. https://doi.org/10.1016/B978-0-12-816398-6.00009-2
Copyright © 2020 Elsevier Inc. All rights reserved.

ticity overtakes deformation in vortex [4–6]. This implies that vorticity and vortex are completely different concepts. In general, direction of the vorticity is different from that of the vortex in three-dimensional vortical flows. The vorticity vector ($\nabla \times \boldsymbol{V}$) can be decomposed into a pure-rotational component ($\boldsymbol{R}$, the vortex vector named 'Rortex' or 'Liutex') and a nonrotational asymmetrical-deformation component ($\boldsymbol{S}$). They declared that, rather than vortex dynamics, R-dynamics (Rortex dynamics or Liutex dynamics) should be used to investigate the mechanisms of turbulence. Based on the new vortex identification method, they revisited the Kolmogorov hypothesis, and declared that turbulent flow has a unique and deterministic solution, which is not governed by vortex breakdown, but vortex build-up. This work provides a unique insight on the mechanism of turbulence. In addition, researchers have also tried to re-examine turbulence under the Lagrangian framework and great efforts were made to understand the kinematic properties of a fluid particle, such as acceleration and velocity [7–13].

Single-phase turbulence has wide applications in life. The turbulent jet is one of the typical examples of turbulent flows with various applications in industry (e.g., free jet, impinging jet, cross jet, slot jet and jet array) [14]. The vortex structure, mixing layers and heat transfer are complex yet very important intrinsic characteristics in jet flows. For example, in the nonisothermal jets, the temperature can be approximately viewed as scalars driven by turbulence with self-diffusive characteristics. Turbulence is intrinsically chaotic or fractal. The fractal measure arises from the fact that the underlying physics of scalar turbulence seems to be closely related to the self-similar multiplicative fragmentation processes. The mixing and heat transfer interfaces are always fractal with a noninteger Hausdorff dimension, which implies the stratified self-similar structure. Fractal dimension could be very important to understand the turbulence. For industrial applications, the heat transfer interface of a large fractal dimension may be beneficial as it can enhance heat transfer efficiency. It is clearly seen that the effect of fluid properties and characteristics of boundary conditions could be very important to enhance mixing and heat transfer, as they may affect the inner interfaces of heat transfer.

A jet flow imparted with a rotational motion about its axis is called a swirling flow (Fig. 1.1). Swirling flows are frequently encountered in nature, such as tornadoes, typhoons, and ocean currents, which have been studied extensively over the past decades. The vortex breakdown, related to the coherent structure of the swirling flows, is considered the most important phenomenon, to which a lot of research has been devoted [15–22]. The most remarkable feature of the vortex breakdown is the abrupt change of the vortex structure with pronounced retardation in the axial flow. Correspondingly, this results in a divergence of the stream surface along the axis [23]. Thus, vortex breakdown can be depicted by three main characteristics: formation of an internal stagnation point along the axis, reversing flows in a limited region, and the sudden expansion of vortex core [24]. The vortex breakdown is considered a major coherent structure in swirling jets, as the well-organized fluid motion is at scales comparable to the flow. However in general, there is no accepted definition for the coherent structure. One attempt is given by Robinson [25] saying that the coherent

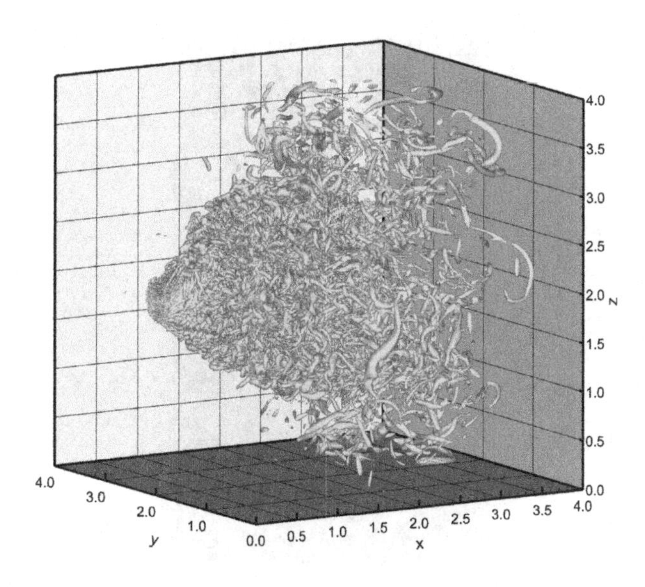

FIGURE 1.1

The single-phase turbulence in the Eulerian frame: swirling flow.

structure is a region over which at least one fundamental flow variable (velocity component, density, etc.) exhibits significant correlation with itself or another one, in time and space significantly larger than the smallest scales of the flow in the region. Besides swirling flow, the vortex breakdown is also very important in aerodynamics and combustion systems. For example, in aeronautics, the vortex breakdown can cause unsteadiness and poor control of the aircraft [26].

In addition, it is necessary to mention that turbulence is not just random as it appears. Instead, it has intrinsic coherent structures, which occur in a wide variety of turbulent flows. It is hoped that the work on coherent structures could lead to efficient methods for prediction of overall features of turbulent flows [27]. The coherent structure in swirling flows has been extensively studied [28], in areas of spatial structure [29], swirling combustion [30] and instability [31]. In particular, for power plant using swirling combustors, the vortex breakdown is extensively studied to stabilize the flame, as the air and fuel drastically mixes in the breakdown region in a result of the there-circulation toward the stagnation point along the jet axis. With regard to the swirl instability, for example, Shtork et al. [28] proposed a distinct modulation of the combustion process by using hydrodynamic instability. Unsteadiness of the flow is caused by the processing of the left-handed spiral (helical) vortical structure. Temporal frequency of the helical mode can be modulated by superposition of its pure rotation and axial translation, weighted with coefficients, which are the azimuthal and axial wavenumbers, respectively.

FIGURE 1.2

The gas-particle flow depicted by Eulerian–Lagrangian approach.

1.1.2 Gas-particle flow

In gas-particle flows, the particle is a small object characterized by finite size, volume and mass (Fig. 1.2). Material of the particle is not limited to solid, for example, a liquid droplet is also viewed as a particle. Usually size of the particle has a finite range that excludes the subatomic scaled particles, such as nucleons and electrons. Microscopic particles of atoms and molecules are also excluded. Gas-particle flows concern most of particles at macroscopic scale, such as powders and granular materials. Therefore, classic mechanics (Newton's laws) can be used to describe the motion of a discrete particle. The gas-solid flow is one type of two-phase flow which frequently appears in natural phenomena and industrial processes. For example, the formation of sand and dust storms, rainfall and avalanche in nature, the process of boiling, flotation, fermentation, coating, spray drying, and particle transport and mixing in pneumatic conveying and powder mixing devices.

In gas-particle flows, the fluid-to-particle, particle-to-particle, and particle-to-wall interactions are the three critical fundamental aspects need to be carefully considered. The forces and torques imposed by fluid-to-particle, and the feedback forces from particle-to-fluid (including the interfacial distributions) are the key focuses of fluid-particle interactions. Accordingly, numerical investigations are classified based on the type of forces considered. If only the fluid-to-particle forces are considered, it is called the one-way coupling approach. When both the fluid-to-particle and particle-to-fluid interactions are considered, it is called the two-way coupling approach. In addition, the particle-to-particle interaction concerns the collision (contact), frictional forces and torques experienced between particles, where the effects of influencing factors such as particle size, shape, density, roughness, etc. are considered. When the particle-to-particle, fluid-to-particle and particle-to-fluid interactions are considered, the method is called four-way coupling. Such an example is the gas-solid flow of sufficiently high particle concentrations (e.g., the dense flow region in fluidized beds), where particle behaviors are not only influenced by the aerodynamic transport of turbulence, but also significantly affected by the inter-particle collisions [32]. In addition, the particle-wall interaction should, at least, include the driving forces from the walls (either stationary or moving), and the effects of wall geometry and roughness should also be considered.

In two-way coupling of gas-particle flow, the research has been mainly focused on two aspects: (1) the modification of turbulence [33–35]; and (2) the characteristics of particle behavior (e.g., particle distribution, dispersion, clustering and coagulation) [36–38]. The main characteristics of particle-laden flow are governed by groups of key parameters, and some of them are elaborated here:

- Size (including particle diameter and the characteristic sizes of the turbulence):
  - Integral length scale: $l_e = \frac{1}{\langle u'u' \rangle} \int_0^\infty \langle u'u'(r) \rangle dr$, where r is the distance between two measurements, and u' is the fluctuation velocity in the same direction. It should be noted that the largest scale always correlates to the most energetic eddies which may obtain energy from the main flow. Therefore, the integral length scale l_e, defined in terms of the normalized two-point flow velocity correlation, is an indication of the largest scales of eddies and velocity fluctuation, which implies a low frequency in the energy spectrum.
  - Taylor microscale λ, corresponding to the inertial subrange, is the intermediate scale between the largest and the smallest eddies. Taylor microscale eddy does not dissipate but transport the energy from the largest to the smallest eddies.
  - Kolmogorov length scale η: it is the smallest scale in turbulence spectrum which corresponds to the range in the viscous sublayer, where energy input from the inertial subrange is balanced with viscous dissipation. These small scale eddies, usually of high frequency, are always considered homogeneous and isotropic.
  - Mean free path of particle l_p: it is the mean path length of the free motion of particles. l_p may start from a collision event and end at the next collision event. According to the kinetic theory, l_p may be affected by the local concentration and velocity of particles, as well as the particle sizes.
- Time:
  - Particle response time τ_p: driven by the fluid-to-particle force, a particle may respond to fluid flow in short or long time duration depending on its physical properties and the interaction of the particle to local turbulence. Usually, the velocity difference between the fluid flow and particle motion decays exponentially, e.g. in the form of $Ce^{-\tau/\tau_p}$ where τ_p is the characteristic time scale of the dynamic response of particle in the fluid, or the relaxation time. It is seen that the relaxation time τ_p, representing the time lapse for the difference of the local velocities between fluid and particle phases, decreases to e^{-1} (herein e is the Napierian base) of the initial value.

- Mean characteristic collision time τ_c: in a stationary or quasi-stationary gas-particle flow, τ_c is the mean characteristic time interval of the free motion for a particle between two consecutive collisions. It is the mean time interval for a particle to move freely starting from prior collision with a partner to the next collision with a different partner.
- The characteristic time scale of fluid turbulence τ_f: turbulence integral time scale can be defined in a similar manner as that of the integral length scale l_e by $\tau_f = \frac{1}{\langle u'u' \rangle} \int_0^\infty \langle u'u'(t) \rangle dt$. This corresponds to the characteristic time of large eddies. In most engineering applications, the characteristic velocity is always determined by the practical inflow boundary conditions, therefore, τ_f can also be derived from the ratio of the characteristic length l_e and characteristic velocity.

- Velocity: including the relative velocity v_r between particles, and the relative velocity between particles and the fluid $(\boldsymbol{v}_p - \boldsymbol{u}_f)$. Frequently, a derived characteristic velocity, such as $v_c = l_e/\tau_f$ or $v_c = Re \cdot \nu/d_p$ is included in the analysis depending on the particular applications.
- Density: including the mass or volumetric loading $\dot{m}_p$ and $\dot{V}_p$, the particle density ρ_p, the fluids density ρ_g, and the bulk density ρ_p^B, which is defined as the ratio of the mass of particles to the volume of the gas-particle mixture.
- Temperature T_c: initial temperature difference is usually used as the characteristic temperature in heat transfer applications.

As the typical applications of gas-particle flows seldom involve electromagnetic fields, or are concerned with the molecular scale variables, the aforementioned characteristic variables are sufficient to describe the most common cases.

As a convention, dimensionless forms of above-mentioned parameters are commonly used to characterize the gas-solid flow, e.g., d_p/l_e, Reynolds number $Re_\lambda = \langle u \rangle \lambda/\nu$, particle Reynolds number $Re_p = |\boldsymbol{v}_p - \boldsymbol{u}_f| d_p/\nu$, mass loading $m_l = \dot{m}_p/\dot{m}_g$, and Stokes number $St = \tau_p/\tau_f$. Among such examples, Re is a key parameter for characterizing the fundamental features of the turbulence, and consequentially it can be used to estimate the needed computer capacity for capturing all scales of the eddies in direct numerical simulation ($\sim Re^{9/4}$ for spatial resolution and $\sim Re^{3/4}$ for temporal resolution). For the particle dynamics in turbulence, Stokes number is defined as the ratio of the particle response time τ_p to the turbulence characteristic time scale τ_f. When $St \ll 1$, particle responds to the local turbulence instantaneously so that it can be regarded as perfect tracer. On the other hand, when $St \gg 1$, particle is hardly driven by the fluid so that its motion is almost independent of that of the turbulence.

Understanding the modulation of turbulence by particle-vortex interaction is of both practical and theoretical interest. The mechanism of turbulence modulation is very complicated as it is related to the various aspects of particles, such as particle inertia [39,40], finite volume or wake effect [41], mass loading and concentration [42,43], and inter-particle spacing [44]. The results of turbulence modification may be different from one case to another; therefore, a large number of two-phase flows are studied to gain knowledge of turbulence modulation under various conditions (e.g., pipe flow [45], mixing layer [46], rotating channel flow [47], and isotropic turbulence flow [35,48]). With regard to the dimensionless parameter for evaluating turbulence modulation, frequently d_p/l_e [33] or Re_p [49] is used. For example, Gore and Crowe [33] suggested to use d_p/l_e, the ratio of particle diameter to the integral length scale (the most energetic eddy), for the demarcation of the turbulence, where augmentation ($d_p/l_e \geq 0.1$) or attenuation ($d_p/l_e < 0.1$) of the turbulence flow by suspended particles can be benchmarked. Using a different approach, Tanaka and Eaton [50] proposed a novel dimensionless number, i.e., the particle moment number *Pa*, to classify the turbulence modulation. The *Re-Pa* relationship was obtained by examining a set of 80 experimental measurements, and this relationship was clustered into three groups representing the attenuation, augmentation and transition regions with two critical particle momentum numbers.

Besides particle properties and fluid turbulence, inter-particle collision is also of great importance and needs to be carefully treated. Many of the early works [51, 36,34,35] assumed that the particle diameter is comparable to or less than the Kolmogorov length scale η. In this context and assuming dilute condition, inter-particle collision and the particle size effects are usually neglected. Thus majority of the early researches neglected the collision dynamics. To extend the regime of dilute gas-particle flow to intermediate dense flow, particle-particle collision should be considered, and appropriate collision model is needed. Tsuji [52] proposed a model to describe the collision of rigid granular particles based on the principle of conservation of the linear and angular momenta. The energy loss in collision was incorporated through a restitution coefficient e. This model is called the hard sphere model (HSM), which can be used to simulate the particle motion and collision deterministically. Another approach, the discrete element method (DEM) – a soft sphere model (SSM) [53] – treats the particle-to-particle collision based upon particle kinetics. The soft sphere model includes three fundamental inter-particle collision mechanisms: elastic collision with viscous damping, inter-particle friction, and sliding trends. These models, together with suitable modifications and extensions, will be discussed extensively in this book.

When the separation of particle centers is small compared to the smallest eddies in the fluid, and at the same time, particles follow the fluid motion completely, then particle collision rates (between small particles in turbulent fluids) approximately depend on the dimension of particles, rate of energy dissipation and the kinematic viscosity of the fluid [54]. On the other hand, if particle-turbulence velocities are independent, in analogy to the kinetic theory, particle collision rate in absence of the external force should follow the suggestions of [55]. The dependence of inter-particle collision on turbulence and particle properties was studied through direct numerical simulation of heavy particle suspension in turbulent flow field [37]. By correlating the collision frequency over several eddy turnover times, two complicated yet well-known effects of intermediate Stokes number on particles were identified, i.e. (1) particles tend to collect in regions of low vorticity (high strain) due to the centrifugal effect; (2) particle pairs are not as strongly correlated with each other, which leads to an increase of relative velocity. These two effects cause the collision rates to increase locally.

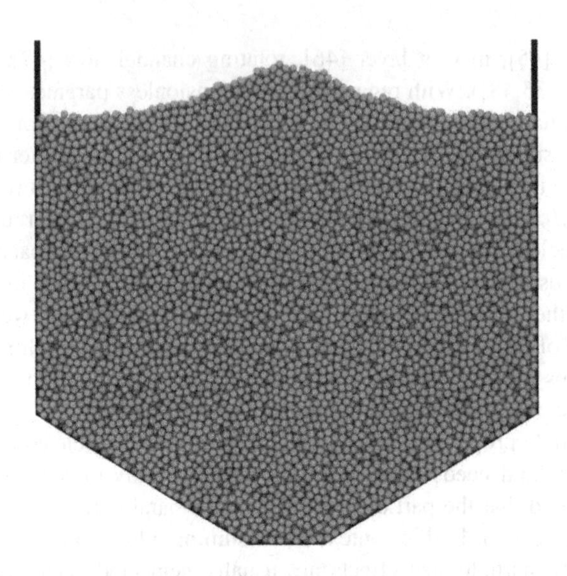

FIGURE 1.3

The granular bed.

1.1.3 Granular flow

A granular system is composed of a large number of granules (e.g. the packed granules in Fig. 1.3), which are abundant in dissipative interactions between particles due to deformation and friction in body contacts. In such systems, the random Brownian motion of particles is always omitted as well as the flow of interstitial gas or liquid (except for the forces caused by them, e.g. the liquid-bridge force). With wide engineering applications, many scientists and engineers have been interested in granular systems, and great progress in fundamentals and the underlying mechanisms have been achieved.

Although generating, transporting and stocking of granules can be traced back to the era of ancient human civilizations, the scientific study of granular flow started from the 18th century when the Coulomb's friction law was proposed. Up to date, the mechanism of granular flow is still poorly understood. It remains one of the central challenges for scientists [56]. Because of the incomplete knowledge, major framework is on developing the constitutive models by using phenomenological parameters in the stress-strain relationship of the granular materials, which is a key practice in engineering applications. As a result, many studies have been focused on the details of the striking mechanics of granules, such as nonlinear behaviors, self-organization, force chain structure, and arch formation.

In engineering, the silos, hoppers, and drums are the most common devices for processing of granular material. Such examples include the mixing, segregating, transporting processes inside a hopper in powder engineering, the crushing, grinding, and blasting processes in mineral and mining engineering, the feeding and coating processes in pharmaceutical engineering, and the dispersion and diffusion processes in nuclear engineering. Another example is the granular flow driven by gravity widely used in agricultural hoppers due to economy, reliability and efficiency. The particle flowability has significant impact on the processes of transportation, storage, and mixing of granular materials. As a result, particle flow dynamics in hoppers have been extensively studied over the past decades. Such topics include the macroscopic empirical relation between the discharge rate and hopper geometry [57–59], the relationship between the discharge rate and the interaction [60], size [61] or shape of particles [62,63], the velocity profiles [64,65] and the jamming mechanism of granular flows [66], the segregation [67,68] and the flow patterns [69,70].

Both experimental and numerical investigations have been conducted to reveal the underlying physics of granular flow in silos (e.g., silos of different shapes [71], the friction effect on flow patterns [72,73], the force analysis with clogging arches [74] and the prediction for discharge rates [75,76]). For the discharge rate, one of the famous correlations was proposed by Beverloo et al. [57] as follows:

$$\dot{m} = C\rho g^{1/2}(D_o - kd)^{2.5} \tag{1.1}$$

where ρ is the bulk density of granular material, g is the gravity acceleration, D_o is the orifice diameter, and d is the particle diameter. The discharge rate is also shown to increase with increasing of the normal stress or solid fraction (flow regime I) before achieving the maximum discharge rate when the flow is choked (flow regime II). For further increase of the solid fraction, the discharge rate becomes independent of the normal stress or the height of the granular materials (flow regime III) [77].

Rotating tumblers are currently employed by chemical industries in a wide range of physical processes, including size reduction, waste reclamation, agglomeration, solid mixing, drying, heating, cooling. The popular application of rotating tumblers is also due to its ability to handle a variety of feedstocks (from slurries to granular materials), and to operate in distinct environments [78]. In addition, particle mixing and segregation in various types of mixers have attracted the interests of scientists and engineers for several decades. Although great progress has been made, further understanding of the fundamentals and underlying mechanisms of the mixing, segregation and heat transfer of granular particles remains in great need. In particular, characteristics of the mixing interface, the microscopic mixing degree and the macroscopic mixing level in combination with the heat transfer of granular materials are still not clearly understood. It is also challenging to specify clear criteria for the design, manufacturing, and operation of the rotating tumblers in the most efficient way under various conditions such as at given energy input. Furthermore, limited empirical knowledge is available on the effect of walls, inner geometries, and fins for optimized operation of mixing, stirring and heat transferring of particles.

1.1.4 Pebble flow

The dense flow of heavy inertia particles with very low velocity is called a pebble flow in nuclear engineering (Fig. 1.4A). The pebbles (Fig. 1.4B) are so large that

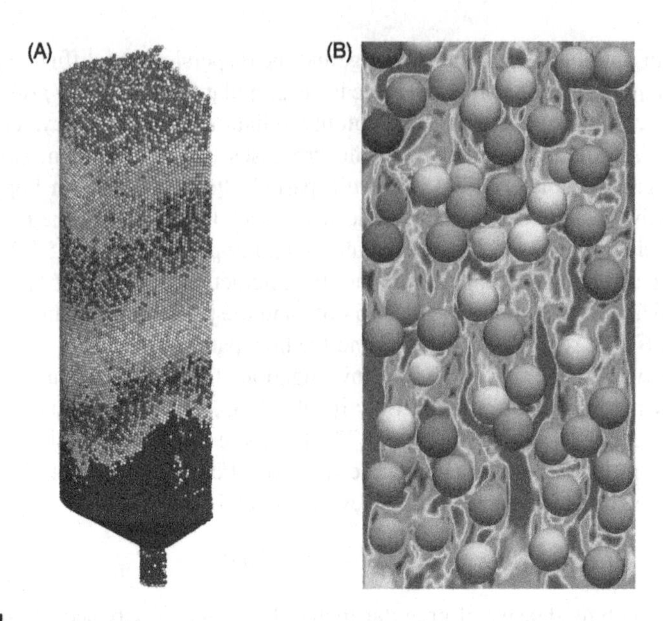

FIGURE 1.4

(A) The pebble bed. (B) Interstitial flow over pebbles.

(1) voids between them permit fluids to pass through at a relatively small pressure drop or low friction loss; (2) drag forces of the particles from the interstitial gas flow are negligible compared to the gravitational force of the particles; (3) only slow motion of the pebbles is allowed in practical applications to reduce/avoid impact, vibration and damage to container and the transporting devices; (4) in particular, in nuclear engineering, flow rate of the pebbles is so slow that the dispersion of fuel pebbles in nuclear reactions should be predictable and controllable.

The pebble bed reactor is becoming the mainstream solution for high temperature gas-cooled reactor (HTGR). The pebble bed (HTGR) is regarded as a promising option for Generation IV advanced reactors. Its excellent performance in inherent safety, modularity, market flexibility, broad applications, and short construction period, has drawn great attention. Other advantages, such as modularized building, high temperature heat process and nonstop reloading and discharging, are also impressive compared with current running Generation II and Generation III reactors [79,80].

Core of the pebble bed HTGR reactor is filled with a large quantity of mono-sized particles of tristructural isotropic (TRISO) fuel, which are moving in-and-out of the reactor recirculatorily. The reactor core of pebble-bed configuration has been used by many facilities, such as PBMR in South Africa [81,82], MPBR in United States [83] and the prototype reactor – AVR in Germany [84,79]. For research purpose, a small pebble-bed high temperature gas-cooled reactor of 10 MW (HTR-10) was built in 2000 at the Institute of Nuclear and New Energy Technology (INET) of Tsinghua university [85]. Currently, a new high temperature reactor pebble-module (HTR-PM) of 2×250 MW is under construction in Shandong province, China [86]. It will be used to investigate and validate the economic features of HTGR while maintaining the inherent safety.

Driven by gravity, pebble flow in the pebble bed reactor is unique and poses considerable challenge to theorists due to body interactions and nonthermal fluctuations of large inertia particles [87]. The discrete graphite coated fuel pebbles have a uniform diameter of several centimeters, much larger than the conventional powders. Thousands of fuel pebbles, loaded from the top and discharged from the bottom, are flowing through the core driven by gravity in the pebble bed. In the actual reactor, pebbles flow at extremely slow rate, about 0.1–1 mm/h, corresponding to a circulating rate of 400 pebbles per day. Therefore, pebble flow in the reactor is quasi-static, with distinctive characteristics from the rapid particle flow. The pebbles remain stationary most of time, however small or large scale pebble avalanches could occur, which are random in nature and cannot be predicted. The phenomenon is related to packing and historic motion of the pebbles and is also called the intermittent pebble flow. Discharging of pebbles in the high temperature gas-cooled reactor (HTGR) is controlled by a discharging facility that removes bottom pebbles one at a time at a controlled rate. This differs from the free outflow conditions in other industrial applications. The controlled constant discharge process at a slow rate is the reason for the very slow pebble flow in the pebble-bed HTGRs.

Pebble flow with a controllable discharging process lays the foundation for the design and operation of pebble-bed reactors. Pebble flow operation conditions are also vital to the efficiency and safety of the pebble-bed HTGRs. Pebbles' behavior should be assured to fulfill the thermal hydraulic rules and radiation safety requirements. In the recirculation process, velocity of the pebbles varies greatly throughout the bed, in relation to bed configurations, the loading method, etc. In general, particles flow fast in the center and slow near the wall. The uniformity of pebble flow is of crucial importance to the performance and safety of the reactors [88]. In general, flow pattern in the reactor core should be uniform and synchronous. Flooding of powders, channeling and jamming should be prevented, and caking, crystallization and forming of stagnant regions should to be minimized. At the same time, pebble recirculation in the reactor core should follow first-in-first-out sequence, so that each fuel pebble reach almost the same burnup level when it is discharged.

Although the pebble flows have drawn great attention, it is still poorly understood. Many investigations, including experimental [89–91] and numerical [92–98], have been carried out. The practical design of HTGRs is to optimize pebble bed configurations, such as the two-region arrangement [99] with the pebble dispersion [100] and the stagnant region [101,88]. For the pebble bed design, modifying the geometry of bed through adjusting the base cone angle [101] and bottom shape [88] has been verified to be important. Introducing well-designed wall structures is also helpful in avoiding near wall pebble caking in large areas [102].

1.2 Flow regimes

1.2.1 Dilute and dense flows

For gas-particle two-phase flows, the flow regimes can be generally categorized by the response time [103], the concentration of gas-particle flow [104] and the spatial scales. When the ratio $\tau_p/\tau_c \ll 1$, the particle can respond quickly to the change of local turbulence in the mean free time of collision, and the flow is called a dilute flow regime. On the contrary, when $\tau_p/\tau_c \gg 1$, it is a dense two-phase flow regime where a particle does not always have enough time to trace the motion of local turbulence before an interruption takes place that is caused by the collision with another particle. When $\tau_p/\tau_c \approx 1$, it can be regarded as an intermediate flow regime accordingly. Because τ_c depends on the local concentration of particles, it is dependent on the rate of mass loading of particles and is affected by the fluid turbulence. If the two-phase flow is not stationary, τ_c may vary largely within a short time, and the flow regime may be an instantaneously dilute or an instantaneously dense. Besides, if the distribution of particle concentration is highly nonuniform, the flow may be regarded as a locally dilute or a locally dense flow regime.

Furthermore, according to the role of particle-particle collision in the flow of different concentrations of particles, a gas-particle flow can be categorized into a collision-free flow (dilute phase flow), a collision-dominated flow (medium concentration flow) and a contact-dominated flow (dense phase flow) [104]. Here the contact-dominated flow is differentiated from the collision-dominated flow by almost continuous and long-time 'many-body' contacts between neighboring particles forming a long-history of deformation, and almost continuous force network between particles, which severely affect the internal structure of particle systems. In contrast, the collision-dominated flow is always characterized by a large number of frequent, instantaneous 'two-body' collision events, in combination with almost clearly separated particle-particle collision affairs. Additionally, in the collision-free flow, particle-particle collision scarcely takes place, and its influence on the gas and particle phases can be neglected.

1.2.2 Inertial and elastic flows

For granular flows, according to Campbell's work [105–107], granular flows can be divided into two main regimes – inertial and elastic. The elastic regime is dominated by force chains which can sustain or resist external forces. It is divided into the elastic-quasi-static regime and the elastic-inertial regime depending on whether there is a noticeable dependence of the stresses on the shear rate. The stresses τ are independent on the shear rate γ in the elastic-quasi-static regime, whereas the stresses are in proportion to the shear rate ($\tau \sim \gamma$) in the elastic-inertial regime. The inertial regime is free of force chains since frequent inter-particle collision takes place in the inertial regime and the force chain cannot exist stably. Inter-particle collision plays the role of momentum transfer instead of the force chains. Because both the momentum transfer and collision rates are dependent on the shear rate γ, which both contribute to the stresses. Therefore, the stresses are scaled with the square of the shear rate ($\tau \sim \gamma^2$). The inertial regime can be divided into the inertial-noncollisional regime and the inertial-collisional (or rapid-flow) regime depending on whether the dominant particle interaction is binary collisions.

Moreover, a dimensionless parameter was proposed by Campbell governing the ratio of the elastic effect ($\sim k/d$) to the inertial one ($\sim \rho d^2\gamma^2$) that is defined by $k^\star = k/(\rho d^3\gamma^2)$, where k, ρ, d, and γ represent inter-particle stiffness, solid density, particle diameter and shear rate, respectively. Campbell drew a flow map of flow regimes of granular materials on the solid concentration ϵ_p–$k^\star$ plane. At solid concentrations ϵ_p larger than 0.6, one cannot leave the elastic regimes by changing $k^\star$ (usually by changing the shear rate γ), and can only have transition between elastic-quasi-static and elastic-inertial regimes. Furthermore, out of a laboratory environment, it is difficult to reach the elastic-inertial regime at high solid concentration, because it mostly requires extremely large shear rate. Therefore, at high solid concentrations, the elastic-quasi-static regime which covers the common useful gravity-driven dense granular flow needs to be investigated.

1.3 Numerical methods

1.3.1 Methods for single-phase flow

Following the conservation laws of mass, momentum and energy, the flows of continuous medium can be fully governed by the well-known Navier–Stokes equation. Therefore, the basic approach for obtaining full features of the flow is through solving the full scale Navier–Stokes equations. Unfortunately, as they are a group of partial differential equations which is totally coupled with strongly nonlinear features, together with enormous practical cases of different initial and boundary conditions, it is almost impossible to analytically solve them by up-to-date mathematical theories. As a result, computational methods become a choice for obtaining the flow behavior characteristics of continuous medium.

In general, the computational methods, following the macroscopic viewpoint and description for continuous mediums by the Navier–Stokes equation in the Eulerian framework, can be divided into three main categories:

- Reynolds Averaged Navier–Stokes equation (RANS): The basic concept of the RANS method is based on the decomposition of fluid velocity into a time-averaged component $\bar{u}$ and fluctuating component u', i.e. $u = \bar{u} + u'$. As $\overline{u_i u_j} = \bar{u}_i\bar{u}_j + \overline{u'_i u'_j}$, Replacing the nonlinear term of $\overline{u_i u_j}$ by $\bar{u}_i\bar{u}_j$ in the left-hand side of the Reynolds Averaged Navier–Stokes equation, the additional term $\overline{u'_i u'_j}$, called the Reynolds stress tensor, appears in the right-hand side. Therefore, the Reynolds stress tensor should be solved or modeled additionally [108], either by the Algebraic stress model (ASM) (Zero-equation model, [109,110]), One-equation model (OEM) [111], Two-equation model (TEM) [112], or the Reynolds stress models (RSM), although none of them is a uniform model of closure.

For example, the zero-, one- and two-equation models are based on the concept of eddy-viscosity ν_T [113]. In the zero-equation model, ν_T is prescribed algebraically, as in the Prandtl's mixing-length theory [114], i.e. $\nu_T = l_m^2 |\frac{d\bar{u}}{dy}|$, where l_m is the 'mixing length' obtained by the analogies between the turbulent length scale and the mean free path in the kinetic theory of gases, and represents the distance traversed by a small lump of fluid before the loss of its momentum. In the one-equation model [115,116], $\nu_T = K_t^{1/2} l_e$ where K_t is named as the turbulent kinetic energy obtained from a modeled exact transport equation. In the two-equation model, two separate modeled transport equations for two independent variables, e.g. $K_t - \epsilon$ (ϵ is the turbulent dissipation rate), $K_t - l_e$ (l_e is the integral length scale), or $K_t - \omega_t$ (ω_t is the reciprocal turbulent time scale $\omega_t = \epsilon / K_t$), are solved for the turbulent length and time scales, e.g. $l_0 \sim K_t^{3/2}/\epsilon$, $\tau_0 \sim K_t/\epsilon$, which gives rise to the eddy-viscosity $\nu_T = l_0^2/\tau_0$.
However, the main deficiencies of the eddy-viscosity models lie in the lacking of uniform models of case configuration independence, failure in consideration of nonlocal and history effects [117] and failure in prediction of anisotropic turbulences, such as the strongly swirling flows, and inability to properly account for the flows with streamline curvature, flow separation, flows with zones of re-circulating flow or flows influenced by mean rotational effects. These deficiencies can be majorly overcome by higher order closures [108]. For example, in the Reynolds stress equation model, the Reynolds Stress Transport equation (RST)

$$\frac{DR_{ij}}{Dt} = \hat{D}_{ij} + \hat{P}_{ij} + \hat{\Pi}_{ij} + \hat{\Omega}_{ij} - \hat{\epsilon}_{ij} \tag{1.2}$$

where $R_{ij} = \overline{u'_i u'_j}$ and $\frac{DR_{ij}}{Dt} = (\frac{\partial}{\partial t} + \boldsymbol{u} \cdot \nabla) R_{ij}$. Eq. (1.2) means the rate of change plus the transport by convection of the Reynolds stress are equal to the high order terms of $\hat{D}_{ij}$ (the diffusion term), $\hat{P}_{ij}$ (production term), $\hat{\Pi}_{ij}$ (pressure-strain correlation term), $\hat{\Omega}_{ij}$ (rotational term), and $\hat{\epsilon}_{ij}$ (dissipation term) respectively [118]. The RSM has advantages over the eddy-viscosity models since it can solve all components of the turbulent transport and consider the effects of buoyancy and curvature on stresses.

- Large Eddy Simulation (LES): Different from the RANS model on averaging the fluid velocity by time scale '‾', the large eddy simulation developed by Smagorinsky [119] is based on filtering the fluid velocity by small spatial scale 'Δ' (or in addition to the time scale '‾' [120]). Therefore, by solving the filtered Navier–Stokes equations, the computational cost is reduced since only the large scale turbulence is resolved and the small scale vortex finer than 'Δ' is filtered out. However, an additional term $\tau_{ij} = \left(\frac{\partial \overline{u'_i u'_j}}{\partial x_j} - \frac{\partial \bar{u}_i \bar{u}_j}{\partial x_j} \right)$ is also generated in the filtering process. Thus, τ_{ij} is considered to be caused by the sub-grid scale (SGS) stress tensor, which needs to be modeled properly. For example, similar to the eddy-viscosity model in RANS, it was assumed that the dissipation of kinetic energy at sub-grid scales are analogous to molecular diffusion, i.e. $\tau_{ij} = \frac{1}{3}\tau_{kk}\delta_{ij} - 2\nu_t \bar{S}_{ij}$, where $\bar{S}_{ij} = \frac{1}{2}\left(\frac{\partial \bar{u}_i}{\partial x_j} + \frac{\partial \bar{u}_j}{\partial x_i}\right)$ is the rate-of-strain tensor and ν_t is the turbulent eddy viscosity. The Smagorinsky–Lilly model [119] and Germano dynamic model [121] are two of the most typical SGS models.

- Direct Numerical Simulation (DNS): Unlike the LES and RANS, DNS tries to solve the full Navier–Stokes equations directly on sufficient fine spatial grids and temporal scales to capture all the spatial scales of turbulence from the smallest dissipative scales (namely the Kolmogorov microscale $\eta = (\nu^3/\epsilon)^{1/4}$ where ν is the kinematic viscosity and ϵ is the rate of kinetic energy dissipation) up to the integral scale l_e, associated with the vortices containing most of the kinetic energies [122]. Suppose a three-dimensional case with a characteristic length L covered by $N_x \times N_y \times N_z$ grids of scale h. As the numerical grids should be finer than η, $Re = \frac{u'L}{\nu}$ and $\epsilon \approx u'^3/L$, the total number of spatial grids will be increased dramatically: $N_x \times N_y \times N_z \sim (L/h)^3 \geq (Re^{3/4})^3$. On the other hand, for time integral, the time step Δt should be fine enough to guarantee the Courant number $Co = \frac{u'\Delta t}{h} < 1$. Therefore, to simulate a time interval of $\Delta T_s = L/u'$, the required time steps would be $N_t = \frac{\Delta T_s'}{\Delta t} > L/h \sim Re^{1/4}$. As a whole, the computational costs in both spatial capacity and temporal integral are proportional to Re^3. Thus, the heavy costs of DNS make it far from the requirements for rapid practical application in industries of high Reynolds flows. Nonetheless, it is regarded as a promising laboratory tool for better understanding of the physics of turbulence, such as providing the benchmark cases for checking the assumptions, correcting the parameters, and validating the LES or RANS models.
- Lattice Boltzmann method (LBM): the former three methods are macroscopic methods based on the Navier–Stokes equation. In addition, it is necessary to mention the lattice Boltzmann method (LBM), a mesoscopic method for fluid flow study. LBM is based on kinetic theory to simulate various hydrodynamic systems and has attracted interest of researchers in computational physics [123,124]. Started from a microscopic viewpoint of collision and migration of discrete fictitious particles on fixed grids that correspond to the discretized spatial-velocity phase space, the Navier–Stokes equations can be derived from the LBE models with particular selection of local equilibrium distribution via the Chapman–Enskog analysis [125,126]. Moreover, LBM has particular suitability for fully parallel computing of complicated boundary conditions and multiphase interfaces [127]. The LBM application is rather diverse and interdisciplinary and it is well accepted that it has become a powerful tool in the modeling of inhomogeneous fluids, such as multiphase or multicomponent fluids [128–130], the flow through porous media [131,132], soil filtration in Earth sciences [133], and the fuel cells in energy sciences [134].

1.3.2 Methods for two-phase flow

Gas-particle flows could cover a large number of scales, e.g. a long pipe for transporting natural gas and oil droplets should be described in kilometers whereas the gas-particle flow through a valve, filter, or seal may be considered in millimeters or microns. Therefore, a variety of numerical methods, including the coupled or multi-scale methods [135], have been developed for describing the gas-particle flow of different scales, to either get insights into the physical principles on microscopic inter-phase interaction or predict the vast applications in engineering scales. These methods can be basically divided into four categories:

- Engineering scale. The characteristic length of this scale is always far larger than the particle size. It is mainly used with regard to the engineering issues, such as pressure drop, mass flow, particle loading rate, mean velocity, and turbulence intensity.
- Cluster/bubble scale. The particle clusters are composed of a large number of particles that appear in local flow regions of relatively high concentration of particles. On the contrary, the bubble scale regards the flow in local regions with relatively low concentration of particles compared to the dense particle concentration surrounding the bubble. The drag force of particle cluster is one of the main issues in this scale.
- Particle scale. It concerns the issues near particle size, in particular the mechanisms on particle scale inter-phase interaction, such as the drag force of single particle and particle-particle collision.
- Sub-particle scale. To better understand the inter-phase interaction, the sub-particle-scale methods are in essence different from the former scales which are all or partially based on empirical or experimental correlations for inter-phase interactions. For example, using the distribution and integral of the gas-phase stress tensor on the particle surface, it is able to directly compute the fluid-to-particle force and particle-to-fluid feedback. To date, the sub-particle scale methods, sometimes including the particle-scale methods, are only for mechanical or physical study in limited testing cases for exploration of the inter-phase interaction models to be incorporated in the even larger scale methods.

Conventionally, the methods for gas-solid flows are following either the Eulerian–Eulerian approach based on the continuum hypothesis for both the gas and solid phases (e.g. the well-known Two-Fluid Model (TFM)), or the Eulerian–Lagrangian approach in which the discrete particle trajectories are simulated through solving the Newton's equation of motion. For most numerical simulations of dilute gas-solid flow in either the Eulerian–Lagrangian or the Eulerian–Eulerian framework, the effect of the particle-particle collision dynamics is always neglected. The particle motion is controlled by local aerodynamic forces in dilute gas-particle flows. But in dense flows, particle motion is also governed by particle-particle collisions, and hence inter-particle collisions become fundamental and dominant. Moreover, in the Eulerian–Lagrangian method, inter-particle collisions are taken into account by either a deterministic or a stochastic way. The former needs to compute simultaneously the trajectories of all particles in the flow domain, and the latter usually uses a random technique, like the Monte Carlo (MC) method, to compute only a bundle of sampling particles to reduce the number of particles for solution. Due to the rapid development and powerful capability of high performance computers, the deterministic way has become a prevalent approach for the modeling study of gas-particle flows. In particular, the discrete particle methods, or the variants, extensions and coupled methods developed from the discrete element method (DEM) [53] have become the most popular methods for exploring the science and engineering applications of particle or gas-particle flows.

The DEM based coupled approaches for gas-particle flows can be basically divided into three main categories:

- The discrete element method – direct numerical simulation (DEM-DNS) coupled method [136–138]: In two-phase flow, DNS is usually referring to a particular high resolution method, termed as fully resolved, to solve the Navier–Stokes equation surrounding the particles directly by high order accurate numerical scheme on a huge number of grids and stepped by high order time integration scheme. It relies severely on the capacity and performance of supercomputers. But, when the turbulent flow is solved in high precision, in combination with suitable inter-particle and inter-phase solution by discrete element method and coupling scheme, the physics and mechanisms on the gas-particle two-phase flow can be captured on the highest solutions. Though its computational costs are heavy, it is still attracting the interest of many scientists on exploring the underlying unknown world of turbulence and particle science.
- The discrete element method – large eddy simulation (DEM-LES) coupled method: With the aid of suitable filtering operators and with suitable modeling on the two-phase sub-grid scale stress tensors, the large eddy simulation method avoids capturing full scales of turbulence in a two-phase flow and only focuses on large scales. In particular, it is assumed that the finest scales in a two-phase flow are isotropic and the modeling of such fine scales are much easier than the full scales. Therefore, LES method meets the common interests of scientists and engineers in the large scale turbulence study and applications. The coupled method of DEM-LES extends the power of LES to the multiphase flows, especially to the dense phase flows in chemical engineering. [139–146].
- The computational fluid dynamics – discrete element method (CFD-DEM), coupled method [147–154]. Here, the CFD-DEM method is different from the DEM-LES or DEM-DNS methods in using conventional computational fluid dynamics methods other than the LES or DNS methods, such as RANS, the inviscous equations and the locally averaged full Navier–Stokes equation (weighted by void fractions) without subgrid turbulence models. It is a powerful method particularly for directly utilizing full advantages of various computational fluid dynamics software.

In addition, many other DEM-based coupled methods, such as DEM-LBM [155–157], DEM-SPH [158–160], or multiple coupled methods, such as LBM-IBM-DEM [161–163], have also been developed. The basic strategy of the immersed-boundary method (IBM) is to account for the gas-particle interaction via keeping the nonslip condition on the gas-phase interface through suitable interpolation operators

[164–169]. By distributing a number of Lagrangian points on the particle surface and compute the velocities on them by interpolation from neighboring fluid nodes, the velocity differences on these Lagrangian points are then used to compute feedback forcing on these points. By re-interpolating this forcing back to nearby fluid nodes, the fluid-particle velocity difference on the gas-particle interface will be reduced. Such a process can be repeated in one time step and step-by-step to keep a perfect nonslip condition. Therefore, it is a promising tool for small scale flow to better understand the physics in gas-particle interaction, with the aid of DNS for capturing the nature of turbulence and DEM to deal with complex particle-particle interactions.

1.4 Summary

In this book, some efforts have been made to extend the DEM models to widen their application in gas-particle flows. The CFD-DEM, DEM-LES, DEM-DNS, DEM-LBM, and multiple coupled methods, like DEM-IBM-LBM, will be introduced in details, particularly by focusing on the fundamentals and numerical techniques and important issues in application, for study of gas-particle and granular flows.

For single-phase flow, particular attention will be made to jet flows, including the planar jet, cross jet and swirling jet flows, since they provide the most common basis for the particle-laden two-phase flows. For the gas-particle two-phase flow, the jet flows with and without swirl, and the fluidized bed, which are the most widely applied dilute and dense flows, will be studied extensively by using suitable DEM-DNS or DEM-LES coupled numerical methods.

The important issues of inter-phase interactions will be particularly taken into account, in combination with the issue of heat transfer. For granular flows, we will mainly address the granular flows in hoppers, silos and rotating drums, focusing on the important issues of numerical modeling of particle-to-particle, particle-to-wall interactions, mixing behaviors, flow patterns, heat transfer characteristics, and the effects of container geometry and particle shapes.

CHAPTER

Discrete particle model

2

2.1 Spherical particle model

The classical discrete (or distinct) element method (DEM) was pioneered by Cundall and Strack [53]. DEM model involves two stages: at first, given the particles' instantaneous positions, the interaction forces are computed by suitable mechanical models when particles slightly inter-penetrate each other. Then, the Newton's second law is used to determine the resulting acceleration of each particle and to find the new positions of particles by time-stepping integration. This process is analogous to the molecular dynamics (MD). Therefore, this approach is based on force displacement formulation since the forces are computed directly and the governing equations of particle dynamics are integrated. It can also be viewed as a smooth contact method or contact-dynamics model.

On the other hand, this approach may not consider possible inter-penetration between particles and unilateral contacts. That means it disregards the contact history between particles, but the post-collisional velocities can be directly obtained from the pre-collisional velocities by the conservation laws. These approaches are referred to as nonsmooth contact methods, such as the event-driven method (EDM) [170] or the hard sphere model [171]. They can also be regarded as kinematic collision models [172,173]. In general, these approaches are considered not suitable for dense particle systems with multiple-body contacts because they treat only each pair of particles at a time.

Particles have been most commonly viewed as spherical shapes since spherical shape is easier than nonspherical shapes for modeling and simulation. A lot of works on spherical particle model development and application have been reported. In this chapter, we will not make a thorough review on the vast majority of spherical shape based models. For readers who are interested in this topic, can refer to the literature [52] for details. We will focus on the models newly developed by our group, which pay particular attention to nonspherical shapes, including both collision dynamics and numerical techniques.

2.2 Generalized hard particle model (GHPM)

2.2.1 Governing equations

Suppose a particle indexed by '(i)' collides with another particle indexed by '(j)' on a pair of contacting points $\boldsymbol{C}'$ and $\boldsymbol{C}''$. The following assumptions are applied to

Gas-Particle and Granular Flow Systems. https://doi.org/10.1016/B978-0-12-816398-6.00010-9
Copyright © 2020 Elsevier Inc. All rights reserved.

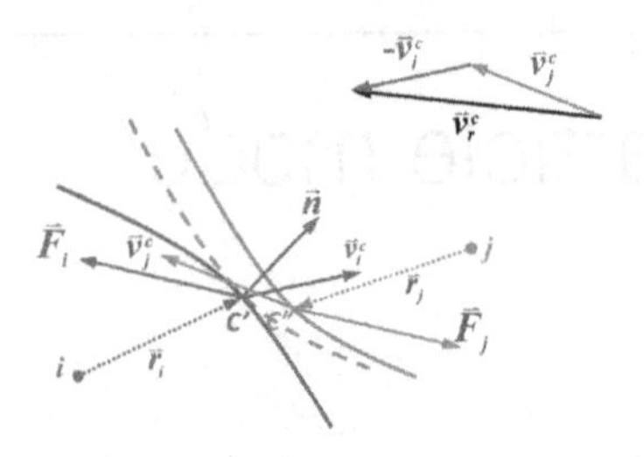

FIGURE 2.1

Sketch of particle-particle collision.

simplify the collision process: (1) The collision is three-dimensional, taking place between rigid particles of arbitrary shapes. (2) The momenta, including translational and angular, are conserved during collision. (3) The Coulomb's law of friction is followed when inter-particle sliding motion on the contact surface takes place. (4) In the normal direction of the contact surface, a normal coefficient of restitution e is defined as the rate of magnitudes of the post-collisional relative velocity to the pre-collisional relative velocity between the pair of colliding points ($\boldsymbol{C}'$ and $\boldsymbol{C}''$), i.e.

$$(v_\alpha^{(j)}(\boldsymbol{C}'',t) - v_\alpha^{(i)}(\boldsymbol{C}',t))n_\alpha = e(v_\beta^{(i,0)}(\boldsymbol{C}',t) - v_\beta^{(j,0)}(\boldsymbol{C}'',t))n_\beta \tag{2.1}$$

where $v_\alpha^{(j)}$ and $v_\beta^{(i)}$ are velocities of the pair of colliding particles '(j)' and '(i)', respectively. Herein the Einstein summation convention is used for the vector and tensor representation, i.e. $\boldsymbol{v} = v_\alpha \boldsymbol{e}_\alpha = v_\beta \boldsymbol{e}_\beta$. The unit orthonormal coordinates $\boldsymbol{e}_\alpha$ are always omitted for simple representation, and $\boldsymbol{v} = v_\alpha = v_\beta$. '$\boldsymbol{C}'$' and '$\boldsymbol{C}''$' denote the contact points and t is time. The superscript '0' denotes the pre-collisional variables. Otherwise, it represents a post-collisional variable. 'n_α' is the normal direction of the contact surface on the contact point $\boldsymbol{C}'$ (Fig. 2.1).

Based on the conservation laws of momenta, the relations between the post- and pre-collisional translational and angular velocities can be expressed as follows

$$m^{(i)} v_\alpha^{(i)} = m^{(i)} v_\alpha^{(i,0)} + \int_{t_0}^{t_0+\delta t} f_\alpha^{(i)} dt = m^{(i)} v_\alpha^{(i,0)} + p_\alpha^{(i)} \tag{2.2}$$

$$m^{(j)} v_\alpha^{(j)} = m^{(j)} v_\alpha^{(j,0)} - \int_{t_0}^{t_0+\delta t} f_\alpha^{(i)} dt = m^{(j)} v_\alpha^{(j,0)} - p_\alpha^{(i)} \tag{2.3}$$

$$\omega_\beta^{(i)} = \omega_\beta^{(i,0)} + I_{\beta\alpha}^{(-1,i)} \int_{t_0}^{t_0+\delta t} \epsilon_{\alpha\gamma\lambda} r_\gamma^{(i)} f_\lambda^{(i)} dt = \omega_\beta^{(i,0)} + I_{\beta\alpha}^{(-1,i)} \epsilon_{\alpha\gamma\lambda} r_\gamma^{(i)} p_\lambda^{(i)} \tag{2.4}$$

$$\omega_\beta^{(j)} = \omega_\beta^{(j,0)} - I_{\beta\alpha}^{(-1,j)} \int_{t_0}^{t_0+\delta t} \epsilon_{\alpha\gamma\lambda} r_\gamma^{(j)} f_\lambda^{(i)} dt = \omega_\beta^{(j,0)} - I_{\beta\alpha}^{(-1,j)} \epsilon_{\alpha\gamma\lambda} r_\gamma^{(j)} p_\lambda^{(i)} \tag{2.5}$$

where v_α, ω_α, m, $I_{\alpha\beta}$ are the translational velocity vector, angular velocity vector, mass and the inertia tensor of three-dimensional particles, respectively. $I_{\beta\alpha}^{(-1)}$ is the inverse tensor of the particle inertia tensor, and

$$I_{\beta\alpha}^{(-1)} I_{\alpha\gamma} = \delta_{\beta\gamma} = \begin{cases} 1, & \text{if } \beta = \gamma \\ 0, & \text{otherwise} \end{cases} \tag{2.6}$$

where $\delta_{\beta\gamma}$ is the Kronecker operator. The superscript '0' denotes the pre-collisional parameters at t^0. $r_\gamma^{(i)}$ and $r_\gamma^{(j)}$ are the position vectors from the centroid of particles (i) and (j) to the contact points C' or C'' respectively (Fig. 2.1). The quantity

$$\epsilon_{\alpha\beta\gamma} = \begin{cases} -1, \text{ odd permutation of } (1, 2, 3) \\ 1, \text{ even permutation of } (1, 2, 3) \\ 0, \text{ if any index is repeated,} \end{cases} \tag{2.7}$$

is the permutation tensor. f_α is the collisional force. p_α is the total impulse on particle '(i)' produced by collision, which can be decomposed into a normal component $p^n = p_\beta n_\beta$ and a tangential component $p^t = p_\beta t_\beta$, respectively, i.e.

$$p_\alpha = p^n n_\alpha + p^t t_\alpha \tag{2.8}$$

The solution of impulse can be obtained by the following procedures:
The relative velocity between the pair of contacting points is given by

$$v_\alpha^{r,C} = v_\alpha^{(j)} + \epsilon_{\alpha\beta\gamma}\omega_\beta^{(j)} r_\gamma^{(j)} - v_\alpha^{(i)} - \epsilon_{\alpha\beta\gamma}\omega_\beta^{(i)} r_\gamma^{(i)} \tag{2.9}$$

Substitute Eqs. (2.2)–(2.5) into Eq. (2.9):

$$\begin{aligned}
v_\alpha^{r,C} &= v_\alpha^{(j)} + \epsilon_{\alpha\beta\gamma}\omega_\beta^{(j)} r_\gamma^{(j)} - v_\alpha^{(i)} - \epsilon_{\alpha\beta\gamma}\omega_\beta^{(i)} r_\gamma^{(i)} \\
&= v_\alpha^{(j,0)} - v_\alpha^{(i,0)} - \frac{p_\alpha^{(i)}}{m^{(j)}} - \frac{p_\alpha^{(i)}}{m^{(i)}} + \epsilon_{\alpha\beta\gamma}\omega_\beta^{(j)} r_\gamma^{(j)} - \epsilon_{\alpha\beta\gamma}\omega_\beta^{(i)} r_\gamma^{(i)} \\
&= v_\alpha^{(r,0)} - \frac{m^{(i)} + m^{(j)}}{m^{(i)} m^{(j)}} p_\alpha^{(i)} + \epsilon_{\alpha\beta\gamma}\omega_\beta^{(j)} r_\gamma^{(j)} - \epsilon_{\alpha\beta\gamma}\omega_\beta^{(i)} r_\gamma^{(i)} \\
&= v_\alpha^{(r,0)} - \frac{p_\alpha^{(i)}}{m^{(r,ij)}} + \epsilon_{\alpha\beta\gamma}\left[\omega_\beta^{(j,0)} - I_{\beta\kappa}^{(-1,j)}\epsilon_{\kappa\theta\sigma} r_\theta^{(j)} p_\sigma^{(i)}\right] r_\gamma^{(j)} \\
&\quad - \epsilon_{\alpha\beta\gamma}\left[\omega_\beta^{(i,0)} + I_{\beta\kappa}^{(-1,i)}\epsilon_{\kappa\theta\sigma} r_\theta^{(i)} p_\sigma^{(i)}\right] r_\gamma^{(i)} \\
&= v_\alpha^{(r,0)} - \frac{p_\alpha^{(i)}}{m^{(r,ij)}} + \epsilon_{\alpha\beta\gamma}\omega_\beta^{(j,0)} r_\gamma^{(j)} - \epsilon_{\alpha\beta\gamma}\omega_\beta^{(i,0)} r_\gamma^{(i)} \\
&\quad - \epsilon_{\alpha\beta\gamma}\left[I_{\beta\kappa}^{(-1,j)}\epsilon_{\kappa\theta\sigma} r_\theta^{(j)} p_\sigma^{(i)}\right] r_\gamma^{(j)} - \epsilon_{\alpha\beta\gamma}\left[I_{\beta\kappa}^{(-1,i)}\epsilon_{\kappa\theta\sigma} r_\theta^{(i)} p_\sigma^{(i)}\right] r_\gamma^{(i)} \\
&= v_\alpha^{(r,C,0)} - \frac{p_\alpha^{(i)}}{m^{(r,ij)}} - \epsilon_{\alpha\beta\gamma}\epsilon_{\kappa\theta\sigma} I_{\beta\kappa}^{(-1,j)} r_\theta^{(j)} p_\sigma^{(i)} r_\gamma^{(j)} - \epsilon_{\alpha\beta\gamma}\epsilon_{\kappa\theta\sigma} I_{\beta\kappa}^{(-1,i)} r_\theta^{(i)} p_\sigma^{(i)} r_\gamma^{(i)}
\end{aligned} \tag{2.10}$$

where $v_\alpha^{(r,0)} = v_\alpha^{(j,0)} - v_\alpha^{(i,0)}$ and $v_\alpha^{(r,C,0)} = v_\alpha^{(r,0)} + \epsilon_{\alpha\beta\gamma}\omega_\beta^{(j,0)} r_\gamma^{(j)} - \epsilon_{\alpha\beta\gamma}\omega_\beta^{(i,0)} r_\gamma^{(i)}$ are the pre-collisional relative velocity of particle centroids and pre-collisional relative velocity of contact points respectively. $m^{(r,ij)} = \frac{m^{(i)}m^{(j)}}{m^{(i)}+m^{(j)}}$ is the reduced mass.

Using the identity

$$\begin{aligned}\epsilon_{\alpha\beta\gamma}\epsilon_{\kappa\theta\sigma} = {} & \delta_{\alpha\kappa}(\delta_{\beta\theta}\delta_{\gamma\sigma} - \delta_{\gamma\theta}\delta_{\beta\sigma}) + \delta_{\beta\kappa}(\delta_{\gamma\theta}\delta_{\alpha\sigma} - \delta_{\alpha\theta}\delta_{\gamma\sigma}) \\ & + \delta_{\gamma\kappa}(\delta_{\beta\sigma}\delta_{\alpha\theta} - \delta_{\beta\theta}\delta_{\alpha\sigma})\end{aligned} \tag{2.11}$$

one concludes

$$\begin{aligned}& \epsilon_{\alpha\beta\gamma}\epsilon_{\kappa\theta\sigma} I_{\beta\kappa}^{(-1,j)} r_\theta^{(j)} p_\sigma^{(i)} r_\gamma^{(j)} \\
&= \delta_{\alpha\kappa}(\delta_{\beta\theta}\delta_{\gamma\sigma} - \delta_{\gamma\theta}\delta_{\beta\sigma}) I_{\beta\kappa}^{(-1,j)} r_\theta^{(j)} p_\sigma^{(i)} r_\gamma^{(j)} \\
&+ \delta_{\beta\kappa}(\delta_{\gamma\theta}\delta_{\alpha\sigma} - \delta_{\alpha\theta}\delta_{\gamma\sigma}) I_{\beta\kappa}^{(-1,j)} r_\theta^{(j)} p_\sigma^{(i)} r_\gamma^{(j)} \\
&+ \delta_{\gamma\kappa}(\delta_{\beta\sigma}\delta_{\alpha\theta} - \delta_{\beta\theta}\delta_{\alpha\sigma}) I_{\beta\kappa}^{(-1,j)} r_\theta^{(j)} p_\sigma^{(i)} r_\gamma^{(j)} \\
&= (\delta_{\beta\theta}\delta_{\gamma\sigma} - \delta_{\gamma\theta}\delta_{\beta\sigma}) I_{\beta\alpha}^{(-1,j)} r_\theta^{(j)} p_\sigma^{(i)} r_\gamma^{(j)} \\
&+ (\delta_{\gamma\theta}\delta_{\alpha\sigma} - \delta_{\alpha\theta}\delta_{\gamma\sigma}) I_{\beta\beta}^{(-1,j)} r_\theta^{(j)} p_\sigma^{(i)} r_\gamma^{(j)} \\
&+ (\delta_{\beta\sigma}\delta_{\alpha\theta} - \delta_{\beta\theta}\delta_{\alpha\sigma}) I_{\beta\gamma}^{(-1,j)} r_\theta^{(j)} p_\sigma^{(i)} r_\gamma^{(j)} \\
&= I_{\beta\alpha}^{(-1,j)} r_\beta^{(j)} p_\gamma^{(i)} r_\gamma^{(j)} - I_{\beta\alpha}^{(-1,j)} r_\gamma^{(j)} p_\beta^{(i)} r_\gamma^{(j)} \\
&+ I_{\beta\beta}^{(-1,j)} r_\gamma^{(j)} p_\alpha^{(i)} r_\gamma^{(j)} - I_{\beta\beta}^{(-1,j)} r_\alpha^{(j)} p_\gamma^{(i)} r_\gamma^{(j)} \\
&+ I_{\beta\gamma}^{(-1,j)} r_\alpha^{(j)} p_\beta^{(i)} r_\gamma^{(j)} - I_{\beta\gamma}^{(-1,j)} r_\beta^{(j)} p_\alpha^{(i)} r_\gamma^{(j)} \\
&= \boldsymbol{r}^{(j)} \cdot \boldsymbol{I}^{(-1,j)} (\boldsymbol{p}^{(i)} \cdot \boldsymbol{r}^{(j)}) - \boldsymbol{p}^{(i)} \cdot \boldsymbol{I}^{(-1,j)} (\boldsymbol{r}^{(j)} \cdot \boldsymbol{r}^{(j)}) \\
&+ tr(\boldsymbol{I}^{(-1,j)}) \boldsymbol{p}^{(i)} (\boldsymbol{r}^{(j)} \cdot \boldsymbol{r}^{(j)}) - tr(\boldsymbol{I}^{(-1,j)}) \boldsymbol{r}^{(j)} (\boldsymbol{p}^{(i)} \cdot \boldsymbol{r}^{(j)}) \\
&+ (\boldsymbol{p}^{(i)} \cdot \boldsymbol{I}^{(-1,j)} \cdot \boldsymbol{r}^{(j)}) \boldsymbol{r}^{(j)} - (\boldsymbol{r}^{(j)} \cdot \boldsymbol{I}^{(-1,j)} \cdot \boldsymbol{r}^{(j)}) \boldsymbol{p}^{(i)}\end{aligned} \tag{2.12}$$

which can be expressed in a similar manner as

$$\begin{aligned}\epsilon_{\alpha\beta\gamma}\epsilon_{\kappa\theta\sigma} I_{\beta\kappa}^{(-1,i)} r_\theta^{(i)} p_\sigma^{(i)} r_\gamma^{(i)} = {} & \boldsymbol{r}^{(i)} \cdot \boldsymbol{I}^{(-1,i)} (\boldsymbol{p}^{(i)} \cdot \boldsymbol{r}^{(i)}) - \boldsymbol{p}^{(i)} \cdot \boldsymbol{I}^{(-1,i)} (\boldsymbol{r}^{(i)} \cdot \boldsymbol{r}^{(i)}) \\
& + tr(\boldsymbol{I}^{(-1,i)}) \boldsymbol{p}^{(i)} (\boldsymbol{r}^{(i)} \cdot \boldsymbol{r}^{(i)}) - tr(\boldsymbol{I}^{(-1,i)}) \boldsymbol{r}^{(i)} (\boldsymbol{p}^{(i)} \cdot \boldsymbol{r}^{(i)}) \\
& + (\boldsymbol{p}^{(i)} \cdot \boldsymbol{I}^{(-1,i)} \cdot \boldsymbol{r}^{(i)}) \boldsymbol{r}^{(i)} - (\boldsymbol{r}^{(i)} \cdot \boldsymbol{I}^{(-1,i)} \cdot \boldsymbol{r}^{(i)}) \boldsymbol{p}^{(i)}\end{aligned} \tag{2.13}$$

Using Eqs. (2.12) and (2.13), Eq. (2.10) can be deduced as follows:

$$\begin{aligned}v_\alpha^{r,C} &= v_\alpha^{(r,C,0)} - \frac{p_\alpha^{(i)}}{m^{(r,ij)}} - \epsilon_{\alpha\beta\gamma}\epsilon_{\kappa\theta\sigma} I_{\beta\kappa}^{(-1,j)} r_\theta^{(j)} p_\sigma^{(i)} r_\gamma^{(j)} - \epsilon_{\alpha\beta\gamma}\epsilon_{\kappa\theta\sigma} I_{\beta\kappa}^{(-1,i)} r_\theta^{(i)} p_\sigma^{(i)} r_\gamma^{(i)} \\
&= v_\alpha^{(r,C,0)} - \frac{p_\alpha^{(i)}}{m^{(r,ij)}} - I_{\beta\alpha}^{(-1,j)} r_\beta^{(j)} p_\gamma^{(i)} r_\gamma^{(j)} + I_{\beta\alpha}^{(-1,j)} r_\gamma^{(j)} p_\beta^{(i)} r_\gamma^{(j)}\end{aligned}$$

$$
\begin{aligned}
&- I_{\beta\beta}^{(-1,j)} r_\gamma^{(j)} p_\alpha^{(i)} r_\gamma^{(j)} + I_{\beta\beta}^{(-1,j)} r_\alpha^{(j)} p_\gamma^{(i)} r_\gamma^{(j)} \\
&- I_{\beta\gamma}^{(-1,j)} r_\alpha^{(j)} p_\beta^{(i)} r_\gamma^{(j)} + I_{\beta\gamma}^{(-1,j)} r_\beta^{(j)} p_\alpha^{(i)} r_\gamma^{(j)} \\
&- I_{\beta\alpha}^{(-1,i)} r_\beta^{(i)} p_\gamma^{(i)} r_\gamma^{(i)} + I_{\beta\alpha}^{(-1,i)} r_\gamma^{(i)} p_\beta^{(i)} r_\gamma^{(i)} \\
&- I_{\beta\beta}^{(-1,i)} r_\gamma^{(i)} p_\alpha^{(i)} r_\gamma^{(i)} + I_{\beta\beta}^{(-1,i)} r_\alpha^{(i)} p_\gamma^{(i)} r_\gamma^{(i)} \\
&- I_{\beta\gamma}^{(-1,i)} r_\alpha^{(i)} p_\beta^{(i)} r_\gamma^{(i)} + I_{\beta\gamma}^{(-1,i)} r_\beta^{(i)} p_\alpha^{(i)} r_\gamma^{(i)}
\end{aligned}
\tag{2.14}
$$

or

$$
\begin{aligned}
\boldsymbol{v}^{r,C} = {} & \boldsymbol{v}^{(j,0)} - \boldsymbol{v}^{(i,0)} + \boldsymbol{\omega}^{(j,0)} \times \boldsymbol{r}^{(j)} - \boldsymbol{\omega}^{(i,0)} \times \boldsymbol{r}^{(i)} - \frac{p_\alpha^{(i)}}{m^{(r,ij)}} \\
& - \boldsymbol{r}^{(j)} \cdot \boldsymbol{I}^{(-1,j)} (\boldsymbol{p}^{(i)} \cdot \boldsymbol{r}^{(j)}) + \boldsymbol{p}^{(i)} \cdot \boldsymbol{I}^{(-1,j)} (\boldsymbol{r}^{(j)} \cdot \boldsymbol{r}^{(j)}) \\
& - tr(\boldsymbol{I}^{(-1,j)}) \boldsymbol{p}^{(i)} (\boldsymbol{r}^{(j)} \cdot \boldsymbol{r}^{(j)}) + tr(\boldsymbol{I}^{(-1,j)}) \boldsymbol{r}^{(j)} (\boldsymbol{p}^{(i)} \cdot \boldsymbol{r}^{(j)}) \\
& - (\boldsymbol{p}^{(i)} \cdot \boldsymbol{I}^{(-1,j)} \cdot \boldsymbol{r}^{(j)}) \boldsymbol{r}^{(j)} + (\boldsymbol{r}^{(j)} \cdot \boldsymbol{I}^{(-1,j)} \cdot \boldsymbol{r}^{(j)}) \boldsymbol{p}^{(i)} \\
& - \boldsymbol{r}^{(i)} \cdot \boldsymbol{I}^{(-1,i)} (\boldsymbol{p}^{(i)} \cdot \boldsymbol{r}^{(i)}) + \boldsymbol{p}^{(i)} \cdot \boldsymbol{I}^{(-1,i)} (\boldsymbol{r}^{(i)} \cdot \boldsymbol{r}^{(i)}) \\
& - tr(\boldsymbol{I}^{(-1,i)}) \boldsymbol{p}^{(i)} (\boldsymbol{r}^{(i)} \cdot \boldsymbol{r}^{(i)}) + tr(\boldsymbol{I}^{(-1,i)}) \boldsymbol{r}^{(i)} (\boldsymbol{p}^{(i)} \cdot \boldsymbol{r}^{(i)}) \\
& - (\boldsymbol{p}^{(i)} \cdot \boldsymbol{I}^{(-1,i)} \cdot \boldsymbol{r}^{(i)}) \boldsymbol{r}^{(i)} + (\boldsymbol{r}^{(i)} \cdot \boldsymbol{I}^{(-1,i)} \cdot \boldsymbol{r}^{(i)}) \boldsymbol{p}^{(i)}
\end{aligned}
\tag{2.15}
$$

where $tr(\boldsymbol{I}^{(-1)}) = I_{\alpha\alpha}^{(-1)}$ is the trace of the inertia tensor. It is noticed that the relative velocity at the contact point is determined by the radius from the centroid to the contact point $\boldsymbol{r}$, the collisional impulse $\boldsymbol{p}$, the momenta of inertia $tr(\boldsymbol{I}^{(-1)})$, as well as the various inner products of them, such as $\boldsymbol{r} \cdot \boldsymbol{I}^{(-1)}$, $\boldsymbol{p} \cdot \boldsymbol{r}$, $\boldsymbol{r} \cdot \boldsymbol{r}$, $\boldsymbol{p} \cdot \boldsymbol{I}^{(-1)} \cdot \boldsymbol{r}$ and $\boldsymbol{r} \cdot \boldsymbol{I}^{(-1)} \cdot \boldsymbol{r}$.

Then, the normal and tangential relative velocities at the contact points can be obtained by taking inner product of $\boldsymbol{v}^{r,C}$ and $\boldsymbol{n}$ or $\boldsymbol{t}$ respectively. The normal relative velocity at the contact point is

$$
\begin{aligned}
\boldsymbol{v}^{r,C} \cdot \boldsymbol{n} = {} & \boldsymbol{v}^{r,C,0} \cdot \boldsymbol{n} - \frac{p_\alpha^{(i)} \cdot \boldsymbol{n}}{m^{(r,ij)}} \\
& + [tr(\boldsymbol{I}^{(-1,j)})(\boldsymbol{r}^{(j)} \cdot \boldsymbol{n}) - (\boldsymbol{r}^{(j)} \cdot \boldsymbol{I}^{(-1,j)} \cdot \boldsymbol{n})](\boldsymbol{p}^{(i)} \cdot \boldsymbol{r}^{(j)}) \\
& + [(\boldsymbol{p}^{(i)} \cdot \boldsymbol{I}^{(-1,j)} \cdot \boldsymbol{n}) - tr(\boldsymbol{I}^{(-1,j)})(\boldsymbol{p}^{(i)} \cdot \boldsymbol{n})](\boldsymbol{r}^{(j)} \cdot \boldsymbol{r}^{(j)}) \\
& - (\boldsymbol{p}^{(i)} \cdot \boldsymbol{I}^{(-1,j)} \cdot \boldsymbol{r}^{(j)})(\boldsymbol{r}^{(j)} \cdot \boldsymbol{n}) - (\boldsymbol{p}^{(i)} \cdot \boldsymbol{I}^{(-1,i)} \cdot \boldsymbol{r}^{(i)})(\boldsymbol{r}^{(i)} \cdot \boldsymbol{n}) \\
& + [tr(\boldsymbol{I}^{(-1,i)})(\boldsymbol{r}^{(i)} \cdot \boldsymbol{n}) - (\boldsymbol{r}^{(i)} \cdot \boldsymbol{I}^{(-1,i)} \cdot \boldsymbol{n})](\boldsymbol{p}^{(i)} \cdot \boldsymbol{r}^{(i)}) \\
& + [(\boldsymbol{p}^{(i)} \cdot \boldsymbol{I}^{(-1,i)} \cdot \boldsymbol{n}) - tr(\boldsymbol{I}^{(-1,i)})(\boldsymbol{p}^{(i)} \cdot \boldsymbol{n})](\boldsymbol{r}^{(i)} \cdot \boldsymbol{r}^{(i)}) \\
& + [\boldsymbol{r}^{(i)} \cdot \boldsymbol{I}^{(-1,i)} \cdot \boldsymbol{r}^{(i)}) + (\boldsymbol{r}^{(j)} \cdot \boldsymbol{I}^{(-1,j)} \cdot \boldsymbol{r}^{(j)})](\boldsymbol{p}^{(i)} \cdot \boldsymbol{n})
\end{aligned}
\tag{2.16}
$$

Using Eq. (2.9) to substitute $\boldsymbol{p}$ by $p^n \boldsymbol{n} + p^t \boldsymbol{t}$, Eq. (2.16) is transformed to

$$
\begin{aligned}
\boldsymbol{v}^{r,C} \cdot \boldsymbol{n} = \boldsymbol{v}^{r,C,0} \cdot \boldsymbol{n} - \frac{p_\alpha^{(n,i)}}{m^{(r,ij)}} \\
&+ p^{(n,i)}[tr(\boldsymbol{I}^{(-1,j)})(\boldsymbol{r}^{(j)} \cdot \boldsymbol{n}) - (\boldsymbol{r}^{(j)} \cdot \boldsymbol{I}^{(-1,j)} \cdot \boldsymbol{n})](\boldsymbol{n} \cdot \boldsymbol{r}^{(j)}) \\
&+ p^{(t,i)}[tr(\boldsymbol{I}^{(-1,j)})(\boldsymbol{r}^{(j)} \cdot \boldsymbol{n}) - (\boldsymbol{r}^{(j)} \cdot \boldsymbol{I}^{(-1,j)} \cdot \boldsymbol{n})](\boldsymbol{t} \cdot \boldsymbol{r}^{(j)}) \\
&+ p^{(n,i)}[(\boldsymbol{n} \cdot \boldsymbol{I}^{(-1,j)} \cdot \boldsymbol{n}) - tr(\boldsymbol{I}^{(-1,j)})](\boldsymbol{r}^{(j)} \cdot \boldsymbol{r}^{(j)}) \\
&+ p^{(t,i)}(\boldsymbol{t} \cdot \boldsymbol{I}^{(-1,j)} \cdot \boldsymbol{n})(\boldsymbol{r}^{(j)} \cdot \boldsymbol{r}^{(j)}) \\
&- p^{(n,i)}[(\boldsymbol{n} \cdot \boldsymbol{I}^{(-1,j)} \cdot \boldsymbol{r}^{(j)})(\boldsymbol{r}^{(j)} \cdot \boldsymbol{n}) + (\boldsymbol{n} \cdot \boldsymbol{I}^{(-1,i)} \cdot \boldsymbol{r}^{(i)})(\boldsymbol{r}^{(i)} \cdot \boldsymbol{n})] \\
&- p^{(t,i)}[(\boldsymbol{t} \cdot \boldsymbol{I}^{(-1,j)} \cdot \boldsymbol{r}^{(j)})(\boldsymbol{r}^{(j)} \cdot \boldsymbol{n}) + (\boldsymbol{t} \cdot \boldsymbol{I}^{(-1,i)} \cdot \boldsymbol{r}^{(i)})(\boldsymbol{r}^{(i)} \cdot \boldsymbol{n})] \\
&+ p^{(n,i)}[tr(\boldsymbol{I}^{(-1,i)})(\boldsymbol{r}^{(i)} \cdot \boldsymbol{n}) - (\boldsymbol{r}^{(i)} \cdot \boldsymbol{I}^{(-1,i)} \cdot \boldsymbol{n})](\boldsymbol{n} \cdot \boldsymbol{r}^{(i)}) \\
&+ p^{(t,i)}[tr(\boldsymbol{I}^{(-1,i)})(\boldsymbol{r}^{(i)} \cdot \boldsymbol{n}) - (\boldsymbol{r}^{(i)} \cdot \boldsymbol{I}^{(-1,i)} \cdot \boldsymbol{n})](\boldsymbol{t} \cdot \boldsymbol{r}^{(i)}) \\
&+ p^{(n,i)}[(\boldsymbol{n} \cdot \boldsymbol{I}^{(-1,i)} \cdot \boldsymbol{n}) - tr(\boldsymbol{I}^{(-1,i)})](\boldsymbol{r}^{(i)} \cdot \boldsymbol{r}^{(i)}) \\
&+ p^{(t,i)}[(\boldsymbol{t} \cdot \boldsymbol{I}^{(-1,i)} \cdot \boldsymbol{n})](\boldsymbol{r}^{(i)} \cdot \boldsymbol{r}^{(i)}) \\
&+ p^{(n,i)}[\boldsymbol{r}^{(i)} \cdot \boldsymbol{I}^{(-1,i)} \cdot \boldsymbol{r}^{(i)}) + (\boldsymbol{r}^{(j)} \cdot \boldsymbol{I}^{(-1,j)} \cdot \boldsymbol{r}^{(j)})] \\
&= \mathcal{A} p^{(n,i)} + \mathcal{B} p^{(t,i)}
\end{aligned}
\tag{2.17}
$$

where

$$
\begin{aligned}
\mathcal{A} = -\frac{1}{m^{(r,ij)}} \\
&+ [tr(\boldsymbol{I}^{(-1,j)})(\boldsymbol{r}^{(j)} \cdot \boldsymbol{n}) - (\boldsymbol{r}^{(j)} \cdot \boldsymbol{I}^{(-1,j)} \cdot \boldsymbol{n}) - (\boldsymbol{n} \cdot \boldsymbol{I}^{(-1,j)} \cdot \boldsymbol{r}^{(j)})](\boldsymbol{n} \cdot \boldsymbol{r}^{(j)}) \\
&+ [tr(\boldsymbol{I}^{(-1,i)})(\boldsymbol{r}^{(i)} \cdot \boldsymbol{n}) - (\boldsymbol{r}^{(i)} \cdot \boldsymbol{I}^{(-1,i)} \cdot \boldsymbol{n}) - (\boldsymbol{n} \cdot \boldsymbol{I}^{(-1,i)} \cdot \boldsymbol{r}^{(i)})](\boldsymbol{n} \cdot \boldsymbol{r}^{(i)}) \\
&+ [(\boldsymbol{n} \cdot \boldsymbol{I}^{(-1,j)} \cdot \boldsymbol{n}) - tr(\boldsymbol{I}^{(-1,j)})](\boldsymbol{r}^{(j)} \cdot \boldsymbol{r}^{(j)}) \\
&+ [(\boldsymbol{n} \cdot \boldsymbol{I}^{(-1,i)} \cdot \boldsymbol{n}) - tr(\boldsymbol{I}^{(-1,i)})](\boldsymbol{r}^{(i)} \cdot \boldsymbol{r}^{(i)}) \\
&+ [(\boldsymbol{r}^{(i)} \cdot \boldsymbol{I}^{(-1,i)} \cdot \boldsymbol{r}^{(i)}) + (\boldsymbol{r}^{(j)} \cdot \boldsymbol{I}^{(-1,j)} \cdot \boldsymbol{r}^{(j)})]
\end{aligned}
\tag{2.18}
$$

$$
\begin{aligned}
\mathcal{B} = [tr(\boldsymbol{I}^{(-1,j)})(\boldsymbol{r}^{(j)} \cdot \boldsymbol{n}) - (\boldsymbol{r}^{(j)} \cdot \boldsymbol{I}^{(-1,j)} \cdot \boldsymbol{n})](\boldsymbol{t} \cdot \boldsymbol{r}^{(j)}) \\
&- (\boldsymbol{t} \cdot \boldsymbol{I}^{(-1,j)} \cdot \boldsymbol{r}^{(j)})(\boldsymbol{r}^{(j)} \cdot \boldsymbol{n}) - (\boldsymbol{t} \cdot \boldsymbol{I}^{(-1,i)} \cdot \boldsymbol{r}^{(i)})(\boldsymbol{r}^{(i)} \cdot \boldsymbol{n}) \\
&+ [tr(\boldsymbol{I}^{(-1,i)})(\boldsymbol{r}^{(i)} \cdot \boldsymbol{n}) - (\boldsymbol{r}^{(i)} \cdot \boldsymbol{I}^{(-1,i)} \cdot \boldsymbol{n})](\boldsymbol{t} \cdot \boldsymbol{r}^{(i)}) \\
&+ (\boldsymbol{t} \cdot \boldsymbol{I}^{(-1,j)} \cdot \boldsymbol{n})(\boldsymbol{r}^{(j)} \cdot \boldsymbol{r}^{(j)}) + (\boldsymbol{t} \cdot \boldsymbol{I}^{(-1,i)} \cdot \boldsymbol{n})(\boldsymbol{r}^{(i)} \cdot \boldsymbol{r}^{(i)})
\end{aligned}
\tag{2.19}
$$

Recall the definition of coefficient of restitution in Eq. (2.1), $\boldsymbol{v}^{r,C} \cdot \boldsymbol{n} = -e\boldsymbol{v}^{r,C,0} \cdot \boldsymbol{n}$. Substituting it into Eq. (2.17) we get

$$
\mathcal{A} p^{(n,i)} + \mathcal{B} p^{(t,i)} + (1+e)(\boldsymbol{v}^{r,C,0} \cdot \boldsymbol{n}) = 0 \tag{2.20}
$$

In a similar manner, the tangential relative velocity at the contact point is expressed by

$$
\begin{aligned}
\boldsymbol{v}^{r,C}\cdot\boldsymbol{t} = {} & \boldsymbol{v}^{r,C,0}\cdot\boldsymbol{t} - \frac{p_{\alpha}^{(t,i)}}{m^{(r,ij)}} \\
& + p^{(n,i)}[tr(\boldsymbol{I}^{(-1,j)})(\boldsymbol{r}^{(j)}\cdot\boldsymbol{t}) - (\boldsymbol{r}^{(j)}\cdot\boldsymbol{I}^{(-1,j)}\cdot\boldsymbol{t})](\boldsymbol{n}\cdot\boldsymbol{r}^{(j)}) \\
& + p^{(t,i)}[tr(\boldsymbol{I}^{(-1,j)})(\boldsymbol{r}^{(j)}\cdot\boldsymbol{t}) - (\boldsymbol{r}^{(j)}\cdot\boldsymbol{I}^{(-1,j)}\cdot\boldsymbol{t})](\boldsymbol{t}\cdot\boldsymbol{r}^{(j)}) \\
& + p^{(n,i)}[(\boldsymbol{n}\cdot\boldsymbol{I}^{(-1,j)}\cdot\boldsymbol{t})](\boldsymbol{r}^{(j)}\cdot\boldsymbol{r}^{(j)}) \\
& + p^{(t,i)}[(\boldsymbol{t}\cdot\boldsymbol{I}^{(-1,j)}\cdot\boldsymbol{t}) - tr(\boldsymbol{I}^{(-1,j)})](\boldsymbol{r}^{(j)}\cdot\boldsymbol{r}^{(j)}) \\
& - p^{(n,i)}[(\boldsymbol{n}\cdot\boldsymbol{I}^{(-1,j)}\cdot\boldsymbol{r}^{(j)})(\boldsymbol{r}^{(j)}\cdot\boldsymbol{t}) + (\boldsymbol{n}\cdot\boldsymbol{I}^{(-1,i)}\cdot\boldsymbol{r}^{(i)})(\boldsymbol{r}^{(i)}\cdot\boldsymbol{t})] \\
& - p^{(t,i)}[(\boldsymbol{t}\cdot\boldsymbol{I}^{(-1,j)}\cdot\boldsymbol{r}^{(j)})(\boldsymbol{r}^{(j)}\cdot\boldsymbol{t}) + (\boldsymbol{t}\cdot\boldsymbol{I}^{(-1,i)}\cdot\boldsymbol{r}^{(i)})(\boldsymbol{r}^{(i)}\cdot\boldsymbol{t})] \\
& + p^{(n,i)}[tr(\boldsymbol{I}^{(-1,i)})(\boldsymbol{r}^{(i)}\cdot\boldsymbol{t}) - (\boldsymbol{r}^{(i)}\cdot\boldsymbol{I}^{(-1,i)}\cdot\boldsymbol{t})](\boldsymbol{n}\cdot\boldsymbol{r}^{(i)}) \\
& + p^{(t,i)}[tr(\boldsymbol{I}^{(-1,i)})(\boldsymbol{r}^{(i)}\cdot\boldsymbol{t}) - (\boldsymbol{r}^{(i)}\cdot\boldsymbol{I}^{(-1,i)}\cdot\boldsymbol{t})](\boldsymbol{t}\cdot\boldsymbol{r}^{(i)}) \\
& + p^{(n,i)}[(\boldsymbol{n}\cdot\boldsymbol{I}^{(-1,i)}\cdot\boldsymbol{t})](\boldsymbol{r}^{(i)}\cdot\boldsymbol{r}^{(i)}) \\
& + p^{(t,i)}[(\boldsymbol{t}\cdot\boldsymbol{I}^{(-1,i)}\cdot\boldsymbol{t}) - tr(\boldsymbol{I}^{(-1,i)})](\boldsymbol{r}^{(i)}\cdot\boldsymbol{r}^{(i)}) \\
& + p^{(t,i)}[\boldsymbol{r}^{(i)}\cdot\boldsymbol{I}^{(-1,i)}\cdot\boldsymbol{r}^{(i)}) + (\boldsymbol{r}^{(j)}\cdot\boldsymbol{I}^{(-1,j)}\cdot\boldsymbol{r}^{(j)})] \\
= {} & \mathcal{C}p^{(n,i)} + \mathcal{D}p^{(t,i)}
\end{aligned}
\tag{2.21}
$$

where

$$
\begin{aligned}
\mathcal{C} = {} & [tr(\boldsymbol{I}^{(-1,j)})(\boldsymbol{r}^{(j)}\cdot\boldsymbol{t}) - (\boldsymbol{r}^{(j)}\cdot\boldsymbol{I}^{(-1,j)}\cdot\boldsymbol{t})](\boldsymbol{n}\cdot\boldsymbol{r}^{(j)}) \\
& + [tr(\boldsymbol{I}^{(-1,i)})(\boldsymbol{r}^{(i)}\cdot\boldsymbol{t}) - (\boldsymbol{r}^{(i)}\cdot\boldsymbol{I}^{(-1,i)}\cdot\boldsymbol{t})](\boldsymbol{n}\cdot\boldsymbol{r}^{(i)}) \\
& - (\boldsymbol{n}\cdot\boldsymbol{I}^{(-1,j)}\cdot\boldsymbol{r}^{(j)})(\boldsymbol{r}^{(j)}\cdot\boldsymbol{t}) - (\boldsymbol{n}\cdot\boldsymbol{I}^{(-1,i)}\cdot\boldsymbol{r}^{(i)})(\boldsymbol{r}^{(i)}\cdot\boldsymbol{t}) \\
& + [(\boldsymbol{n}\cdot\boldsymbol{I}^{(-1,j)}\cdot\boldsymbol{t})](\boldsymbol{r}^{(j)}\cdot\boldsymbol{r}^{(j)}) + [(\boldsymbol{n}\cdot\boldsymbol{I}^{(-1,i)}\cdot\boldsymbol{t})](\boldsymbol{r}^{(i)}\cdot\boldsymbol{r}^{(i)})
\end{aligned}
\tag{2.22}
$$

$$
\begin{aligned}
\mathcal{D} = {} & -\frac{1}{m^{(r,ij)}} \\
& + [tr(\boldsymbol{I}^{(-1,j)})(\boldsymbol{r}^{(j)}\cdot\boldsymbol{t}) - (\boldsymbol{r}^{(j)}\cdot\boldsymbol{I}^{(-1,j)}\cdot\boldsymbol{t}) - (\boldsymbol{t}\cdot\boldsymbol{I}^{(-1,j)}\cdot\boldsymbol{r}^{(j)})](\boldsymbol{t}\cdot\boldsymbol{r}^{(j)}) \\
& + [tr(\boldsymbol{I}^{(-1,i)})(\boldsymbol{r}^{(i)}\cdot\boldsymbol{t}) - (\boldsymbol{r}^{(i)}\cdot\boldsymbol{I}^{(-1,i)}\cdot\boldsymbol{t}) - (\boldsymbol{t}\cdot\boldsymbol{I}^{(-1,i)}\cdot\boldsymbol{r}^{(i)})](\boldsymbol{t}\cdot\boldsymbol{r}^{(i)}) \\
& + [(\boldsymbol{t}\cdot\boldsymbol{I}^{(-1,j)}\cdot\boldsymbol{t}) - tr(\boldsymbol{I}^{(-1,j)})](\boldsymbol{r}^{(j)}\cdot\boldsymbol{r}^{(j)}) \\
& + [(\boldsymbol{t}\cdot\boldsymbol{I}^{(-1,i)}\cdot\boldsymbol{t}) - tr(\boldsymbol{I}^{(-1,i)})](\boldsymbol{r}^{(i)}\cdot\boldsymbol{r}^{(i)}) \\
& + [\boldsymbol{r}^{(i)}\cdot\boldsymbol{I}^{(-1,i)}\cdot\boldsymbol{r}^{(i)}) + (\boldsymbol{r}^{(j)}\cdot\boldsymbol{I}^{(-1,j)}\cdot\boldsymbol{r}^{(j)})]
\end{aligned}
\tag{2.23}
$$

In an actual collision process, particle-particle friction can take place between the two contacting points 'C''' and 'C'', no matter whether the inter-particle sliding motion takes place or not. Therefore, the tangential velocity can be zero when the sliding motion doesn't occur during collision. By $\boldsymbol{v}^{r,C}\cdot\boldsymbol{t}=0$, Eq. (2.21) gives rise to

$$\mathcal{C}p^{(n,i)} + \mathcal{D}p^{(t,i)} + (\boldsymbol{v}^{r,C,0} \cdot \boldsymbol{t}) = 0 \tag{2.24}$$

Otherwise, the tangential force can be given by the Coulomb's law of friction:

$$p^{(t,i)} = \mu_f p^{(n,i)} \tag{2.25}$$

where μ_f is the coefficient of friction. Combining Eqs. (2.20), (2.24) and (2.25), the following solutions of $p^{(n,i)}$ and $p^{(t,i)}$ can be obtained:

- For collision without inter-particle sliding motion: If the relative motion on the contact point is absent, it is called nonsliding collision, which implies $\boldsymbol{v}^{r,C} \cdot \boldsymbol{t} = 0$. In this case, Eqs. (2.20) and (2.24) are combined together to solve the collisional impulse

$$\begin{cases} \mathcal{A}p^{(n,i)} + \mathcal{B}p^{(t,i)} + (1+e)(\boldsymbol{v}^{r,C,0} \cdot \boldsymbol{n}) = 0 \\ \mathcal{C}p^{(n,i)} + \mathcal{D}p^{(t,i)} + (\boldsymbol{v}^{r,C,0} \cdot \boldsymbol{t}) = 0 \end{cases} \tag{2.26}$$

  and the solutions are

$$\begin{cases} p^{(n,i)} = \dfrac{\mathcal{B}(\boldsymbol{v}^{r,C,0} \cdot \boldsymbol{t}) - \mathcal{D}(1+e)(\boldsymbol{v}^{r,C,0} \cdot \boldsymbol{n})}{\mathcal{AD} - \mathcal{BC}} \\ p^{(t,i)} = \dfrac{\mathcal{C}(1+e)(\boldsymbol{v}^{r,C,0} \cdot \boldsymbol{n}) - \mathcal{A}(\boldsymbol{v}^{r,C,0} \cdot \boldsymbol{t})}{\mathcal{AD} - \mathcal{BC}} \end{cases} \tag{2.27}$$

- For collision with inter-particle sliding motion: the impulse components can be obtained by solving

$$\begin{cases} \mathcal{A}p^{(n,i)} + \mathcal{B}p^{(t,i)} + (1+e)(\boldsymbol{v}^{r,C,0} \cdot \boldsymbol{n}) = 0 \\ p^{(t,i)} = \mu_f p^{(n,i)} \end{cases} \tag{2.28}$$

  and the solutions are

$$\begin{cases} p^{(n,i)} = \dfrac{-(1+e)(\boldsymbol{v}^{r,C,0} \cdot \boldsymbol{n})}{\mathcal{A} + \mu_f \mathcal{B}} \\ p^{(t,i)} = \dfrac{-\mu_f(1+e)(\boldsymbol{v}^{r,C,0} \cdot \boldsymbol{n})}{\mathcal{A} + \mu_f \mathcal{B}} \end{cases} \tag{2.29}$$

- For the inter-particle sliding criterion, it can be quantitatively determined by

$$\mu_\theta = \left| \frac{\mathcal{C}(1+e)(\boldsymbol{v}^{r,C,0} \cdot \boldsymbol{n}) - \mathcal{A}(\boldsymbol{v}^{r,C,0} \cdot \boldsymbol{t})}{\mathcal{B}(\boldsymbol{v}^{r,C,0} \cdot \boldsymbol{t}) - \mathcal{D}(1+e)(\boldsymbol{v}^{r,C,0} \cdot \boldsymbol{n})} \right| \begin{cases} > \mu_f, \text{ sliding} \\ \leq \mu_f, \text{ nonsliding} \end{cases} \tag{2.30}$$

To say conclusively, the collisional impulse can be obtained by

$$p^{(i)}=\begin{cases}\dfrac{\mathcal{B}v^{r,C,0}\cdot t-\mathcal{D}(1+e)v^{r,C,0}\cdot n}{\mathcal{AD}-\mathcal{BC}}n \\ +\dfrac{\mathcal{C}(1+e)v^{r,C,0}\cdot n-\mathcal{A}v^{r,C,0}\cdot t}{\mathcal{AD}-\mathcal{BC}}t,\ \text{if } \mu_\theta\le\mu_f \\ \dfrac{-(1+e)(v^{r,C,0}\cdot n)}{\mathcal{A}+\mu_f\mathcal{B}}(n+\mu_f t),\ \text{if } \mu_\theta>\mu_f\end{cases} \tag{2.31}$$

where $v^{r,C,0}=v^{(j,0)}+\omega^{(j,0)}\times r^{(j)}-v^{(i,0)}-\omega^{(i,0)}\times r^{(i)}$.

2.2.2 Consistency with the hard sphere model

Above are generalized solutions of the particle-particle collision of arbitrary shapes as no information on particle shape is involved. As conventional solutions for spherical particle collisions were available from prior studies (see the standard hard sphere model [52,174,175]), it is useful to degenerate the solutions to special cases for spherical particles to validate the correctness and consistency of the present model.

For uniform spheres, the aforementioned general conditions are simplified as follow:

- The inertia tensor is a scalar value for spheres $I^{(i)}=I^{(j)}=IE=\frac{2}{5}mr^2E$, where $r^{(i)}=r^{(j)}=r$ (uniform), $I=\frac{2}{5}mr^2$, and $I^{(-1,i)}=I^{(-1,j)}=\frac{1}{I}E$, and E is the identity matrix, namely $E_{\alpha\beta}=\delta_{\alpha\beta}$.
- The pair of contact points are on the line between the centers of mass of the colliding particles. We have $r^{(i)}=rn$, $r^{(j)}=-rn$, and $r^{(j)}\cdot t=r^{(i)}\cdot t=0$.
- $\frac{1}{m^{(r,ij)}}=\frac{m_i+m_j}{m_im_j}=\frac{2}{m}$, $r^{(j)}\cdot n=-rn\cdot n=-r$, $r^{(i)}\cdot n=rn\cdot n=r$
- $n\cdot I^{(-1,j)}\cdot r^{(j)}=r^{(j)}\cdot I^{(-1,j)}\cdot n=\frac{1}{I}n\cdot r^{(j)}=-\frac{r}{I}$
- $n\cdot I^{(-1,i)}\cdot r^{(i)}=r^{(i)}\cdot I^{(-1,i)}\cdot n=\frac{1}{I}n\cdot r^{(i)}=\frac{r}{I}$
- $r^{(j)}\cdot I^{(-1,j)}\cdot r^{(j)}=r^{(i)}\cdot I^{(-1,i)}\cdot r^{(i)}=\frac{r^2}{I}$
- $n\cdot I^{(-1,i)}\cdot n=n\cdot I^{(-1,j)}\cdot n=\frac{1}{I}$
- $tr I^{(-1,i)}=tr I^{(-1,j)}=\frac{3}{I}$
- $t\cdot I^{(-1,i)}\cdot n=t\cdot I^{(-1,j)}\cdot n=\frac{1}{I}t\cdot n=0$
- $t\cdot I^{(-1,i)}\cdot r^{(i)}=t\cdot I^{(-1,j)}\cdot r^{(j)}=\frac{1}{I}t\cdot r^{(j)}=0$
- $t\cdot I^{(-1,j)}\cdot t=\frac{1}{I}$
- $t\cdot I^{(-1,i)}\cdot t=\frac{1}{I}$
- $\mathcal{A}=-\frac{1}{m^{(r,ij)}}-[\frac{-3r}{I}+\frac{2r}{I})]r+[\frac{3r}{I}-\frac{2r}{I}]r+[\frac{1}{I}-\frac{3}{I}]r^2+[\frac{1}{I}-\frac{3}{I}]r^2+[\frac{r^2}{I}+\frac{r^2}{I}]=-\frac{2}{m}$
- $\mathcal{B}=\mathcal{C}=0$
- $\mathcal{D}=-\frac{1}{m^{(r,ij)}}+[\frac{1}{I}-\frac{3}{I}]r^2+[\frac{1}{I}-\frac{3}{I}]r^2+[\frac{r^2}{I}+\frac{r^2}{I}]=-\frac{2}{m}-\frac{2r^2}{I}=-\frac{7}{m}$

- $\mu_\theta = \left|\frac{2(\boldsymbol{v}^{r,C,0}\cdot\boldsymbol{t})}{7(1+e)(\boldsymbol{v}^{r,C,0}\cdot\boldsymbol{n})}\right| \begin{cases} > \mu_f, \text{ sliding} \\ \leq \mu_f, \text{ nonsliding} \end{cases}$
- $\boldsymbol{v}^{r,C,0}\cdot\boldsymbol{t} = \boldsymbol{v}^{(j,0)}\cdot\boldsymbol{t} + \boldsymbol{\omega}^{(j,0)}\times\boldsymbol{r}^{(j)}\cdot\boldsymbol{t} - \boldsymbol{v}^{(i,0)}\cdot\boldsymbol{t} - \boldsymbol{\omega}^{(i,0)}\times\boldsymbol{r}^{(i)}\cdot\boldsymbol{t}$
  $= (\boldsymbol{v}^{(j,0)} - \boldsymbol{v}^{(i,0)})\cdot\boldsymbol{t} - r^{(j)}\boldsymbol{\omega}^{(j,0)}\times\boldsymbol{n}\cdot\boldsymbol{t} - r^{(i)}\boldsymbol{\omega}^{(i,0)}\times\boldsymbol{n}\cdot\boldsymbol{t}$
- $\boldsymbol{v}^{r,C,0}\cdot\boldsymbol{n} = \boldsymbol{v}^{(j,0)}\cdot\boldsymbol{n} + \boldsymbol{\omega}^{(j,0)}\times\boldsymbol{r}^{(j)}\cdot\boldsymbol{n} - \boldsymbol{v}^{(i,0)}\cdot\boldsymbol{n} - \boldsymbol{\omega}^{(i,0)}\times\boldsymbol{r}^{(i)}\cdot\boldsymbol{n}$
  $= (\boldsymbol{v}^{(j,0)} - \boldsymbol{v}^{(i,0)})\cdot\boldsymbol{n}$
- $p^{(i,n)} = \frac{-(1+e)}{A}(\boldsymbol{v}^{r,C,0}\cdot\boldsymbol{n}) = \frac{m(1+e)}{2}(\boldsymbol{v}^{r,C,0}\cdot\boldsymbol{n})$
- $p^{(i,t)} = \min\{\frac{-(\boldsymbol{v}^{r,C,0}\cdot\boldsymbol{t})}{D}, \mu_f p^{(i,n)}\} = \min\{\frac{m}{7}(\boldsymbol{v}^{r,C,0}\cdot\boldsymbol{t}), \mu_f p^{(i,n)}\}$
- $\boldsymbol{p}^{(i)} = \begin{cases} \frac{m(1+e)(\boldsymbol{v}^{r,C,0}\cdot\boldsymbol{n})}{2}\boldsymbol{n} + \frac{m(\boldsymbol{v}^{r,C,0}\cdot\boldsymbol{t})}{7}\boldsymbol{t}, \text{ if } \mu_\theta \leq \mu_f \\ \frac{m(1+e)(\boldsymbol{v}^{r,C,0}\cdot\boldsymbol{n})}{2}(\boldsymbol{n} + \mu_f\boldsymbol{t}), \text{ if } \mu_\theta > \mu_f \end{cases}$

The solution of $\boldsymbol{p}^{(i)}$ is exactly the same as in the standard hard sphere models, as described in Refs. [52,174,175]. Thus, the present generalized solution correctly reproduces the classical hard sphere model.

2.2.3 Numerical procedures for implementation

Numerical procedures to implement the present model to perform numerical simulation of particle flows are illustrated in Fig. 2.2. The main procedures are composed of sub-procedures called 'free motion', 'collision detection', 'collision solution', and 'post-velocity updating'. 'Free motion' tracks the particle motion by solving the Newton's 2nd law, whereas the 'collision solution' solves the post-velocities by using the current model as depicted above.

In the collision detection procedure, every case of 'vertex-vertex', 'vertex-edge', 'vertex-face' and 'edge-edge' collisions is checked according to the positions and orientations of the particles. For example, as shown in Fig. 2.3, the 'vertex-face' collision is checked by computing the distance between vertex 'A' and surface 'α' as follows:

$$d_{A\to a} = \frac{|c_1x_A + c_2y_A + c_3z_A + c_4|}{\sqrt{c_1^2 + c_2^2 + c_3^2}} < \epsilon \tag{2.32}$$

where $c_1x + c_2y + c_3z + c_4 = 0$ is the surface 'α', and (x_A, y_A, z_A) is the vertex A. $\epsilon = 0.01l$ is a small threshold value, and l is the edge length of particle. If the projection of the vertex 'A' onto the face 'α' is inside the face 'α', a vertex-face collision is detected. The 'vertex-vertex' and 'vertex-edge' collisions can be checked in similar ways.

Moreover, for the 'edge-edge' collision, assume the edges '$\overline{AB}$' and '$\overline{CD}$' are formulated as $\boldsymbol{x}_A + k_{\overline{AB}}\boldsymbol{e}_{\overline{AB}}$ and $\boldsymbol{x}_C + k_{\overline{CD}}\boldsymbol{e}_{\overline{CD}}$ respectively, where $\boldsymbol{x}_A$ and $\boldsymbol{x}_C$ are the position vectors of the vertices 'A' and 'C' respectively, and '$\boldsymbol{e}_{\overline{AB}}$' and '$\boldsymbol{e}_{\overline{CD}}$' are unit vectors from vertex 'A' to 'B' and from vertex 'C' to 'D' respectively. If the

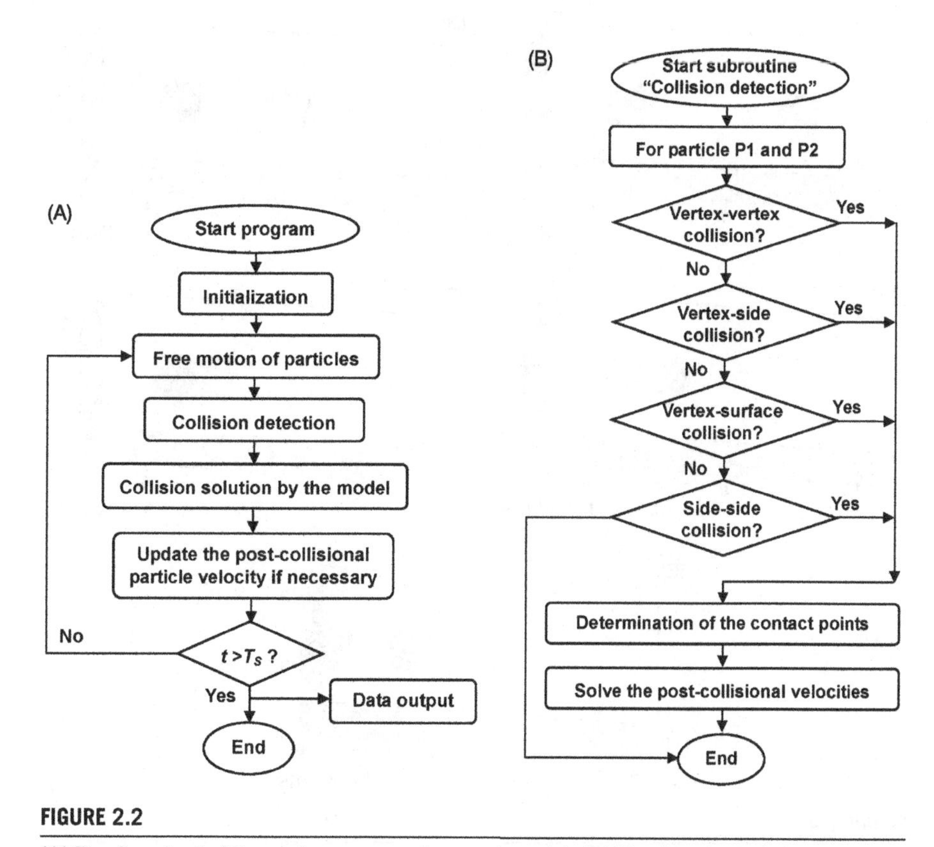

FIGURE 2.2

(A) The flowchart of the main program of current model. (B) The flowchart of the collision detection procedure.

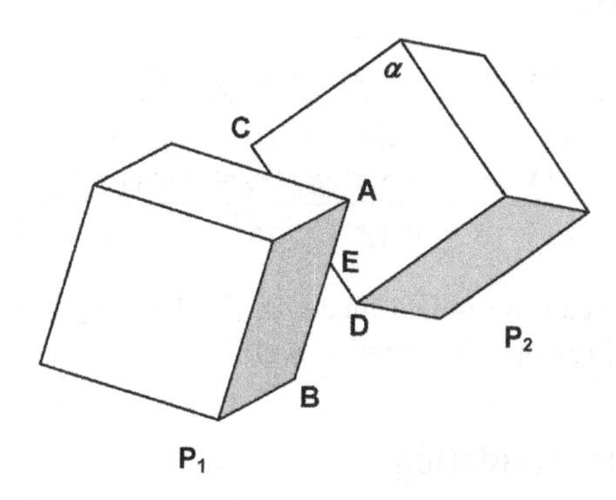

FIGURE 2.3

A schematic diagram of detection of the cubic-to-cubic particle collision.

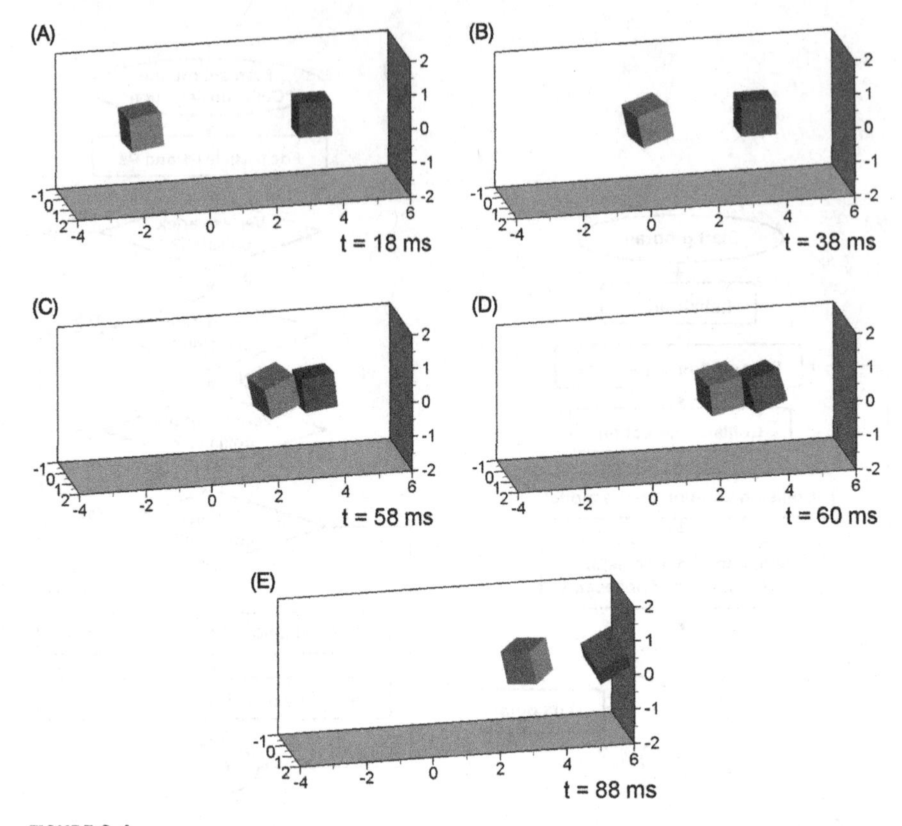

FIGURE 2.4

The procedure of collision between a pair of cubic particles.

scalar parameters satisfy

$$\begin{aligned} k_{\overline{AB}} &= \frac{(\boldsymbol{x}_C - \boldsymbol{x}_A) \times \boldsymbol{e}_{\overline{CD}} \cdot (\boldsymbol{e}_{\overline{AB}} \times \boldsymbol{e}_{\overline{CD}})}{|\boldsymbol{e}_{\overline{AB}} \times \boldsymbol{e}_{\overline{CD}}|^2} \in (0, 1) \\ k_{\overline{CD}} &= \frac{(\boldsymbol{x}_A - \boldsymbol{x}_C) \times \boldsymbol{e}_{\overline{AB}} \cdot (\boldsymbol{e}_{\overline{CD}} \times \boldsymbol{e}_{\overline{AB}})}{|\boldsymbol{e}_{\overline{AB}} \times \boldsymbol{e}_{\overline{CD}}|^2} \in (0, 1) \end{aligned} \tag{2.33}$$

and the distance between the points meets $|\boldsymbol{x}_A + k_{\overline{AB}}\boldsymbol{e}_{\overline{AB}} - \boldsymbol{x}_C - k_{\overline{CD}}\boldsymbol{e}_{\overline{CD}}| < \epsilon$, an 'edge-edge' collision can be detected as well.

2.2.4 GHPM model validation

Theoretical validation

Following the above numerical procedures, a representative collision between a pair of cubic particles is simulated by the GHPM model. As shown in Fig. 2.4, at the beginning ($t = 0$ s), the west cubic particle at $\boldsymbol{x}_w^0 = (0, -3.3641, 0)$ mm is moving in the y-direction with velocities of $\boldsymbol{v}_w^0 = (0, 100, 0)$ mm/s and $\boldsymbol{\omega}_w^0 = (2\pi, \pi, 2\pi)$ rad/s directly toward the east rest cubic particle at $\boldsymbol{x}_e^0 = (0, 3.3641, 0)$ mm with $\boldsymbol{v}_e^0 = 0$ and $\boldsymbol{\omega}_e^0 = 0$. They have the same size ($l = 1$ mm^3), same density ($\rho_p = 1000$ kg/m^3), and the same mass $m = 10^{-6}$ kg and inertia tensor $\boldsymbol{I}_w^0 = \boldsymbol{I}_e^0 = 1.667 \times 10^{-13} \begin{pmatrix} 1 & 0 & 0 \\ 0 & 1 & 0 \\ 0 & 0 & 1 \end{pmatrix}$ kg·m^2. In collision, the restitution coefficient is set as $e = 0.9$ and the friction coefficient is set as $\mu = 0.3$.

A total time of 120 ms is simulated using a fine simulation time step of $\delta t = 0.1$ ms. Particle-particle collisions take place from $t = 58$ ms to 60 ms. At about $t = 58$ ms, the first collision takes place between a lower vertex of the west particle and a lower point on the face of the east particle. Then, they rotate immediately and the next collision between them takes place. At about $t = 60$ ms, an edge-to-edge collision takes place. The series of particle-particle collision involve three momentum transfers between the pair of particles for both the translational and rotational motions.

After the collision (e.g. $t = 0.12$ s), the east particle obtains translational and rotational momenta from the collision ($\boldsymbol{v}_w^0 = (2.9116, 28.41, 0.4165)$ mm/s, $\boldsymbol{\omega}_w^0 = (55.526, 20.1, 17.041)$ rad/s). Meanwhile, the west particle loses a part of translational and rotational momenta ($\boldsymbol{v}_e^0 = (-2.9116, 71.59, -0.4165)$ mm/s, $\boldsymbol{\omega}_e^0 = (-48.35, -17.273, -33.333)$ rad/s). The pre-collisional translational and post-collisional momenta are

$$\boldsymbol{T}^0 = \begin{pmatrix} T_x^0 \\ T_y^0 \\ T_z^0 \end{pmatrix} = \begin{pmatrix} \Sigma m v_x^0 \\ \Sigma m v_y^0 \\ \Sigma m v_z^0 \end{pmatrix} = \begin{pmatrix} 0 \\ 100m \\ 0 \end{pmatrix} \tag{2.34}$$

$$\boldsymbol{T} = \begin{pmatrix} T_x \\ T_y \\ T_z \end{pmatrix} = \begin{pmatrix} m v_{x,w} + m v_{x,e} \\ m v_{y,w} + m v_{y,e} \\ m v_{z,w} + m v_{z,e} \end{pmatrix} = \begin{pmatrix} (2.9116 - 2.9116)m \\ (71.59 + 28.41)m \\ (0.4165 - 0.4165)m \end{pmatrix} = \begin{pmatrix} 0 \\ 100m \\ 0 \end{pmatrix} \tag{2.35}$$

respectively, where $m = 10^{-6}$ kg is the mass of particle. With $\boldsymbol{T}^0 = \boldsymbol{T}$, the translational momentum is conserved in collision. On the other hand, the pre-collisional and post-collisional angular momenta with respect to the original point are

$$\begin{aligned} \boldsymbol{L}^0 &= \Sigma(\boldsymbol{r}^0 \times m\boldsymbol{v}^0 + \boldsymbol{I}\boldsymbol{\omega}^0) = \begin{vmatrix} \boldsymbol{e}_x & \boldsymbol{e}_y & \boldsymbol{e}_z \\ 0 & 3.3641 & 0 \\ 0 & 100m & 0 \end{vmatrix} + \begin{vmatrix} \boldsymbol{e}_x & \boldsymbol{e}_y & \boldsymbol{e}_z \\ 0 & -3.3641 & 0 \\ 0 & 0 & 0 \end{vmatrix} \\ &+ \frac{m}{6} \begin{pmatrix} 2\pi \\ \pi \\ 2\pi \end{pmatrix} + \frac{m}{6} \begin{pmatrix} 0 \\ 0 \\ 0 \end{pmatrix} \\ &= \begin{pmatrix} 1.04719 \\ 0.52360 \\ 1.04719 \end{pmatrix} \end{aligned} \tag{2.36}$$

$$
\begin{aligned}
\boldsymbol{L} = \Sigma(\boldsymbol{r} \times m\boldsymbol{v} + \boldsymbol{I}\boldsymbol{\omega}) &= \begin{vmatrix} \boldsymbol{e}_x & \boldsymbol{e}_y & \boldsymbol{e}_z \\ 0.1664 & 4.1201 & 0.03282 \\ 2.9116 & 28.41 & 0.4165 \end{vmatrix} \\
&+ \begin{vmatrix} \boldsymbol{e}_x & \boldsymbol{e}_y & \boldsymbol{e}_z \\ -0.1664 & 7.8799 & -0.03282 \\ -2.9116 & 71.59 & -0.4165 \end{vmatrix} + \frac{m}{6}\begin{pmatrix} 55.526 \\ 20.1 \\ 17.041 \end{pmatrix} + \frac{m}{6}\begin{pmatrix} -48.350 \\ -17.273 \\ -33.333 \end{pmatrix} \\
&= \begin{pmatrix} 1.0471 \\ 0.52367 \\ 1.04655 \end{pmatrix}
\end{aligned}
\tag{2.37}
$$

respectively. We found $\boldsymbol{L}^0 = \boldsymbol{L}$. Thus, the angular momenta are also conserved in collision.

In addition, the translational and rotational kinetic energies are defined as $E_{tr} = \frac{1}{2}m\boldsymbol{V} \cdot \boldsymbol{V}$ and $E_{ro} = \frac{1}{2}m\boldsymbol{\omega} \cdot \boldsymbol{I} \cdot \boldsymbol{\omega}$ respectively. In simulation, the inertia tensors at $t = 0.12$ s are

$$
\boldsymbol{I}_w = \begin{pmatrix} 1.667 \times 10^2 & 2.511 \times 10^{-12} & 9.920 \times 10^{-12} \\ 2.540 \times 10^{-12} & 1.667 \times 10^2 & 6.233 \times 10^{-12} \\ 9.908 \times 10^{-12} & 6.291 \times 10^{-12} & 1.667 \times 10^2 \end{pmatrix} \times 10^{-15} \tag{2.38}
$$

$$
\boldsymbol{I}_e = \begin{pmatrix} 1.667 \times 10^2 & 4.541 \times 10^{-12} & -6.370 \times 10^{-12} \\ 4.517 \times 10^{-12} & 1.667 \times 10^2 & 2.697 \times 10^{-12} \\ -6.352 \times 10^{-12} & 2.677 \times 10^{-12} & 1.667 \times 10^2 \end{pmatrix} \times 10^{-15} \tag{2.39}
$$

Thus, the pre-collisional and post-collisional kinetic energies are ($E^0_{tr,w} = 50 \times 10^{-10}$ J, $E^0_{ro,w} = 0.074 \times 10^{-10}$ J, $E^0_{tr,e} = E^0_{ro,e} = 0$) and ($E_{tr,w} = 4.079 \times 10^{-10}$ J, $E_{ro,w} = 3.148 \times 10^{-10}$ J, $E_{ro,e} = 3.123 \times 10^{-10}$ J, $E_{tr,e} = 25.67 \times 10^{-10}$ J) respectively.

Fig. 2.5 shows the variations of the kinetic energies in collision. At $t = 58$ ms, the west particle lost a large part of its translational kinetic energy, which is partially transferred into the translational kinetic energy of the east particle by the repulsive impulse, and partially transformed into the rotational kinetic energies of the two particles caused by the collisional torque.

At $t = 60$ ms, another collision takes place between the upper edges of the two particles. The collisional impulse reduces the translational kinetic energy of the west particle again, and increases the translational kinetic energy of the east particle. But the collisional torque resists the rotational motion of each particle. Thus, the rotational energies of the pair of particles are also decreased. As a result, the translational energy of the west particle and the rotational energies of the two particles are transformed into the translational energy of the east particle. However, no matter how many kinetic energies are transformed or transferred from one to the other. The total kinetic energy is always decreasing in every collision, which is caused by the incomplete elastic collision nature ($e = 0.9$). Thus, the results follow the laws of energy.

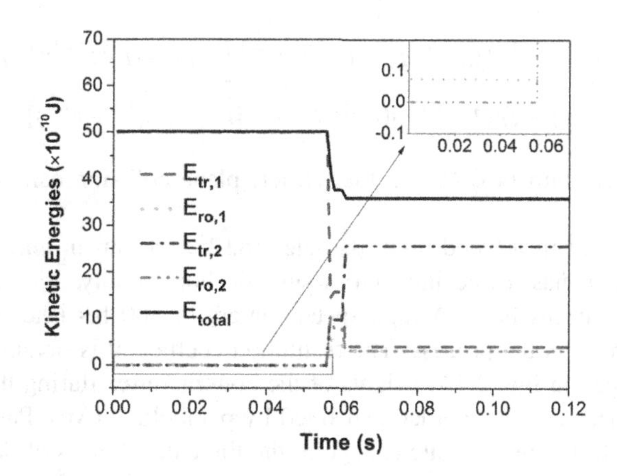

FIGURE 2.5

The variation of kinetic energies in this collision.

In conclusion, the total translational and angular momenta are conserved in collision, and the energy laws are followed. To this regard, the current model has been validated by theoretical laws.

Experimental validation

In this section, an experiment is performed to further validate the model. In this model as well as in other existing DEM models, the restitution coefficient is an input related to a material property which cannot be predicted theoretically in advance. With the restitution coefficient known, the post-collision velocities and motion trajectories can be accurately determined by the current model.

To simplify the validation, let one particle be infinitely large, such as the size of the Earth. The particle-particle collision becomes the particle-face collision, where the face is an infinite stationary flat plane. Let $m^{(i)} = \infty$, $I^{(-1,i)} = 0$, the coefficients $\mathcal{A}$, $\mathcal{B}$, $\mathcal{C}$ and $\mathcal{D}$ are reduced to $\mathcal{A}'$, $\mathcal{B}'$, $\mathcal{C}'$ and $\mathcal{D}'$ respectively

$$
\begin{aligned}
\mathcal{A}' = & -\frac{1}{m^{(j)}} \\
& + [tr(\boldsymbol{I}^{(-1,i)})(\boldsymbol{r}^{(i)} \cdot \boldsymbol{n}) - (\boldsymbol{r}^{(i)} \cdot \boldsymbol{I}^{(-1,i)} \cdot \boldsymbol{n}) - (\boldsymbol{n} \cdot \boldsymbol{I}^{(-1,i)} \cdot \boldsymbol{r}^{(i)})](\boldsymbol{n} \cdot \boldsymbol{r}^{(i)}) \\
& + [(\boldsymbol{n} \cdot \boldsymbol{I}^{(-1,i)} \cdot \boldsymbol{n}) - tr(\boldsymbol{I}^{(-1,i)})](\boldsymbol{r}^{(i)} \cdot \boldsymbol{r}^{(i)}) + [(\boldsymbol{r}^{(i)} \cdot \boldsymbol{I}^{(-1,i)} \cdot \boldsymbol{r}^{(i)})] \\
\mathcal{B}' = & -(\boldsymbol{t} \cdot \boldsymbol{I}^{(-1,i)} \cdot \boldsymbol{r}^{(i)})(\boldsymbol{r}^{(i)} \cdot \boldsymbol{n}) \\
& + [tr(\boldsymbol{I}^{(-1,i)})(\boldsymbol{r}^{(i)} \cdot \boldsymbol{n}) - (\boldsymbol{r}^{(i)} \cdot \boldsymbol{I}^{(-1,i)} \cdot \boldsymbol{n})](\boldsymbol{t} \cdot \boldsymbol{r}^{(i)}) \\
& + (\boldsymbol{t} \cdot \boldsymbol{I}^{(-1,i)} \cdot \boldsymbol{n})(\boldsymbol{r}^{(i)} \cdot \boldsymbol{r}^{(i)}) \\
\mathcal{C}' = & [tr(\boldsymbol{I}^{(-1,i)})(\boldsymbol{r}^{(i)} \cdot \boldsymbol{t}) - (\boldsymbol{r}^{(i)} \cdot \boldsymbol{I}^{(-1,i)} \cdot \boldsymbol{t})](\boldsymbol{n} \cdot \boldsymbol{r}^{(i)}) - (\boldsymbol{n} \cdot \boldsymbol{I}^{(-1,i)} \cdot \boldsymbol{r}^{(i)})(\boldsymbol{r}^{(i)} \cdot \boldsymbol{t}) \\
& + [(\boldsymbol{n} \cdot \boldsymbol{I}^{(-1,i)} \cdot \boldsymbol{t})](\boldsymbol{r}^{(i)} \cdot \boldsymbol{r}^{(i)})
\end{aligned}
$$

$$\mathcal{D}' = -\frac{1}{m^{(j)}} + [tr(\boldsymbol{I}^{(-1,i)})(\boldsymbol{r}^{(i)} \cdot \boldsymbol{t}) - (\boldsymbol{r}^{(i)} \cdot \boldsymbol{I}^{(-1,i)} \cdot \boldsymbol{t}) - (\boldsymbol{t} \cdot \boldsymbol{I}^{(-1,i)} \cdot \boldsymbol{r}^{(i)})](\boldsymbol{t} \cdot \boldsymbol{r}^{(i)})$$
$$+ [(\boldsymbol{t} \cdot \boldsymbol{I}^{(-1,i)} \cdot \boldsymbol{t}) - tr(\boldsymbol{I}^{(-1,i)})](\boldsymbol{r}^{(i)} \cdot \boldsymbol{r}^{(i)}) + [\boldsymbol{r}^{(i)} \cdot \boldsymbol{I}^{(-1,i)} \cdot \boldsymbol{r}^{(i)})] \tag{2.40}$$

Substituting them into Eqs. (2.31), the particle-plane collision can be determined readily.

In the experiment, a wood-brick particle, 58, 28, 15 mm in length, width and height, is used. It has a free fall onto a smooth, horizontally oriented, flat wood surface and rebounces back. A high speed camera of 500 fps (the right insets of Fig. 2.6A–E) records the process. The restitution coefficient is obtained by the experiment. As seen in Fig. 2.7A, heights of the bottom vertex during the falling and rebouncing processes are recorded and fitted by parabolic curves. Particle velocity can be derived from the time rate change of the finite differences of the height. Using the pre- and post-collisional velocities, the restitution coefficient is calculated as $e = -\frac{v_{ib}}{v_{tf}}\Big|_{t=t_0} = -\frac{3538.42 \text{ Pixel}}{-6905.61 \text{ Pixel}}\Big|_{t_0=1.02s} = 0.51$, where v_{ib} is the initial velocity of the post-collisional rebounding process at $t = t_0^+$ and v_{tf} is the terminal velocity of the pre-collisional falling process at $t = t_0^-$. Substitute $e = 0.51$ into the present model, the rebouncing process can be obtained and displayed in the left insets of Fig. 2.6A–E. General agreements are obtained by a phenomenological comparison between the numerical and experimental results as seen in Fig. 2.6. Furthermore, for quantitative comparison, simulated heights of the vertex are displayed in Fig. 2.7B, together with the experimental measurements. The heights in the falling process are used to validate the consistence of pre-collisional conditions between simulation and experiment. Simulated heights of the rebouncing process are obtained using the current model. It is observed from Fig. 2.7B that the simulation results are consistent with the experimental results. Thus, it is verified that the current model is correct and can be used for the simulation of collision of nonspherical particles.

2.2.5 Application in a lifting hopper

An experimental test [176] is used for comparison with the simulation, where a hopper filled with tetrahedral particles is lifted gradually and the particles are discharged from orifice at the bottom of the hopper. The geometry, size, main parameters and the lift velocity of the hopper in the simulation are set to those of the experiment (particle size $s_p = 7.4$ mm, density $\rho_p = 1410$ kg/m^3, friction coefficient $\mu_p = 0.27$, lift speed $w_l = 4.2$ mm/s). To save the simulation time, a 2D hopper bed model and triangular particles are used.

The particle discharging processes in simulation and experiment are shown in Fig. 2.8 for comparison. It is seen from the figure that particle distribution is generally similar with experimental results, although there are slight differences. Phenomenological comparisons between them are generally acceptable.

Quantitative comparison of the heights of particle piles between numerical and experimental results is shown in Fig. 2.9. In addition, the numerical results obtained by using a stochastic soft DEM model [176] are also plotted for comparison. It is

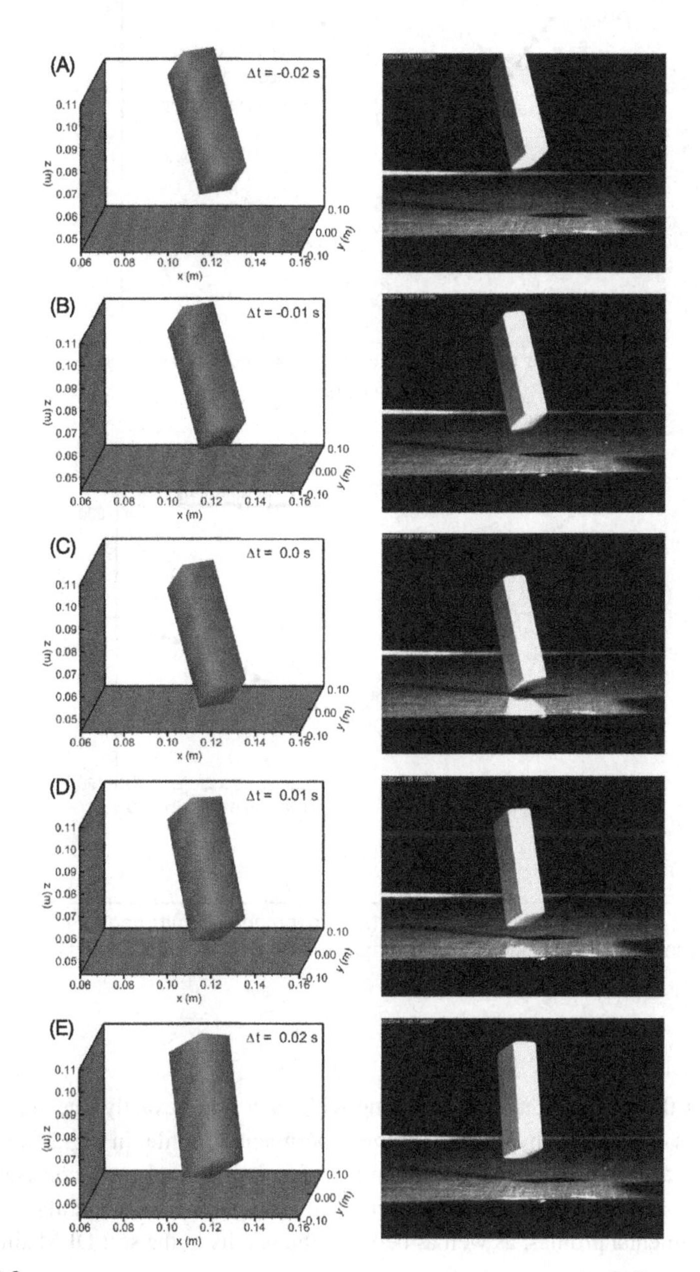

FIGURE 2.6

Comparison between simulation and experimental results for the collision process.

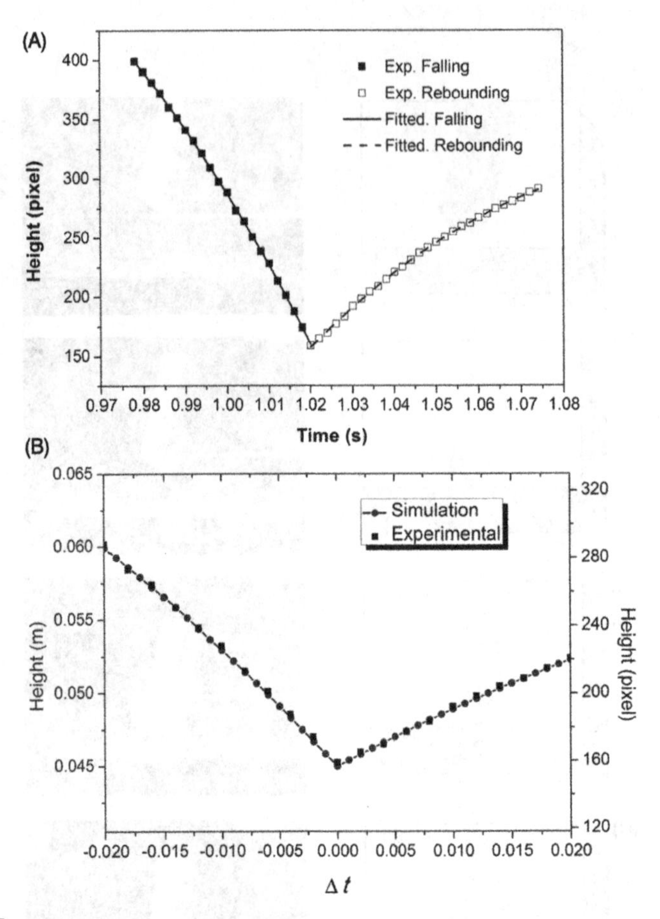

FIGURE 2.7

Variation of height of the left bottom vertex: (A) Experimental results and fitted curves for obtaining the restitution coefficient. (B) Comparison between simulation and experimental results.

noted that these profiles are not matching with each other exactly. Inevitable slight differences exist between each case of the experimental profile. In Fig. 2.9, a sample experimental profile is displayed. Generally, simulation results are consistent with each other, although slight differences can still be found between simulation results and experimental profiles, as well as between the results in the soft DEM simulation and the experiments.

In conclusion, both qualitative and quantitative comparisons are satisfactory, demonstrating successful application of the GHPM model for granular flow simulation.

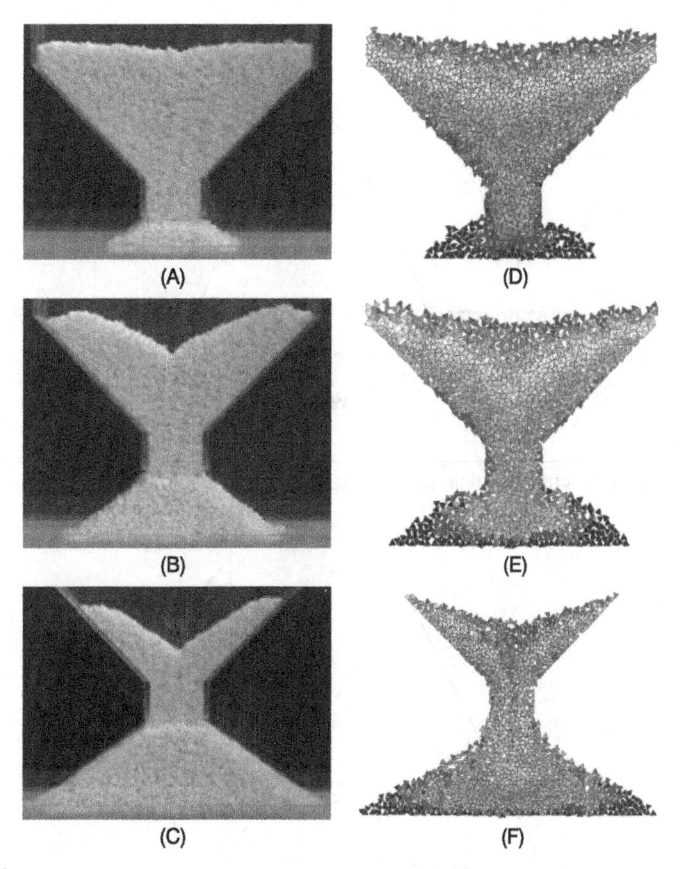

FIGURE 2.8

Comparison of distribution of particles in a gently lifted hopper between the current simulation and the experiment in [176] ($t = 10$ s, 18 s and 23 s, respectively).

2.2.6 Application for the particle-wall collision

In general, the particle-to-wall collision generates a force and a torque on a particle. A typical particle-to-wall collision process is sketched in Fig. 2.10. Post-collisional velocities, including both the translational and rotational ones, are changed following the linear and angular impulse theory. Assumptions are made to simplify the solution process. There are also the limitations for particle-to-wall collision in the GHPM model:

- The collision duration Δt is small.
- The materials of particle and wall are rigid, where the deformations of them are neglected.

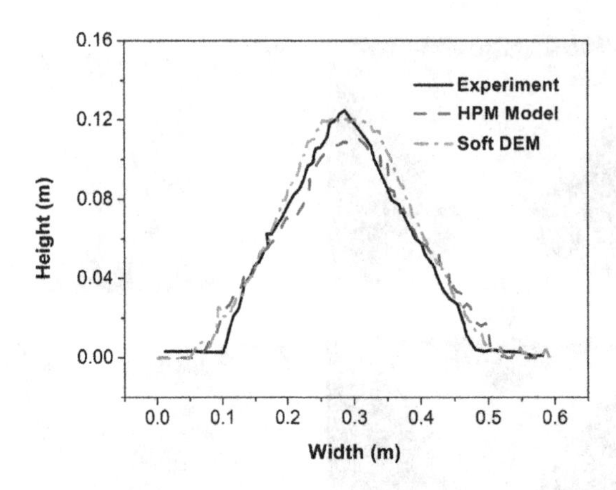

FIGURE 2.9

Comparison of the height of particle piles between the current simulation and the experiment.

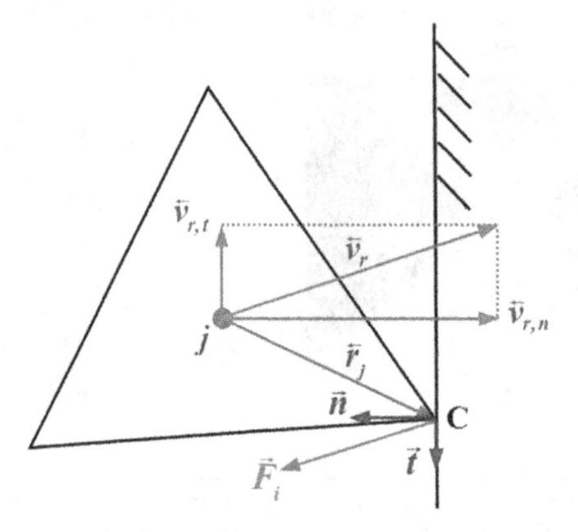

FIGURE 2.10

A sketch of the particle-to-wall collision process. (As a simplest sketch, the triangular shape of particle is used here. The diagrams for other shapes are similar).

- Without deformation, the instantaneous force $\boldsymbol{F}(\boldsymbol{t})$ does not need to be computed. Instead, an integral vector $\boldsymbol{P} = \int_{t_0}^{t_0+\Delta t} \boldsymbol{F} dt$, the total impulse caused by collision over time duration Δt is needed for the solution.
- The total kinetic energy of the system is always decreasing because of the irreversible condition and dissipation of the kinetic energy in the actual collisions. For ideal case without dissipation, total kinetic energy should be conserved.
- The theorem of impulse determines the changes of momentum and velocity of the particle after collision, where the effect of gravity is neglected. In addition, the gravity force can be added independently in the right-hand of the governing equation, which, together with the collisional force, determines the movements of particles.

The Coulomb's law of friction is used since particle may slide on the wall during collision,

$$m_j \boldsymbol{v}_j = m_j \boldsymbol{v}_j^0 + \int_{t_0}^{t_0+\Delta t} \boldsymbol{F}_j dt = m_j \boldsymbol{v}_j^0 + \boldsymbol{P}_j \tag{2.41}$$

$$[\boldsymbol{I}_j] \cdot \boldsymbol{\omega}_j = [\boldsymbol{I}_j] \cdot \boldsymbol{\omega}_j^0 + \int_{t_0}^{t_0+\Delta t} \boldsymbol{r}_j \times \boldsymbol{F}_j dt = [\boldsymbol{I}_j] \cdot \boldsymbol{\omega}_j^0 + \boldsymbol{r}_j \times \boldsymbol{P}_j \tag{2.42}$$

where m_j, $[\boldsymbol{I}_j]$, $\boldsymbol{v}_j$, $\boldsymbol{\omega}_j$, $\boldsymbol{F}_j$, $\boldsymbol{r}_j$, are the jth particle's mass, inertia tensor, velocity vector, angular velocity vector, force vector, and the position vector from the center of mass to the contact point on the particle surface. The subscript '0' denotes the pre-collisional variables. Otherwise, it is a post-collisional variable.

Relative velocity between the contact vertex of the jth particle and the contact point 'C' on the wall is defined as

$$\boldsymbol{v}_r^C = \boldsymbol{v}_j + \boldsymbol{\omega}_j \times \boldsymbol{r}_j - \boldsymbol{v}_w \tag{2.43}$$

where $\boldsymbol{v}_r^C$ denotes the relative velocity vector of the jth particle on the contact point 'C', and $\boldsymbol{v}_j$ is the velocity vector of the center of mass of the particle. $\boldsymbol{v}_w$ is the velocity vector of the wall. Using Eqs. (2.41) and (2.42), after some effort, $\boldsymbol{v}_r^C$ can be given by

$$\begin{aligned}\boldsymbol{v}_r^C =& \boldsymbol{v}_r^0 + \boldsymbol{\omega}_j^0 \times \boldsymbol{r}_j + \frac{\boldsymbol{P}_j}{m_j} \\ &+ (\boldsymbol{r}_j \cdot \boldsymbol{P}_j)(\boldsymbol{r}_j \cdot [\boldsymbol{I}_j^{-1}]) - tr([\boldsymbol{I}_j^{-1}])\boldsymbol{r}_j) + (\boldsymbol{P}_j \cdot [\boldsymbol{I}_j^{-1}] \cdot \boldsymbol{r}_j)\boldsymbol{r}_j \\ &- (\boldsymbol{r}_j \cdot \boldsymbol{r}_j)(\boldsymbol{P}_j \cdot [\boldsymbol{I}_j^{-1}] - tr([\boldsymbol{I}_j^{-1}])\boldsymbol{P}_j) - (\boldsymbol{r}_j \cdot [\boldsymbol{I}_j^{-1}] \cdot \boldsymbol{r}_j)\boldsymbol{P}_j\end{aligned} \tag{2.44}$$

where $\boldsymbol{v}_r^0 = \boldsymbol{v}_j^0 - \boldsymbol{v}_w$, and $[\boldsymbol{I}_j^{-1}]$ is the inverse of the inertia tensor, i.e.

$$[\boldsymbol{I}_j] \cdot [\boldsymbol{I}_j^{-1}] = \begin{pmatrix} 1 & 0 & 0 \\ 0 & 1 & 0 \\ 0 & 0 & 1 \end{pmatrix}.$$

$\mathrm{tr}([\boldsymbol{I}_j^{-1}]) = I_{j,kk}^{-1}$ is the matrix trace. Using Eq. (2.44), the post-collisional relative velocity can be computed from the pre-collisional velocities and the impulses. The inner product of Eq. (2.44) and the unit normal vector $\boldsymbol{n}$ of the wall can be written as

$$\begin{aligned}\boldsymbol{v}_r^C \cdot \boldsymbol{n} =& \boldsymbol{v}_r^{C,0} \cdot \boldsymbol{n} + \frac{\boldsymbol{P}_j \cdot \boldsymbol{n}}{m_j} \\ &+ (\boldsymbol{P}_j \cdot \boldsymbol{n})[(\boldsymbol{r}_j \cdot \boldsymbol{n})(\boldsymbol{r}_j \cdot [\boldsymbol{I}_j^{-1}] \cdot \boldsymbol{n} - tr([\boldsymbol{I}_j^{-1}])(\boldsymbol{r}_j \cdot \boldsymbol{n})) \\ &- (\boldsymbol{r}_j \cdot \boldsymbol{r}_j)(\boldsymbol{n} \cdot [\boldsymbol{I}_j^{-1}] \cdot \boldsymbol{n} - tr([\boldsymbol{I}_j^{-1}])) - (\boldsymbol{r}_j \cdot \boldsymbol{t})(\boldsymbol{t} \cdot [\boldsymbol{I}_j^{-1}] \cdot \boldsymbol{r}_j)] \\ &+ (\boldsymbol{P}_j \cdot \boldsymbol{t})[(\boldsymbol{r}_j \cdot \boldsymbol{t})(\boldsymbol{r}_j \cdot [\boldsymbol{I}_j^{-1}] \cdot \boldsymbol{n} - tr([\boldsymbol{I}_j^{-1}])(\boldsymbol{r}_j \cdot \boldsymbol{n})) \\ &- (\boldsymbol{r}_j \cdot \boldsymbol{r}_j)(\boldsymbol{t} \cdot [\boldsymbol{I}_j^{-1}] \cdot \boldsymbol{n}) + (\boldsymbol{r}_j \cdot \boldsymbol{n})(\boldsymbol{t} \cdot [\boldsymbol{I}_j^{-1}] \cdot \boldsymbol{r}_j)].\end{aligned} \tag{2.45}$$

In a similar manner, the inner product of Eq. (2.44) and the tangential unit vector $\boldsymbol{t}$ of the wall is given by

$$\begin{aligned}\boldsymbol{v}_r^C\cdot\boldsymbol{t}=\boldsymbol{v}_r^{C,0}\cdot\boldsymbol{t}+\frac{\boldsymbol{P}_j\cdot\boldsymbol{t}}{m_j}\\+(\boldsymbol{P}_j\cdot\boldsymbol{n})[(\boldsymbol{r}_j\cdot\boldsymbol{n})(\boldsymbol{r}_j\cdot[\boldsymbol{I}_j^{-1}]\cdot\boldsymbol{t}-tr([\boldsymbol{I}_j^{-1}])(\boldsymbol{r}_j\cdot\boldsymbol{t}))\\-(\boldsymbol{r}_j\cdot\boldsymbol{r}_j)(\boldsymbol{n}\cdot[\boldsymbol{I}_j^{-1}]\cdot\boldsymbol{t})+(\boldsymbol{r}_j\cdot\boldsymbol{t})(\boldsymbol{n}\cdot[\boldsymbol{I}_j^{-1}]\cdot\boldsymbol{r}_j)]\\+(\boldsymbol{P}_j\cdot\boldsymbol{t})[(\boldsymbol{r}_j\cdot\boldsymbol{t})(\boldsymbol{r}_j\cdot[\boldsymbol{I}_j^{-1}]\cdot\boldsymbol{t}-tr([\boldsymbol{I}_j^{-1}])(\boldsymbol{r}_j\cdot\boldsymbol{t}))\\-(\boldsymbol{r}_j\cdot\boldsymbol{r}_j)(\boldsymbol{t}\cdot[\boldsymbol{I}_j^{-1}]\cdot\boldsymbol{t}-tr([\boldsymbol{I}_j^{-1}]))-(\boldsymbol{r}_j\cdot\boldsymbol{n})(\boldsymbol{n}\cdot[\boldsymbol{I}_j^{-1}]\cdot\boldsymbol{r}_j)]\end{aligned}\tag{2.46}$$

The restitution coefficient e is defined as the ratio of post- to pre-collision relative velocities on the contact point, i.e.

$$\boldsymbol{v}_r^C\cdot\boldsymbol{n}=-e(\boldsymbol{v}_r^{C,0}\cdot\boldsymbol{n})\tag{2.47}$$

Combining Eqs. (2.45) and (2.46) and substituting Eq. (2.47) into Eq. (2.45), the normal and tangential components of the impulse, i.e. $\boldsymbol{P}_j\cdot\boldsymbol{n}$ and $\boldsymbol{P}_j\cdot\boldsymbol{t}$, can be solved by

$$\begin{cases}\alpha(\boldsymbol{P}_j\cdot\boldsymbol{n})+\beta(\boldsymbol{P}_j\cdot\boldsymbol{t})=-(1+e)(\boldsymbol{v}_r^{C,0}\cdot\boldsymbol{n}) & (2.48)\\ \gamma(\boldsymbol{P}_j\cdot\boldsymbol{t})+\delta(\boldsymbol{P}_j\cdot\boldsymbol{n})=(\boldsymbol{v}_r^C\cdot\boldsymbol{t})-(\boldsymbol{v}_r^{C,0}\cdot\boldsymbol{t}) & (2.49)\end{cases}$$

where

$$\begin{aligned}\alpha=\frac{1}{m_j}+(\boldsymbol{r}_j\cdot\boldsymbol{n})(\boldsymbol{r}_j\cdot[\boldsymbol{I}_j^{-1}]\cdot\boldsymbol{n}-tr([\boldsymbol{I}_j^{-1}])(\boldsymbol{r}_j\cdot\boldsymbol{n}))\\-(\boldsymbol{r}_j\cdot\boldsymbol{r}_j)(\boldsymbol{n}\cdot[\boldsymbol{I}_j^{-1}]\cdot\boldsymbol{n}-tr([\boldsymbol{I}_j^{-1}]))\\-(\boldsymbol{r}_j\cdot[\boldsymbol{I}_j^{-1}]\cdot\boldsymbol{r}_j)+(\boldsymbol{r}_j\cdot\boldsymbol{n})(\boldsymbol{n}\cdot[\boldsymbol{I}_j^{-1}]\cdot\boldsymbol{r}_j)\end{aligned}\tag{2.50}$$

$$\begin{aligned}\beta=(\boldsymbol{r}_j\cdot\boldsymbol{t})(\boldsymbol{r}_j\cdot[\boldsymbol{I}_j^{-1}]\cdot\boldsymbol{n}-tr([\boldsymbol{I}_j^{-1}])(\boldsymbol{r}_j\cdot\boldsymbol{n}))\\-(\boldsymbol{r}_j\cdot\boldsymbol{r}_j)(\boldsymbol{t}\cdot[\boldsymbol{I}_j^{-1}]\cdot\boldsymbol{n})+(\boldsymbol{r}_j\cdot\boldsymbol{n})(\boldsymbol{t}\cdot[\boldsymbol{I}_j^{-1}]\cdot\boldsymbol{r}_j)\end{aligned}\tag{2.51}$$

$$\begin{aligned}\gamma=\frac{1}{m_j}+(\boldsymbol{r}_j\cdot\boldsymbol{t})(\boldsymbol{r}_j\cdot[\boldsymbol{I}_j^{-1}]\cdot\boldsymbol{t}-tr([\boldsymbol{I}_j^{-1}])(\boldsymbol{r}_j\cdot\boldsymbol{t}))\\-(\boldsymbol{r}_j\cdot\boldsymbol{r}_j)(\boldsymbol{t}\cdot[\boldsymbol{I}_j^{-1}]\cdot\boldsymbol{t}-tr([\boldsymbol{I}_j^{-1}]))\\-(\boldsymbol{r}_j\cdot[\boldsymbol{I}_j^{-1}]\cdot\boldsymbol{r}_j)+(\boldsymbol{r}_j\cdot\boldsymbol{t})(\boldsymbol{t}\cdot[\boldsymbol{I}_j^{-1}]\cdot\boldsymbol{r}_j)\end{aligned}\tag{2.52}$$

$$\begin{aligned}\delta=(\boldsymbol{r}_j\cdot\boldsymbol{n})(\boldsymbol{r}_j\cdot[\boldsymbol{I}_j^{-1}]\cdot\boldsymbol{t}-tr([\boldsymbol{I}_j^{-1}])(\boldsymbol{r}_j\cdot\boldsymbol{t}))\\-(\boldsymbol{r}_j\cdot\boldsymbol{r}_j)(\boldsymbol{n}\cdot[\boldsymbol{I}_j^{-1}]\cdot\boldsymbol{t})+(\boldsymbol{r}_j\cdot\boldsymbol{t})(\boldsymbol{n}\cdot[\boldsymbol{I}_j^{-1}]\cdot\boldsymbol{r}_j)\end{aligned}\tag{2.53}$$

It should be noted that the coefficients α, β, γ, δ are determined by the material property of the particle and the pre-collisional geometry between particle and the wall. Only one term in the right-hand side of Eq. (2.49) is unknown before collision, namely, $\boldsymbol{v}_r^C \cdot \boldsymbol{t}$. It is the post-collisional relative tangential velocity on the wall. For a collision with static friction, the $\boldsymbol{v}_r^C \cdot \boldsymbol{t}$ must be zero. In this case, the governing equations of collision are

$$\begin{cases} \alpha(\boldsymbol{P}_j \cdot \boldsymbol{n}) + \beta(\boldsymbol{P}_j \cdot \boldsymbol{t}) = -(1+e)(\boldsymbol{v}_r^{C,0} \cdot \boldsymbol{n}) & (2.54) \\ \gamma(\boldsymbol{P}_j \cdot \boldsymbol{t}) + \delta(\boldsymbol{P}_j \cdot \boldsymbol{n}) = -(\boldsymbol{v}_r^{C,0} \cdot \boldsymbol{t}). & (2.55) \end{cases}$$

The solution of the impulse from the above equations is

$$\begin{cases} \boldsymbol{P}_j \cdot \boldsymbol{n} = [\beta(\boldsymbol{v}_r^{C,0} \cdot \boldsymbol{t}) - \gamma(1+e)(\boldsymbol{v}_r^{C,0} \cdot \boldsymbol{n})]/(\alpha\gamma - \beta\delta) \\ \boldsymbol{P}_j \cdot \boldsymbol{t} = [\delta(1+e)(\boldsymbol{v}_r^{C,0} \cdot \boldsymbol{n}) - \alpha(\boldsymbol{v}_r^{C,0} \cdot \boldsymbol{t})]/(\alpha\gamma - \beta\delta). \end{cases} \tag{2.56}$$

For the collision with sliding friction on the wall, it is not necessary to use Eq. (2.49) to calculate the component of post-collisional relative sliding velocity. Instead, according to the Coulomb's law of friction, the normal and tangential components of the impulse should follow the relation of

$$\boldsymbol{P}_j \cdot \boldsymbol{t} = \mu_d(\boldsymbol{P}_j \cdot \boldsymbol{n}) \tag{2.57}$$

where μ_d is the coefficient of dynamic (kinetic) friction. Therefore, it is more convenient to use Eqs. (2.45) and (2.57) together to solve for the collision with sliding motion, i.e.

$$\begin{cases} \alpha(\boldsymbol{P}_j \cdot \boldsymbol{n}) + \beta(\boldsymbol{P}_j \cdot \boldsymbol{t}) = -(1+e)(\boldsymbol{v}_r^{C,0} \cdot \boldsymbol{n}) & (2.58) \\ \boldsymbol{P}_j \cdot \boldsymbol{t} = \mu_d(\boldsymbol{P}_j \cdot \boldsymbol{n}) & (2.59) \end{cases}$$

and the solutions are

$$\begin{cases} \boldsymbol{P}_j \cdot \boldsymbol{n} = -(1+e)(\boldsymbol{v}_r^{C,0} \cdot \boldsymbol{n})/(\alpha + \mu_d\beta) \\ \boldsymbol{P}_j \cdot \boldsymbol{t} = -\mu_d(1+e)(\boldsymbol{v}_r^{C,0} \cdot \boldsymbol{n})/(\alpha + \mu_d\beta) \end{cases} \tag{2.60}$$

The collision can be categorized based on the consistency of the characteristics between the static and sliding frictions. For the static solutions, if ratio of the tangential impulse to normal impulse is smaller or equal to the friction coefficient (i.e. $P_{j,t} \le \mu_s P_{j,n}$), the tangential impulse is too weak to cause sliding movement on the wall; otherwise, it causes sliding on the wall. According to the static collision described by Eq (2.56), the criterion is

$$\lambda = \frac{\left|\delta(1+e)(\boldsymbol{v}_r^{C,0} \cdot \boldsymbol{n}) - \alpha(\boldsymbol{v}_r^{C,0} \cdot \boldsymbol{t})\right|}{\left|\beta(\boldsymbol{v}_r^{C,0} \cdot \boldsymbol{t}) - \gamma(1+e)(\boldsymbol{v}_r^{C,0} \cdot \boldsymbol{n})\right|} = \begin{cases} > \mu_s \text{ sliding friction;} \\ < \mu_s \text{ static friction.} \end{cases} \tag{2.61}$$

It is worth to mention that μ_s is the maximum coefficient of the static friction. In general, it is slightly larger than the coefficient of the dynamic friction and only used to categorize the collision. When the collision is dynamic, the dynamic friction coefficient μ_d will be used in solution. For simplicity, the difference between the dynamic and static friction coefficient can be neglected.

In conclusion, the impulse from wall to particle during the collision can be computed as

$$\boldsymbol{P}_j = \begin{cases} \frac{\beta(\boldsymbol{v}_r^{C,0}\cdot\boldsymbol{t})-\gamma(1+e)(\boldsymbol{v}_r^{C,0}\cdot\boldsymbol{n})}{(\alpha\gamma-\beta\delta)}\boldsymbol{n} + \frac{\delta(1+e)(\boldsymbol{v}_r^{C,0}\cdot\boldsymbol{n})-\alpha(\boldsymbol{v}_r^{C,0}\cdot\boldsymbol{t})}{(\alpha\gamma-\beta\delta)}\boldsymbol{t}, & \text{if } \lambda < \mu_s \\ \frac{-(1+e)(\boldsymbol{v}_r^{C,0}\cdot\boldsymbol{n})}{(\alpha+\mu_d\beta)}\boldsymbol{n} + \frac{-\mu_d(1+e)(\boldsymbol{v}_r^{C,0}\cdot\boldsymbol{n})}{(\alpha+\mu_d\beta)}\boldsymbol{t}, & \text{if } \lambda > \mu_s. \end{cases} \tag{2.62}$$

With the impulse known, post-collisional velocities and angular velocities can be readily obtained by substituting them in Eqs. (2.41) and (2.42). Then, particle trajectory can be calculated by integrating the motion equations

$$\dot{\boldsymbol{x}}_j = \boldsymbol{v}_j \tag{2.63}$$

$$\dot{\boldsymbol{\theta}}_j = \boldsymbol{\omega}_j \tag{2.64}$$

$$\dot{\boldsymbol{v}}_j = \frac{1}{m_j \Delta t}\boldsymbol{P}_j - \boldsymbol{g} \tag{2.65}$$

$$\dot{\boldsymbol{\omega}}_j = [\boldsymbol{I}_j^{-1}]\cdot(\frac{1}{\Delta t}\boldsymbol{r}_j \times \boldsymbol{P}_j) \tag{2.66}$$

where $\boldsymbol{x}$, $\boldsymbol{\theta}$ are the translational and angular displacements of particles, and $\boldsymbol{g}$ is the gravity acceleration.

2.2.7 Numerical procedure for multiple contacts

As the particles are rigid and without deformation, the particle-wall collision can be categorized into three types: the vertex-to-wall collision, edge-to wall collision, and face-to-wall collision. In reality, there exists a continuous distribution of collision forces and torques throughout the edges and faces. However the equivalent resulting force $\boldsymbol{P}_{j,tot}^{equ}$ and torque $\boldsymbol{T}_{j,tot}^{equ}$ can be used. In the current model, it is assumed that the edge-to-wall or face-to-wall collision takes action through selected critical contact points, such as C_1, C_2 on the side, and C_1, C_2, C_3, C_4 on the face (Fig. 2.11). Thus, to simplify the numerical procedures, the edge-to-wall and face-to-wall collision can be transformed into a combination of two or four vertex-to-wall collisions respectively. Taking a particle-face collision for example, the collision force $\boldsymbol{P}_{j,i}$ on each contact point C_i is solved by

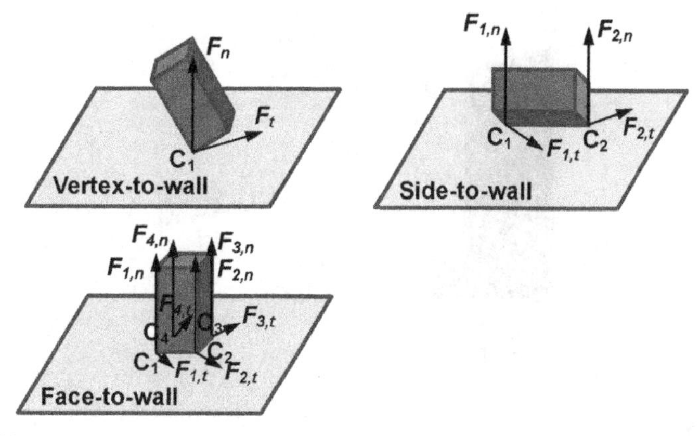

FIGURE 2.11

Sketches of three basic types of the particle-to-wall collision.

$$\begin{cases} m_j \boldsymbol{v}_j = m_j \boldsymbol{v}_j^0 + \boldsymbol{P}_{j,1} & \text{(a)} \\ m_j \boldsymbol{v}_j = m_j \boldsymbol{v}_j^0 + \boldsymbol{P}_{j,2} & \text{(b)} \\ m_j \boldsymbol{v}_j = m_j \boldsymbol{v}_j^0 + \boldsymbol{P}_{j,3} & \text{(c)} \\ m_j \boldsymbol{v}_j = m_j \boldsymbol{v}_j^0 + \boldsymbol{P}_{j,4} & \text{(d)} \end{cases} \tag{2.67}$$

respectively. By summing up Eqs. (2.67)(a)–(d), the total equivalent collisional force on the particle is

$$\boldsymbol{P}_{j,tot}^{equ} = \frac{1}{n_j^c} \sum_{i=1}^{n_j^c} \boldsymbol{P}_{j,i} \tag{2.68}$$

where $n_j^c = 4$ is the number of contact points. Similarly, the collision torque can be expressed as

$$\boldsymbol{T}_{j,tot}^{equ} = \frac{1}{n_j^c} \sum_{i=1}^{n_j^c} \boldsymbol{T}_{j,i} \tag{2.69}$$

By this approach, the edge-to-wall or face-to-wall collision can be simplified as the vertex-to-wall collision.

To detect the contact, the distance between each vertex and the wall should be examined at each time step. If the distance is less than an infinitesimal value, e.g. $\epsilon = 0.001 d_p$, the collision occurs.

2.2.8 Simulation test and experimental validation

As explained, the edge-to-wall and face-to-wall collision can be transformed to the vertex-to-wall collision. In this section, an experiment of a brick particle-to-wall

FIGURE 2.12

Sketches of the particle-to-wall collision experiment.

Table 2.1 Comparison of parameters used in simulation and experiment.

Parameters		Experiment	Simulation
Brick length, width, height (mm)		(58, 28, 15)	(58, 28, 15)
Brick density ρ (kg/m^{-3})		500	500
Friction coefficient μ		–	0.3
Restitution coefficient e		0.6	0.6
Initial height of the brick $H_{b,0}$ (mm)	case 1:	91.07	91.07
	case 2:	114.09	114.09
Initial angle of the brick θ_0	case 1:	0.275π	0.275π
	case 2:	0.276π	0.276π
Initial velocity of the brick,	case 1:	(0, 0, −0.667)	(0, 0, −0.667)
$(v_{x,0}, v_{y,0}, v_{z,0})$ (m/s)	case 2:	(0, 0, −0.153)	(0, 0, −0.153)
Initial rotating velocity of the brick,	case 1:	(0, 0, 0)	(0, 0, 0)
$(\omega_{x,0}, \omega_{y,0}, \omega_{z,0})$ (/s)	case 2:	–	$(-0.22, -0.16, -0.20)\pi$
Height of the desk wall $H_{d,0}$ (mm)	case 1:	44.72	44.72
	case 2:	44.26	44.26

collision is performed and a validation simulation under the same conditions are conducted and compared. A wood brick of 5.8 cm × 2.8 cm × 1.5 cm in length, width and height, falls onto a wall (a laboratory desk wrapped with smooth steel surface placed horizontally) freely under gravity (Fig. 2.12). The collision process is recorded by a high speed camera at a rate of 500 fps (Norpix FR-625, with the highest resolution of 1280×1024 and a maximum frame rate of 5000 fps). The same experimental and simulation parameters are used (see in Table 2.1). Comparisons between the experimental and numerical results of the brick-to-wall collision processes are shown in Fig. 2.13, Fig. 2.14 and Fig. 2.15.

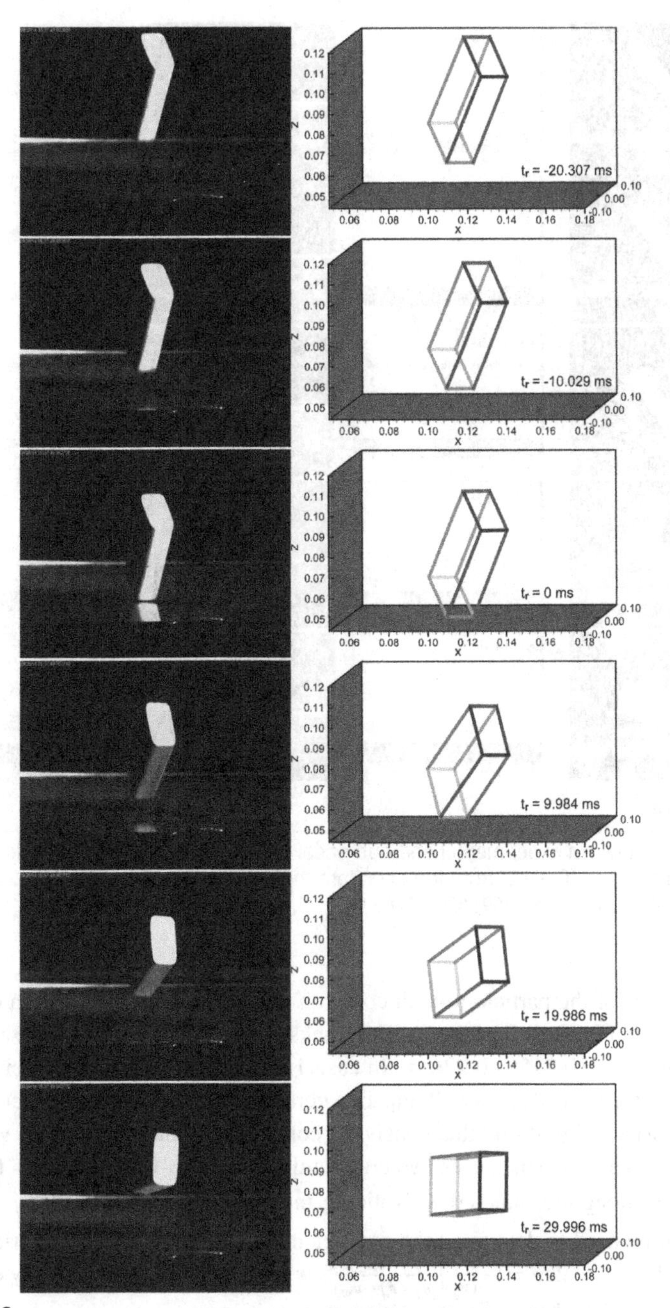

FIGURE 2.13

Comparison between experimental and numerical snapshots of the brick particle-to-wall collision processes. (Case 1: from top to bottom the time points are $t = 922.689$, 932.697, 942.726, 952.710, 962.712, 972.722 ms respectively).

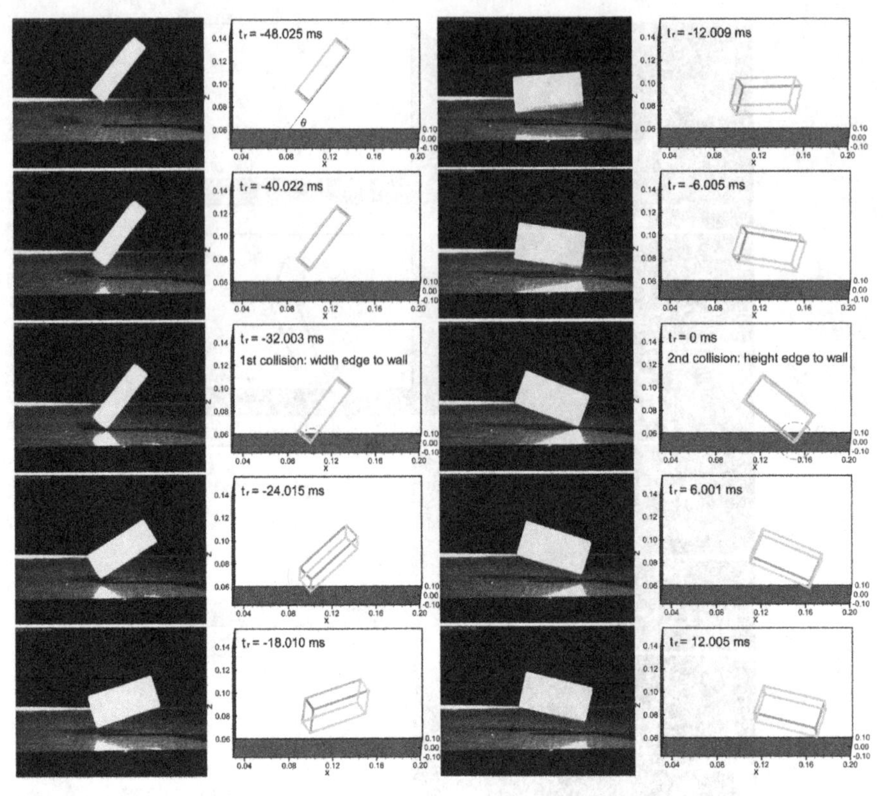

FIGURE 2.14

Comparison between experimental and numerical snapshots of the brick particle-to-wall collision processes. (Case 2: from top to bottom the time points are $t = 21.953$, 29.956, 37.975, 45.964, 51.968, 57.969, 63.973, 69.978, 75.979, 81.983 ms respectively).

Two cases of the particle-to-wall collision are recorded by the camera (Fig. 2.13 for case 1 and Fig. 2.14 for case 2). The bricks in the two cases have the same initial orientation: $\theta_0 = (0.275 - 0.276)\pi$. In case 1, the initial rotational velocity is zero, which causes a right edge-to-wall impact where all the points on the edge collide with the wall surface almost simultaneously. In contrast, the initial rotational velocity in case 2 is nonzero, which causes two consecutive vertex-to-wall collisions for two of the vertices along the same edge. Noticing that $e = -\frac{(\boldsymbol{v}_j - \boldsymbol{v}_w)_n}{(\boldsymbol{v}_j - \boldsymbol{v}_w)_n^0}$ when $\boldsymbol{\omega}_j = 0$, thus, case 1 can be used to determine the coefficient of restitution of the material. On the other hand, in case 2, $e = -\frac{(\boldsymbol{v}_j + \boldsymbol{\omega}_j \times \boldsymbol{r}_j - \boldsymbol{v}_w)_n}{(\boldsymbol{v}_j + \boldsymbol{\omega}_j \times \boldsymbol{r}_j - \boldsymbol{v}_w)_n^0}$ when $\boldsymbol{\omega}_j \neq 0$, which is newly defined in the current model when nonzero rotation motion is incorporated.

It is seen from Fig. 2.13 that the qualitative comparison agrees satisfactorily. The brick particle moves downward onto the wall and collides with it at 942.726 ms. When the collision takes place, the entire height-edge of the particle is in contact

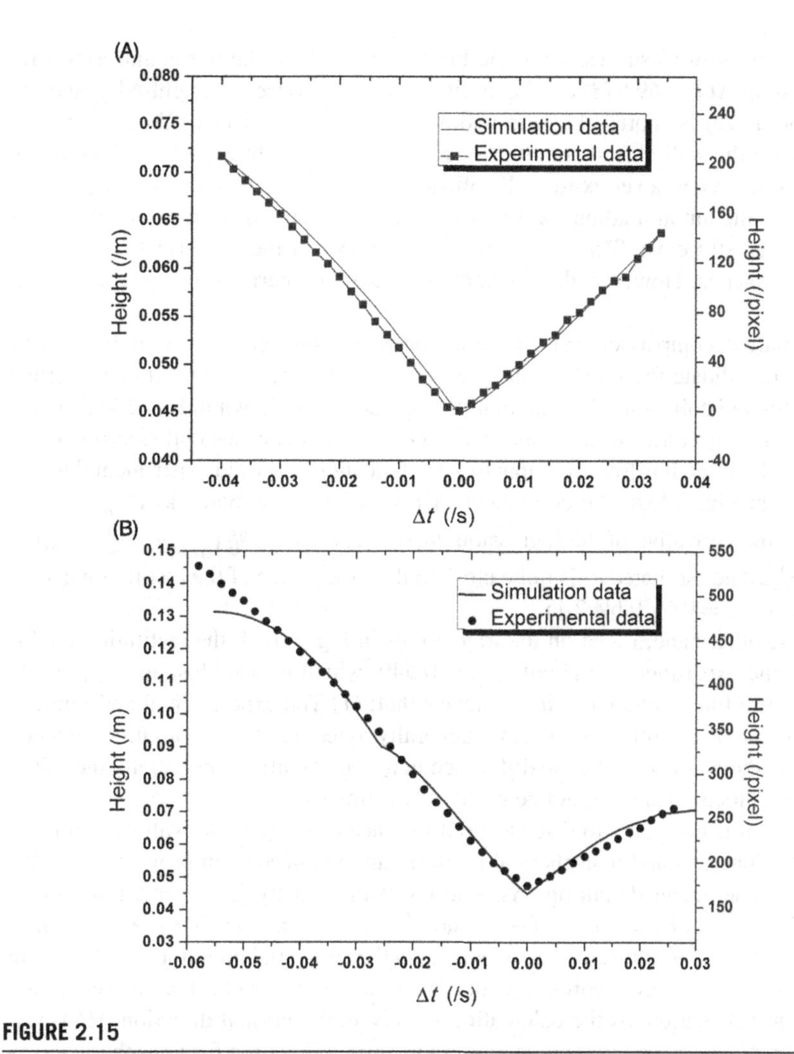

FIGURE 2.15

Comparison of temporal variation of the heights of the contact points for case 1 (A) and case 2 (B) respectively. (The abscissa axis is $\Delta t = t - t_{col}$, where t_{col} is the collision time point. Therefore, $\Delta t < 0$ means pre-collisional time and $\Delta t > 0$ means post-collisional time).

with the wall, including the two vertices at both ends of the height-edge. After the collision, the particle rotates around the axis which is parallel to the height direction. It is worth to elaborate the technique to perform the edge-to-wall collision. It is the averaged solutions of the two vertex-to-wall collisions. Multiple edge-to-edge or surface-to-surface collisions can follow a similar strategy.

It is noted from Fig. 2.14 that two of the brick-to-wall collisions take place at $t = 37.975$ and 69.978 ms, respectively. The collision at $t = 37.975$ ms is between a width-edge of the brick and the wall. Because the two vertices of the width-edge collide with the wall successively, the brick rotates about the horizontal axis after the collision. At $t = 69.978$ ms, the collision occurs between a height-edge and the wall. The height is shorter than the width, and the two vertices of the height-edge collide with the wall almost simultaneously. Thus, the height-edge-to-wall collision can be considered as a vertex-to-wall collision. As it is difficult to determine parameters such as the initial rotational velocity $\boldsymbol{\omega}_0$ and the orientation of the brick $\boldsymbol{\theta}_0$, slight uncertainties still exist. This causes difference between the numerical and experimental snapshots. However, the phenomenological comparison in Fig. 2.14 is still acceptable.

Qualitative comparisons of the collision snapshots are shown in Figs. 2.13 and 2.14 to validate feasibility of the present model to simulate a rigid nonspherical particle-to-wall collision. The quantitative comparison as shown in Fig. 2.15A and B displays the trajectories of the contact vertices in the above two collision processes (i.e. cases 1 and 2). It is found out that trends of the numerical and experimental tracks agree. From Fig. 2.15A, the coefficient of restitution is estimated as $e^*_{exp} = 0.609$ through time derivative of the regression curve, i.e. $e^* = -(\frac{dh}{dt}|_{\Delta t>0})/(\frac{dh}{dt}|_{\Delta t<0}) = -\boldsymbol{v}_{r,n}/\boldsymbol{v}^0_{r,n}$. The estimated $e^\star$ is quite close to the exact value of the restitution coefficient, i.e. $e_{num} = 0.6$ (Table 2.1).

On the other hand, based on the trajectories in Fig. 2.15B, the restitution coefficients in the experiment are about $e^*_{exp} = 0.549$, which is also close to $e_{num} = 0.6$, the value used in simulation. This indicates that: (1) The experimental and numerically simulated restitution coefficients generally agree. (2) Slight variation between the two methods is caused by the difference in initial conditions, especially the initial rotational velocity which cannot be accurately estimated.

Finally, it is necessary to discuss the difference of the particle-wall collision between the spherical and nonspherical particles and the need of improvement for the restitution coefficient definition. Assume the wall velocity is $\boldsymbol{v}_w = 0$. For spherical particles, $\boldsymbol{\omega}_j \times \boldsymbol{r}_j = r_j\boldsymbol{\omega}_j \times (-\boldsymbol{n})$, and $\boldsymbol{v}^C_r \cdot \boldsymbol{n} = \boldsymbol{v}_j \cdot \boldsymbol{n} + \boldsymbol{\omega}_j \times \boldsymbol{r}_j \cdot \boldsymbol{n} = \boldsymbol{v}_j \cdot \boldsymbol{n} - r_j\boldsymbol{\omega}_j \times \boldsymbol{n} \cdot \boldsymbol{n} = \boldsymbol{v}_j \cdot \boldsymbol{n}$. It means that $\boldsymbol{\omega}_j \times \boldsymbol{r}_j$ is in the tangential direction, and therefore $\boldsymbol{\omega}_j \times \boldsymbol{r}_j \cdot \boldsymbol{n} = 0$. In other words, the rotational motion does not affect the restitution coefficient calculation, as the calculation is only in the normal direction. However, $\boldsymbol{\omega}_j \times \boldsymbol{r}_j$ can have a nonzero component in the normal direction for nonspherical particles, and therefore, the rotational motion and the normal restitution coefficient (and the post-collisional rebound motion) are correlated. This indicates the complexity of the nonspherical particle-to-wall collision mechanism and the necessity to use the present definition for the restitution coefficient.

2.2.9 Application in the particle-wall collision

Case 1: Multiple particle-wall collisions

GHPM is applied to other cases to demonstrate its applicability for more complex particle-wall collisions (e.g., the multiple particle-wall collisions without considering gravity). As seen in Fig. 2.16, an array of cubic particles with different initial velocities (e.g. first inset of Fig. 2.16 and Table 2.2) move southward and impact

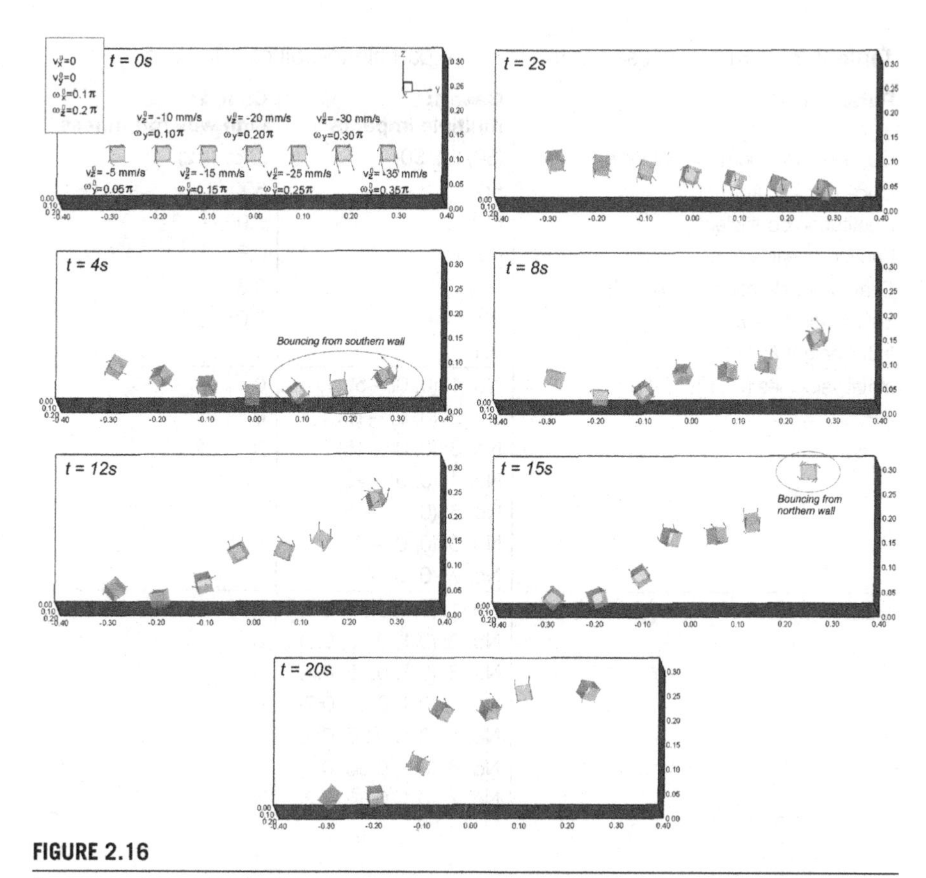

FIGURE 2.16

Demonstration of the application. Case 1: the multiple particle-to-wall collision, i.e. the collision process (from $t = 0$ s to $t = 20$ s) between seven cubic particles and a smooth wall with different initial velocities and without gravity.

on the southern wall one by one. Without gravity, after collision, the particles are rebounded toward the north and impact on the northern wall.

The translational and rotational kinetic energy ($E_{tra} = \frac{1}{2} m\boldsymbol{v} \cdot \boldsymbol{v}$ and $E_{rot} = \frac{1}{2}\boldsymbol{\omega} \cdot \boldsymbol{I} \cdot \boldsymbol{\omega}$) may decrease or increase in each collision (Fig. 2.17). Taking particle No. 5 as an example (Fig. 2.17E), E_{tra} is decreased from about 4×10^{-6} (J) to 0 and E_{rot} is increased from 1.2×10^{-6} (J) to 2.4×10^{-6} (J) after the first collision. The sum of them is decreased from about 5.2×10^{-6} (J) to 2.4×10^{-6} (J). The decrease of the total kinetic energy is caused by irreversible dissipation of kinetic energies in each collision. As a result, total kinetic energy is always decreasing in all of the collisions. This does not violate the fundamental laws of energy conservation.

Case 2: With wall roughness

The second case shows particle-to-wall collisions with wall roughness under gravity. An artificial rough wall is constructed as seen in Fig. 2.18. Five small brick

Table 2.2 Parameters used in the multiple particle-to-wall collision case.

Parameters	Case 1: multiple impacts	Case 2: with wall roughness
Cube length, width, height (mm)	(30, 30, 30)	(0.5, 1, 2)
Cube density (kg/m^3)	500	500
Restitution coefficient e	0.95	0.95
Friction coefficient μ	0.3	0.3
Gravity acceleration g (kg/m^{-2})	0	9.8
Simulation time t_s (s)	20	0.07
Initial height h_z (m)	0.1	0.01
Initial velocities (v_x^0, v_y^0, v_z^0) (mm/s)	No. 1: (0, 0, −5)	0
	No. 2: (0, 0, −10)	0
	No. 3: (0, 0, −15)	0
	No. 4: (0, 0, −20)	0
	No. 5: (0, 0, −25)	0
	No. 6: (0, 0, −30)	
	No. 7: (0, 0, −35)	
Initial angular velocities $(\omega_x^0, \omega_y^0, \omega_z^0)$ (π/s)	No. 1: (0.1, 0.05, 0.2)	0
	No. 2: (0.1, 0.10, 0.2)	0
	No. 3: (0.1, 0.15, 0.2)	0
	No. 4: (0.1, 0.20, 0.2)	0
	No. 5: (0.1, 0.25, 0.2)	0
	No. 6: (0.1, 0.30, 0.2)	
	No. 7: (0.1, 0.35, 0.2)	

particles are positioned stationary above the wall with the same initial height of h_z before falling vertically down onto the rough wall. Compared to the former case, the kinetic energy E_{tra} is continuously increasing as a result of the decrease of the potential energy $E_{pot} = mgh$. Most of the bricks impact on the inclined rough wall consecutively. Under the effect of gravity and with low initial height ($h_z = 0.01$ m), the particles don't rebound as evidently as in the former case in absence of gravity. As a consequence, they are more likely to be "entrapped" by the roughnesses. However, no matter what the collision patterns (e.g. direct impact vs. inclined impact) and the post-collisional behaviors are (e.g. direct rebounding with absence of gravity vs. without evident rebounce due to gravity and structured roughness), the total mechanical energies, including E_{tra}, E_{rot} and the potential energy E_{pot}, are always decreasing (Fig. 2.19). This also follows the basic laws of energy.

In conclusion, the two cases demonstrate the feasibility of using GHPM model for multiple particle-to-wall collisions with/without gravity onto smooth/rough walls.

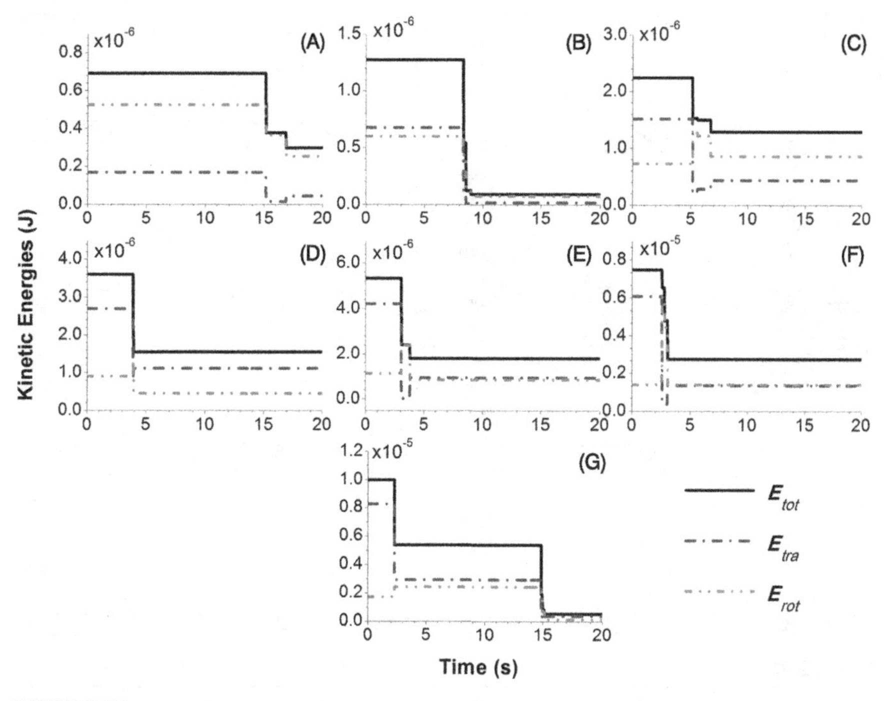

FIGURE 2.17

The variations of kinetic energies in the multiple particle-to-wall collision case.

2.3 SIPHPM model

2.3.1 Description of the model basis

In this section, we introduce the SIPHPM model. It describes smooth particle-particle/wall contacts. Without loss of generality, a rectangular shape is chosen to interpret the SIPHPM model. The model procedure contains several steps:

STEP 1: Boundary discretization (BD)

The boundary of any nonspherical particle (or the "host" particle) is covered by a series of soft subspheres with finite volumes, e.g. the sides of the host particles are covered by N_s (e.g. $N_s = 40$ for squares in Fig. 2.20 and $N_s = 60$ for cubes in Fig. 2.21) subspheres equally and continuously distributed. This is the "boundary-discretization" process or "subsphere imbedding" process.

STEP 2: Collision of subspheres (CS)

In the existing "multisphere" models (or "clumped particle", "sphere-gluing", "bonded particle" method (BPM)), "bonding" interaction between the sub-elements of the host particle is always present. However, in the SIPHPM model, the interactions between the subspheres covering the host particle are neglected. As a result, the computational cost of solving the inner interactions of the host particle is reduced.

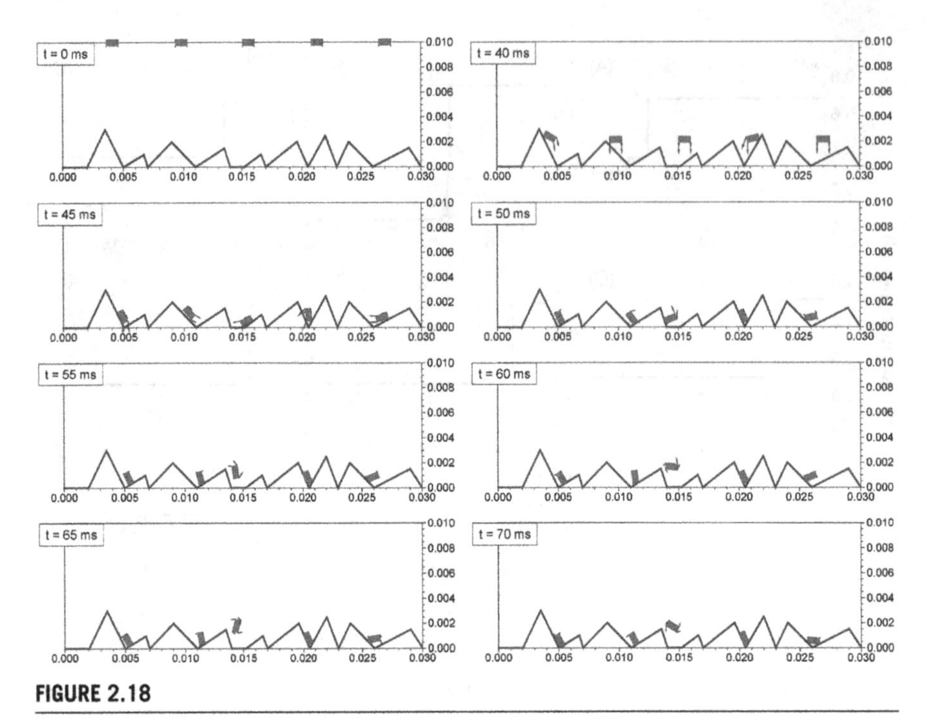

FIGURE 2.18

Demonstration of the application. Case 2: the particle-rough-wall collision, i.e. the collision process (from $t = 0$ ms to $t = 70$ ms) between five brick particles and a rough wall by a pure sedimentation process under gravity.

However, collision between the subspheres covering different host particles are still needed. In this study, the conventional DEM model is utilized to simulate the collision between the sub-elements of different host particles.

STEP 3: Motion prediction (MP, an integration process)

After the collision of subspheres is solved, the motion of subspheres can be integrated by a prediction process. This causes the subspheres to depart from the equilibrium position, and this generates forces and torques acting on the host particles. Similar to DEM, a coupled damping-spring system is employed to model the forces and torques applied on the host particles. Meanwhile, the forces and torques from all the imbedded subspheres are summed up to determine the movement of the host particle.

After the motion of the host particle is resolved, the new equilibrium positions of the imbedded subspheres can be determined. It is noted that, as the host particle is regarded as pseudo-rigid, displacements between the imbedded subsphere and the new equilibrium location can be determined by the positions and orientations of the host particle.

STEP 4: Motion correction (MC, a relaxation process)

During a relaxation process, a restoring force is generated, similar to the coupled damping spring system, where motions of the imbedded subspheres are re-

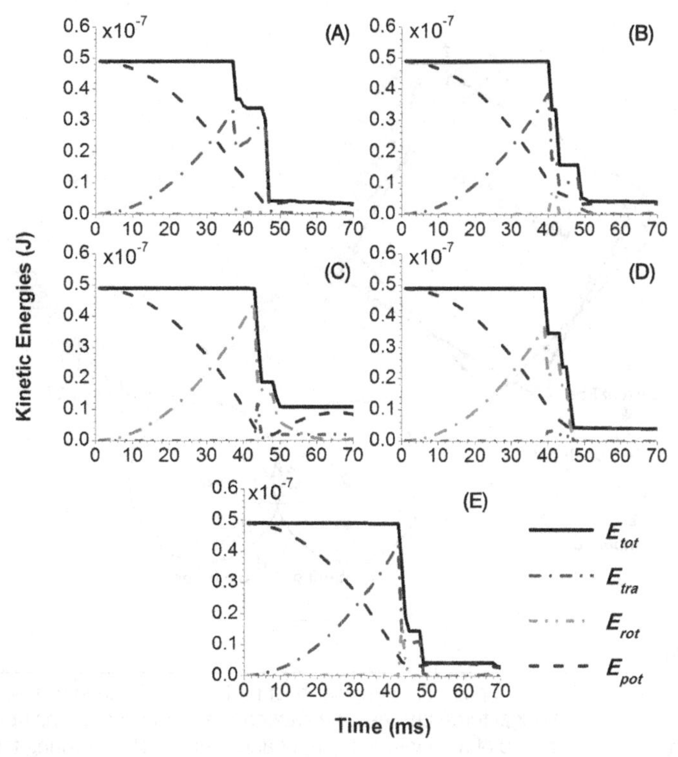

FIGURE 2.19

The variations of kinetic energies in the multiple particle-to-wall collision case.

stored based on the displacements of the current locations to the new equilibrium positions.

In the above procedures, particle is viewed as a material with the follow features:

1. The host particle is pseudo-"rigid", and deformation of the pseudo-rigid host particle can be viewed from two perspectives: On one hand, the imbedded subspheres are not allowed to move far from the "equilibrium" location, and can only oscillate about the equilibrium location, just as an "atom" with only slight fluctuations about its equilibrium position inside a molecule. Only small relative displacement is allowed between the position of subsphere and its equilibrium position. On the other hand, the subspheres distributed along the boundary of the host particle are soft materials, allowing only small micro-deformation of the host particle, since the equilibrium positions of the subspheres are determined by the globally undeformable property of the host particle. As a result, it is called a pseudo-hard particle.
2. The computational cost is moderate and acceptable. In the "multisphere" approach, there is always a conflict between precision and efficiency. On one hand,

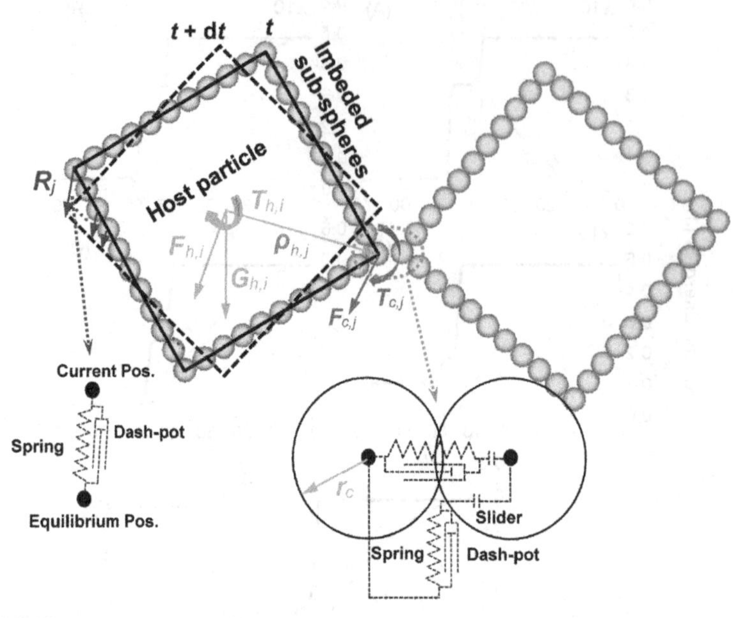

FIGURE 2.20

The sketch of the collision of a pair of rectangular host particles, imbedded with subspheres on the borders. The collisional force between the subspheres at the contact point is solved by the DEM model composed of a spring, a dash-pot and a slider. The restoring force of the subspheres generated in the relaxation process is modeled by a spring and a dash-pot toward the equilibrium position on the border of the host particle.

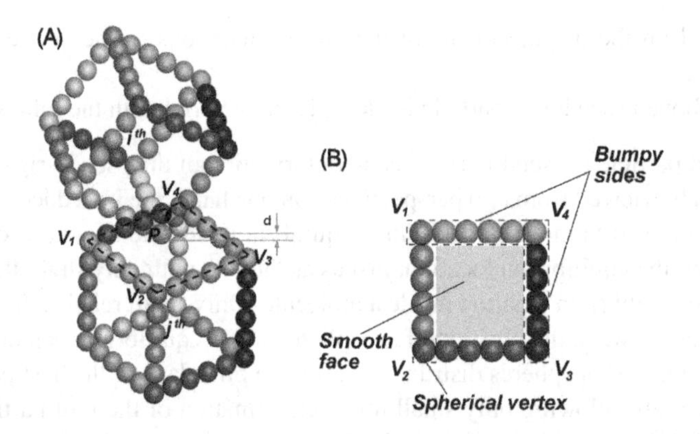

FIGURE 2.21

(A) The sketch of collision between a subelement of one host particle and a surface of another host particle in 3D case. (B) The models of the surface V_{1-4} of the host particle, composed of spherical vertex, bumpy sides and smooth face (named as V_{1-4} too).

it is advantageous to use as many and fine subspheres as possible to model a nonspherical shape accurately. On thc other hand, it is desirable to use as few subspheres as possible to save the computational cost. In current model, as there is no need to calculate the inner interactions within the host particle, the computational cost is not as large as in the "multisphere" approaches.

Finally, it is needed to compare the current approach to the multisphere models to see the main differences, pros and cons of the methods:

- In the clump method, a clump consists of rigid bodies of overlapping spheres. The spheres displace independently from each other and interact only at the contact points [177]. Moreover, the spheres are bonded together at contact points and the bonds can break when a predefined bond strength is reached. A parallel bond between two balls is just like a set of elastic springs uniformly distributed over a circular disk lying on the contact plane, which can transmit forces and moments between particles [177,178]. With the bonds, the mechanical behaviors of clumps, e.g. friction, compression, rupture, can be studied. For simplicity, two-ball and three-ball clumps are frequently used [177,179,180]. It can generate complex force chains or networks within the clump if more spheres are used [181]. This increases the computational costs for solution. It is still possible to neglect the interaction between the spheres of a clump. In this way, the spheres are permanently bonded together and the clump is a rigid assembly without deformation, inter-displacement or rupture. The transmit of forces and torques between spheres are not considered. To this regard, it is not a good approach for practical use if the bonds are unchangeable.
- In the SIPHPM model, solution of the interaction network between the subspheres within the clump is not required. As a consequence, the interactions between the subspheres are not transmitted through the force and moment networks within the clump. Instead, they are taken into account through the interactions between every subelement and the host particle by the relaxation process. With this method, further extensions can be made. The inter-displacement between subspheres and the deformation or rupture of the host particle can also be considered if needed. For example, when the impact energy between subspheres is too large, the restoring force of the relaxation process may not be sufficient to bring back the subsphere to its equilibrium position. Then, the host particle is ruptured at that instant. These types of extensions are left to the respective researchers for further applications.
- From point of view of the interaction between subspheres, the multisphere method is based on a 'net'-structure whereas the present method is based on a 'tree'-structure. Root of the tree is the host particle, and all interactions between the subspheres are through the 'root'. According to theoretical consideration, the 'net'-structure has $O(N_e^2)$ (N_e is the number of subspheres) interactions whereas the 'tree'-structure only has $O(N_e)$ interactions. Thus, the present model is cheaper to perform than the multisphere method, especially with a large number of subspheres. As the computational cost increases linearly with the number of subspheres, it can be conveniently extended to any large three-dimensional clump

of arbitrary shapes. In addition, the present model has a better view and control of the movements of the host particle.

- At last, limitations of these models are that they require a very fine time step in simulation, especially for the present model. To achieve a better approximation of the nonspherical shape, more subspheres with finer size are employed in the present model. In contrast, the clump method can use less subspheres for large size and coarse shape approximation. The contradiction between simulation efficiency and accuracy cannot be avoided. Because of the reduced complexity $O(N_e)$, the present model prefers to use a large number of subspheres to get more precise resemblance of a brick host particle.

2.3.2 Governing equation

The soft sphere approach of discrete element method is utilized in the procedure, which is formulated as [182]:

$$\boldsymbol{F}_c = -k\boldsymbol{\delta x} - \beta\boldsymbol{\delta\dot{x}}, \text{ if } |\langle \boldsymbol{F}_c, \boldsymbol{t}\rangle| > \mu\,|\langle \boldsymbol{F}_c, \boldsymbol{n}\rangle|, \text{ then } \langle \boldsymbol{F}_c, \boldsymbol{t}\rangle = \mu\langle \boldsymbol{F}_c, \boldsymbol{n}\rangle \tag{2.70}$$

$$\boldsymbol{T}_c = \boldsymbol{r}_c \times \boldsymbol{f}_c \tag{2.71}$$

where $\boldsymbol{F}_c$, $\boldsymbol{T}_c$, $\boldsymbol{r}_c$, $\boldsymbol{\delta x}$, $\boldsymbol{n}$, $\boldsymbol{t}$ are the contact force, contact torque, contact radius (the radius of the subspheres), inter-displacement, normal and tangential unit vectors respectively. k, β, μ are the stiffness factor, damping and friction coefficients. The "$\langle\cdot\rangle$" is the inner product operator, "$\dot{}$" is the derivative operator, and the subscript "c" means collisional.

Relaxation means that the subspheres are not always on the equilibrium positions. Once particle deviates from the equilibrium position, a restoring force is induced to return the particle back to the equilibrium positions and orientations. Suppose the displacement between the equilibrium and current position of the subspheres is $\boldsymbol{\delta\xi}$, a damping-spring system is used to model the restoring force $\boldsymbol{R}$ toward the equilibrium position as

$$\boldsymbol{R} = -k_r\boldsymbol{\delta\xi} - \beta_r\boldsymbol{\delta\dot{\xi}} \tag{2.72}$$

where k_r and β_r are the coefficients of stiffness and damping of the restoring force respectively.

2.3.3 Equation of motion

The forces and torques of each subsphere, including both the collisional and restoring forces and torques, are summed up to compute that of the host particle:

$$\boldsymbol{F}_h = \sum_{j=1}^{N_c} (\boldsymbol{F}_{c,j} + \boldsymbol{R}_j) \tag{2.73}$$

$$\boldsymbol{T}_h = \sum_{j=1}^{N_c} \boldsymbol{\rho}_{h,j} \times (\boldsymbol{F}_{c,j} + \boldsymbol{R}_j) + \sum_{j=1}^{N_c} \boldsymbol{T}_{c,j} \tag{2.74}$$

where $\boldsymbol{F}_h$, $\boldsymbol{T}_h$, are the total contact force and torque of the host particle. $\rho_{h,j}$ is the distance from the jth subsphere to centroid of the host particle. N_c is the total number of subspheres and $\boldsymbol{R}_j$ is the restoring force of the jth subsphere in the relaxation process.

With the total force and torque known, the host particle's motion is governed by

$$\ddot{\boldsymbol{x}}_h = \boldsymbol{F}_h/m_h - \boldsymbol{g} \tag{2.75}$$

$$\ddot{\boldsymbol{\theta}}_h = \langle \boldsymbol{I}_h^{-1}, \boldsymbol{T}_h \rangle \tag{2.76}$$

where $\boldsymbol{x}_h$, $\boldsymbol{\theta}_h$, m_h, $\boldsymbol{I}_h$, $\boldsymbol{g}$ are the translational and rotational displacements, mass and moment of inertia, and gravitational acceleration of the host particle. By time integration of Eqs. (2.75) and (2.76), position and orientation of the host particle can be recovered. Note that the host particle is regarded as a pseudo-hard material. Therefore, the relative position and orientation between any subsphere and the centroid of the host particle are fixed. After updating the host particle's position and orientation, the equilibrium position and velocity of the subspheres can be obtained accordingly.

The equation of motion of the subspheres can be determined by the collisional and restoring forces generated in the CS and MC procedures as

$$\ddot{\boldsymbol{x}}_e = (\boldsymbol{F}_c + \boldsymbol{R})/m_e - \boldsymbol{g} \tag{2.77}$$

$$\ddot{\boldsymbol{\theta}}_e = \boldsymbol{T}_c/I_e \tag{2.78}$$

2.3.4 Numerical procedures and techniques

Collision detection

For collision detection, many strategies have been developed with the analytical expressions of particle shape given, such as [183] or, with quaternions utilized, [184]. The conventional detection strategies between local pair on the boundaries of different host particles are utilized for the current study. Thus, the need of particle shape, in the governing equations is avoided.

Three-dimensional extension

As the border of host particle in a two-dimensional case is one-dimensional, covering the subelements is relatively easy, and the computational cost is of the order $O(N_s \times n_e)$, where N_s and n_e are the number of borders and subelements for each side respectively. However, in a three-dimensional case, boundary of the host particle is a two-dimensional surface, and requires a large number of subelements to cover it. The computational cost is in the order of $O(N_s \times n_e^2)$. A simplification strategy has to be employed.

Polyhedral elements are used to model the host particles of arbitrary shape. For a 3D polyhedron, it is not necessary to cover surfaces of the host particle completely. Instead, sides of the host particle should give the full description of the particle shape. Second, a particle-surface collision needs to be employed to tackle the collision process between a subelement of one host particle and the surface of another host particle. As sketched in Fig. 2.21A and B, each surface of the host particle may be modeled as a smooth face enclosed by four sides, and only the sides are covered by subspheres. For the particle-surface interaction, collision is detected when distance between one subelement P of the ith host particle and the face V_{1-4} of the jth particle satisfy $d < r_e$, where r_e is the radius of the subelements. The cost of such process is of the order of $O(N_s \times n_e) + O(N_f \times n_e)$, where N_f is the number of surfaces. Since $O(N_s \times n_e) + O(N_f \times n_e) \sim O(N_s \times n_e) \ll O(N_s \times n_e^2)$, the computational cost is greatly reduced by using such scheme, i.e. it reduces to the computational cost of a three-dimensional nonspherical shape of almost the same order as the cost of a two-dimensional case.

Is there any other, more simplified scheme? Such an example of further simplification is to distribute the subelements on vertices of the polyhedron, and therefore reduce the computational cost to $O(N_e)$. It is possible to use such an extremely simplified distribution scheme for detecting collision between subelements. However, in reality, the computational cost depends on other conditions in the simulation. The rationales are: (1) For such a simplified scheme, in addition to the vertex-vertex collision, detection schemes for vertex-side, side-side and side-surface collision must be incorporated. This significantly increases the computational expense (with the orders of $O(N_e)$, $O(N_s^2)$, $O(N_s \times N_f)$ respectively) and requires complicated numerical techniques. (2) As an alternative scheme, we would like to distribute the subelements covering all the sides of polyhedra. Then, the solution for the side-side collision is incorporated into (or transferred to) the vertex-vertex collision. There is a balance of the computational cost between the two schemes, i.e. one for using subelements imbedded on sides and one for using subelements on the vertexes only. In fact, we have already implemented the first scheme for solving the collision of fully rigid particles in prior works [185]. Therefore, in current work, we would like to use the second scheme for collision detection of pseudo-rigid particles.

2.3.5 SIPHPM model validation

In this section, five validation cases are illustrated by using the SIPHPM model.

2.3.5.1 Case 1: Pair collision

In the first case, impact between a pair of square particles is investigated by using the SIPHPM model. The results are compared to those obtained by using the generalized hard particle model (GHPM). The impact processes and correlated parameters are shown in Fig. 2.22A and B, including the position coordinates x, y, and the velocity components u, v and the rotational velocity ω. The parameters and conditions are also listed in Table 2.3.

As seen from Fig. 2.22A, the simulated collision processes of the pair of particles by GHPM and SIPHPM show good agreement. Agreement is also observed for positions, velocities, and angular velocities (Fig. 2.22B), especially for the position coordinates. However, tiny differences are observed in the post-collisional velocity components. It should be noted that the relaxation process requires only approximate

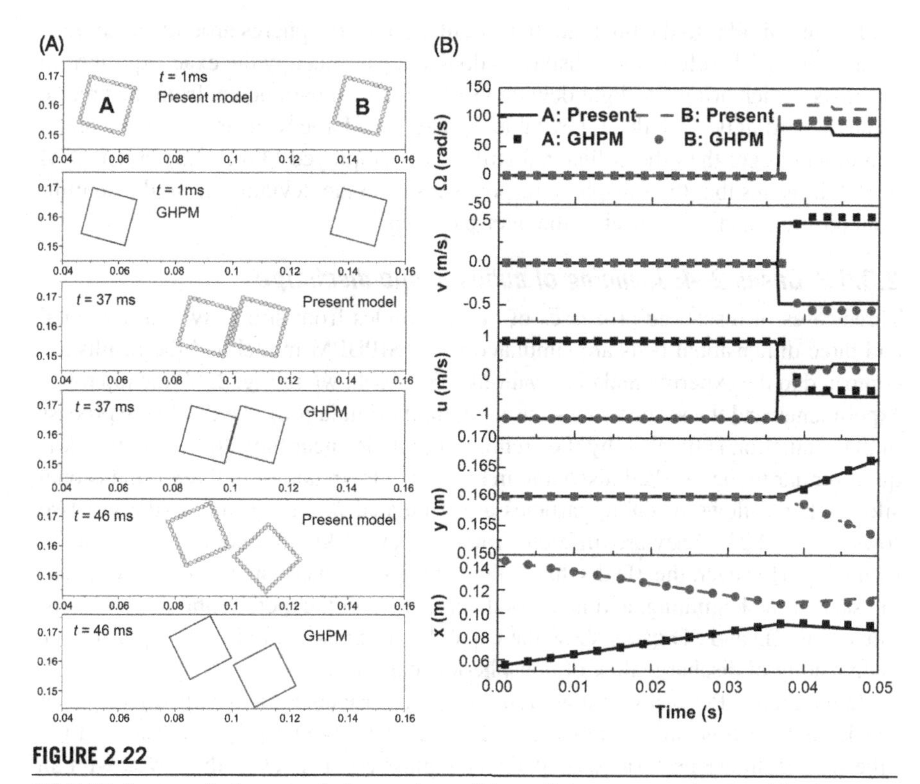

FIGURE 2.22

(A) Validation results for case 1: Comparison of the collision process simulated by the present model and the GHPM accurate solution. (B) Comparison of the positions, velocities and angular velocity of the pair of host particles.

Table 2.3 Parameters used in the four cases for validation of SIPHPM.

Parameters	Case 1: impacts	Case 2: silo discharge
Cube size (s) (mm)	16.1 / 16.1	16.1
Cube density (kg/m^3)	1455 / 1455	1455
Collisional stiffness k_c, (N/m)	– / 10^4	10^4
Restoring stiffness k_r, (N/m)	– / 10^6	10^6
Restitution coefficient e	0.8 / 0.8	0.5
Friction coefficient μ	0.232 / 0.232	0.232
Poisson ratio ν	– / 0.377	0.377
Gravity acceleration g (kg/m^{-2})	0 / 0	9.8
Simulation time t_s (s)	0.05 / 0.05	8
Filling height h_z (m)	–	0.4
Initial velocities (v_x^0, v_y^0, ω) (m/s, m/s, rad/s)	particle A: (1, 0, 0) particle B: (−1, 0, 0)	– –

predictions, similar to the fluctuation or oscillation of subspheres around the equilibrium position. Therefore, the subspheres do not always occupy the exact equilibrium positions, which brings in slight deviations of collisional parameters from the precise values accordingly. To reduce inevitable deviations, a large restoring stiffness, about two-orders larger than the collisional stiffness, is employed (Table 2.3) in SIPHPM model. It means that the subsphere responds faster to the deviation from the equilibrium position than the impact of the inter-particles.

2.3.5.2 Cases 2–4: Jamming of cubes in silo discharge

Three cases of discharge processes of brick particles from silo in two-dimensional and three-dimensional beds are simulated using SIPHPM model and the results are compared to the experimental observations [63]. As shown in Fig. 2.23 (the top row – experiments, and the second row – case 2 with 2D simulation), the discharge process in 2D simulation is blocked by the arching of particles near the silo orifice, which is quite similar to the blocked discharge in experiments. In addition, three-dimensional silo discharge flows of cubic particles are simulated in case 3 and 4 (the last two rows of Fig. 2.23). They are different from each other due to random packing at the beginning. However, the 3D discharge flow are also blocked, independently of packing state at the beginning, and is in good agreement with experimental observations. Thus, both 2D and 3D cases show the capability of present model for capturing the main feature of discharge flow of nonspherical particles.

In regard to CPU consumption (Table 2.4), 3D case takes an averaged 1.85 μs per subelement per time step, whereas the 2D case takes about 1 μs per subelement per time step. Both are performed on personal computers. It is clear that, even with an additional subsolution for the vertex-surface collision (as indicated in Fig. 2.21B), the mean computational efficiency is about the same. On the other hand, using the present 3D scheme, the number of subelements for covering the 3D particle is about 1.5 times as large as the number of subelements of 2D case. Therefore, the mean efficiency for 3D case is about 2/3 ($\approx 67.5\%$) of that of 2D case. As indicated the overall computational costs are acceptable. In conclusion, the SIPHPM model is a good candidate for simulation of complex nonspherical particle flows.

2.3.5.3 Case 5: Free discharge of cubes in silos

In this section, validation on free discharge of cubic particles is performed following the experimental work by [186]. 200 cubes of 12.5 mm in length are initially inside a silo of 500 (height) $\times$ 200 (width) $\times$ 30 (depth) mm, and are discharged through the orifice (width $B_o = 60, 80, 116$ mm) at the bottom center. It is noticed that the exact positions and velocities of all particles cannot be obtained in experiment. Therefore, we used random initial particle velocities in the simulation, and let particles fall freely from some distances from the bottom when the orifice is closed to build similar initial packing states. After that, the orifice is opened, and the discharge starts. Validation of parameters in the simulation are listed in Table 2.5.

The snapshots of discharge at $t = 0.1$ s for $B_o = 60, 80, 116$ mm are illustrated in Fig. 2.24A–C. When the orifice size is different, the number of particles mov-

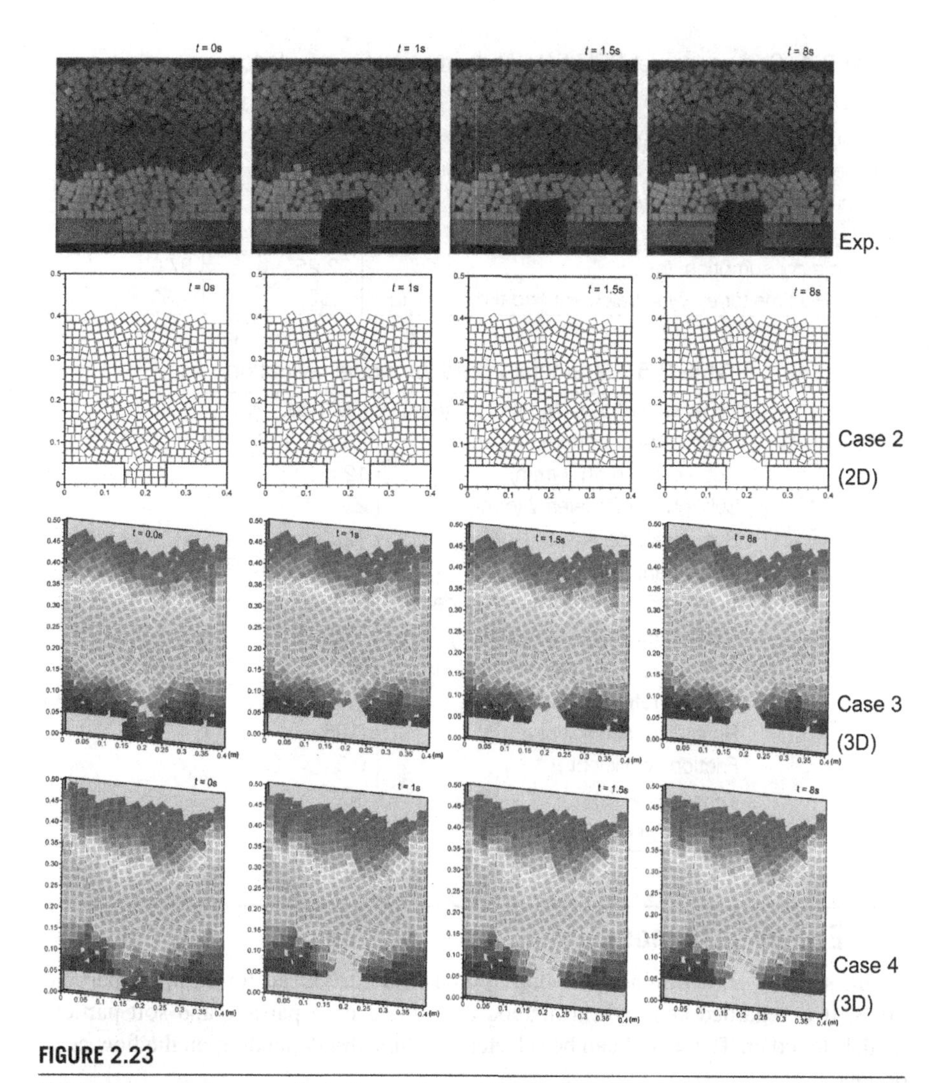

FIGURE 2.23

Validation results for cases 2–4: Comparison of the discharge process within a silo between the SIPHPM simulation (the 2nd row is for 2D and the 3th and 4th rows are for 3D cases) and the experimental results (the 1st row) in [63].

ing downward are different. The corresponding number flow rates are computed (Fig. 2.25) and compared to the experimental data. It is noted that although the initial packing states are different, the simulated number flow rates are close to the experimental data. Once again, the simulation validates the current model for simulating cubic particle motions.

Table 2.4 Comparison of CPU time consumption for simulation.

	Case 2 (2D)	Case 3 (3D)
Total number of host particles	405	406
Number of subelements for covering each host particle	40	60
Total number of subelements	16200	24360
Number of time steps	4×10^6	3×10^6
CPU time consumption, (h)	33.25	37.03
Mean CPU time for every subelement and time step, (μs)	1.82	1.85

Table 2.5 Parameters used for model validation.

Dimension of silo, (H_s, W_s, D_s) (mm)	(500, 200, 30)
Orifice size at bottom center, B_o (mm)	60, 80, 116
Cube side length, l (mm)	12.5
Subsphere diameter, d (mm)	2.5
Subsphere density, ρ_s (kg/m^3)	1200
Cubic particle number N_c	200
Subsphere number in each cube N_s	56
Total number of subspheres N_p	11200
Stiffness factor k_c, k_r (N/m)	10^3, 10^5
Poisson ratio ν	0.3
Restitution coefficient e	0.95
Friction coefficient μ	0.3
Time step δt (s)	10^{-6}
Number of simulation step N_t	2.5×10^6

2.4 EHPM-DEM model

In this section, an additional coupling approach is introduced for complex particle flows. It is regarded as a hybrid method combining hard particle and soft particle models together. The model can be selected dynamically depending on the flow conditions.

2.4.1 Extended general hard particle model (EHPM)

The general hard particle model is derived from the hard sphere model. The main assumptions are the same, i.e. (1) the particles are quasi-rigid, and particle deformation is neglected; (2) the collision is binary and quasi-instantaneous, which means that the particle-particle collision takes place during a short duration. The fundamental laws are as follows:

$$m_i \boldsymbol{v}_i = m_i \boldsymbol{v}_i^0 + \boldsymbol{P}_i \tag{2.79}$$

$$m_j \boldsymbol{v}_j = m_j \boldsymbol{v}_j^0 - \boldsymbol{P}_i \tag{2.80}$$

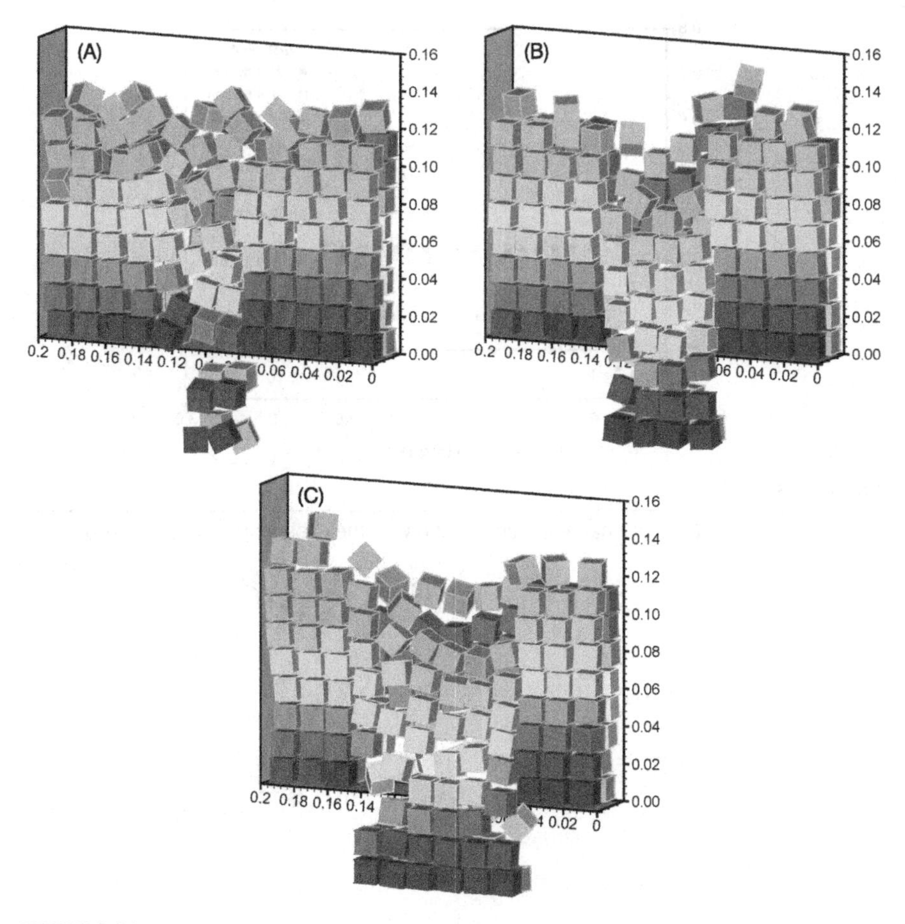

FIGURE 2.24

The snapshots ($t = 0.1$ s) of cubic particle discharge from an orifice at the bottom of a silo. (A)–(C) Orifice size $B_o = 60$, 80 and 116 mm, respectively [186].

$$I_i \boldsymbol{\omega}_i = I_i \boldsymbol{\omega}_i^0 + \boldsymbol{r}_i \times \boldsymbol{P}_i \tag{2.81}$$

$$I_j \boldsymbol{\omega}_j = I_j \boldsymbol{\omega}_j^0 - \boldsymbol{r}_j \times \boldsymbol{P}_i \tag{2.82}$$

where the subscripts 'i' and 'j' are the indices of the pair of colliding particles. m, I, $\boldsymbol{v}$, $\boldsymbol{\omega}$, $\boldsymbol{r}$, $\boldsymbol{P}$ are the particle mass, inertia, velocity, angular velocity, the position vector from the center of mass to the contact point on the particle surface, and the impulse generated in collision, respectively. The superscript '0' denotes the parameter before collision. Other parameters without superscript '0' are post-collisional parameters. In Eqs. (2.79)–(2.82), five post-collisional parameters, i.e. $\boldsymbol{v}_i$, $\boldsymbol{\omega}_i$, $\boldsymbol{v}_j$, $\boldsymbol{\omega}_j$ and $\boldsymbol{P}$, are the unknowns. Therefore, an additional equation is required to close the governing

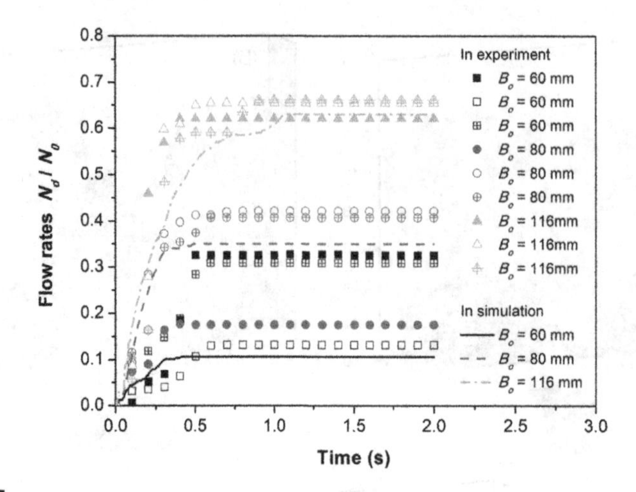

FIGURE 2.25

The number flow rates of discharge in comparison with the experimental data of [186].

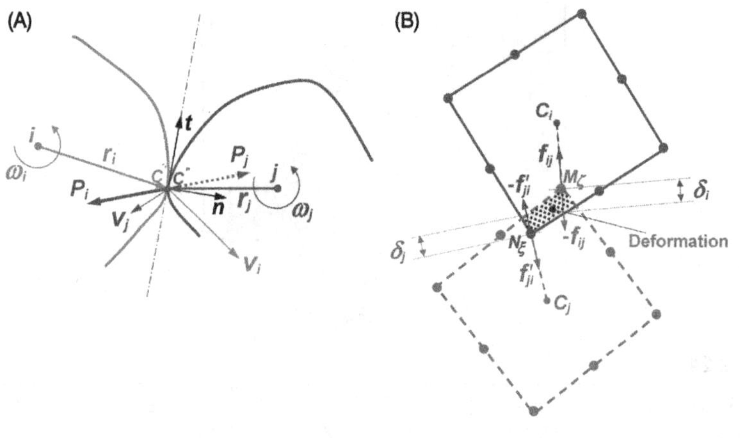

FIGURE 2.26

The sketches of the particle-particle collision processes in the EHPM model (A) and DEM model (B).

equations. The restitution coefficient is formulated as:

$$\boldsymbol{v}_r^C \cdot \boldsymbol{n} = -e(\boldsymbol{v}_r^{C,0} \cdot \boldsymbol{n}) \tag{2.83}$$

where $\boldsymbol{v}_r^C = \boldsymbol{v}_j - \boldsymbol{v}_i + \boldsymbol{\omega}_j \times \boldsymbol{r}_j - \boldsymbol{\omega}_i \times \boldsymbol{r}_i$ is the relative velocity between the pair of contact points (C' and C''), and $\boldsymbol{n}$ is the normal vector of the contact surface (Fig. 2.26A). Notice that the definition of the restitution coefficient in Eq. (2.83) is a general form, which is suitable for particles of arbitrary shape. The differences between the present definition of the restitution coefficient for general shape particles and the conventional definition for spherical particles are: (1) For spherical particles, the normal vector is pointing from the center of mass from one particle to another, whereas for nonspherical particles, the normal vector is the one on the contact surface. For spherical particles, the vector from the center of one particle to another superimposes the normal vector of the contact surface, whereas they are different for nonspherical particles (Fig. 2.26A). (2) For spherical particles, the relative velocity used to define the restitution coefficient is the ratio of incident velocities of the centroids of spherical particles. However, for nonspherical particles, it is the ratio of the velocities at the pair of contact points. Both definitions result in the same value for spherical particles and different for nonspherical particles. In fact, the latter is an extension of the former (Fig. 2.26A).

Based on Eqs. (2.79)–(2.83), the solutions of impulse after collision are as follows:

$$\boldsymbol{P}_i = \begin{cases} \frac{\beta(\boldsymbol{v}_r^{C,0}\cdot\boldsymbol{t})+\gamma(1+e)(\boldsymbol{v}_r^{C,0}\cdot\boldsymbol{n})}{(\alpha\gamma-\beta^2)}\boldsymbol{n} + \frac{\beta(1+e)(\boldsymbol{v}_r^{C,0}\cdot\boldsymbol{n})+\alpha(\boldsymbol{v}_r^{C,0}\cdot\boldsymbol{t})}{(\alpha\gamma-\beta^2)}\boldsymbol{t}, \text{ nonsliding collision} \\ \frac{(1+e)(\boldsymbol{v}_r^{C,0}\cdot\boldsymbol{n})}{(\alpha-\mu\beta)}(\boldsymbol{n}+\mu\boldsymbol{t}), \text{ sliding collision} \end{cases} \tag{2.84}$$

where the coefficients α, β, γ are expressed, respectively, as:

$$\begin{cases} \alpha = \dfrac{m_i+m_j}{m_i m_j} + \dfrac{(\boldsymbol{r}_i\cdot\boldsymbol{t})^2}{I_i} + \dfrac{(\boldsymbol{r}_j\cdot\boldsymbol{t})^2}{I_j}, & \text{(a)} \\ \beta = \dfrac{(\boldsymbol{r}_i\cdot\boldsymbol{n})(\boldsymbol{r}_i\cdot\boldsymbol{t})}{I_i} + \dfrac{(\boldsymbol{r}_j\cdot\boldsymbol{n})(\boldsymbol{r}_j\cdot\boldsymbol{t})}{I_j}, & \text{(b)} \\ \gamma = \dfrac{m_i+m_j}{m_i m_j} + \dfrac{(\boldsymbol{r}_i\cdot\boldsymbol{n})^2}{I_i} + \dfrac{(\boldsymbol{r}_j\cdot\boldsymbol{n})^2}{I_j}, & \text{(c)} \end{cases} \tag{2.85}$$

and $\boldsymbol{t}$ is the tangential vector on the contact surface. In addition, collisions with or without sliding can be determined by

$$\lambda = \frac{\left|\alpha(\boldsymbol{v}_r^{C,0}\cdot\boldsymbol{t})+\beta(1+e)(\boldsymbol{v}_r^{C,0}\cdot\boldsymbol{n})\right|}{\left|\beta(\boldsymbol{v}_r^{C,0}\cdot\boldsymbol{t})+\gamma(1+e)(\boldsymbol{v}_r^{C,0}\cdot\boldsymbol{n})\right|} \begin{cases} > \mu \text{ sliding collision} \\ < \mu \text{ nonsliding collision} \end{cases} \tag{2.86}$$

where inter-particle sliding does not occur on the surface in the nonsliding collision, and vice versa. Finally, with the impulse known, post-collisional velocities and angular velocities can be readily obtained by substituting $\boldsymbol{P}_i$ into Eqs. (2.79)–(2.82).

2.4.2 Extended discrete element method

In general, the discrete element method is characterized by three basic mechanisms of the particle-particle interaction, parameterized respectively by an elastic stiffness factor k, a damping coefficient η, and a friction coefficient μ. In order to couple DEM with the EHPM model, some simplifications are needed (as sketched in Fig. 2.26):

- The elastic force is only exerted on the vertices and sample representative points on the particle surface (Fig. 2.26B), if they are inside another particle. Totally 8 vertices on the border of particle surface are considered.
- The damping force depends on the relative velocity between the pair of particles on the vertices.
- The friction is not considered since it has already been considered in the EHPM model.
- The edge-edge collision is transformed into a pair of vertex-edge collisions (when the edges are parallel to each other with one vertex of each particle contacting the edge of the other) or a pair of vertex-vertex collision (when the two edges are parallel and superposed completely). With solutions to the two collisions solved, the two post-collisional impulses are averaged, while the post-collisional torques are added.

Based on these assumptions, the particle-particle interaction force can be modeled as follows:

$$\boldsymbol{f}_{ij}^{\zeta} = -k\boldsymbol{\delta}_i^{\zeta} - \eta \boldsymbol{v}_{r,ij}^{\zeta} \tag{2.87}$$

$$\boldsymbol{f}_{ji}^{\xi} = -k\boldsymbol{\delta}_j^{\xi} - \eta \boldsymbol{v}_{r,ji}^{\xi} \tag{2.88}$$

$$\boldsymbol{F}_{c,ij} = \sum_{\zeta} \boldsymbol{f}_{ij}^{\zeta} - \sum_{\xi} \boldsymbol{f}_{ji}^{\xi} \tag{2.89}$$

$$\boldsymbol{M}_{c,ij} = \sum_{\zeta} (\overline{C_i M_{\zeta}} \times \boldsymbol{f}_{ij}^{\zeta}) - \sum_{\xi} (\overline{C_i N_{\xi}} \times \boldsymbol{f}_{ji}^{\xi}) \tag{2.90}$$

where ξ and ζ $(= 1, \cdots, 8)$ denote the 8 vertices on the surfaces of particles 'i' and 'j' respectively. $\boldsymbol{f}_{ij}^{\zeta}$ is the force from particle 'j' to particle 'i' through the 'ζ' vertex. $\boldsymbol{\delta}_i^{\zeta}$ represents the deformation of particle 'i' caused by the impact of vertex 'ζ' of particle 'j'. $\boldsymbol{v}_{r,ij}^{\zeta} = \boldsymbol{v}_j - \boldsymbol{v}_i + \boldsymbol{\omega}_j \times \overline{C_j M_{\zeta}} - \boldsymbol{\omega}_i \times \overline{C_i M_{\zeta}}$ is the relative velocity on the points where the vertex 'M_{ζ}' is located. $\boldsymbol{F}_{c,ij}$ is the total force of particle 'i' induced by all the vertices inside the other particle.

The stiffness factor, k, and damping coefficient, η, are used according to the works [187], [52], and [188], i.e.

$$\begin{cases} k = \dfrac{\sqrt{2r_p}E}{3(1-\nu^2)} \\ \eta = -\dfrac{2\ln e}{\sqrt{\pi^2 + \ln^2 e}}\sqrt{m_p k} \end{cases} \tag{2.91}$$

where E, ν, r_p and m_p are the Young's modulus, Poisson ratio, particle radius (from center to the vertex) and mass, respectively.

In this model, the inter-particle collision force due to the deformations (characterized by the continuous overlapping between particle 'i' and 'j', Fig. 2.26B) of the pair of particles is simplified as the forces acting on the vertices. Therefore, the

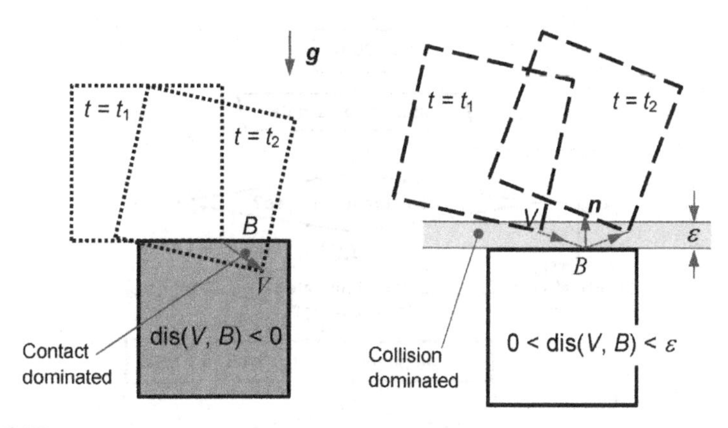

FIGURE 2.27

The sketch of the collision type determination in the contact dominated (left inset) and collision dominated flows (right inset).

detailed calculation of the distributed deformation forces over the region is avoided. The material is assumed to be near rigid, i.e. the deformation region is fairly small. The rationality for this assumption is that in an ideal case, the deformation region should be reduced to a segment of the particle border. In addition, the force distribution throughout the segment should be linear and characterized by the forces on the two ends of the segment.

2.4.3 EHPM-DEM coupling strategy

As mentioned previously, the EHPM model is suitable for a collision dominated flow and the DEM model is suitable for a contact dominated flow. However, for numerical implementation, it is difficult to quantitatively distinguish the flow regimes. In this section, a simple but useful method is proposed to determine which model to use for a specified condition. As sketched in Fig. 2.27, since the contact dominated flow is characterized by continuous inter-particle contact, which leads to particle deformation and inter-particle geometry overlap, at least one vertex of a particle is inside another particle. In contrast, in the collision dominated flow, particles are separated with a nonzero relative incident velocity before collision, i.e. $v^0_{r,n} < 0$, which leads to the particle-particle collision and consequentially causes separation after collision, i.e. $v_{r,n} > 0$. This occurs outside the particles but within a very short region.

For numerical simulation, a distance function $dis(V, B)$ is used to quantify the distance of any vertex 'V' of one particle to the border 'B' of other particles. Then, the inter-particle collision within the contact dominated flow regime can be determined by $dis(V, B) < 0$. In this case, the DEM model is employed to solve the particle-particle collision force. Otherwise the flow is collision dominated and inter-particle collision may take place when the distance between any vertex of one particle and the border of another particle is less than a very small threshold, i.e.

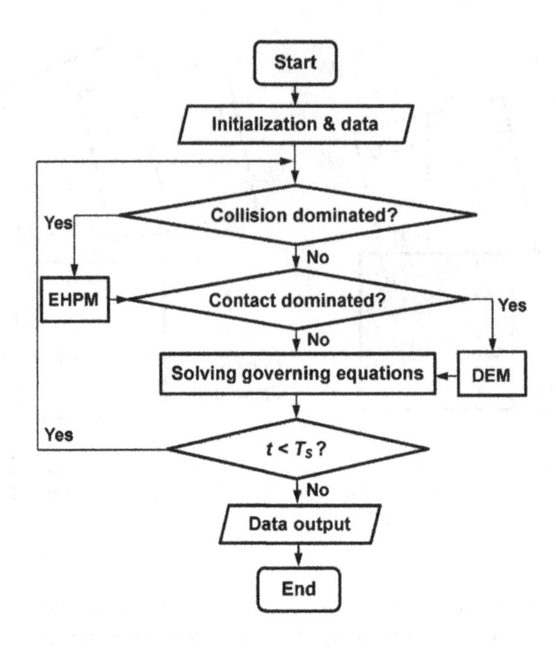

FIGURE 2.28

The flow chart of the whole EHPM-DEM coupled approach.

$0 < dis(V, B) < \varepsilon$. In this case, the EHPM model is utilized to solve the collisional impulse.

To summarize, the programming flowchart is shown in Fig. 2.28. In the main cycle, the $dis(V, B)$ function is calculated first to determine the collision type. If $dis(V, B) < 0$, the DEM model is used for the particle-particle contact; if $0 < dis(V, B) < \varepsilon$, the EHPM model is used to solve the particle-particle collision. Then, governing equations of the entire motion are solved to trace the particle motion. This process is repeated in the next time step to solve the temporal evolution of the particle flow.

2.4.4 Collision detection strategy

The particle-particle collision detection strategy is important for numerical implementation of the current model. In this work, in context of the coupled EHPM and DEM approach, the particle-particle collision is strictly detected by every edge-edge contact, where vertex-vertex and vertex-edge collisions are also incorporated.

As sketched in Fig. 2.29, it is assumed that A (x_a, y_a) and B (x_b, y_b) are ends of one edge of particle P_1, and C (x_c, y_c) and D (x_d, y_d) are ends of an edge of particle P_2. If $\overline{AB}$ and $\overline{CD}$ intersects at (x_p, y_p), they can be computed as follows:

$$\begin{cases} x_p = x_a + \tau_{ab}(x_b - x_a) = x_c + \tau_{cd}(x_d - x_c) \\ y_p = y_a + \tau_{ab}(y_b - y_a) = y_c + \tau_{cd}(y_d - y_c) \end{cases} \tag{2.92}$$

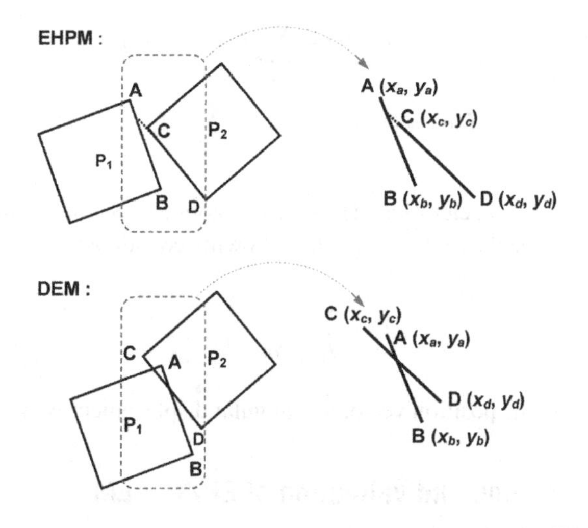

FIGURE 2.29

The sketch of the collision detection in EHPM and DEM.

which are equivalent to:

$$\begin{cases} (x_b - x_a)\tau_{ab} - (x_d - x_c)\tau_{cd} = (x_c - x_a) \\ (y_b - y_a)\tau_{ab} - (y_d - y_c)\tau_{cd} = (y_c - y_a) \end{cases} \tag{2.93}$$

If the solutions of Eq. (2.93) satisfy the following conditions,

$$\begin{cases} \text{if } \tau_{ab} \text{ or } \tau_{cd} \in [-\epsilon, 0] \text{ or } [1, 1+\epsilon], \text{ when } 0 < dis(V, B) < \varepsilon, \text{ using EHPM} \\ \text{if } 0 < \tau_{ab}, \tau_{cd} < 1, \text{ when } dis(V, B) < 0, \text{ using DEM} \end{cases} \tag{2.94}$$

where ϵ is a small tolerance, a collision between P_1 and P_2 at (x_p, y_p) is detected and either the EHPM or DEM model is used to solve the collision process.

Differently from the contact detection scheme based on the common plane (CP method) [189,190], the key point for the current strategy is the combination of collision detection for EHPM and DEM. By incorporation of the DEM method, a slight overlap of the particles can occur, and the edge-edge cross is allowed. This method enables a fast searching of the edge-edge cross by solving Eq. (2.93) directly. As mentioned before, the result of this equation determines which particle-particle collision model should be selected.

2.4.5 Governing equations of motion

With the collisional forces/impulses known, the Newton's law of motion is followed to trace the particle motions:

$$m_i \dot{\boldsymbol{v}}_i = (\boldsymbol{P}_i/\Delta t + \sum_j \boldsymbol{F}_{c,ij}) - m_i \boldsymbol{g} \tag{2.95}$$

$$I_i \dot{\boldsymbol{\omega}}_i = (\boldsymbol{r}_i \times \boldsymbol{P}_i)/\Delta t + \sum_j \boldsymbol{M}_{c,ij} \tag{2.96}$$

where $\boldsymbol{g}$ is the gravity acceleration. The particle motion trajectories and angular displacements are obtained by integrating the following equations:

$$\dot{\boldsymbol{x}}_i = \boldsymbol{v}_i \tag{2.97}$$

$$\dot{\boldsymbol{\theta}}_i = \boldsymbol{\omega}_i \tag{2.98}$$

where $\boldsymbol{x}_i$ and $\boldsymbol{\theta}_i$ are the position vector and angular displacement vector, respectively.

2.4.6 Demonstration and validation of EHPM-DEM

Case 1: Pair collision

In this section, a typical particle-particle collision is solved by the EHPM model and validated by to the fundamental laws of momentum and mechanical energy. The following case is considered: using restitution coefficient $e = 0.9$ and frictional coefficient $\mu = 0.3$, Fig. 2.30 shows the process of collision between a pair of equilateral triangular particles (the basic nonspherical particle shape with the same edge length $l = \sqrt{3}$ cm). Particle 1 is moving toward particle 2 at a constant velocity, and collides with it at $t = 5$ ms on location $\delta h = 5$ mm from the bottom of the vertical edge of particle 2. After collision, particle 2 moves forward and rotates, and particle 1 loses most of its translational kinetic energy. Particle 1 gains a weak rotational energy due to the frictional torque.

For quantitative analysis, after collision, the $V_{x,1}$ is changed from 1 to 0.335, $V_{x,2}$ from 0 to 0.665, $V_{y,1}$ from 0 to −0.095 and $V_{y,2}$ from 0 to 0.095. Thus, the translational momentum in either the x or the y direction is conserved. Meanwhile, the rotational velocity ω_1 is decreased from 0 to −38 rad/s and ω_2 is increased from 0 to 114 rad/s. Take the collisional point between particle 1 and 2 as a reference point, the angular momentum with respect to the reference point is zero before the collision. After the collision, the total angular momentum with respect to the reference point is

$$\begin{aligned}
L_{tot,c} &= \sum_{i=1,2} \{r_{p,i} \times m_{p,i} v_{p,i} + I_{p,i}\omega_{p,i}\} \\
&= m_p\{V_{x,2}\delta h + V_{y,2}\frac{l}{3}sin(\frac{\pi}{3}) - V_{y,1}\frac{2l}{3}sin(\frac{\pi}{3}) + \frac{l^2}{12}(\omega_1 + \omega_2)\} \\
&= m_p * 10^{-3} * \{-0.665 * 5 + 0.095 * \frac{17.32}{3} * 0.866 + \\
&\quad 0.095 * \frac{17.32 * 2}{3} * 0.866 + \frac{17.32^2 * 10^{-3}}{12} * (114 - 38)\} \\
&\approx 0
\end{aligned}$$

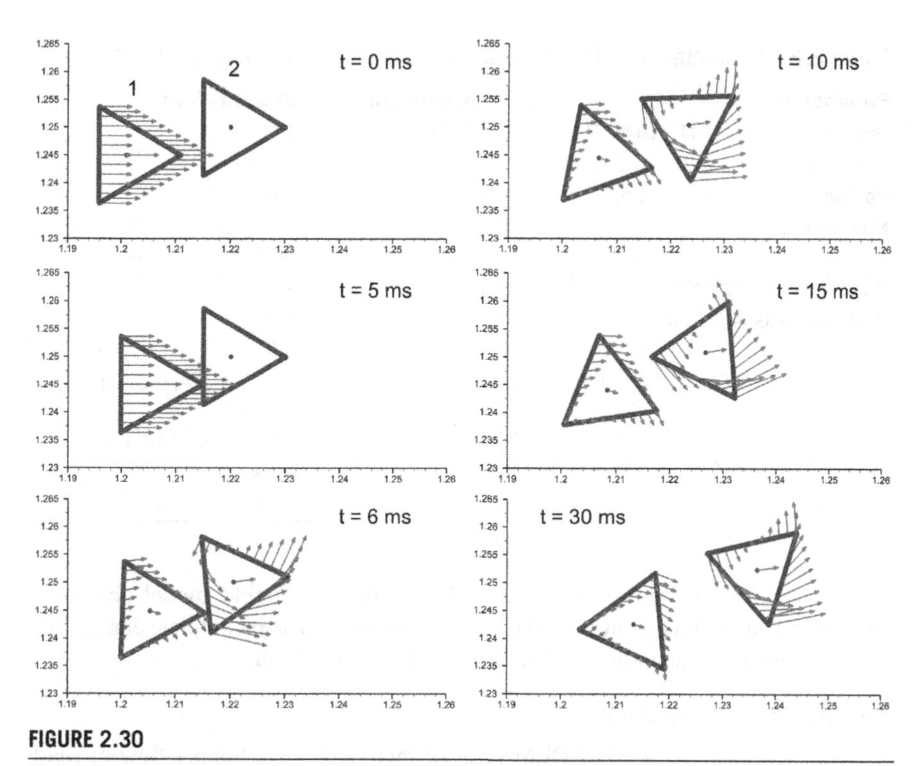

FIGURE 2.30

The collision process between a pair of triangular particles (/mm).

where $m_p = m_{p,i}$ is the mass, and $I_{p,i} = \frac{1}{12} m_{p,i} l^2$ is the momentum of inertia with respect to the centroid. Thus, the total angular momentum is also conserved during the collision.

For energy analysis, assume the mass is equal to 1, then the translational, rotational, and total kinetic energies can be expressed as

$$K_{tr} = \frac{1}{2} \boldsymbol{v} \cdot \boldsymbol{v} \tag{2.99}$$

$$K_{ro} = \frac{1}{24} \boldsymbol{\omega} \cdot \boldsymbol{\omega} \tag{2.100}$$

$$K_{total} = K_{tr} + K_{ro}. \tag{2.101}$$

Then, the following results are obtained: (1) Particle 1 loses most of its kinetic energy, and particle 2 only obtains a small portion of the kinetic energy from particle 1. (2) Meanwhile, the rotational kinetic energy $K_{ro,1}$ is increased slightly by friction, and $K_{ro,2}$ is increased by both collision impulse and friction. (3) The total kinetic energy is decreased, partially caused by the incomplete elastic collision with the restitution coefficient $e = 0.9 < 1$, and partially caused by the frictional motion on the particle edge.

Table 2.6 Parameters used in the EHPM-DEM simulation and experiment.

Parameters	In experiment	In simulation
Bed width W, depth D, height H (mm)	(200, 200, 600)	(200, –, 600)
Wedge angle φ (°), orifice size s_o (mm)	60, 40	60, 40
Particle shape, equivalent size s_p (mm)	hexahedron, 6.5	square, 6.5
Number of particles N_p	–	1000, 1500, 2000, 2500, 3000, 3500, 4000
Particle density (kg·m^{-3})	778	778
Young's modulus E (Pa))	–	1.0×10^6
Poisson ratio ν	–	0.3
Restitution coefficient e	–	0.2, 0.3, 0.4, 0.5, 0.6, 0.7, 0.8, 0.9
Friction coefficient μ	0.12	0.01, 0.06, 0.12, 0.20
Time step Δt (s)	–	1.0×10^{-5}
Small threshold value ε	–	$0.01 s_p$

In conclusion, the above simulation validates that the EHPM model obeys the conservation laws of momentum and angular momentum. The incomplete elastic collision and the nonideal collision with friction are both included.

Case 2: Hopper discharge

In order to validate the EHPM-DEM coupled methodology, a hopper flow of rigid square particles is simulated following an existing experiment [191]. In the experiment, the hopper bed is three-dimensional and the particle is hexahedron (hopper B in [191]). For simplification, a two-dimensional bed filled with square particles is simulated. The detailed parameters used in the experimental apparatus and in the present simulation are listed in Table 2.6.

Initially, the square particles are distributed in a regular array and lifted in certain heights. They fall into the wedge of the bed under the effect of gravity. After a short time, an initial packing state is built, e.g. Fig. 2.31A at the packing state of $t = 1.0$ s after the falling process. After inspection, it shows that some particles are overlapped within compactly packed regions – a contact dominated state with continuous inter-particle contact. During the falling process, majority of the moving particles frequently experience the inter-particle collision – a collision dominated state. This step acts as a pre-processing step to obtain the packing state for the discharging hopper flow.

At $t = 1$ s, the bottom of the hopper is opened to let the particles discharge freely. According to the experiment [191], three thin layers of colored particles are used as tracer particles. They are separated every 150 mm in the vertical direction. It is convenient to compare the flow patterns in the hopper by the tracer particles. As shown in Fig. 2.31B, the three thin layers of particles are placed horizontally at the beginning ($t = 5$ ms immediately after the natural packing state (Fig. 2.31A). At $t = 0.4$ s, the three layers of tracer particles move down in a shape that the two ends

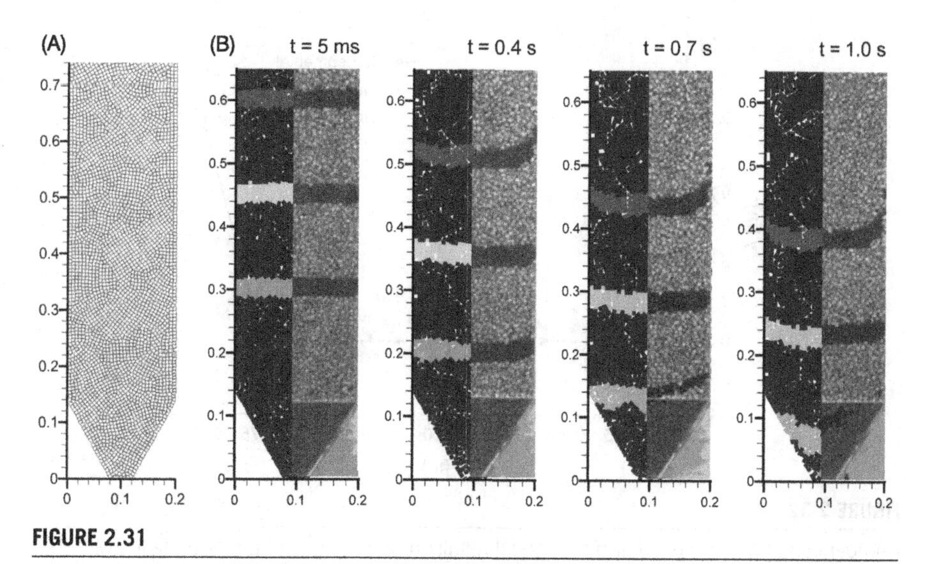

FIGURE 2.31

(A) The distribution of particles by falling into the wedge. (B) Comparison of the flow patterns in the hopper flow from $t = 0.005$ s to $t = 1.0$ s. In each inset, the left part is a numerical result and the right part is an experimental result.

of the top layer are higher than the center due to the resistance of friction of the wall. Similar trends can be seen for the middle and lowest layer of particles. Although slight difference exists between the experimental and numerical results especially for the bottom layer of particles, the simulation results show the main features of the flow pattern of the drainage hopper flow of hexahedron particles successfully. The flow pattern is characterized by a straight-line-shaped pattern in the central core, similar to that in typical plug flow, where all the particles flow at almost the same velocity except for the particles near the wall [191]. This flow pattern was different from other cases of hopper flows using other kinds of particle shapes. Based on the flow pattern comparison, the EHPM-DEM model is capable of capturing the main features of particle flows.

Quantitative comparison of the heights of the stripes of tracer particles are shown in Fig. 2.32. The left ends of the stripes are kept at initial height of 300 mm. Then, different restitution coefficients are used in the simulation to find the best match. Fig. 2.32 shows that the simulation data agree well with the experimental data. Although, there are slight differences between them, especially at the right ends. It is seen that the particles are located almost linearly on the right wall, which could result in a faster downward movement compared to that near the left wall. That is why the heights in the right end are lower than those at the left end. It is generally seen that profiles of the heights are concave upward, with a flat bottom at about 0.24 m.

In conclusion, although the simulation conditions are not exactly the same, the main features of flow patterns have been reproduced successfully and general quan-

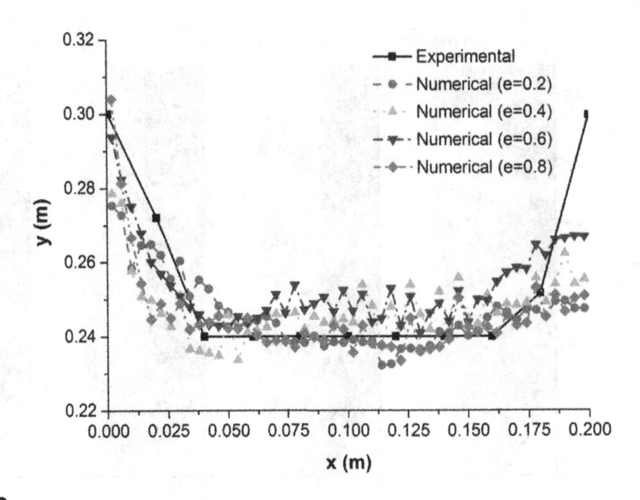

FIGURE 2.32

Comparison of experimental and numerical results of the heights of tracer particles with the left ends at $H = 0.3$ m.

titative agreements have been obtained. The EHPM-DEM coupled approach for simulating both contact dominated and collision dominated particle flows is reliable.

2.4.7 Simulation efficiency

It is well known that searching for collisions and determining the contact status are the most expensive procedures in performing the EHPM and the soft DEM simulation. One of the main advantages of EHPM over the soft DEM is the relatively higher efficiency. In this section, the relative efficiency for searching and solving the collision by either the EHPM or the soft DEM in the EHPM-DEM method is compared. The total efficiency for the EHPM-DEM coupled method and the pure soft DEM method is compared too.

2.4.7.1 Case 1: Hopper discharge

The first case is a discharging flow of $N_p = 4000$ square particles (Fig. 2.33), with restitution coefficient $e = 0.6$ and friction coefficient $\mu = 0.12$ (Table 2.6). The EHPM and DEM models are employed to simulate the freely discharging flow from the bottom hole. A small threshold value $\varepsilon = 0.01 s_p$ of the distance function is used which determines the selection of collision schemes, either by EHPM or by soft DEM. The process and main features of the discharge flow (from $t = 0.1$ s to 0.6 s) in the hopper are successfully simulated (Fig. 2.33). At $t = 0.7$ s, all the particles are almost discharged out of the hopper.

2.4.7.2 Case 2: Particle deposition

The second case is a deposition of $N_p = 100$ square particles solved by both the EHPM-DEM coupled method (Fig. 2.34A–C) and by the pure DEM method

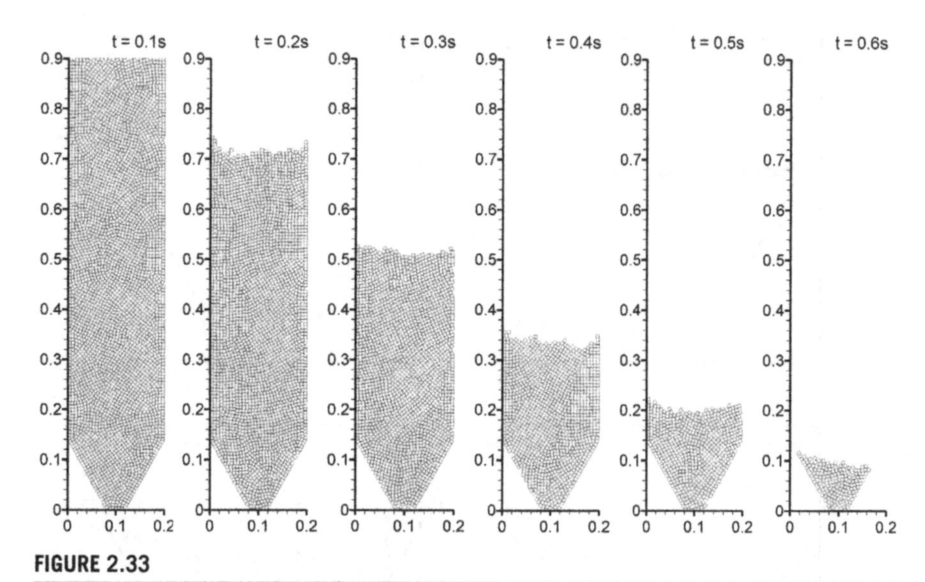

FIGURE 2.33

The process of discharge flow of the first case.

(Fig. 2.34D–F) with the same restitution coefficient $e = 0.6$ and friction coefficient $\mu = 0.12$ as in the first case. In the pure DEM solution, the damping elastic force is dominant, whereas the impulse is considered dominant in the EHPM model. Therefore, motion of particles in the pure DEM solution is more elastic than that in the EHPM-DEM solution. However, the main processes of deposition are reproduced successfully by both solutions. As many-body granular system is nonlinear (e.g. dicing is hard to be predicted or determined accurately), only qualitative agreements between the two solutions can be obtained.

The total and mean CPU time for the EHPM ($T_{t,EHPM}$ and $T_{m,EHPM}$), DEM ($T_{t,DEM}$ and $T_{m,DEM}$), and the total and mean CPU time for searching collision ($T_{t,SC}$ and $T_{m,SC}$) are shown in Fig. 2.35A and B to compare the efficiency. The time steps for the EHPM and the DEM model are the same ($\Delta t = 10^{-5}$ s). It is seen that the total time consumption for EHPM is smaller than that of DEM. The total consumption time divided by the number of collisions (the inset of Fig. 2.35A) gives the mean computational time for each collision in EHPM which is always less than that in DEM. Thus, EHPM is more advantageous than DEM for the collision solutions. For the collision searching process, as the discharge continues, the mean CPU time for searching is decreased as the number of particles is reduced.

Fig. 2.35C shows the total computational time in the EHPM-DEM solution and the pure DEM solution. Both models can simulate the particle deposition process. The only difference is that the pure soft DEM model requires a finer time step, namely $\Delta t = 10^{-6}$ s. It shows that the efficiency of the EHPM-DEM solution is higher than for the pure DEM method.

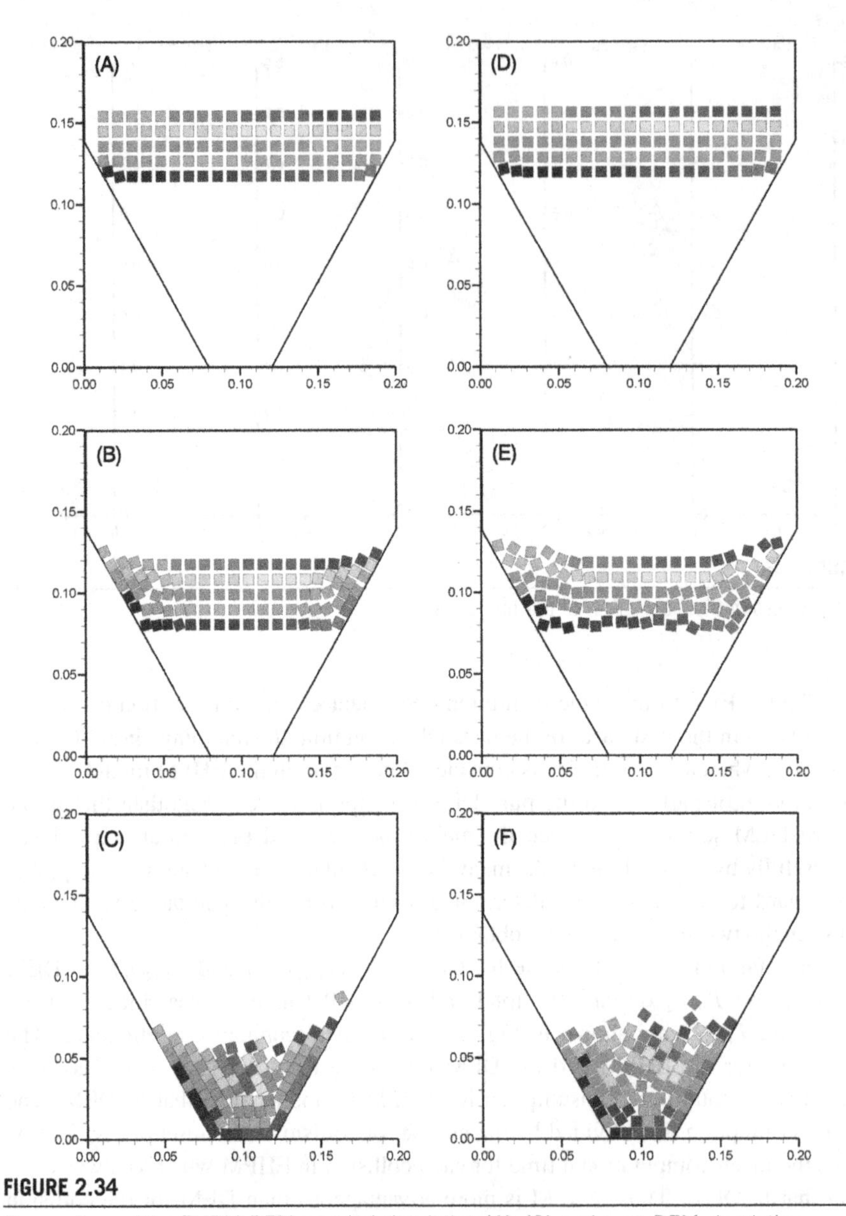

FIGURE 2.34

Comparison of the EHPM-DEM coupled simulation (A)–(C) and pure DEM simulation (D)–(F) of the deposition process of an array of square particles.

In conclusion, the EHPM model is computationally more efficient than the pure DEM approach from both the microscopic point of view to solve individual collisions, and from the macroscopic point of view to simulate a complex granular flow. Therefore, the EHPM-DEM coupled method is a good choice for fast simulation of

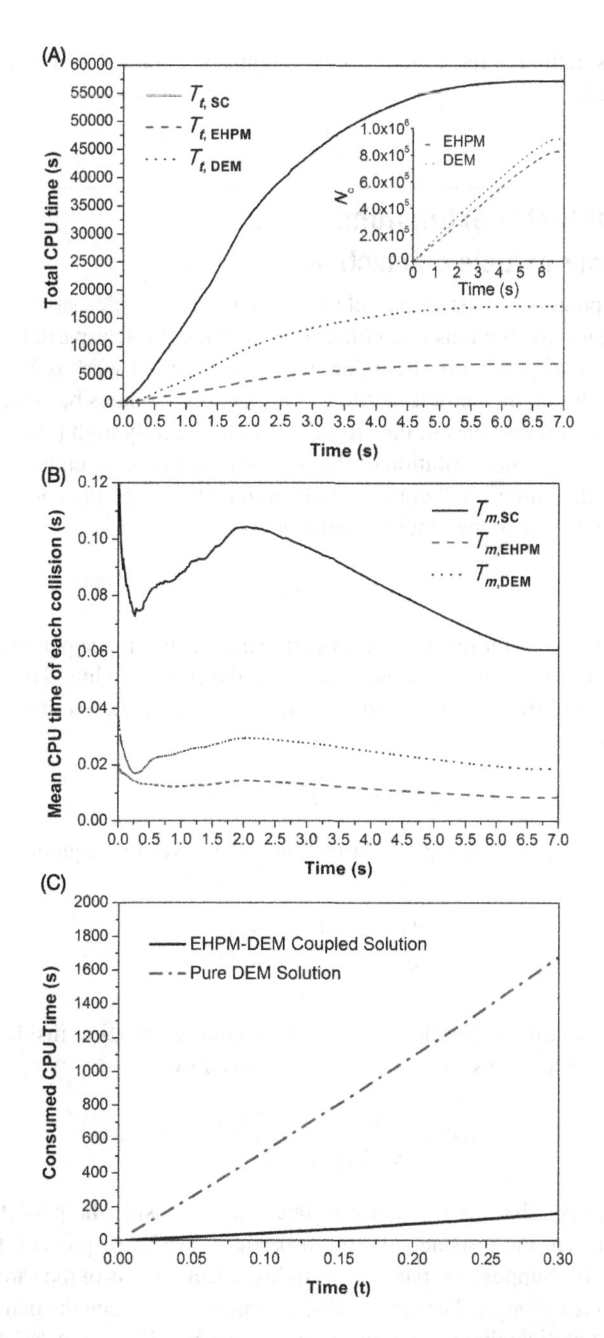

FIGURE 2.35

The total (A) and mean (B) computational CPU time in the collision detection, EHPM and DEM solution in the EHPM-DEM coupled model; the inset in (A) shows the number of collisions. (C) The comparison of the CPU time consumed in the EHPM-DEM coupled solution and the pure DEM solution.

complex flows, although the contact forces cannot be obtained as accurate as in the pure DEM model.

2.5 Heat transfer extensions

2.5.1 Particle-particle conduction

Heat transfer processes exist in particle flows, e.g. the particle-particle conduction, the particle-fluids conductions and convection, and the particle-particle radiation. If the fluid phase is air, the ratio of conductivity in the present work is $\lambda_p/\lambda_f > 2000$. Under this condition, the particle-particle conduction dominates because the ratio of conductivity of steel particles to interstitial gas is sufficiently high [192].

Based on the analytical solution of thermal conduction between two smooth elastic particles with a finite small contact area obtained by [192], the conductance from one particle centerline to the other is predicted by

$$\dot{q}_{c,ij} = 2\lambda_{p,ij} r_{c,ij}, \tag{2.102}$$

where $\dot{q}_{c,ij}$ is the heat transported per unit temperature difference per unit time. $\lambda_{p,ij}$ is the thermal conductivity of particle, and $r_{c,ij}$ is the contact radius. The total amount of heat transported from the j$^t h$ particle to the i$^t h$ particle across their contact area per unit time is

$$\dot{Q}_{ij} = \dot{q}_{c,ij}(T_j - T_i), \tag{2.103}$$

where T_i and T_j are particle temperatures. Thus, the governing equation of energy of particle 'i' is

$$\frac{dT_i}{dt} = \frac{1}{c_p m_p} \sum_j \dot{Q}_{ij}, \tag{2.104}$$

where c_p is the particle specific heat. The temperature variation inside the particle volume is neglected in this case, which is guaranteed by

$$Bi^\star = \frac{\dot{q}_{c,ij}}{\lambda_p A_p / r_p} = \frac{2}{\pi} \frac{r_{c,ij}}{r_p} \ll 1. \tag{2.105}$$

Eq. (2.105) implies that the resistance to heat transfer inside the particle is considerably smaller than the resistance between particles [193]. At present, the stiffness factor is $k_s = 10^4$. Suppose the particle-particle contact force is of the same rate as the gravity force, $k_s \delta x \approx m_p g$. The ratio of displacement δx between the pair of colliding particles to the particle diameter is about $\delta x / d_p < 5.8 \times 10^{-4}$, and $Bi^\star \approx 0.02 \ll 1$. Eq. (2.105) is always satisfied. Thus, the assumption of a uniform temperature distribution within the volume of any particle at all time is appropriate. The thermal conduction process inside the particle volume is neglected, i.e., a fast thermal re-

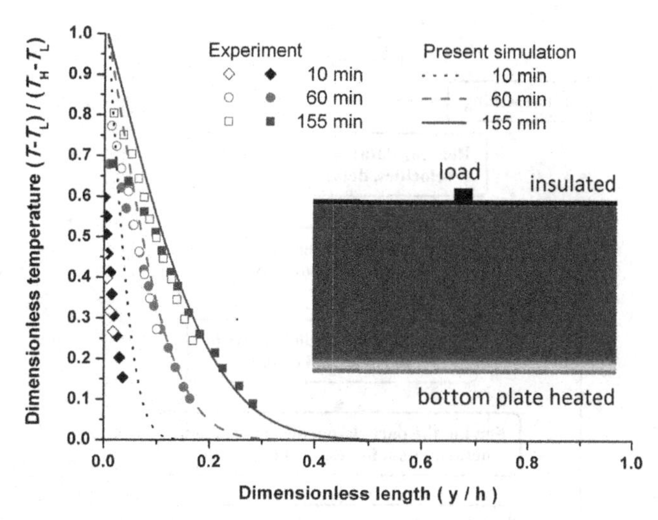

FIGURE 2.36

Comparison of particle temperature distributions within a packed bed between numerical and experimental results.

sponse or relaxation process inside the particle volume always dominates. Therefore, it is reasonable to neglect the temperature variation inside the particle volume.

It is noticed that this thermal conduction model has been used and validated by many research groups [194,195,182,196], and is generally well accepted. In addition, code validation for using this thermal conduction model has been performed in previous studies [197,182]. As seen in Fig. 2.36, a packed bed is composed of regularly arranged 200 × 100 particles of 304 stainless steel. The bottom plate is uniformly heated at a high temperature and the top is insulated. A 5 kg load is applied on the bed to have the particles packed and ensure them contacted with each other. The physical properties of the particle are $\rho_p = 7930$ kg/m^3, $c_p = 502$ J/(kg·K) and $\lambda_p = 12.1$ W/(m·K). No freely adjustable parameters were used in the model. However, as the bed was packed in a perfect hexagonal lattice, the stress chains might be slightly different from the experiment, and the stress inhomogeneity was not considered.

The simulation results in Fig. 2.36 show that: (1) The numerical results are smoother and more ideal than the experimental results because of the perfectly hexagonal packing of the particles. (2) With the exception of the results of 10 min, the simulated results agree very well with the experimental data, although visible differences between them still exist. (3) The difference of 10 min may have resulted from the difference in stress chains and stress inhomogeneities between simulation and the experiment. That is, the "fault" of the stress chains in the experiment may have led to the slower response of thermal conduction, especially at the early stage.

In conclusion, the present simulation is validated by a comparison of the numerical and experimental results.

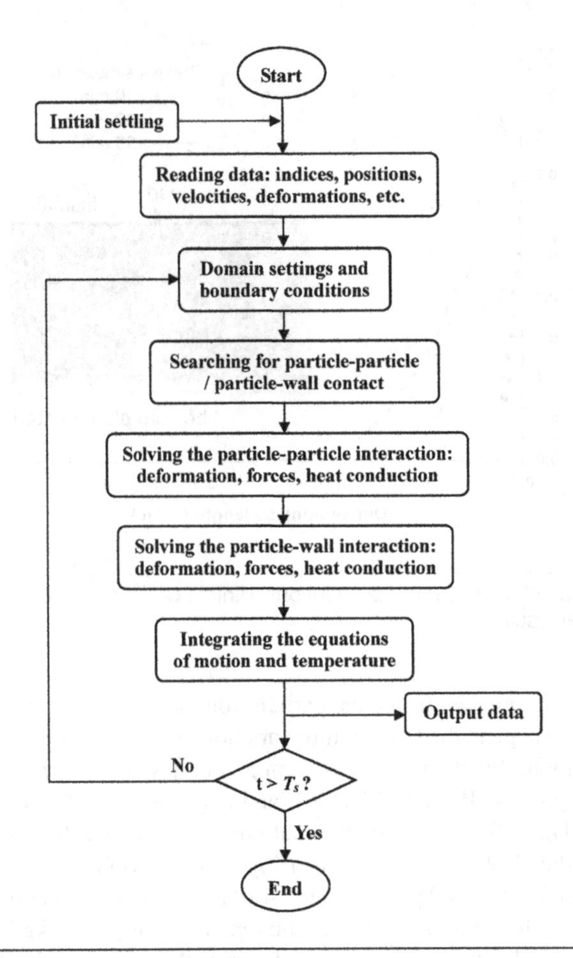

FIGURE 2.37

Flowchart of thermal discrete element method.

The concept of the thermal conduction model is to some degree similar to the pipe-network model [195], which is a simplified version of a more specific DTEM model based on the element thermal conductivity matrix incorporating the distribution of temperatures throughout the particle volume and the contact zone. The current model is advantageous in calculating the amount of heat flux and solving the energy equation, but limited to the conditions of small Bi^* numbers, where the temperature gradient inside the particle volume is negligible.

At last, the flowchart of thermal DEM procedure is presented in Fig. 2.37.

2.6 Summary

In this chapter, we have showed three collision models, which are suitable for particle flows of different flow regimes. GHPM is a pure kinematic two-body collision model of general particle shapes. It also belongs to the event-driven approaches, suitable for dilute flows and suitable for coupled CFD method, which will be discussed in the following chapter, for dilute gas-nonspherical particle flows. The SIPHPM method is a dynamical collision model, which considers the smooth contact history between particles. It is capable to deal with dense particle flows where multiple contacts are present. However, the deformation of particles is limited, and therefore this approach is regarded as pseudo-rigid. The EHPM-DEM model is a hybrid approach of GHPM and DEM model, which can select the collision model between particle dynamically depending on the local flow regimes and conditions. It may partially accelerate the simulation and improve efficiency in complex particle flows.

With the particle model developed, we will investigate gas-particle flows, using coupled particle models for the discrete phase and CFD models for the continuum phase, and treat the inter-phase interaction by suitable forces.

CHAPTER

Coupled methods

3

3.1 LES-DEM coupled methods

3.1.1 Governing equations

In general, the governing equations of the gas phase are the cell-averaged Navier–Stokes equations, which follow the law of conservation of momentum as in what follows [140]

$$\frac{\partial \epsilon_g}{\partial t} + \frac{\partial \epsilon_g \widetilde{u}_{i,g}}{\partial x_i} = 0 \tag{3.1}$$

$$\frac{\partial \epsilon_g \widetilde{u}_{i,g}}{\partial t} + \frac{\partial \epsilon_g \widetilde{u}_{i,g} \widetilde{u}_{j,g}}{\partial x_j} = -\frac{\epsilon_g}{\rho_g}\frac{\partial \widetilde{p}_g}{\partial x_i} + \epsilon_g g_i + \frac{1}{\rho_g}\frac{\partial \epsilon_g \widetilde{\tau}_{ij,g}}{\partial x_j} - \frac{f_{D,i}}{\rho_g} + \frac{1}{\rho_g}\frac{\partial \epsilon_g \widetilde{\Gamma}_{ij,g}}{\partial x_j} \tag{3.2}$$

where ϵ_g, $\widetilde{u}_{i,g}$, $\widetilde{p}_g$, $\widetilde{\tau}_{ij,g}$, $\widetilde{\Gamma}_{ij,g}$ are the void fraction, filtered variables of velocity, pressure, stress tensor, and subgrid stress tensor of the gas phase respectively. ρ_g, g_i are the density of gas and the gravity acceleration component respectively. The '$\widetilde{\cdot}$' is the top-hat filtering operator. The subscript 'i' and 'j' are the coordinates of dimension, and 'g' means the gas phase. The Smagorinsky model is used here for the subgrid scale stress tensor

$$\widetilde{\Gamma}_{ij,g} = \widetilde{u}_i \widetilde{u}_j - \widetilde{u_i u_j} = 2\nu_t \widetilde{S}_{ij} + \frac{1}{3}\widetilde{\Gamma}_{kk}\delta_{ij}, \tag{3.3}$$

where $\widetilde{S}_{ij} = \frac{1}{2}\left(\frac{\partial \widetilde{u}_i}{\partial x_j} + \frac{\partial \widetilde{u}_j}{\partial x_i}\right)$, $\nu_t = (C_s\Delta)^2|\widetilde{S}|$, $|\widetilde{S}| = (2\widetilde{S}_{ij}\widetilde{S}_{ij})^{1/2}$ and Δ are the deformation tensor of the filtered field, the turbulent eddy viscosity, the local strain rate and the subgrid characteristic length (the mesh size here), respectively. δ_{ij} is the Kronecker delta function, and $C_s = 0.1$ is a constant [198].

The drag force $\boldsymbol{f}_D = \beta_D(\boldsymbol{u}_g - \boldsymbol{v}_p)$, where the coefficient β_D is formulated by [199]

$$\beta_D = \begin{cases} \frac{150(1-\epsilon)^2\mu}{\epsilon d_p^2} + \frac{1.75(1-\epsilon)\rho|\boldsymbol{v}_g - \boldsymbol{v}_p|}{d_p}, & \text{if } \epsilon < 0.8 \\ \frac{0.75C_d\epsilon(1-\epsilon)\rho|\boldsymbol{v}_g - \boldsymbol{v}_p|\epsilon^{-2.65}}{d_p}, & \text{if } \epsilon \geq 0.8 \end{cases} \tag{3.4}$$

Gas-Particle and Granular Flow Systems. https://doi.org/10.1016/B978-0-12-816398-6.00011-0
Copyright © 2020 Elsevier Inc. All rights reserved.

$$C_d = \begin{cases} \frac{24(1+0.15Re_p^{0.687})}{Re_p}, \text{ if } Re_p < 1000 \\ 0.43, \text{ if } Re_p \geq 1000 \end{cases} \tag{3.5}$$

where $\boldsymbol{v}_g$, $\boldsymbol{v}_p$, d_p, C_d, $Re_p = \frac{\epsilon \rho |\boldsymbol{v}_g - \boldsymbol{v}_p| d_p}{\mu}$ are the gas velocity, the particle velocity, the particle diameter, the drag coefficients, and the particle Reynolds number respectively. Moreover, it can also be obtained from the correlation of [200], which is regarded to be more suitable for large particles than others [201].

$$\begin{cases} \beta_D = \frac{18\mu_g \epsilon_g^2 \epsilon_p}{d_p^2}\left[F_0(\epsilon_p) + \frac{1}{2}F_3(\epsilon_p)Re_p\right] \\ F_0(\epsilon_p) = \begin{cases} \frac{1+3\sqrt{\epsilon_p/2}+\frac{135}{64}\epsilon_p \ln(\epsilon_p)+16.14\epsilon_p}{1+0.681\epsilon_p-8.48\epsilon_p^2+8.16\epsilon_p^3}, \text{ if } \epsilon_p < 0.4 \\ \frac{10\epsilon_p}{\epsilon_g^3}, \text{ if } \epsilon_p \geq 0.4 \end{cases} \\ F_3(\epsilon_p) = 0.0673 + 0.212\epsilon_p + \frac{0.0232}{\epsilon_g^5} \end{cases} \tag{3.6}$$

where μ_g is the viscosity of gas, and ϵ_p $(= 1 - \epsilon_g)$ and $Re_p = \frac{\epsilon_g \rho_g |\boldsymbol{u}_g - \boldsymbol{v}_p| d_p}{\mu_g}$ are the void fraction of particle phase and the particle Reynolds number respectively. The negative drag force which appears in the right term of the momentum equation plays the role of feedback force upon the gas phase.

3.1.2 Discrete element method (DEM)

The discrete element method is characterized by three basic mechanisms of the particle-particle interaction, parameterized respectively by an elastic stiffness factor k, a damping coefficient η, and a friction coefficient μ. The particle-particle interaction force can be modeled as follows

$$\boldsymbol{f}_{ij} = -k\boldsymbol{\delta}_{ij} - \eta \boldsymbol{v}_{r,ij}, \text{ if } (\boldsymbol{f}_{ij} \cdot \boldsymbol{t}_{ij}) > \mu(\boldsymbol{f}_{ij} \cdot \boldsymbol{n}_{ij}), (\boldsymbol{f}_{ij} \cdot \boldsymbol{t}_{ij}) = \mu(\boldsymbol{f}_{ij} \cdot \boldsymbol{n}_{ij}) \tag{3.7}$$

where $\boldsymbol{f}_{ij}$ is the force from particle 'j' to particle 'i'. $\boldsymbol{\delta}_{ij}$ represents the deformation of particle 'i' caused by the impact of particle 'j'. $\boldsymbol{v}_{r,ij}$ is the relative velocity on the contact points when the particle is deformed in collision. The k, η and μ are the stiffness factor, damping coefficient and friction coefficient, respectively.

The motion of particle is governed by the Newton's law of motion:

$$m_i \ddot{\boldsymbol{x}}_{i,p} = \boldsymbol{f}_D + \sum_j \boldsymbol{f}_{ij} - m_i \boldsymbol{g} \tag{3.8}$$

$$I_i \ddot{\boldsymbol{\theta}}_{i,p} = \sum_j \boldsymbol{r}_{i,p} \times \boldsymbol{f}_{ij} \tag{3.9}$$

where $\boldsymbol{x}_p$ and $\boldsymbol{\theta}_p$ are the translational and rotational displacements respectively. I_i and $\boldsymbol{r}_p$ are the moment of inertia and radius of particle. The particle motion trajectories and angular rotations are obtained by integrating Eqs. (3.8)–(3.9).

3.1.3 Approach A: conventional method

In the CFD-DEM model, particles are treated as discrete elements solved by DEM and the fluid is treated as a continuum and solved by CFD. The conventional CFD-DEM methods [202,203,201,204–212] are based on the local averaged Navier–Stokes equations, where the concept of cell-averaged void fraction of gas (or equivalently phase volume fraction) is taken into account [213]. This coupled approach is valid and widely and successfully applied in many applications, including gas-particle interaction, heat transfer and combustion [214–218].

However in the conventional CFD-DEM method, the void fraction is usually defined as $\epsilon_g = 1 - \sum_{i=1}^{n} \frac{V_{p,i}}{\Delta V_{cell}}$ [219], where $V_{p,i}$ is the volume or volume segment of the ith particle located in the cell of volume V_{cell}. Therefore, using such a concept of local void fraction, the particle size should be finer than the cell size. Due to such a limitation, when the CFD-DEM method is used for large particles, the governing equation has to be solved on even larger grids. As a result, the grids for the gas phase are too coarse to draw out any meaningful information.

Recently, some models have been developed to treat the conditions for large particles. The most popular model is the fully resolved direct numerical simulation (DNS) through the immersed boundary method (IBM), which is regarded as a promising method to solve this problem [220–223]. However, the direct numerical simulation by IBM must use a mesh scale far smaller than the particle scale (e.g. the particle size is always one order of magnitude larger than the mesh scale), and a large number of Lagrangian points are needed to cover the whole surface of the immersed body to guarantee a precise nonslip boundary condition. Therefore, it requires huge computational time and capacity, severely restricts the practical application of the IBM method for any real scale gas-particle system in engineering and industry. In addition, Kuang et al. have developed new modeling techniques, e.g. using suitable periodic boundary condition and spherical cell as well as the least-square interpolation of fluid properties, to improve numerical efficiency and stability of CFD-DEM. [224–227]. Other methods, such as the fictitious particle method [228], were also proposed to treat the coexistence of fine and large particles.

It is necessary to mention that the drag force $\boldsymbol{f}_D$ in Eq. (3.2) is a filtered or mesh volume-averaged variable, which is a sum of the drag forces experienced by the particles in the mesh volume, i.e. $\boldsymbol{f}_D = \sum_{i=1}^{n} \boldsymbol{f}_{D,i}$, where n is the number of particles within the mesh volume. It is based on the assumption that the sum of all the drag forces in a control volume acts on the fluid of the control volume. For numerical implementation, the feed-back of the drag force on the position occupied by any particle is interpolated to the mesh grids by the volume-averaged interpolation scheme.

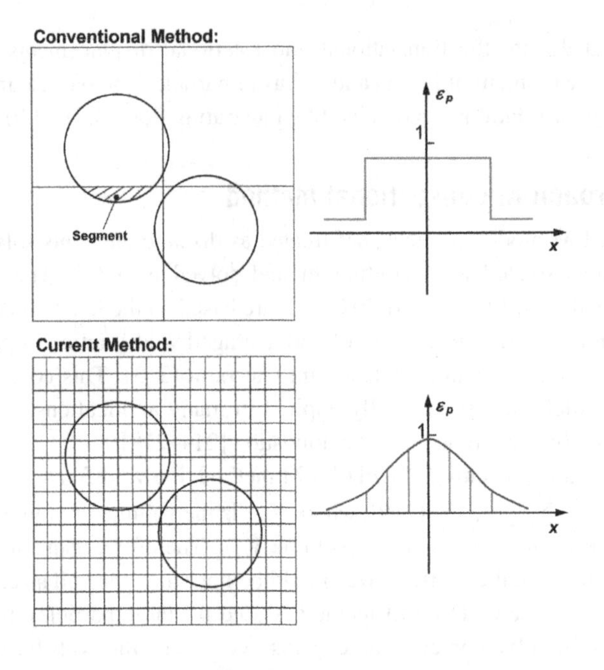

FIGURE 3.1

Schematic drawings of mesh scales and void fraction functions ϵ_p for large particles in the conventional CFD-DEM (A) and the current virtual void fraction method (B).

3.1.4 Approach B: smoothed void fraction method

3.1.4.1 Virtual void fraction model

Suppose there is a three-dimensional distribution function of particle mass as follows:

$$M(r) = \int_{|\xi|=0}^{r} m(\xi)d\xi \tag{3.10}$$

where $\xi = x - p$ is a radial vector from the centroid of particle p to a spatial location x, and $m(\xi)$ is the density of particle at ξ. In a real case, it is

$$m_T(\xi) = \begin{cases} \rho_p, \text{ if } |\xi| \le R \\ 0, \text{ if } |\xi| > R \end{cases} \tag{3.11}$$

where R and ρ_p are the radius and density of particle, respectively. The subscript T denotes the true distribution.

In the conventional CFD-DEM method (Fig. 3.1A), $m_T(\xi)$ is discretized by a large mesh size Δ (Δ denotes a mesh scale larger than the particle diameter hereafter) to compute the void fraction of gas

$$\epsilon_g = 1 - \epsilon_p = 1 - \frac{1}{V_{cell}} \sum_i \int_{|\xi| \in V_{cell}} \frac{1}{\rho_p} m_{i,T}(\xi) d\xi \tag{3.12}$$

where $V_{cell} = \Delta^D$ is the cell volume of 'D' dimension, and 'i' is the particle index. Notice that $\Delta > d_p$ and $V_p < V_{cell}$, and the local minimum void fraction is restricted by the packing density. Therefore, it may lead to a rigorous relation of $0 \leq \epsilon_p < 1$ and $0 < \epsilon_g \leq 1$. However, when this method is used for a large size particle of $d_p > \delta$ (hereafter δ denotes a subparticle mesh scale), it may result in $\epsilon_g = 0$ if the whole cell is occupied by a particle. When $\epsilon_g = 0$, the governing equations (3.1) and (3.2) are not meaningful. So are the subgrid scale stress tensor in Eq. (3.3) and the drag force in Eqs. (3.6). As a result, the conventional CFD-DEM method is infeasible when $d_p > \delta$. On the other hand, the simulation results of the gas phase obtained from the conventional CFD-DEM may be inadequate to obtain meaningful results under the condition of using coarser meshes for large particles, especially when the particle scale is comparable to the bed scale.

Therefore, in this section, a virtual void fraction method is proposed to improve the conventional CFD-DEM method and extend its application for a subparticle-diameter scale simulation of large particles. As sketched in Fig. 3.1B, the fine cells of $\delta < d_p$ should be here to capture the detailed information of the gas phase. An appropriate virtual mass distribution function (VMDF) of $m_V(\xi)$ should be used to avoid the condition of $\epsilon_g = 0$ (when $r < R - \delta$) and still make valid the cell-averaged Navier–Stokes equations (3.1)–(3.2), the Smagorinsky subgrid scale stress tensor in Eq. (3.3), and the drag force in Eqs. (3.6).

The virtual mass distribution function should follow the criteria as follow:

- The virtual volume fraction distribution should avoid zero of ϵ_g, i.e. $\epsilon_g > 0$ and $\epsilon_p = 1 - \epsilon_g < 1$.
- The integral of $m_V(\xi)$ should converge and be finite, i.e.

$$\lim_{r \to +\infty} M_V(r) = \int_{|\xi|=0}^{r} m_V(\xi) d\xi < \infty,$$

  where the subscript V denotes the virtual distribution. In particular, there should exist a finite difference $R_f < +\infty$, which leads to $\lim_{r \to +\infty} M_V(r) = \lim_{r \to +R_f} M_V(r)$
- The virtual mass distribution $m_V(\xi)$ should be conservative to give the same whole mass as the true particle, i.e.

$$M_T(\infty) = \int_{|\xi|=0}^{\infty, R_f} m_T(\xi) d\xi = \int_{|\xi|=0}^{\infty, R_f} m_V(\xi) d\xi = M_V(\infty),$$

  where $\int^{\infty, R_f}$ means the upper limit of the integral can be ∞ or R_f.
- The $m_V(\xi)$ should have a truncation radial distance $R_{trun} \ll R_f$ to save computational cost, i.e.

$$\int_{|\xi|=0}^{\infty, R_f} m_V(\xi) d\xi = \int_{|\xi|=0}^{R_{trun}} m_V(\xi) d\xi.$$

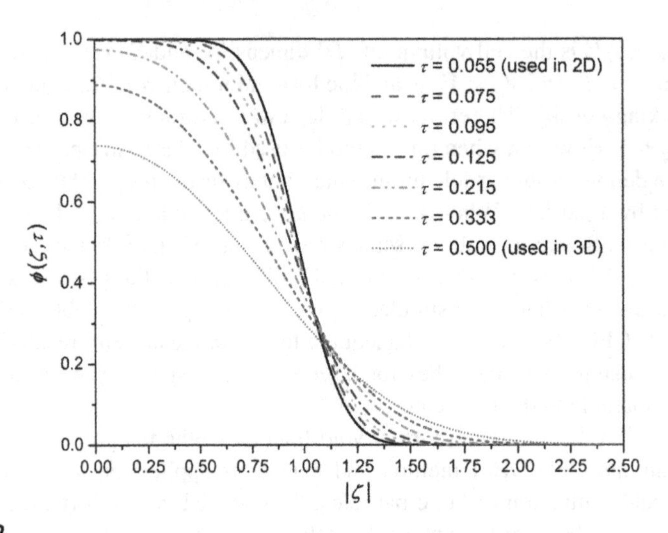

FIGURE 3.2

The function $\phi(\boldsymbol{\zeta}, \tau)$ used for the virtual mass distribution function.

- The virtual distribution $m_V(\boldsymbol{\xi})$ should be smoothing to restrict abrupt change of mass distribution on the particle surface to avoid numerical instability in simulation. It is desirable to have the smoothing adaptable.

To ensure the conservation of m_V, the linear partial differential equations

$$\begin{cases} \dfrac{\partial \phi(\zeta, \tau)}{\partial t} = \lambda^2 \nabla^2 \phi(\zeta, \tau) \\ \phi(\zeta, \tau)|_{\tau=0} = \begin{cases} 1, \text{ when } |\zeta| \leq 1 \\ 0, \text{ when } |\zeta| > 1 \end{cases} \end{cases} \tag{3.13}$$

should have the solution

$$\phi(\zeta, \tau) = -\frac{1}{2}\left[\operatorname{erf}(\alpha) - \operatorname{erf}(\beta)\right] - \frac{\lambda\sqrt{\tau}}{\sqrt{\pi}|\zeta|}\left[e^{-\alpha^2} - e^{-\beta^2}\right] \tag{3.14}$$

where $\alpha = \frac{|\zeta|-1}{2\lambda\sqrt{\tau}}$, $\beta = \frac{|\zeta|+1}{2\lambda\sqrt{\tau}}$, and $\operatorname{erf}(x) = \frac{2}{\sqrt{\pi}}\int_0^x e^{-s^2} ds$ is the error function. Herein ζ can be viewed as a position parameter $\zeta \propto |\boldsymbol{r}|$, and τ can be viewed as a time parameter. For example, let $\lambda = 0.5$, the values of $\phi(\boldsymbol{\zeta}, \tau)$ at $\tau = 0.055$ to 0.5 are shown in Fig. 3.2 (e.g. to demonstrate the wide application of τ, $\tau = 0.055$ is used for the 2D case and $\tau = 0.5$ is used for the 3D case in this work). It can be seen that $0 < \phi(\zeta) < 1$, $\lim_{|\zeta| \to +\infty} \phi(\zeta) = 0$, and $\int_{|\zeta|=0}^{+\infty} \phi(\zeta, \tau)\, d\zeta = \frac{4}{3}\pi$. Then, we can define the virtual mass distribution function (VMDF) $m_V(\boldsymbol{\xi}, \tau)$ as below

$$m_V(\boldsymbol{\xi}, \tau) = \rho_p \phi\left(\frac{\boldsymbol{\xi}}{R}, \tau\right) \tag{3.15}$$

It is found that $M_V(\infty) = \int_{|\xi|=0}^{+\infty} m_V(\xi,\tau)d\xi = \int_{|\xi|=0}^{+\infty} \rho_p \phi\left(\frac{\xi}{R},\tau\right)d\xi = \frac{4\pi}{3}\rho_p R^3 = M_T(\infty)$. In addition, it is seen from Fig. 3.2 that it is appropriate to regard $\phi(\zeta,\tau)$ having a truncation radius of $R_{trun,\phi} = 2$, $R_{trun,mv} = 2R$.

With $m_V(\xi,\tau)$, the local void distribution function at any mesh scale δ can be computed as

$$\epsilon_{g,\tau}(\boldsymbol{x}) = 1 - \frac{1}{\delta^3}\int_{|x-\xi|\le\frac{1}{2}\delta} \frac{1}{\rho_p} m_V(\xi,\tau)d\xi \tag{3.16}$$

To say conclusively, VMDF defined by Eq. (3.15) meets all the criteria as mentioned before, and Eq. (3.16) gives the virtual distribution of void fraction at any mesh scale, especially for $\delta < d_p$. τ in Eq. (3.15) and (3.16) can be regarded as a parameter to adapt the shape of VMDF to meet additional boundary conditions, properties, and operating requirements in any particular application.

VMDF is an analytical function which can be computed by integration within fluid cells. Therefore, VMDF is theoretically independent of mesh size. However, it always has truncation errors caused by the discretization of VMDF using finite mesh size in simplified integration over fluid cells to save time in computation, especially when the mesh is too coarse. Two kinds of mesh size are considered in this research work to show the sufficiency of accuracy of the discretized VMDF. In particular, for multi-size particles, the accuracy of the discretized VMDF is mainly dependent on the coarsest mesh to cover the finest particle, i.e. the ratio of $\frac{\delta}{d_p}$. Based on the demonstrative results, it is suggested that $\frac{\delta}{d_p} < 1/6$ should be enforced to guarantee the numerical accuracy. The VMDF herein is a radial distribution function of mass for spherical shapes, and it could not be used for nonspherical particles.

On the other hand, for a fluid cell, the contribution to the volume fraction of solids from nearby particle is determined by the range of VMDF. As seen from Fig. 3.2, when the cell is within a range of about $2R$ from the centers of neighboring particles, VMDF of these particles should be summed up to obtain the total mass of particles within the cell.

It is worth to mention that the concept of 'virtual mass distribution' of particle is for computation of the subparticle scale void distribution only. It is not substantial for the particle-particle collision. Therefore, the virtual distribution function is not considered in DEM model for the particle-particle collision.

3.1.4.2 Drag force computation

The drag force is computed from (Eq. (3.6)) [200]. In the conventional DEM, the mesh size was larger than the particle diameter, i.e. $\Delta > d_p$. For computation of the drag force of one particle, it may correspond to one cell where the centroid of particle locates, and consequentially have only one ϵ_g and Re_p. But, when the mesh size $\delta < d_p$, one particle may occupy several cells and have several ϵ_g and Re_p. As Eq. (3.6) is used to compute the drag force of one particle, which uses a single value of the void fraction or fluid velocity, the ϵ_g or $\boldsymbol{u}_g$ in the cells covered by the particle should be locally averaged to avoid multivaluedness. Herein, the fluid void fractions within the truncation radius from the centroid of the particle are all averaged to compute a mean void fraction, i.e.

$$\overline{\varphi(x)} = \frac{1}{\Omega}\int_{|x-\xi|\le\frac{1}{2}R_{trun}} \varphi(\xi)d\xi, \text{ where } \Omega = \int_{|x-\xi|\le\frac{1}{2}R_{trun}} d\xi \tag{3.17}$$

where φ can be ϵ_g or $\boldsymbol{u}_g$.

By using this strategy of averaging, it deploys the inherent feature and the concept of the existing correlation of the drag force without tremendous modification. Then, we can focus our primary attention on the accuracy of the virtual void fraction model as proposed above.

3.1.4.3 Feedback force and the subparticle scale four-way coupling

Differently from the averaging technique for the drag force computation, the feedback force from a particle to the gas phase must be distributed onto subparticle grids of $\delta < d_p$ to define the source term $-f_D^\delta$ in the governing equation. To be conservative, we still use the distribution function $\phi\left(\frac{\xi}{R}, \tau\right)$ to distribute the feedback force of the whole particle onto the subparticle cell of δ^3 as follows:

$$-f_D^\delta(\xi) = -\frac{1}{V_p} f_D^\Delta(x)\phi\left(\frac{x-\xi}{R}, \tau\right) \tag{3.18}$$

where $f_D^\Delta = \beta(\boldsymbol{u}_g - \boldsymbol{v}_p)$ is the drag force of particle computed from the parameters obtained by using the averaging technique as mentioned in the above section, and V_p is the particle volume. It is noticed that $\int_{|x-\xi|\le R_{trun}} f_D^\delta(\xi)d\xi = \int_{|x-\xi|\le R_{trun}} \frac{f_D^\Delta(x)}{V_p}\phi\left(\frac{x-\xi}{R}, \tau\right)d\xi = \frac{f_D^\Delta(x)}{V_p}\int_{|x-\xi|\le R_{trun}} \phi\left(\frac{x-\xi}{R}, \tau\right)d\xi = f_D^\Delta(x)$. Thus, such a distribution function is conservative for interphase momentum transfer between the scales of δ and Δ.

Eq. (3.18) ensures the conservation of the two-way coupling momentum transfer in the interphase interaction. Combined with the DEM for the particle-particle interaction, the present model gives a full description of the four-way coupling for the interphase interaction on subparticle scales. Such a conservative interaction between the superparticle scale Δ and the subparticle scale δ cannot be achieved in conventional CFD-DEM approaches.

3.2 DNS-DEM coupled methods

3.2.1 Immersed boundary method

By using mesh grids finer than the size of immersed bodies, the immersed boundary method is a fully resolved method which can capture full scales of turbulence and subparticle fine scales of the gas-particle interaction. Suppose the dimensionless governing equations for incompressible flows in the entire computational domain D_c are

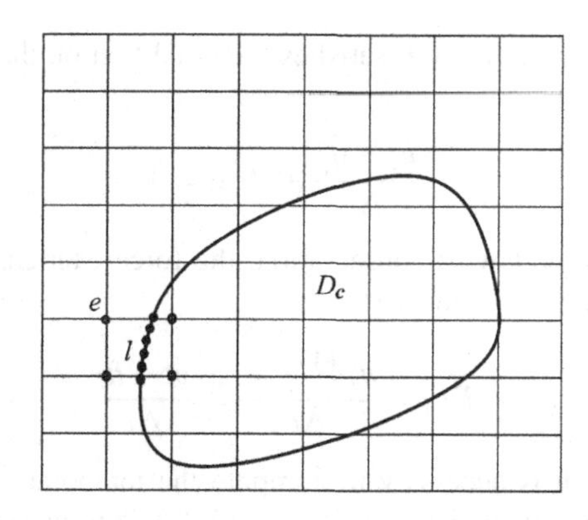

FIGURE 3.3

Sketch of the immersed boundary with embedded Lagrangian points and covered by Eulerian grids.

$$\nabla \cdot \boldsymbol{u} = 0 \tag{3.19}$$

$$\frac{\partial \boldsymbol{u}}{\partial t} + (\boldsymbol{u} \cdot \nabla)\boldsymbol{u} = -\nabla p + \frac{1}{Re}\nabla \boldsymbol{u} + \boldsymbol{F}_b \tag{3.20}$$

where p, $\boldsymbol{u}$, $\boldsymbol{F}_b$ are the fluid pressure, velocity and feedback force respectively. ∇ is the gradient operator. $Re = Ud/\nu$ is the Reynolds number, where the U, d and ν are the mean velocity, characteristic length scale of flow, and kinematic viscosity respectively.

In the immersed boundary method, the mutual interaction force $\boldsymbol{F}_{b,e}$ between fluid and immersed boundary on the Eulerian points (Fig. 3.3) is expressed as follows:

$$\boldsymbol{F}_{b,e}(\boldsymbol{x}) = \int_{D_c} \boldsymbol{F}_{b,l}(\boldsymbol{x}_l)\delta(\boldsymbol{x}_e - \boldsymbol{x}_l)d\boldsymbol{x}_l \tag{3.21}$$

where $\delta(\boldsymbol{x}_e - \boldsymbol{x}_l)$ is the Dirac delta function, $\boldsymbol{x}_l$ is the position of Lagrangian points at the immersed boundary. $\boldsymbol{x}$ is the position of computational Eulerian mesh points and $\boldsymbol{F}_{b,l}(\boldsymbol{x}_l)$ is the force exerted on the Lagrangian points $\boldsymbol{x}_l$.

In order to ensure the no-slip boundary condition on the Lagrangian points of the immersed boundary, a forcing $\boldsymbol{F}_{b,l}(\boldsymbol{x}_l)$ is imposed to keep the velocity equal to the solid velocity ($\boldsymbol{v}_l$) on the immersed boundary. The forcing $\boldsymbol{F}_{b,l}(\boldsymbol{x}_l)$ is determined by the following procedures:

- Discretize Eq. (3.20) in time to get

$$\begin{aligned}\boldsymbol{F}_{b,e} &= \frac{\partial \boldsymbol{u}}{\partial t} + (\boldsymbol{u} \cdot \nabla)\boldsymbol{u} + \nabla p - \frac{1}{Re}\nabla \boldsymbol{u} \\ &= \frac{\boldsymbol{u}^{n+1} - \boldsymbol{u}^n}{\Delta t} - \text{Rhs}\end{aligned} \tag{3.22}$$

  where $\text{Rhs} = -(\boldsymbol{u} \cdot \nabla)\boldsymbol{u} - \nabla p + \frac{1}{Re}\nabla \boldsymbol{u}$.

- Suppose a temporary velocity $\hat{\boldsymbol{u}}$ satisfies the condition on the Lagrangian point, that is

$$\frac{\hat{\boldsymbol{u}}_l - \boldsymbol{u}_l^n}{\Delta t} - \text{Rhs}_l = 0 \tag{3.23}$$

Then, based upon the law of conservation, the force exerted on the Lagrangian points at the immersed boundary is

$$\boldsymbol{F}_{b,l} = \frac{\boldsymbol{u}_l^{n+1} - \hat{\boldsymbol{u}}_l}{\Delta t} = \frac{\boldsymbol{v}_l - \hat{\boldsymbol{u}}_l}{\Delta t} \tag{3.24}$$

- To compute $\boldsymbol{F}_{b,l}$, it is necessary to compute the temporary velocity $\hat{\boldsymbol{u}}_l(\boldsymbol{x}_l)$ on the immersed boundary which satisfies Eq. (3.23) from its surrounding Eulerian grids $\boldsymbol{x}_e$. The Dirac delta function $\delta_h(\boldsymbol{x}_l - \boldsymbol{x}_e)$ can be applied for interpolation between Eulerian grids $\boldsymbol{x}$ and Lagrangian points $\boldsymbol{x}_l$.

$$\hat{\boldsymbol{u}}_l = \sum_{x \in D_c} \hat{\boldsymbol{u}}_e \Delta_h \delta_h(\boldsymbol{x}_l - \boldsymbol{x}_e) \tag{3.25}$$

where $\hat{\boldsymbol{u}}_e$ is the temporary fluid velocity on the Eulerian grids. h is the scale of the Eulerian grid, and Δ_h is the volume of Eulerian cells.
- Inversely, to compute $\boldsymbol{F}_{b,e}$, it is needed to interpolate the forcing $\boldsymbol{F}_{b,l}$ on the Lagrangian points onto the Eulerian grids backwardly as follows:

$$\boldsymbol{F}_{b,e}(\boldsymbol{x}) = \sum_{n=1}^{N_l} \boldsymbol{F}_{b,l} \delta_h(\boldsymbol{x}_e - \boldsymbol{x}_l) \Delta_l \tag{3.26}$$

where N_l is the number of Lagrangian points, and Δ_l is the discrete volume occupied by each Lagrangian point.
- With regard to the Dirac delta function $\delta_h(\boldsymbol{x}_l - \boldsymbol{x}_e)$, it can be chosen as that of [229]

$$\delta_h(\boldsymbol{x}_e - \boldsymbol{x}_l) = \frac{1}{h^D} \left(\delta_i \left(\frac{x_{e,i} - x_{l,i}}{h} \right) \right)^D \tag{3.27}$$

where $\delta_i(x)$ can be of various forms. For example, six types of $\delta_i(x)$ are expressed as

$$\delta_1(x) = \begin{cases} 1 - \frac{5|x|^2}{2} + \frac{3|x|^3}{2}, \ |x| < 1.0 \\ \frac{(1-|x|)(2-|x|)^2}{2}, \ 1 \le |x| < 2 \\ 0, \ 2 \le |x| \end{cases}$$

$$\delta_2(x) = \begin{cases} 1 - \frac{|x|}{2} - |x|^2 + \frac{|x|^3}{2}, \ |x| < 1.0 \\ 1 - \frac{11|x|}{6} + |x|^2 - \frac{|x|^3}{6}, \ 1 \le |x| < 2 \\ 0, \ 2 \le |x| \end{cases}$$

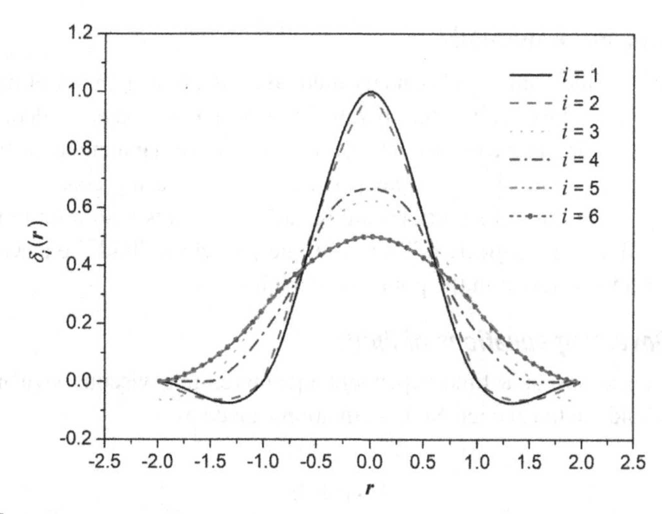

FIGURE 3.4

$\delta_i(x)$ functions with $i = 1, \ldots, 6$.

$$\delta_3(x) = \begin{cases} \frac{61}{112} - \frac{11|x|}{42} - \frac{11|x|^2}{56} + \frac{|x|^3}{12} \\ +\frac{\sqrt{3}\sqrt{(243+1584|x|-748|x|^2-1560|x|^3+500|x|^4+336|x|^5-112|x|^6)}}{336}, & |x| < 1.0 \\ \frac{21}{16} + \frac{7|x|}{16} - \frac{7|x|^2}{8} + \frac{|x|^3}{6} - \frac{3}{2}\delta_3(|x|-1), \ 1 \le |x| < 2 \\ \frac{9}{8} - \frac{23|x|}{12} + \frac{3|x|^2}{4} - \frac{|x|^3}{12} - \frac{1}{2}\delta_3(|x|-2), \ 2 \le |x| < 3 \\ 0, \ 3 \le |x| \end{cases}$$

$$\delta_4(x) = \begin{cases} \frac{(1+\sqrt{-3|x|^2+1})}{3}, \ |x| < 0.5 \\ \frac{(5-3|x|-\sqrt{-3(1-|x|)^2+1})}{6}, \ 0.5 \le |x| < 1.5 \\ 0, \ 1.5 \le |x| \end{cases}$$

$$\delta_5(x) = \begin{cases} \frac{1}{4}(1+\cos(\frac{\pi|x|}{2})), \ |x| \le 2 \\ 0, \ 2 < |x| \end{cases}$$

$$\delta_6(x) = \begin{cases} \frac{1}{8}(3-2|x|+\sqrt{1+4|x|-4x^2}, \ 0 \le |x| < 1 \\ \frac{1}{8}(5-2|x|-\sqrt{-7+12|x|-4x^2}, \ 1 \le |x| < 2 \\ 0, \ 2 \le |x| \end{cases} \quad (3.28)$$

and their profiles are shown in Fig. 3.4.

When the forces are determined, usually numerical schemes are used, such as finite difference/volume/element methods or the lattice Boltzmann method, to solve the Navier–Stokes equations of (3.19) and (3.20).

3.2.2 Point-force method

As the meshes of the immersed boundary method need a huge number of mesh grids to simulate a laboratory-scale setup, parallel computing is always needed, which is restricted by the computer capacity. Therefore, the point-force method is still a good choice to simulate industrial scale gas-particle flows. By using coarse grids larger than the size of particles, the particles are regarded as points with a finite mass and the volume effects are neglected. Therefore, the particle-to-fluid feedback force is simplified as a force vector in the point-force method.

3.2.2.1 Governing equations of fluids

Governing equations of a time-dependent incompressible viscous swirling jet of Newtonian fluid are the Navier–Stokes equations, given as

$$\nabla \cdot \boldsymbol{u} = 0 \tag{3.29}$$

$$\frac{\partial \boldsymbol{u}}{\partial t} + (\boldsymbol{u} \cdot \nabla)\boldsymbol{u} = -\nabla p + \frac{1}{Re}\nabla \boldsymbol{u} + \boldsymbol{F}_b \tag{3.30}$$

$$\frac{\partial T}{\partial t} + (\boldsymbol{u} \cdot \nabla)T = \frac{1}{RePr}\left(\nabla^2 T\right) \tag{3.31}$$

where p, t, T are the fluid pressure, time and temperature. $\boldsymbol{u} = (u, v, w)$ is the fluid velocity. ∇ is the gradient operator. The Reynolds number is defined as $Re = Ud/\nu$, where U and d are the mean axial (streamwise) velocity and characteristic length scale of the flow, respectively. ν is the kinematic viscosity. $Pr = \nu/a$ is the Prandtl number, where a is the thermal diffusion coefficient. $\boldsymbol{F}_b$ is the back-forward force from particle to fluid. Conceptually, it is a sum of the drag forces $\boldsymbol{f}_d$ experienced by the particle in the control volume, i.e. $\boldsymbol{F}_b = -\sum_{Vol} \boldsymbol{f}_d$. In the numerical implementation, the drag forces are interpolated from the positions of the particles to the grid nodes of the fluid by using a multiple linear interpolation scheme. In Eq. (3.31), the temperature can be viewed as passive self-diffusive scalars.

To solve the governing equations, the finite difference method is applied. The upwind compact schemes [230] are used to discretize the convection term. Although the upwind has dispersion, whereas the central compact schemes are dispersion-free, the upwind schemes are better for discretization of the convective term because the spurious waves can be suppressed. Sixth order compact difference schemes [231] are applied for space derivatives and the pressure-gradient terms:

$$\frac{1}{3}\left(f'_{i-1} + f'_i + f'_{i+1}\right) = \frac{14}{9}\frac{f_{i+1} + f_{i-1}}{2h} + \frac{1}{9}\frac{f_{i+2} + f_{i-2}}{4h} \tag{3.32}$$

where f denotes the flow variable. The third-order explicit schemes are used to treat the boundary points, and to maintain the global fourth-order spatial accuracy:

$$
\begin{aligned}
f_1' &= \frac{1}{720}(-1764f_1 + 4320f_2 - 5400f_3 + 4800f_4 - 2700f_5 + 864f_6 - 120f_7) \\
f_2' &= \frac{1}{720}(-120f_1 - 924f_2 + 1800f_3 - 1200f_4 + 600f_5 - 180f_6 + 24f_7)
\end{aligned} \tag{3.33}
$$

here the subscripts denote the number of nodes, and h is the mesh spacing. The fourth-order Runge–Kutta schemes [232] are used for time integration. Let

$$
\text{rhs} = -(\boldsymbol{u} \cdot \nabla \boldsymbol{u} + \nabla p - \frac{1}{Re}\nabla^2 \boldsymbol{u}) \tag{3.34}
$$

and we have

$$
\begin{aligned}
\hat{\boldsymbol{u}}^{(1)} &= \boldsymbol{u}^n + \frac{\delta t}{4}\text{rhs}(\boldsymbol{u}^n) \\
\hat{\boldsymbol{u}}^{(2)} &= \boldsymbol{u}^n + \frac{\delta t}{3}\text{rhs}(\boldsymbol{u}^{(1)}) \\
\hat{\boldsymbol{u}}^{(3)} &= \boldsymbol{u}^n + \frac{\delta t}{2}\text{rhs}(\boldsymbol{u}^{(2)}) \\
\hat{\boldsymbol{u}}^{n+1} &= \boldsymbol{u}^n + \delta t \cdot \text{rhs}(\boldsymbol{u}^{(3)})
\end{aligned} \tag{3.35}
$$

where $\boldsymbol{u}^{(i)}$ is the temporary velocity of intermediate flow fields. The pressure-Poisson equation is solved by using the fourth-order finite difference method [233].

As a direct numerical simulation requires the capture of all ranges of length scales of turbulence, the largest scale is in the order of flow width and the smallest scale is the Kolmogorov length scale η. Assume the Kolmogorov scale is estimated in the order of $\eta \sim (\frac{\nu^3}{\epsilon})^{\frac{1}{4}}$ where $\epsilon = \nu\left(\frac{\partial u_i}{\partial x_k}\frac{\partial u_i}{\partial u_k}\right)$ is the kinematic viscosity and ν is the dissipation rate of the turbulent kinetic energy. All scales of turbulence can be captured. According to Ref. [234], it is sufficient to meet the requirement of spatial resolution of the direct numerical simulation.

3.2.2.2 Motion equations of particles

In general, the motion equation of a particle in nonuniform and unsteady flows can be expressed as [235]

$$
\begin{aligned}
\frac{d\boldsymbol{v}_p}{dt} = \frac{1}{m_p}((&\boldsymbol{f}_D + \boldsymbol{f}_G + \boldsymbol{f}_Y + \boldsymbol{f}_P + \boldsymbol{f}_A + \boldsymbol{f}_B + \boldsymbol{f}_M + \boldsymbol{f}_S + \boldsymbol{f}_{Br} \\
&+ \boldsymbol{f}_{Th} + \boldsymbol{f}_{Tb} + \boldsymbol{f}_C + \cdots)
\end{aligned} \tag{3.36}
$$

$$
\dot{\boldsymbol{\omega}} = \frac{1}{I}(\boldsymbol{T}_C + \boldsymbol{T}_g) \tag{3.37}
$$

where the right-hand terms of Eqs. (3.36) and (3.37) correspond to the drag force $\boldsymbol{f}_D$ [236], the gravity force $\boldsymbol{f}_G$, the buoyant force $\boldsymbol{f}_Y$, the pressure gradient force $\boldsymbol{f}_P$, the visual (added) mass force $\boldsymbol{f}_A$ [237], the Basset (history) force $\boldsymbol{f}_B$ [238], the Saffman force $\boldsymbol{f}_S$ [239–242], the Magnus force $\boldsymbol{f}_M$ [243] (including the slip-shear lift force

$\boldsymbol{f}_{LS}$ [241,239] and the slip-rotation list force $\boldsymbol{f}_{LR}$ [244–246]), the Brownian motion force $\boldsymbol{f}_{Br}$, the turbulent fluctuation force $\boldsymbol{f}_{Tb}$ [247–252], the thermophoresis force $\boldsymbol{f}_{Th}$ [253–255], the particle-particle collisional force $\boldsymbol{f}_{C}$, the particle-particle collisional torque $\boldsymbol{T}_{C}$, and the fluid-particle interaction torque $\boldsymbol{T}_{g}$ [246], respectively. They are expressed as:

$$\boldsymbol{f}_D = \frac{1}{2}\rho_g A_p C_D |\boldsymbol{u}_g - \boldsymbol{v}_p|(\boldsymbol{u}_g - \boldsymbol{v}_p) \tag{3.38}$$

$$\boldsymbol{f}_G + \boldsymbol{f}_Y = (\rho_p - \rho_g)V_p \boldsymbol{g} \tag{3.39}$$

$$\boldsymbol{f}_P = -V_p \nabla p_g \tag{3.40}$$

$$\boldsymbol{f}_A = \frac{1}{2}\rho_p V_p \frac{d}{dt}(\boldsymbol{u}_g - \boldsymbol{v}_p + \frac{d_p^2}{40}\nabla^2 \boldsymbol{u}_g) \tag{3.41}$$

$$\boldsymbol{f}_B = \frac{3}{2} d_p^2 \sqrt{\pi \mu_g \rho_g} \int_0^t \frac{1}{(t-\tau)^{0.5}} \frac{d}{d\tau}\left(\boldsymbol{u}_g - \boldsymbol{v}_p + \frac{d_p^2}{24}\nabla^2 \boldsymbol{u}_g\right) d\tau \tag{3.42}$$

$$\boldsymbol{f}_M = \begin{cases} \boldsymbol{f}_{LR} = \frac{1}{2}\rho_g A_p C_M |\boldsymbol{u}_g - \boldsymbol{v}_p| \frac{(\boldsymbol{\omega}_g - \boldsymbol{\omega}_p)\times(\boldsymbol{u}_g - \boldsymbol{v}_p)}{|\boldsymbol{\omega}_g - \boldsymbol{\omega}_p|} \\ \boldsymbol{f}_{LS} = \frac{\rho_g}{2} d_p A_p C_S (\boldsymbol{u}_g - \boldsymbol{v}_p) \times \boldsymbol{\omega}_g \end{cases} \tag{3.43}$$

$$\boldsymbol{f}_S = \begin{cases} \left(\frac{1.615 d_p^2 \sqrt{\rho_g \mu_g}}{\sqrt{|\frac{d\boldsymbol{u}_g}{dy}|}}(\boldsymbol{u}_g - \boldsymbol{v}_p) \times \frac{d\boldsymbol{u}_g}{dy}\right) \times \\ \left((1 - 0.3314\sqrt{\frac{Re_s}{2Re_p}})\exp(-\frac{Re_p}{10}) + 0.3314\sqrt{\frac{Re_s}{2Re_p}}\right), Re_p < 40 \\ \left(\frac{1.615 d_p^2 \sqrt{\rho_g \mu_g}}{\sqrt{|\frac{d\boldsymbol{u}_g}{dy}|}}(\boldsymbol{u}_g - \boldsymbol{v}_p) \times \frac{d\boldsymbol{u}_g}{dy}\right)\left(0.0524\sqrt{\frac{Re_s}{2}}\right), Re_p > 40 \end{cases} \tag{3.44}$$

$$\boldsymbol{f}_{Br} = 6\pi \mu_g d_p k_B T \tag{3.45}$$

$$\boldsymbol{f}_{Th} = -3\pi \mu_g^2 d_p k_T \frac{\nabla T}{\rho_g T_\infty} \tag{3.46}$$

$$\boldsymbol{f}_C = \frac{1}{\delta t}\boldsymbol{J} \tag{3.47}$$

$$\boldsymbol{T}_C = \frac{1}{\delta t}\boldsymbol{r} \times \boldsymbol{J} \tag{3.48}$$

$$\boldsymbol{T}_g = \frac{\rho_g}{2}(\frac{d_p}{2})^5 C_R |\boldsymbol{\omega}_g - \boldsymbol{\omega}_p|(\boldsymbol{\omega}_g - \boldsymbol{\omega}_p) \tag{3.49}$$

where m_p, ρ_p, d_p, A_p $(= \frac{\pi}{4}d_p^2)$, V_p, I, $\boldsymbol{v}$, $\boldsymbol{\omega}_p$ are the particle mass, density, diameter, cross-sectional area, volume, moment of inertia, particle's velocity and angular velocity, respectively. ρ_g, μ_g, p_g, $\boldsymbol{u}_g$, $\boldsymbol{\omega}_g$ $(= \frac{1}{2}\nabla \times \boldsymbol{u}_g)$ are the gas phase density, viscosity, pressure, velocity, and angular velocity respectively. $\boldsymbol{J}$ is the collisional impulse over a small time interval Δt. $\boldsymbol{r}$ is a vector from the center of particle to the collision point with its length equal to the radius of the rigid sphere. k_B is the Boltzmann constant, and k_T is a parameter related to the fluid property. The coefficients are [236,256–259,241,245,244,239,246]:

$$
\begin{aligned}
&C_D = \frac{24 f_D}{Re_p}, Re_p = \frac{\rho_p d_p |\boldsymbol{u} - \boldsymbol{v}|}{\mu},\\
&f_D = 1 + 0.15 Re_p^{0.687} + 0.0175 \frac{Re_p}{1 + 42500 Re_p^{-1.16}}\\
&C_D(I_R, Re_p)_{\text{(Turb. Drag)}}\\
&= \begin{cases}
162 I_R^{1/3} Re_p^{-1}, Re_p < 50, 0.05 < I_R < 0.5;\\
0.133(1 + \frac{150}{Re_p})^{1.565} + 4 I_R\\
0.3(\frac{Re_p}{Re_c})^{-3}, 0.9 Re_c < Re_p < Re_m\\
0.3(\frac{Re_p}{Re_M})^{0.45+20 I_R}, Re_m < Re_p < Re_M\\
3990 Re_p^{-6.10} - 4.47 \times 10^5 I_R^{-0.97} Re^{-1.8}, Re_M < Re_p < 3 \times 10^4\\
\lg Re_c = 5.562 - 16.4 I_R, I_R \le 0.15\\
\lg Re_c = 3.371 - 1.75 I_R, I_R > 0.15\\
\lg Re_M = 6.878 - 23.2 I_R, I_R \le 0.15\\
\lg Re_M = 3.663 - 1.8 I_R, I_R > 0.15
\end{cases}\\
&C_M = \begin{cases}
\frac{d_p |\boldsymbol{\omega}_g - \boldsymbol{\omega}_p|}{\boldsymbol{u}_g - \boldsymbol{v}_p} = \frac{Re_r}{Re_p},\ Re_p \le 1\\
0.45 + \left(\frac{Re_r}{Re_p} - 0.45\right) \exp(-0.05684 Re_r^{0.4} Re_p^{0.3}),\ 1 < Re_p < 140
\end{cases}\\
&C_S = \frac{4.1126}{Re_s^{0.5}} f(Re_p, Re_s)\\
&f(Re_p, Re_s) = \begin{cases}
(1 - 0.3314 \beta^{\frac{1}{2}}) \exp(\frac{-Re_p}{10}) + 0.3314 \beta^{\frac{1}{2}},\ Re_p \le 40\\
0.0524 (\beta Re_p)^{\frac{1}{2}},\ Re_p > 40
\end{cases}\\
&C_R = \begin{cases}
\frac{64\pi}{Re_r},\ Re_r \le 32\\
\frac{12.9}{Re_r^{0.5}} + \frac{128.4}{Re_r},\ 32 < Re_p < 1000
\end{cases}\\
&k_T = \begin{cases}
\frac{3}{4(1+\pi f/8)}, \text{when } K_N \gg 1\\
\begin{cases}
\frac{2C_1(\lambda_g + 2\lambda_p K_N)(1 + 2K_N(1.2 + 0.41\exp(-0.44/K_N))}{(1+6C_2 K_N)(2\lambda_g + \lambda_p + 4\lambda_p C_3 K_N)}\\
\frac{1}{1+6K_N} \frac{\lambda_g + 4.4\lambda_p K_N}{\lambda_p + 2\lambda_g + 8.8\lambda_p K_N}
\end{cases}, \text{when } K_N > 0.1\\
\frac{2C_1(\lambda_g + 2\lambda_p K_N)}{(1+4C_2 K_N)(2\lambda_g + \lambda_p + 4\lambda_p C_3 K_N)}, \text{when } K_N < 0.1
\end{cases}
\end{aligned}
\tag{3.50}
$$

where C_D is the drag coefficient, and Re_p is the particle Reynolds number. $I_R = \frac{\sqrt{\boldsymbol{u}_g'^2}}{|\boldsymbol{u}_g - \boldsymbol{v}_p|}$ is the dimensionless turbulent intensity. Re_c, Re_m, Re_M are the transitional particle Reynolds number, the minimum drag Reynolds number and the critical particle Reynolds number with $C_D(Re_M) = 0.3$, respectively. $Re_r = \frac{\rho_g d_p^2 |\omega_g - \omega_p|}{\mu_g}$ is the so-called rotational Reynolds number of a particle. $\beta = \frac{d_p|\omega_g|}{2|u_g - v_p|} = \frac{1}{2}\frac{Re_s}{Re_p}$, where $Re_s = \frac{\rho_g d_p^2 |\omega_g|}{\mu_g}$ is the shear Reynolds number. K_N is the Knudsen number. λ_g and λ_p are conductivity coefficients of the gas and the particle respectively. The parameters are $f = 0.9$ [253], $C_1 = 1.17$, $C_2 = 1.14$, $C_3 = 2.18$ [254,255,260].

Based on Eqs. (3.38)–(3.46), the equation of motion can be integrated as follows:

$$\begin{aligned}
\frac{d\boldsymbol{v}_p}{dt} &= \frac{3}{4}\frac{\rho_g}{\rho_p d_p} C_D (\boldsymbol{u}_g - \boldsymbol{v}_p)|\boldsymbol{u}_g - \boldsymbol{v}_p| + \frac{3}{4}\frac{\rho_g}{\rho_p} C_S (\boldsymbol{u}_g - \boldsymbol{v}_p) \times \boldsymbol{\omega}_g \\
&+ \frac{3}{4}\frac{\rho_g}{\rho_p d_p} C_M |\boldsymbol{u}_g - \boldsymbol{v}_p| \frac{(\boldsymbol{\omega}_g - \boldsymbol{\omega}_p) \times (\boldsymbol{u}_g - \boldsymbol{v}_p)}{|\boldsymbol{\omega}_g - \boldsymbol{\omega}_p|} + \boldsymbol{g} \\
&= \frac{f_D}{\tau_p} C_D (\boldsymbol{u}_g - \boldsymbol{v}_p) + \frac{3}{4}\frac{\rho_g}{\rho_p} C_S ((\boldsymbol{u}_g - \boldsymbol{v}_p) \times \boldsymbol{\omega}_g) \\
&+ \frac{3}{4}\frac{\rho_g}{\rho_p}\frac{Re_p}{Re_r} C_M (\boldsymbol{\omega}_g - \boldsymbol{\omega}_p) \times (\boldsymbol{u}_g - \boldsymbol{v}_p) + \boldsymbol{g} \\
\frac{d\boldsymbol{\omega}_p}{dt} &= \frac{15}{16}\frac{\rho_g}{\rho_p \pi} C_R |\boldsymbol{\omega}_g - \boldsymbol{\omega}_p| (\boldsymbol{\omega}_g - \boldsymbol{\omega}_p)
\end{aligned} \tag{3.51}$$

For the gas-particle flows considered in this book, the drag force is usually the dominating force from fluids to particle. Sometimes the slip-rotation and slip-shear forces are considered. Then, the motion equations of particle, Eq. (3.51), can the dimensionless form:

$$\begin{cases}
\dfrac{d\boldsymbol{v}_p}{dt} = \dfrac{f_D}{St}(\boldsymbol{u}_g - \boldsymbol{v}_p) + \dfrac{3}{4}\dfrac{\rho_g}{\rho_p} C_S ((\boldsymbol{u}_g - \boldsymbol{v}_p) \times \boldsymbol{\omega}_g) \\
\qquad + \dfrac{3}{4}\dfrac{\rho_g}{\rho_p}\dfrac{Re_p}{Re_r} C_M (\boldsymbol{\omega}_g - \boldsymbol{\omega}_p) \times (\boldsymbol{u}_g - \boldsymbol{v}_p) + \dfrac{\boldsymbol{g}}{Fr} \\
\dfrac{d\boldsymbol{\omega}_p}{dt} = \dfrac{15}{16}\dfrac{\rho_g}{\rho_p \pi} C_R |\boldsymbol{\omega}_g - \boldsymbol{\omega}_p| (\boldsymbol{\omega}_g - \boldsymbol{\omega}_p)
\end{cases} \tag{3.52}$$

where $St = \frac{\tau_p}{\tau_g} = \frac{\rho_p d_p^2/(18\mu)}{L^*/U^*}$ is the Stokes number, which is the ratio of the particle response time to the characteristic time of fluid (L^*/U^*). $Fr = \frac{U^2}{gl}$ is the Froude number.

3.2.2.3 Particle-particle collision model

For the particle-particle collision, fundamental assumptions are: (1) The particles are rigid spheres and with uniform densities and diameters. (2) The particle-particle collision follows the law of conservation of momentum. (3) The loss of kinetic energy due to inelastic collision and damping friction is taken into account by two empirical parameters, i.e. the restitution coefficient e and the friction coefficient γ respectively.

(4) Each particle-particle collision is solved deterministically without introducing any statistical treatment.

However, the hard sphere model has some intrinsic shortcomings, they are: (1) It is a two-body-based collision model, and cannot handle a multiple-collision at the same time. (2) Consequently, it is not suitable for dense gas-solid flows. (3) Although this model uses the least number of parameters for the particle-particle collision process, the results depend greatly on collision parameters, especially the restitution coefficient. As the present study focuses on the effects of fluid turbulence and particle dynamical properties on the particle-particle collision distribution in swirling jets, not the effect of collision parameters on particle behavior, a nearly ideal restitution coefficient can be used. As a result, particle coagulation or coalescence is unlikely to occur. For this reason, the particle-cluster collision with different sizes is not accounted, and the particle-particle collision is simplified by considering particles with the uniform size.

In the hard sphere model, based on the law of conservation of momentum, the relationship between the pre- and post-collision variables is formulated as:

$$\begin{cases} \boldsymbol{v}_i = \boldsymbol{v}_i^0 + \boldsymbol{J}/m_i \\ \boldsymbol{v}_j = \boldsymbol{v}_j^0 - \boldsymbol{J}/m_j \\ \boldsymbol{\omega}_i = \boldsymbol{\omega}_i^0 + \boldsymbol{r}_i \times \boldsymbol{J}/I_i \\ \boldsymbol{\omega}_j = \boldsymbol{\omega}_j^0 - \boldsymbol{r}_j \times \boldsymbol{J}/I_j \end{cases} \tag{3.53}$$

where the subscripts 'i' and 'j' indicate the member of the pair of colliding particles. The impulse is decomposed as $\boldsymbol{J} = J_n\boldsymbol{n} + J_t\boldsymbol{t}$ where $\boldsymbol{n} = \boldsymbol{r}_i/|\boldsymbol{r}|$ is the unit normal vector and $\boldsymbol{t}$ is the unit tangential vector. With the normal restitution coefficient $e = (\boldsymbol{v}_i - \boldsymbol{v}_j)_n/(\boldsymbol{v}_j^0 - \boldsymbol{v}_i^0)_n$, the components of the impulse are obtained as:

$$\begin{aligned} J_n &= \Gamma_\mu(1+e)(\boldsymbol{v}_i^0 - \boldsymbol{v}_j^0)_n \\ J_t &= \min\{-\gamma J_n, -\frac{2}{7}\Gamma_\mu|\boldsymbol{v}_j^0 - \boldsymbol{v}_i^0 - \boldsymbol{\omega}_i^0 \times \boldsymbol{r}_i - \boldsymbol{\omega}_j^0 \times \boldsymbol{r}_i|\} \end{aligned} \tag{3.54}$$

where $\Gamma_\mu = \frac{m_i m_j}{m_i + m_j}$ is the reduced mass. In Eq. (3.54), the static friction and the sliding friction are all considered. The deduction for the tangential impulse is based on the assumption that the particles are spherical and solid.

Several steps need to be followed to realize the coupling of the hard sphere model and the direct numerical simulation. The steps are: (1) Solve the motion equation of particle interpolating the particle-to-fluid forces. (2) Check if any particle-particle collision is taking place. If yes, solve the particle-particle collision equation and update particle velocities and angular velocities. (3) Solve the Navier–Stokes equation of fluid (Eqs. (3.29)–(3.31)), taking into account the particle-to-fluid forces. (4) Go to step (1). Following the steps, the fluid and particle motions are correlated due to the two-way coupling. It is also necessary to mention that the interpolation scheme is linear, which means the particle-to-fluid force is commonly undertaken by the imme-

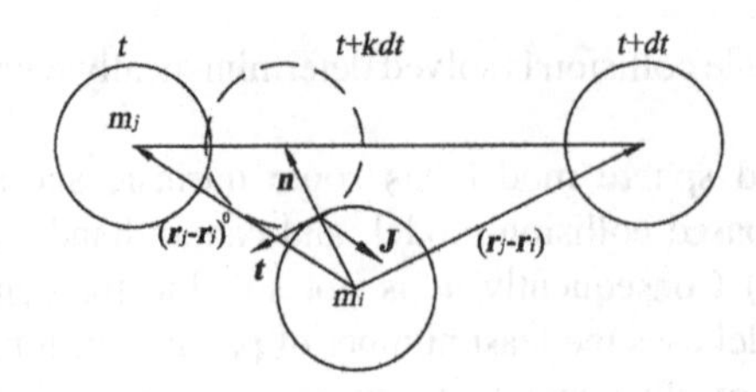

FIGURE 3.5

A sketch for the process of the interparticle collision detection.

diate surrounding fluids through a linearly proportional division. It is reasonable for direct numerical simulation since the grid is sufficiently fine.

3.2.2.4 Particle-particle collision detection

The inter-particle collision, depicted as above, needs to be implemented by using several numerical procedures. As all particles are traced, the occurrence of inter-particle collision, at each time step, can be detected by a geometric solution of relative displacements of any pair of particles, potential to collide with each other. As shown in Fig. 3.5, suppose particles 'm_i' and 'm_j' are a random pair, and $(\boldsymbol{r}_j^0 - \boldsymbol{r}_i^0)$ and $(\boldsymbol{r}_j - \boldsymbol{r}_i)$ are the pre-collisional (at time t) and post-collision (at time $t + dt$) distances between them, the detection of collision can be carried out by the following method. When the pair of particles collides, the equation

$$|(\boldsymbol{r}_j^0 - \boldsymbol{r}_i^0) + k\left((\boldsymbol{r}_j - \boldsymbol{r}_i) - (\boldsymbol{r}_j^0 - \boldsymbol{r}_i^0)\right)| = d_p \tag{3.55}$$

has one or two real roots with $0 \le k_1 \le k_2 < 1$ where k_1 is the required solution. If k_1 is obtained, the virtual position of the particle 'm_j' (Fig. 3.5) is known and the normal and tangential vectors are consequentially calculated. Then Eq. (3.54) is solved to calculate the impulse, and Eq. (3.55) is solved to calculate the post-collisional velocities.

In the above procedures, an overall scanning of all the 'N_p' traced particles at each time step for detecting inter-particle collision consumes computational cost of order $O(N_p^2)$ [261] to solve Eq. (3.55). Hence it is an exhaustion of the CPU time. Alternatively, an estimation of the maximum moving distance for any particle in a time step in a localized domain, e.g. several cells surrounding the particle, is sufficient for collision scanning, and consumes computation of order only about $O(N_p^2)$.

3.3 LBM-DEM coupled methods

3.3.1 Recovery of governing equation

3.3.1.1 Chapman–Enskog analysis

Let's start from the discrete lattice Boltzmann equation in the BGK formulation:

$$f_i(\boldsymbol{x} + \boldsymbol{\xi}_i\delta_t, t + \delta_t) - f_i(\boldsymbol{x}, t) = -\frac{1}{\tau}[f_i(\boldsymbol{x}, t) - f_i^{(eq)}(\boldsymbol{x}, t)] \tag{3.56}$$

with $\rho = \sum_i f_i$, $\rho \boldsymbol{u} = \sum_i \boldsymbol{\xi}_i f_i$. The ρ is fluid density and $\boldsymbol{u}$ is fluid velocity. f_i is the density distribution function in the ith direction which represents the ith component of the discretized velocity, $\boldsymbol{\xi}_i$.

Let $\hat{D}_i = \left(\frac{\partial}{\partial t} + \boldsymbol{\xi}_i \cdot \nabla\right)$, then perform the Taylor expansion of f_i at $(\boldsymbol{x}, t)$ as follows:

$$
\begin{aligned}
f_i(\boldsymbol{x} + \boldsymbol{\xi}_i \delta_t, t + \delta_t) &= f_i(\boldsymbol{x}, t) + \hat{D}_i f_i(\boldsymbol{x}, t)\delta t + \frac{1}{2}\hat{D}_i^2 f_i(\boldsymbol{x}, t)\delta t^2 + o(\delta t^2) \\
&= f_i(\boldsymbol{x}, t) + \frac{\partial f_i(\boldsymbol{x}, t)}{\partial t}\delta t + \nabla f_i(\boldsymbol{x}, t) \cdot \boldsymbol{\xi}_i \delta t \\
&+ \left(\frac{1}{2}\frac{\partial^2 f_i(\boldsymbol{x}, t)}{\partial t^2} + (\boldsymbol{\xi}_i \cdot \nabla)\frac{\partial}{\partial t} f_i(\boldsymbol{x}, t) + \frac{1}{2}\boldsymbol{\xi}_i \boldsymbol{\xi}_i : \nabla\nabla f_i(\boldsymbol{x}, t)\right)\delta t^2 + o(\delta t^2) \\
&= f_i(\boldsymbol{x}, t) - \frac{1}{\tau}[f_i(\boldsymbol{x}, t) - f_i^{(eq)}(\boldsymbol{x}, t)]
\end{aligned}
\tag{3.57}
$$

which leads to

$$
\begin{aligned}
&\frac{\partial f_i}{\partial t} + \boldsymbol{\xi}_i \cdot \nabla f_i + \left(\frac{1}{2}\frac{\partial^2 f_i}{\partial t^2} + (\boldsymbol{\xi}_i \cdot \nabla)\frac{\partial f_i}{\partial t} + \frac{1}{2}\boldsymbol{\xi}_i \boldsymbol{\xi}_i : \nabla\nabla f_i\right)\delta t \\
&= -\frac{1}{\tau \delta t}[f_i(\boldsymbol{x}, t) - f_i^{(eq)}(\boldsymbol{x}, t)]
\end{aligned}
\tag{3.58}
$$

Suppose $\frac{\partial}{\partial t} = \epsilon \frac{\partial}{\partial t_1} + \epsilon^2 \frac{\partial}{\partial t_2}$, $\nabla = \epsilon \nabla_1$, and the distribution function $f_i = f_i^{eq} + \epsilon f_i^{neq}$, where $f_i^{neq} = f_i^{(1)} + \epsilon f_i^{(2)} + O(\epsilon^2)$ is the nonequivalent part of f_i.

To guarantee the conservation of local macroscopic variables ρ and $\rho \boldsymbol{u}$ by

$$
\rho = \sum_i f_i = \sum_i f_i^{eq} \tag{3.59}
$$

$$
\rho \boldsymbol{u} = \sum_i \boldsymbol{\xi}_i f_i = \sum_i \boldsymbol{\xi}_i f_i^{eq}, \tag{3.60}
$$

the constraints

$$
\sum_i f_i^{(k)} = 0, \quad \sum_i \boldsymbol{\xi}_i f_i^{(k)} = 0, \; k = 1, 2, \tag{3.61}
$$

should be followed.

Then, perform the Chapman–Enskog expansion of Eq. (3.58) like the following:

$$
\begin{aligned}
&(\frac{\partial}{\partial t_1}+\epsilon\frac{\partial}{\partial t_2})(f_i^{eq}+\epsilon f_i^{(1)}+\epsilon^2 f_i^{(2)})+\boldsymbol{\xi}_i\cdot\boldsymbol{\nabla}_1(f_i^{eq}+\epsilon f_i^{(1)}+\epsilon^2 f_i^{(2)})\\
&+\frac{1}{2}\epsilon\delta t\left(\frac{\partial}{\partial t_1}+\epsilon\frac{\partial}{\partial t_2}\right)^2(f_i^{eq}+\epsilon f_i^{(1)}+\epsilon^2 f_i^{(2)})\\
&+(\frac{\partial}{\partial t_1}+\epsilon\frac{\partial}{\partial t_2})(\boldsymbol{\xi}_i\cdot\epsilon\boldsymbol{\nabla}_1)(f_i^{eq}+\epsilon f_i^{(1)}+\epsilon^2 f_i^{(2)})\delta t\\
&+\frac{1}{2}\boldsymbol{\xi}_i\boldsymbol{\xi}_i:\epsilon\boldsymbol{\nabla}_1\boldsymbol{\nabla}_1(f_i^{eq}+\epsilon f_i^{(1)}+\epsilon^2 f_i^{(2)})\delta t\\
&=-\frac{1}{\tau\delta t}\frac{(f_j-f_j^{eq})}{\epsilon}=-\frac{1}{\tau\delta t}(f_j^{neq})\\
&=-\frac{1}{\tau\delta t}(f_i^{(1)}+\epsilon f_i^{(2)})
\end{aligned}
\tag{3.62}
$$

Now separating the terms of order ϵ^0 and of order ϵ^1 from Eq. (3.62).
Order ϵ^0:

$$
\frac{\partial}{\partial t_1}f_i^{eq}+\boldsymbol{\xi}_i\cdot\boldsymbol{\nabla}_1 f_i^{eq}=-\frac{f_i^{(1)}}{\tau\delta t}
\tag{3.63}
$$

or

$$
\hat{D}_{i1}f_i^{eq}=-\frac{f_i^{(1)}}{\tau\delta t}
\tag{3.64}
$$

where $\hat{D}_{i1}=(\frac{\partial}{\partial t_1}+\boldsymbol{\xi}_i\cdot\boldsymbol{\nabla}_1)$.
Order ϵ^1:

$$
\begin{aligned}
&\frac{\partial f_i^{(1)}}{\partial t_1}+\boldsymbol{\xi}_i\cdot\boldsymbol{\nabla}_1 f_i^{(1)}+\frac{\partial f_i^{eq}}{\partial t_2}+\frac{1}{2}\frac{\partial^2 f_i^{eq}}{\partial t_1^2}\delta t+(\boldsymbol{\xi}_i\cdot\boldsymbol{\nabla}_1)\frac{\partial f_i^{eq}}{\partial t_1}\delta t+\frac{1}{2}\boldsymbol{\xi}_i\boldsymbol{\xi}_i:\boldsymbol{\nabla}_1\boldsymbol{\nabla}_1 f_i^{eq}\delta t\\
&=-\frac{1}{\tau\delta t}f_i^{(2)}
\end{aligned}
\tag{3.65}
$$

or

$$
\frac{\partial f_i^{eq}}{\partial t_2}+\hat{D}_{i1}f_i^{(1)}+\frac{1}{2}\hat{D}_{i1}^2 f_i^{eq}\delta t=-\frac{1}{\tau\delta t}f_i^{(2)}
\tag{3.66}
$$

Substituting Eq. (3.64) into Eq. (3.66) gives

$$
\frac{\partial f_i^{eq}}{\partial t_2}+\hat{D}_{i1}f_i^{(1)}+\frac{1}{2}\hat{D}_{i1}\left(-\frac{f_i^{(1)}}{\tau\delta t}\right)\delta t=-\frac{1}{\tau\delta t}f_i^{(2)}
\tag{3.67}
$$

or

$$
\frac{\partial f_i^{eq}}{\partial t_2}+(1-\frac{1}{2\tau})\hat{D}_{i1}f_i^{(1)}=-\frac{1}{\tau\delta t}f_i^{(2)}
\tag{3.68}
$$

To make it clearly, we rewrite Eqs. (3.64) and (3.68) as follows:

$$(*)\begin{cases}\hat{D}_{i1} f_i^{eq} = -\dfrac{f_i^{(1)}}{\tau\delta t} \\ \dfrac{\partial f_i^{eq}}{\partial t_2} + (1-\dfrac{1}{2\tau})\hat{D}_{i1} f_i^{(1)} = -\dfrac{1}{\tau\delta t} f_i^{(2)}\end{cases} \tag{3.69}$$

Then, taking the zeroth moments of Eq. (3.69) and summing up over all i gets

$$\begin{cases}\sum_i \hat{D}_{i1} f_i^{eq} = -\dfrac{\sum_i f_i^{(1)}}{\tau\delta t} \\ \dfrac{\partial \sum_i f_i^{eq}}{\partial t_2} + (1-\dfrac{1}{2\tau})\sum_i \hat{D}_{i1} f_i^{(1)} = -\dfrac{1}{\tau\delta t}\sum_i f_i^{(2)}.\end{cases} \tag{3.70}$$

Notice

$$\sum_i \boldsymbol{\xi}_i \cdot \nabla_1 f_i^{eq} = \sum_i (\nabla_1 \cdot (f_i^{eq}\boldsymbol{\xi}_i) - f_i^{eq}\nabla_1 \cdot \boldsymbol{\xi}_i) = \nabla_1 \cdot \sum_i (f_i^{eq}\boldsymbol{\xi}_i) = \nabla_1 \cdot (\rho \boldsymbol{u}) \tag{3.71}$$

With Eq. (3.61), Eq. (3.70) is simplified to

$$\frac{\partial \rho}{\partial t_1} + \nabla_1 \cdot (\rho \boldsymbol{u}) = 0 \tag{3.72}$$

$$\frac{\partial \rho}{\partial t_2} = 0 \tag{3.73}$$

The expression Eq. (3.72) multiplied by ϵ plus Eq. (3.73) multiplied by ϵ^2 results in

$$\epsilon\frac{\partial \rho}{\partial t_1} + \epsilon^2\frac{\partial \rho}{\partial t_2} + \epsilon\nabla_1 \cdot (\rho \boldsymbol{u}) = 0 \tag{3.74}$$

or

$$\frac{\partial \rho}{\partial t} + \nabla \cdot (\rho \boldsymbol{u}) = 0 \tag{3.75}$$

which is a continuum equation in the group of governing equations of fluids.

Moreover, taking the first moment of velocity of Eq. (3.69) and summing up over i gets

$$\sum_i \hat{D}_{i1} f_i^{eq}\boldsymbol{\xi}_i = -\frac{\sum_i f_i^{(1)}\boldsymbol{\xi}_i}{\tau\delta t} \tag{3.76}$$

$$\frac{\partial \sum_i f_i^{eq}\boldsymbol{\xi}_i}{\partial t_2} + (1-\frac{1}{2\tau})\sum_i \hat{D}_{i1} f_i^{(1)}\boldsymbol{\xi}_i = -\frac{1}{\tau\delta t}\sum_i f_i^{(2)}\boldsymbol{\xi}_i. \tag{3.77}$$

Using Eq. (3.61) to simplify Eq. (3.76) and Eq. (3.77) one gets:

$$\sum_i \hat{D}_{i1} f_i^{eq} \boldsymbol{\xi}_i = 0 \tag{3.78}$$

$$\frac{\partial \sum_i f_i^{eq} \boldsymbol{\xi}_i}{\partial t_2} + (1 - \frac{1}{2\tau}) \sum_i \hat{D}_{i1} f_i^{(1)} \boldsymbol{\xi}_i = 0. \tag{3.79}$$

Notice

$$\begin{cases} \boldsymbol{\xi}_i \cdot \nabla_1 (f_i^{eq} \boldsymbol{\xi}_i) = \boldsymbol{\xi}_i \cdot \nabla_1 (f_i^{eq} \boldsymbol{\xi}_i) + (f_i^{eq} \boldsymbol{\xi}_i) \nabla_1 \cdot \boldsymbol{\xi}_i = \nabla_1 \cdot (f_i^{eq} \boldsymbol{\xi}_i \boldsymbol{\xi}_i), \\ \sum_i \hat{D}_{i1} f_i^{eq} \boldsymbol{\xi}_i = \sum_i \left(\frac{\partial}{\partial t_1} + \boldsymbol{\xi}_i \cdot \nabla_1 \right) f_i^{eq} \boldsymbol{\xi}_i = \sum_i \frac{\partial}{\partial t_1} f_i^{eq} \boldsymbol{\xi}_i + \sum_i \boldsymbol{\xi}_i \cdot \nabla_1 f_i^{eq} \boldsymbol{\xi}_i \\ \qquad = \sum_i \frac{\partial}{\partial t_1} f_i^{eq} \boldsymbol{\xi}_i + \nabla_1 \cdot \sum_i (f_i^{eq} \boldsymbol{\xi}_i \boldsymbol{\xi}_i) = \sum_i F_i^{(1)} \boldsymbol{\xi}_i \\ \sum_i \hat{D}_{i1} f_i^{(1)} \boldsymbol{\xi}_i = \sum_i \left(\frac{\partial}{\partial t_1} + \boldsymbol{\xi}_i \cdot \nabla_1 \right) f_i^{(1)} \boldsymbol{\xi}_i = \sum_i \frac{\partial}{\partial t_1} f_i^{(1)} \boldsymbol{\xi}_i + \sum_i \boldsymbol{\xi}_i \cdot \nabla_1 f_i^{(1)} \boldsymbol{\xi}_i \\ \qquad = \frac{\partial}{\partial t_1} \sum_i f_i^{(1)} \boldsymbol{\xi}_i + \nabla_1 \cdot \sum_i (f_i^{(1)} \boldsymbol{\xi}_i \boldsymbol{\xi}_i) = \nabla_1 \cdot \sum_i (f_i^{(1)} \boldsymbol{\xi}_i \boldsymbol{\xi}_i) \end{cases} \tag{3.80}$$

and assume

$$\sum_i \xi_{i\alpha} \xi_{i\beta} f_i^{eq} = p\delta_{\alpha\beta} + \rho \boldsymbol{u}_\alpha \boldsymbol{u}_\beta \tag{3.81}$$

Eq. (3.78) can be reformulated as follows

$$\sum_i \hat{D}_{i1} f_i^{eq} \boldsymbol{\xi}_i = \frac{\partial}{\partial t_1} (\rho \boldsymbol{u}_\alpha) + \nabla_{1\beta} \cdot \sum_i (f_i^{eq} \xi_{i\alpha} \xi_{i\beta}) = 0 \tag{3.82}$$

or

$$\frac{\partial}{\partial t_1} (\rho \boldsymbol{u}_\alpha) + \nabla_{1\beta} \cdot \left(p\delta_{\alpha\beta} + \rho \boldsymbol{u}_\alpha \boldsymbol{u}_\beta \right) = 0 \tag{3.83}$$

On the other hand, performing $\epsilon \times$ Eq. (3.78) $+ \epsilon^2 \times$ Eq. (3.79), we obtain

$$\epsilon \frac{\partial (f_i^{eq} \boldsymbol{\xi}_i)}{\partial t_1} + \epsilon^2 \frac{\partial (f_i^{eq} \boldsymbol{\xi}_i)}{\partial t_2} + \epsilon \nabla_1 \cdot (f_i^{eq} \boldsymbol{\xi}_i \boldsymbol{\xi}_i) + (1 - \frac{1}{2\tau}) \epsilon \hat{D}_{i1} \epsilon f_i^{(1)} \boldsymbol{\xi}_i = 0 \tag{3.84}$$

Summing up over i

$$\frac{\partial (\sum_i f_i^{eq} \boldsymbol{\xi}_i)}{\partial t} + \nabla \cdot (\sum_i f_i^{eq} \boldsymbol{\xi}_i \boldsymbol{\xi}_i) + (1 - \frac{1}{2\tau}) \nabla \cdot (\sum_i \epsilon f_i^{(1)} \boldsymbol{\xi}_i \boldsymbol{\xi}_i) = 0 \tag{3.85}$$

or

$$\frac{\partial(\rho \boldsymbol{u})}{\partial t} + \nabla \cdot \left(p\delta_{\alpha\beta} + \rho \boldsymbol{u}_\alpha \boldsymbol{u}_\beta\right) + (1 - \frac{1}{2\tau})\nabla \cdot (\sum_i \epsilon f_i^{(1)} \boldsymbol{\xi}_i \boldsymbol{\xi}_i) = 0 \tag{3.86}$$

Therefore,

$$\frac{\partial(\rho \boldsymbol{u})}{\partial t} + \nabla \cdot \boldsymbol{\Pi} = 0 \tag{3.87}$$

where

$$\begin{cases} \boldsymbol{\Pi} = (\sum_i f_i^{eq} \boldsymbol{\xi}_i \boldsymbol{\xi}_i) + (1 - \frac{1}{2\tau})(\sum_i \epsilon f_i^{(1)} \boldsymbol{\xi}_i \boldsymbol{\xi}_i) = \boldsymbol{\Pi}'^{(0)} + \epsilon \boldsymbol{\Pi}'^{(1)} \\ \boldsymbol{\Pi}^{(0)} = \sum_i \xi_{i\alpha} \xi_{i\beta} f_i^{eq} = p\delta_{\alpha\beta} + \rho \boldsymbol{u}_\alpha \boldsymbol{u}_\beta \\ \boldsymbol{\Pi}^{(1)} = (1 - \frac{1}{2\tau}) \sum_i \boldsymbol{\xi}_i \boldsymbol{\xi}_i f_i^{(1)} \end{cases} \tag{3.88}$$

With Eq. (3.63) and $f_i^{(1)} = -\tau\delta t \left(\frac{\partial}{\partial t_1} f_i^{eq} + \boldsymbol{\xi}_i \cdot \nabla_1 f_i^{eq}\right)$, one has

$$\begin{aligned} \boldsymbol{\Pi}'^{(1)} &= (1 - \frac{1}{2\tau}) \sum_i \boldsymbol{\xi}_i \boldsymbol{\xi}_i f_i^{(1)} \\ &= -\frac{\delta t(2\tau - 1)}{2} \sum_i \boldsymbol{\xi}_i \boldsymbol{\xi}_i \left(\left(\frac{\partial}{\partial t_1} f_i^{eq} + \boldsymbol{\xi}_i \cdot \nabla_1 f_i^{eq}\right)\right) \\ &= -\frac{\delta t(2\tau - 1)}{2} \sum_i \left(\frac{\partial}{\partial t_1} f_i^{eq} \boldsymbol{\xi}_i \boldsymbol{\xi}_i + (\boldsymbol{\xi}_i \cdot \nabla_1) f_i^{eq} \boldsymbol{\xi}_i \boldsymbol{\xi}_i\right) \\ &= -\frac{\delta t(2\tau - 1)}{2} \left(\frac{\partial}{\partial t_1} \sum_i f_i^{eq} \boldsymbol{\xi}_i \boldsymbol{\xi}_i + \nabla_1 \cdot \sum_i f_i^{eq} \boldsymbol{\xi}_i \boldsymbol{\xi}_i \boldsymbol{\xi}_i\right) \\ &= -\frac{\delta t(2\tau - 1)}{2} \left(\frac{\partial(\rho u_\alpha u_\beta + \rho c_s^2 \delta_{\alpha\beta})}{\partial t_1} + \frac{\partial \sum_i f_i^{eq} \xi_{i\gamma} \xi_{i\alpha} \xi_{i\beta}}{\partial x_{1\gamma}}\right) \end{aligned} \tag{3.89}$$

In addition, with Eqs. (3.72) and (3.83), we have

$$\begin{aligned} \frac{\partial \rho}{\partial t_1} &= \nabla_1 \cdot (-\rho \boldsymbol{u}) = -\frac{\partial \rho u_\gamma}{\partial x_{1,\gamma}} \\ \frac{\partial \rho u_\alpha}{\partial t_1} &= -\nabla_{1\beta} \cdot \left(p\delta_{\alpha\beta} + \rho \boldsymbol{u}_\alpha \boldsymbol{u}_\beta\right) = -c_s^2 \delta_{\alpha\beta} \frac{\partial \rho}{\partial x_{1,\beta}} - \frac{\partial \rho u_\alpha u_\beta}{\partial x_{1,\beta}} \\ \frac{\partial(\rho u_\alpha u_\beta)}{\partial t_1} &= u_\beta \frac{\partial(\rho u_\alpha)}{\partial t_1} + \rho u_\alpha \frac{\partial u_\beta}{\partial t_1} = u_\beta \frac{\partial(\rho u_\alpha)}{\partial t_1} + u_\alpha \left(\frac{\partial \rho u_\beta}{\partial t_1} - u_\beta \frac{\partial \rho}{\partial t_1}\right) \\ &= u_\beta \frac{\partial(\rho u_\alpha)}{\partial t_1} + u_\alpha \frac{\partial(\rho u_\beta)}{\partial t_1} - u_\alpha u_\beta \frac{\partial \rho}{\partial t_1} \end{aligned} \tag{3.90}$$

then

$$\begin{aligned}
&\frac{\partial(\rho u_\alpha u_\beta+\rho c_s^2\delta_{\alpha\beta})}{\partial t_1}\\
&=\frac{\partial}{\partial t_1}(\rho u_\alpha u_\beta)+c_s^2\delta_{\alpha\beta}\frac{\partial\rho}{\partial t_1}=\frac{\partial(\rho u_\alpha u_\beta)}{\partial t_1}+c_s^2\delta_{\alpha\beta}\boldsymbol{\nabla}_{\mathbf{1}}\cdot(-\rho\boldsymbol{u})\\
&=u_\beta\frac{\partial(\rho u_\alpha)}{\partial t_1}+u_\alpha\frac{\partial(\rho u_\beta)}{\partial t_1}-u_\alpha u_\beta\frac{\partial\rho}{\partial t_1}+c_s^2\delta_{\alpha\beta}\left(-\frac{\partial\rho u_\gamma}{\partial x_{1,\gamma}}\right)\\
&=u_\beta\left(-\frac{\partial\rho u_\alpha u_\gamma}{\partial x_{1,\gamma}}-\frac{\partial\rho c_s^2\delta_{\alpha\gamma}}{\partial x_{1,\gamma}}\right)+u_\alpha\left(-\frac{\partial\rho u_\beta u_\gamma}{\partial x_{1,\gamma}}-\frac{\partial\rho c_s^2\delta_{\beta\gamma}}{\partial x_{1,\gamma}}\right)\\
&-u_\alpha u_\beta\left(-\frac{\partial\rho u_\gamma}{\partial x_{1,\gamma}}\right)+c_s^2\delta_{\alpha\beta}\left(-\frac{\partial\rho u_\gamma}{\partial x_{1,\gamma}}\right)\\
&=\left(-u_\beta\frac{\partial\rho u_\alpha u_\gamma}{\partial x_{1,\gamma}}-u_\alpha\frac{\partial\rho u_\beta u_\gamma}{\partial x_{1,\gamma}}\right)-c_s^2\left(u_\beta\frac{\partial\rho}{\partial x_{1,\alpha}}+u_\alpha\frac{\partial\rho}{\partial x_{1,\beta}}\right)\\
&+u_\alpha u_\beta\frac{\partial\rho u_\gamma}{\partial x_{1,\gamma}}-c_s^2\delta_{\alpha\beta}\frac{\partial\rho u_\gamma}{\partial x_{1,\gamma}}
\end{aligned}\tag{3.91}$$

If we assume

$$\sum_i\xi_{i\alpha}\xi_{i\beta}\xi_{i\gamma}f_i^{eq}=\frac{\rho}{3}\left(u_\alpha\delta_{\beta\gamma}+u_\beta\delta_{\gamma\alpha}+u_\gamma\delta_{\alpha\beta}\right)+\rho u_\alpha u_\beta u_\gamma,\tag{3.92}$$

then

$$\begin{aligned}
\frac{\partial\sum_i\xi_{i\alpha}\xi_{i\beta}\xi_{i\gamma}f_i^{eq}}{\partial x_{1,\gamma}}&=\frac{\partial}{\partial x_{1,\gamma}}\left[\rho c_s^2\left(u_\alpha\delta_{\beta\gamma}+u_\beta\delta_{\alpha\gamma}+u_\gamma\delta_{\alpha\beta}\right)\right]+\frac{\partial}{\partial x_{1,\gamma}}\left[\rho u_\alpha u_\beta u_\gamma\right]\\
&=\frac{\partial(\rho c_s^2u_\alpha)}{\partial x_{1,\beta}}+\frac{\partial(\rho c_s^2u_\beta)}{\partial x_{1,\alpha}}+\frac{\partial(\rho c_s^2u_\gamma\delta_{\alpha\beta})}{\partial x_{1,\gamma}}+\frac{\partial}{\partial x_{1,\gamma}}\left[\rho u_\alpha u_\beta u_\gamma\right]\\
&=\rho c_s^2\left(\frac{\partial u_\alpha}{\partial x_{1,\beta}}+\frac{\partial u_\beta}{\partial x_{1,\alpha}}\right)+c_s^2\left(u_\alpha\frac{\partial\rho}{\partial x_{1,\beta}}+u_\beta\frac{\partial\rho}{\partial x_{1,\alpha}}\right)+\\
&+c_s^2\delta_{\alpha\beta}\frac{\partial\rho u_\gamma}{\partial x_{1,\gamma}}+\frac{\partial\rho u_\alpha u_\beta u_\gamma}{\partial x_{1,\gamma}}
\end{aligned}\tag{3.93}$$

$$\begin{aligned}
&\frac{\partial}{\partial x_{1,\gamma}}\sum_i\xi_{i\alpha}\xi_{i\beta}\xi_{i\gamma}f_i^{eq}+\frac{\partial}{\partial t_1}(\rho u_\alpha u_\beta+\rho c_s^2\delta_{\alpha\beta})=\\
&=\rho c_s^2\left(\frac{\partial u_\alpha}{\partial x_{1,\beta}}+\frac{\partial u_\beta}{\partial x_{1,\alpha}}\right)+c_s^2\left(u_\alpha\frac{\partial\rho}{\partial x_{1,\beta}}+u_\beta\frac{\partial\rho}{\partial x_{1,\alpha}}\right)+c_s^2\delta_{\alpha\beta}\frac{\partial\rho u_\gamma}{\partial x_{1,\gamma}}\\
&+\frac{\partial\rho u_\alpha u_\beta u_\gamma}{\partial x_{1,\gamma}}+\left(-u_\beta\frac{\partial\rho u_\alpha u_\gamma}{\partial x_{1,\gamma}}-u_\alpha\frac{\partial\rho u_\beta u_\gamma}{\partial x_{1,\gamma}}\right)-c_s^2\left(u_\beta\frac{\partial\rho}{\partial x_{1,\alpha}}+u_\alpha\frac{\partial\rho}{\partial x_{1,\beta}}\right)\\
&-c_s^2\delta_{\alpha\beta}\frac{\partial\rho u_\gamma}{\partial x_{1,\gamma}}+u_\alpha u_\beta\frac{\partial\rho u_\gamma}{\partial x_{1,\gamma}}
\end{aligned}$$

$$
\begin{aligned}
&= \rho c_s^2\left(\frac{\partial u_\alpha}{\partial x_{1,\beta}}+\frac{\partial u_\beta}{\partial x_{1,\alpha}}\right)+\frac{\partial \rho u_\alpha u_\beta u_\gamma}{\partial x_{1,\gamma}}+\left(-u_\beta\frac{\partial \rho u_\alpha u_\gamma}{\partial x_{1,\gamma}}-u_\alpha\frac{\partial \rho u_\beta u_\gamma}{\partial x_{1,\gamma}}\right)\\
&+u_\alpha u_\beta\frac{\partial \rho u_\gamma}{\partial x_{1,\gamma}}\\
&= \rho c_s^2\left(\frac{\partial u_\alpha}{\partial x_{1,\beta}}+\frac{\partial u_\beta}{\partial x_{1,\alpha}}\right)+u_\beta\frac{\partial \rho u_\alpha u_\gamma}{\partial x_{1,\gamma}}+\rho u_\alpha u_\gamma\frac{\partial u_\beta}{\partial x_{1,\gamma}}-u_\beta\frac{\partial \rho u_\alpha u_\gamma}{\partial x_{1,\gamma}}\\
&-u_\alpha\frac{\partial \rho u_\beta u_\gamma}{\partial x_{1,\gamma}}+u_\alpha u_\beta\frac{\partial \rho u_\gamma}{\partial x_{1,\gamma}}\\
&= \rho c_s^2\left(\frac{\partial u_\alpha}{\partial x_{1,\beta}}+\frac{\partial u_\beta}{\partial x_{1,\alpha}}\right)+\rho u_\alpha u_\gamma\frac{\partial u_\beta}{\partial x_{1,\gamma}}-u_\alpha\frac{\partial \rho u_\beta u_\gamma}{\partial x_{1,\gamma}}+u_\alpha u_\beta\frac{\partial \rho u_\gamma}{\partial x_{1,\gamma}}\\
&= \rho c_s^2\left(\frac{\partial u_\alpha}{\partial x_{1,\beta}}+\frac{\partial u_\beta}{\partial x_{1,\alpha}}\right)+u_\alpha\left(\rho u_\gamma\frac{\partial u_\beta}{\partial x_{1,\gamma}}-\frac{\partial \rho u_\beta u_\gamma}{\partial x_{1,\gamma}}\right)+u_\alpha u_\beta\frac{\partial \rho u_\gamma}{\partial x_{1,\gamma}}\\
&= \rho c_s^2\left(\frac{\partial u_\alpha}{\partial x_{1,\beta}}+\frac{\partial u_\beta}{\partial x_{1,\alpha}}\right)+u_\alpha\left(\rho u_\gamma\frac{\partial u_\beta}{\partial x_{1,\gamma}}-\rho u_\gamma\frac{\partial u_\beta}{\partial x_{1,\gamma}}-u_\beta\frac{\partial \rho u_\gamma}{\partial x_{1,\gamma}}\right)\\
&+u_\alpha u_\beta\frac{\partial \rho u_\gamma}{\partial x_{1,\gamma}}\\
&= \rho c_s^2\left(\frac{\partial u_\alpha}{\partial x_{1,\beta}}+\frac{\partial u_\beta}{\partial x_{1,\alpha}}\right)+u_\alpha\left(-u_\beta\frac{\partial \rho u_\gamma}{\partial x_{1,\gamma}}\right)+u_\alpha u_\beta\frac{\partial \rho u_\gamma}{\partial x_{1,\gamma}}\\
&= \rho c_s^2\left(\frac{\partial u_\alpha}{\partial x_{1,\beta}}+\frac{\partial u_\beta}{\partial x_{1,\alpha}}\right)\\
&= \rho c_s^2\left(\nabla_{1\beta}\cdot \boldsymbol{u}_\alpha+\nabla_{1\alpha}\cdot \boldsymbol{u}_\beta\right)
\end{aligned}
\tag{3.94}
$$

Therefore, to sum up, with Eqs. (3.89)–(3.94), we have

$$
\begin{aligned}
&\boldsymbol{\Pi}^{(1)}=-\frac{\delta t(2\tau-1)}{2}\rho c_s^2\left(\frac{\partial u_\alpha}{\partial x_{1,\beta}}+\frac{\partial u_\beta}{\partial x_{1,\alpha}}\right)=-\frac{\rho\delta t(2\tau-1)}{6}\left(\frac{\partial u_\alpha}{\partial x_{1,\beta}}+\frac{\partial u_\beta}{\partial x_{1,\alpha}}\right)\\
&-\boldsymbol{\nabla}_\beta\cdot\epsilon\boldsymbol{\Pi}^{(1)}=\mu\boldsymbol{\nabla}_\beta\cdot(\nabla_\alpha\boldsymbol{u}_\beta+\nabla_\beta\boldsymbol{u}_\alpha)
\end{aligned}
\tag{3.95}
$$

Let $\mu=\frac{\delta t(2\tau-1)}{6}\rho$, or

$$
\nu=\frac{\delta t(2\tau-1)}{6}, \tag{3.96}
$$

combined with Eq. (3.88),

$$
\begin{aligned}
\boldsymbol{\nabla}_\beta\cdot\boldsymbol{\Pi}_{\alpha\beta}&=\boldsymbol{\nabla}\cdot(\boldsymbol{\Pi}^{(0)}_{\alpha\beta}+\epsilon\boldsymbol{\Pi}^{(1)}_{\alpha\beta})\\
&=\boldsymbol{\nabla}_\beta\cdot(\rho\boldsymbol{u}_\alpha\boldsymbol{u}_\beta)+\boldsymbol{\nabla}_\alpha p-\boldsymbol{\nabla}_\beta\cdot(\mu(\boldsymbol{\nabla}_\alpha\boldsymbol{u}_\beta+\boldsymbol{\nabla}_\beta\boldsymbol{u}_\alpha))\\
&=\boldsymbol{\nabla}_\beta\cdot(\rho\boldsymbol{u}_\alpha\boldsymbol{u}_\beta)+\boldsymbol{\nabla}_\alpha p-\mu(\boldsymbol{\nabla}_\beta\cdot\boldsymbol{\nabla}_\beta)\boldsymbol{u}_\alpha-\mu\boldsymbol{\nabla}_\alpha(\boldsymbol{\nabla}_\beta\cdot\boldsymbol{u}_\beta)
\end{aligned}
\tag{3.97}
$$

For incompressible fluids, $\nabla_\beta \cdot \boldsymbol{u}_\beta = 0$, then Eq. (3.97) is simplified to

$$\nabla_\beta \cdot \boldsymbol{\Pi}_{\alpha\beta} = \nabla_\beta \cdot (\rho \boldsymbol{u}_\alpha \boldsymbol{u}_\beta) + \nabla_\alpha p - \mu(\nabla_\beta \cdot \nabla_\beta)\boldsymbol{u}_\alpha \tag{3.98}$$

In conclusion, combining Eqs. (3.75), (3.87), (3.97) or (3.98), one can build the Navier–Stokes equation as follows:

$$\begin{cases} \dfrac{\partial \rho}{\partial t} + \nabla \cdot (\rho \boldsymbol{u}) = 0 \\ \dfrac{\partial (\rho \boldsymbol{u})}{\partial t} + \nabla \cdot (\rho \boldsymbol{u}\boldsymbol{u}) = -\nabla p + \mu \nabla^2 \boldsymbol{u} \end{cases} \tag{3.99}$$

3.3.1.2 Inperfect equilibrium distribution function

It is noted that, up to now, we still have not mentioned the detailed expressions of the equilibrium density distribution function $f^{(eq)}$. But, in the aforementioned analysis, the equilibrium density distribution function is assumed to be subject to the limitation and assumptions

$$\begin{cases} \sum_i \boldsymbol{\xi}_i f_i^{eq} = \rho \boldsymbol{u} \\ \sum_i \boldsymbol{\xi}_i \boldsymbol{\xi}_i f_i^{eq} = p\delta_{\alpha\beta} + \rho u_\alpha u_\beta \\ \sum_i \xi_{i\alpha}\xi_{i\beta}\xi_{i\gamma} f_i^{eq} = \dfrac{\rho}{3}\left(u_\alpha \delta_{\beta\gamma} + u_\beta \delta_{\gamma\alpha} + u_\gamma \delta_{\alpha\beta}\right) + \rho u_\alpha u_\beta u_\gamma \end{cases} \tag{3.100}$$

Unfortunately, the general form of f_i^{eq},

$$\begin{aligned} f_i^{eq} &= \rho w_i \left[1 + \frac{\boldsymbol{\xi}_i \cdot \boldsymbol{u}}{c_s^2} + \frac{\boldsymbol{u}\boldsymbol{u} : (\boldsymbol{\xi}_i \boldsymbol{\xi}_i - c_s^2 \boldsymbol{I})}{2c_s^4}\right] \\ &= \rho w_i [1 + 3(\boldsymbol{\xi}_i \cdot \boldsymbol{u}) + \frac{9}{2}(\boldsymbol{\xi}_i \cdot \boldsymbol{u})^2 - \frac{3}{2}u^2] \end{aligned} \tag{3.101}$$

only satisfies

$$\begin{cases} \sum_i f_i^{eq} = \rho, \\ \sum_i \boldsymbol{\xi}_i f_i^{eq} = \rho \boldsymbol{u}, \\ \sum_i \xi_{i\alpha}\xi_{i\beta}\xi_{i\gamma} f_i^{eq} = \dfrac{\rho}{3}\left(u_\alpha \delta_{\beta\gamma} + u_\beta \delta_{\gamma\alpha} + u_\gamma \delta_{\alpha\beta}\right) \end{cases} \tag{3.102}$$

In other words, it cannot rebuild the full Navier–Stokes equations precisely. Consequently $\nabla_\beta \nabla_\gamma : (\rho \boldsymbol{u}_\beta \boldsymbol{u}_\gamma \boldsymbol{u}_\alpha)$ must be moved to the additional force source term, i.e.

$$\begin{aligned}\check{F}_\alpha &= -\frac{\delta t(2\tau-1)}{2}\nabla_\beta\nabla_\gamma:(\rho u_\beta u_\gamma u_\alpha) = -\delta t(\tau-\frac{1}{2})\rho u_\alpha \nabla\nabla:(uu)\\ &= -\delta t(\tau-\frac{1}{2})\rho u_\alpha(\nabla\cdot u)^2\end{aligned} \tag{3.103}$$

For incompressible fluids with $\nabla \cdot u = 0$, this term can be neglected $\check{F} = 0$.

3.3.1.3 Considering the interphase force

With the additional force term on the right-hand side, the discrete lattice Boltzmann equation in the BGK formulation can be formulated as

$$f_i(x+\xi_i\delta t, t+\delta t) - f_i(x,t) = -\frac{1}{\tau}[f_i(x,t) - f_i^{(eq)}(x,t)] + \delta t F_i(x,t) \tag{3.104}$$

with $\rho = \sum_i f_i$, $\rho u = \sum_i \xi_i f_i + \chi_b F_b \delta t$. F_b is the interphase force or the feedback force from discrete phase to fluids. χ_b is an unknown parameter to be determined later. Herein F_i is related to the contribution of the interphase force F_b on the component f_i, which, in multiphase flow, can play the role of interphase coupling on the component of density distribution function f_i.

Let $\hat{D}_i = \left(\frac{\partial}{\partial t} + \xi_i \cdot \nabla\right)$, and suppose $\frac{\partial}{\partial t} = \epsilon\frac{\partial}{\partial t_1} + \epsilon^2\frac{\partial}{\partial t_2}$, $\nabla = \epsilon\nabla_1$, $F_i = \epsilon F_i^{(1)}$, $f_i = f_i^{eq} + \epsilon f_i^{neq}$, where $f_i^{neq} = f_i^{(1)} + \epsilon f_i^{(2)} + O(\epsilon^2)$ is the nonequivalent part of f_i, performing the Chapman–Enskog analysis similar to the former section, χ_b is required to be $\frac{1}{2}$. Moreover, assuming the fluids are incompressible and the external forces are free of divergence, which are subject to the restrictions of

$$\nabla \cdot u = 0 \tag{3.105}$$

$$\nabla \cdot B = 0 \tag{3.106}$$

We have

$$\begin{cases}\dfrac{\partial\rho}{\partial t} + \nabla\cdot(\rho u) = 0\\ \dfrac{\partial(\rho u)}{\partial t} + \nabla\cdot(\rho uu) = -\nabla p + \mu\nabla^2\tilde{u} + B\end{cases} \tag{3.107}$$

which are near the full Navier–Stokes equations with external forces, subjected to the restrictions of Eqs. (3.105) and (3.106).

3.3.2 Coupled with heat transfer

Taking D2G9 model for example [265], the discrete velocities are formulated as

$$e_\alpha = \begin{cases}(0,0), \ \alpha = 0,\\ \left(\cos(\alpha-1)\frac{\pi}{2}, \sin(\alpha-1)\frac{\pi}{2}\right)c, \ \alpha = 1,2,3,4,\\ \sqrt{2}\left(\cos(2\alpha-1)\frac{\pi}{4}, \sin(2\alpha-1)\frac{\pi}{4}\right)c, \ \alpha = 5,6,7,8,\end{cases} \tag{3.108}$$

where $c = \delta x/\delta t = 1$ is the lattice velocity, δx and δt are the mesh spacing and time step, respectively. The density equilibrium function is defined as

$$f_\alpha^{eq} = \begin{cases} -\frac{5p}{3c^2} + s_\alpha(\boldsymbol{u}), \ \alpha = 0, \\ \frac{p}{3c^2} + s_\alpha(\boldsymbol{u}), \ \alpha = 1, 2, 3, 4, \\ \frac{p}{12c^2} + s_\alpha(\boldsymbol{u}), \ \alpha = 5, 6, 7, 8, \end{cases} \tag{3.109}$$

where $s_\alpha(\boldsymbol{u}) = w_\alpha \left(\frac{(\boldsymbol{e} \cdot \boldsymbol{u})}{c_s^2} + \frac{(\boldsymbol{e} \cdot \boldsymbol{u})^2}{2c_s^4} - \frac{u^2}{2c_s^2} \right)$. $c_s = \frac{c}{\sqrt{3}}$ is the lattice sound speed, and w_α are weighting coefficients $\left(w_0 = \frac{4}{9}, w_{1-4} = \frac{1}{9}, w_{5-8} = \frac{1}{36} \right)$. p is the dimensionless pressure. The evolution of the density distribution functions are divided into the collision process and free-streaming process. Including the feedback force [266,267], they are expressed as

$$\begin{cases} \tilde{f}_\alpha(\boldsymbol{x}_\alpha, t) = f_\alpha(\boldsymbol{x}_\alpha, t) - \frac{1}{\tau}(f_\alpha(\boldsymbol{x}_\alpha, t) - f_\alpha^{eq}(\boldsymbol{x}_\alpha, t)) + \frac{\boldsymbol{f}_b \cdot (\boldsymbol{e}_\alpha - \boldsymbol{u}_f)}{c_s^2} f_\alpha^{eq} \delta_t \\ \tilde{f}_\alpha(\boldsymbol{x}_\alpha + \boldsymbol{e}_\alpha \delta_t, t + \delta_t) = \tilde{f}_\alpha(\boldsymbol{x}_\alpha, t), \end{cases} \tag{3.110}$$

where τ is the relaxation time which is related to the kinetic diffusivity $\nu = \frac{1}{3}c^2(\tau - 0.5)\delta t$. The macroscopic variables of density, velocity and pressure are

$$\boldsymbol{u}^\star = \frac{1}{\rho} \sum_{\alpha=0}^{8} \boldsymbol{e}_\alpha f_\alpha, \text{ where } \rho = \sum_{\alpha=0}^{8} f_\alpha \tag{3.111}$$

$$p = \frac{3\rho c^2}{5} (\sum_{i \neq 0} f_\alpha + s_0(\boldsymbol{u})) \tag{3.112}$$

Finally, for the boundary condition of f_i, nonequilibrium extrapolation scheme is applied:

$$f_\alpha(B, t) = f_\alpha^{eq}(B, t) + (f_\alpha(N, t) - f_\alpha^{eq}(N, t)) \tag{3.113}$$

where B is the node on boundary and N is the node adjacent to boundary.

It is common to use the temperature distribution functions to include coupling of the heat transfer process of the gas-particle flow. For example, the temperature distribution function can be expressed as

$$T_i^{eq} = \frac{T}{4}\left(1 + 2\frac{\boldsymbol{e} \cdot \boldsymbol{u}}{c^2}\right), i = 1, 2, 3, 4. \tag{3.114}$$

The evolution processes of density and temperature distribution functions are classified into collision and free-streaming steps, where the gas-solid convection heat transfer are considered as follows, [262].

$$T_i(\boldsymbol{x}+\boldsymbol{e}_i\delta t, t+\delta t) = T_i(\boldsymbol{x},t) - \frac{1}{\tau_T}\left(T_i(\boldsymbol{x},t) - T_i^{eq}(\boldsymbol{x},t)\right) + \frac{Q_s}{4}, \tag{3.115}$$

where τ_T is the temperature relaxing time, and Q_s is the amount of the gas-particle convective heat transfer computed by the area-weighted method.

With regard to the boundary condition of the heat transfer, the nonequilibrium extension scheme of distribution function is utilized for treating boundary conditions. Assume 'O' is the boundary node and the node 'B' is immediately adjacent to the node 'O', the relation can be formulated as:

$$T_i(O,t) = T_i^{eq}(O,t) + (T_i(B,t) - T_i^{eq}(B,t)). \tag{3.116}$$

The macro-variables are computed by

$$\begin{aligned} T &= \sum_i T_i. \\ \alpha &= \frac{1}{2}c^2(\tau_T - \frac{1}{2})\delta t. \end{aligned} \tag{3.117}$$

If the temperature is low, the radiation heat transfer can be neglected. Only the thermal conduction between particles and the convective heat transfer between gas and the particle phases are considered. Based on the conservation law [262], and writing the conduction coefficient and thermal capacity of particles as k_p and c_p respectively, the governing equation of energy for particle 'i' is expressed as

$$m_p c_p \frac{T_{p,i}}{dt} = \sum_{j=1}^{k} q_{cd,ij} + q_{cv,ij} \tag{3.118}$$

where $q_{cd,ij}$ is the amount of thermal conduction from a number of particles $(1-k$, where k is the number of particles contacting particle 'i') to the particle 'i', and $q_{cd,ij}$ is the amount of heat transferred from gas to particle by convection. They are expressed as:

$$\begin{aligned} q_{cd,ij} &= 2k_p\sqrt{\delta_{ij}d_p - \delta_{ij}^2}\left(T_{p,j} - T_{p,i}\right) \\ q_{cv,ij} &= \pi d_i^2 h_i \left(T_{f,i} - T_{p,i}\right) \end{aligned} \tag{3.119}$$

where δ_{ij} is the overlap or deformation between particles 'i' and 'j'. $T_{f,i}$ is the temperature of ambient gas phase around particle 'i'. $h_i \frac{Nu_{p,i}k_f}{d_i}$ is the convective coefficient, and $Nu_{p,i}$ is the Nusselt number of particle 'i', formulated as:

$$Nu_{p,i} = \begin{cases} 2 + 0.6\epsilon_i^{3.5} Re_{p,i}^{0.5} Pr^{1/3}, \text{ if } Re_{p,i} < 200, \\ 2 + \epsilon_i^{3.5}\left(0.5Re_{p,i}^{0.5} + 0.02Re_{p,i}^{0.8})Pr^{1/3}\right), \text{ if } 200 < Re_{p,i} < 1500, \\ 2 + 4.5\times 10^{-5}\epsilon_i^{3.5} Re_{p,i}^{1.8}, \text{ if } Re_{p,i} > 1500 \end{cases} \tag{3.120}$$

where ϵ_i is the void fraction of the control volume of particle 'i'. The feedback convective heat transfer from particle to gas phase is computed by an interpolation method based on the area-weighted approach.

In order to validate the numerical model and heat transfer codes, the TD2G9 model is used to simulate the laminar flow between two parallel plates, with the analytical solution [263]. This is an incompressible two-dimensional steady state laminar flow. The distance between the plates is $H = 0.01$ m, and the length is $L = 20$, $H = 0.2$ m. The fluid is air, with density $\tau = 1.0$ kg/m^3, viscosity $\mu = 2.0 \times 10^{-5}$ Pa·s, and thermal capacity $c_p = 1250$ J/(kg·K). The Reynolds number is 150. The velocity and temperature distribution on the inlet boundary are uniform, with $U = 0.3$ m/s, $T = 300$ K. The outlet condition is set as fully developed free outflow.

The simulation results are shown in Fig. 3.6A and 3.6B. Agreements are met between the numerical results and analytical solutions, both for velocity and temperature fields. Thus, the numerical model and codes of the present simulation are validated. On the other hand, validation for free jets can be found in the previous work of our group [264].

3.3.3 Multiple schemes LBM-IBM-DEM

In this section, we will show an example of using multiple schemes of the coupled lattice Boltzmann equation (LBM), immersed boundary method (IBM), and discrete element method (DEM) in simulation of the gas-particle system. The LBM is used to simulate the fluid phase, and the discrete element method is used to simulate the particle phase. The immersed boundary method is used to deal with the momentum coupling between the gas and particle phase at fine scales less than the particle diameter.

In the quasi-two-dimensional test facility of our laboratory [89], the pebbles are recirculated in the air atmosphere. For this apparatus, the gas phase is the stationary air and herein simulated by the lattice Boltzmann method (LB), which consists of three components: the discrete velocity model, the equilibrium distribution function and the evolution equation. We simply employed the D2G9 model [265] to simulate the stationary gas atmosphere based on the Boussinesq hypothesis.

The pebble motion is solved by the discrete element method (DEM) [268]. The key features of DEM are the fundamental contact mechanisms modeled by the elastic collision, viscous damping, and friction, which are formulated by

$$\begin{cases} f_{p,n}^c = -k_n x_n - \beta_n \dot{x}_n \\ f_{p,t}^c = -k_t x_t - \beta_t \dot{x}_t \\ \text{if } |f_{p,t}^c| \geq \mu_f |f_{p,n}^c|, \text{ then } |f_{p,t}^c| = \mu_f |f_{p,n}^c| \\ T_p^c = r_p \times f_{p,t}^c \end{cases} \tag{3.121}$$

where $f_{p,n}^c$, $f_{p,t}^c$, x_n, x_t and T_p^c are the contact forces, interparticle displacements in the normal and tangential directions, and the collisional torque. k, β and μ_f are the

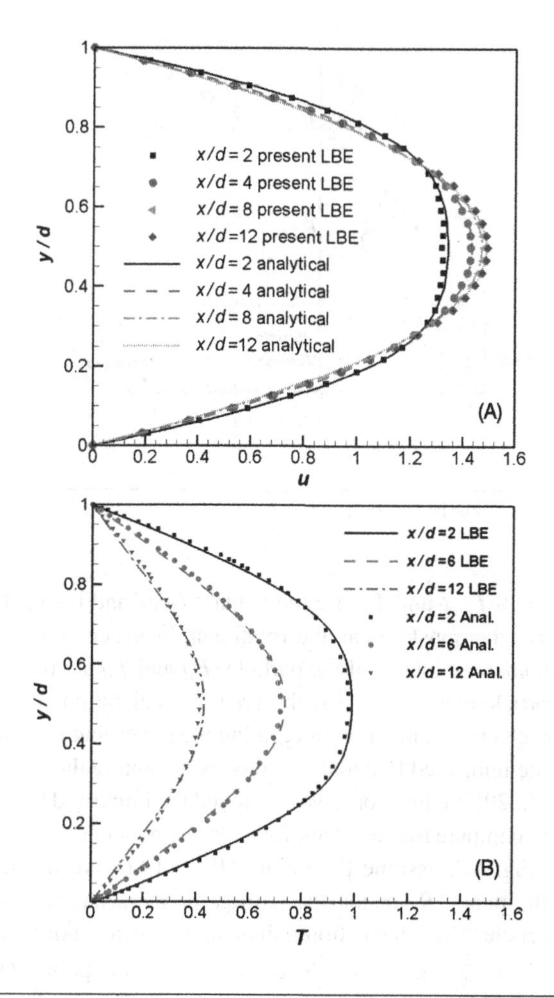

FIGURE 3.6

Comparison of simulation results and analytical solutions. (A) Velocity. (B) Temperature (shown in dimensionless values).

stiffness factor, damping coefficient and friction coefficient respectively. An over-dot denotes the time derivative. The subscripts t and n denote the tangential and normal directions between a pair of colliding particles, respectively.

The translational and rotational motions of particles are governed by the Newton laws of motion as

$$\begin{cases} \ddot{\boldsymbol{x}}_p = \dfrac{\boldsymbol{F}_p^c + \boldsymbol{F}_b + \boldsymbol{F}_d}{m_p} + \boldsymbol{g} \\ \ddot{\boldsymbol{\theta}}_p = \dfrac{\boldsymbol{T}_p^c + \boldsymbol{T}_d}{I_p} \end{cases} \tag{3.122}$$

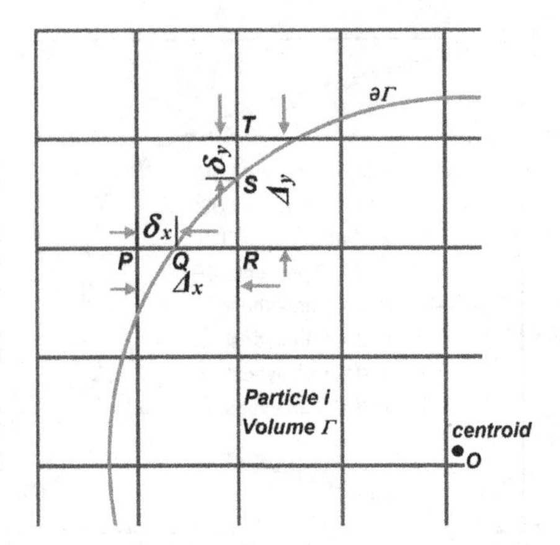

FIGURE 3.7

Sketch of the scheme of the IB method.

where $\boldsymbol{F}^c_p = f^c_{p,n}\boldsymbol{n} + f^c_{p,t}\boldsymbol{t}$ and $\boldsymbol{F}_b$ are the contact force and buoyant force. $\boldsymbol{x}_p$, $\boldsymbol{\theta}_p$, m_p, $\boldsymbol{T}^c_p$, and I_p are the translational and rotational displacements, mass, collisional torque, and the moment of inertia of the particle. $\boldsymbol{F}_d$ and $\boldsymbol{T}_d$ are the force and torque due to fluid-to-particle interaction. $\boldsymbol{g}$ is the gravity acceleration. The particle movement and rotation can be obtained by integrating the governing equations.

The basis of the immersed boundary method is to capture the interphase momentum coupling [164,220]. In this work, we employed the immersed boundary approach similar to [269] to compute the feedback force on the pebble boundary.

As shown in Fig. 3.7, assume the points 'P' and 'T' are in the fluid domain, which are near the point 'R' inside the particle domain, and the point 'O' is the centroid of the particle. The intersection points on the particle border are 'Q' and 'S' respectively. Let $\lambda_x = \frac{\delta_x}{\Delta_x}$ and $\lambda_y = \frac{\delta_y}{\Delta_y}$, the velocity at the points Q and P can be expressed as

$$\begin{cases} u^I_{\alpha,Q} = u^F_{\alpha,P}(1-\lambda_x) + \lambda_x v^S_{\alpha,R} \\ u^I_{\alpha,S} = u^F_{\alpha,T}(1-\lambda_y) + \lambda_y v^S_{\alpha,R} \end{cases} \tag{3.123}$$

where the superscripts F, S, I denote the fluid phase, the solid phase and the interface between them respectively. Therefore, $v^F_{i,P}$ and $v^F_{i,T}$ in the flow domain can be solved by the LBM, and $\boldsymbol{v}^S_R = \boldsymbol{v}^S_O + \dot{\boldsymbol{\theta}}^S_O \times (\boldsymbol{x}^S_R - \boldsymbol{x}^S_O)$ is determined by the motion of the solid particles. The nonslip condition on the immersed particle boundary gives rise to the feedback force from particle to fluid in the form of

$$\boldsymbol{f}_b(\boldsymbol{x}) = \int_{\xi\in\Lambda} \left(\boldsymbol{v}^I_p(\xi) - \boldsymbol{u}^\star_g(\xi)\right)\delta(\boldsymbol{x}-\xi)d\xi \tag{3.124}$$

where $\boldsymbol{v}_p^I(\boldsymbol{\xi})$ is the particle velocity on the border, and $\boldsymbol{u}_g^\star(\boldsymbol{\xi})$ is the fluid velocity on the border of the particle by interpolation using Eq. (3.123). In Eq. (3.124), $\Lambda = \Gamma \cap \Delta(\boldsymbol{x})$ where Γ is the set of the intersection points (such as 'Q' and 'S' in Fig. 3.7) between the border of particles and the meshing grids for the fluids, $\Delta(\boldsymbol{x})$ is the neighborhood of $\boldsymbol{x}$, i.e. $\boldsymbol{\xi} \in \Delta(\boldsymbol{x})$ means $\Delta = \{\boldsymbol{\xi} : |\boldsymbol{x} - \boldsymbol{\xi}| \leq \Delta/2\}$. $\delta(\boldsymbol{x} - \boldsymbol{\xi})$ is a tri-linear three-dimensional interpolation function, which has the form of

$$\delta(\boldsymbol{x} - \boldsymbol{\xi}) = \frac{1}{\Delta^D} \prod_{j=1}^{D} \delta_j(\frac{x_j - \xi_j}{\Delta}) \tag{3.125}$$

in every direction, where D is the flow dimension. The interpolation function $\delta_j(\zeta)$ has the form [270]

$$\delta_j(\zeta) = \begin{cases} \frac{1+\cos(\frac{1}{2}\pi\zeta)}{4}, & \text{if} |\zeta| \leq 2 \\ 0, & \text{otherwise} \end{cases} \tag{3.126}$$

Based on the formula of the feedback force on node $\boldsymbol{x}$ in Eq. (3.124), the fluid-to-particle force and torque can be obtained following the Newton third law of motion:

$$\begin{cases} \boldsymbol{f}_d = \frac{\rho_g}{\rho_p N(\Lambda)} \sum_{\boldsymbol{x} \in \Lambda} \boldsymbol{f}_b(\boldsymbol{x}) \\ \boldsymbol{T}_d = \boldsymbol{r}_p \times \boldsymbol{f}_d \end{cases} \tag{3.127}$$

where $N(\Lambda)$ is the number of nodes $\boldsymbol{x}$ inside the set $\Lambda = \Delta(\boldsymbol{x}) \cap \Gamma\ (\neq \emptyset)$.

By Eq. (3.127), the fluid-to-particle force and torque are computed without explicit empirical correlations. Although following the fundamental law of the Newton motion equation, it still needs validation before application. Therefore, a simple case of sedimentation of one particle is simulated here using the present interphase coupling algorithm.

A single sphere of diameter $d_p = 15$ mm fell in the silicon oil of various densities and viscosities, in a box [271]. Four cases of different density and viscosities of silicon oil were simulated according to the experimental condition (see Table 1 for case E1–E4 in [271]). The snapshots of particle sedimentation processes are shown in Fig. 3.8A–D. The particles were falling under the driving force of gravity and against the resistance forces of the buoyancy and fluid viscosity (Fig. 3.8A–C). After colliding with the bottom wall (Fig. 3.8D), the particle rebounded back. This process went on for a period until the mechanical energy is fully dissipated. During this process, the fluid was disturbed and pairing vortices were induced (Fig. 3.8).

The histories of variation of the particle position and velocity for the four cases are shown in Fig. 3.9A and B. It is seen that the position and velocity of particles are generally in good agreement with the experimental measurements. In Fig. 3.9A, positions of the particle centroid are always above $0.5d_p$ and finally stay at $0.5d_p$, which is reasonable although slightly different from the experimental measurements. In conclusion, based on the comparison in Fig. 3.9, the current method can predict the particle motion in fluid in acceptable accuracy, although with some slight differences.

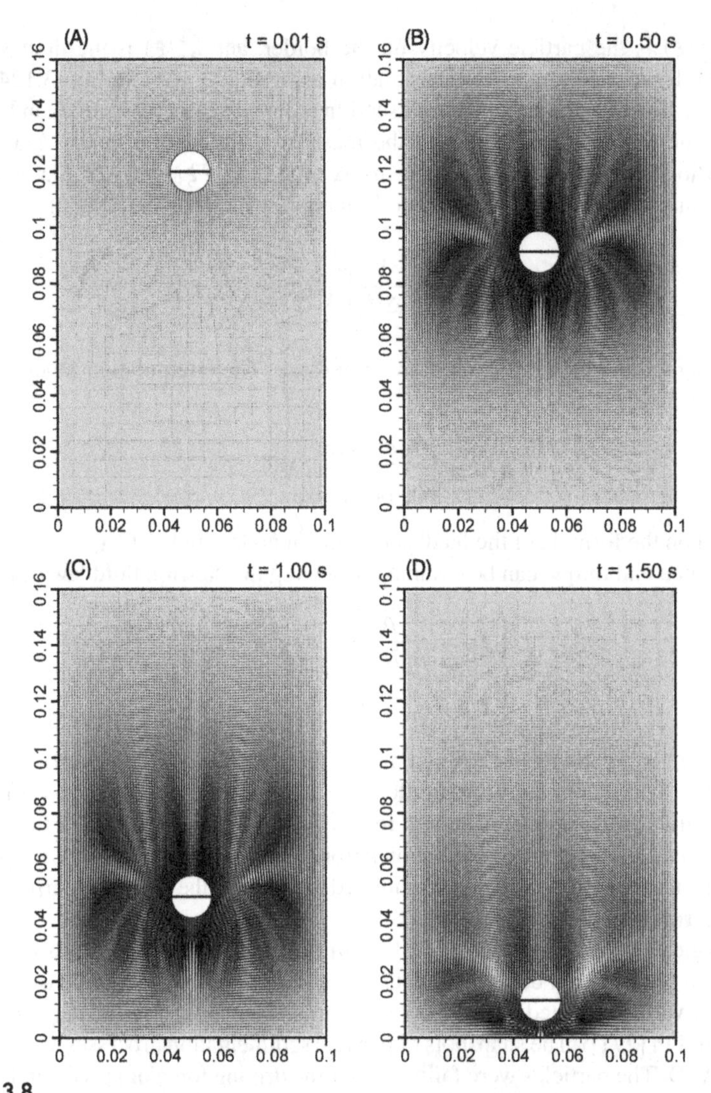

FIGURE 3.8

Snapshots of the velocity fields for the particle sedimentation case.

3.4 Summary

This chapter introduced the details of coupled schemes within the Eulerian-Lagrangian framework, including multiple schemes with or without heat conduction, which have been widely used for simulation of gas-particle flows. In general, the DNS based coupled schemes are powerful tools for studying the interphase interaction mechanisms of gas-particle flows. The LES based coupled schemes are applicable to

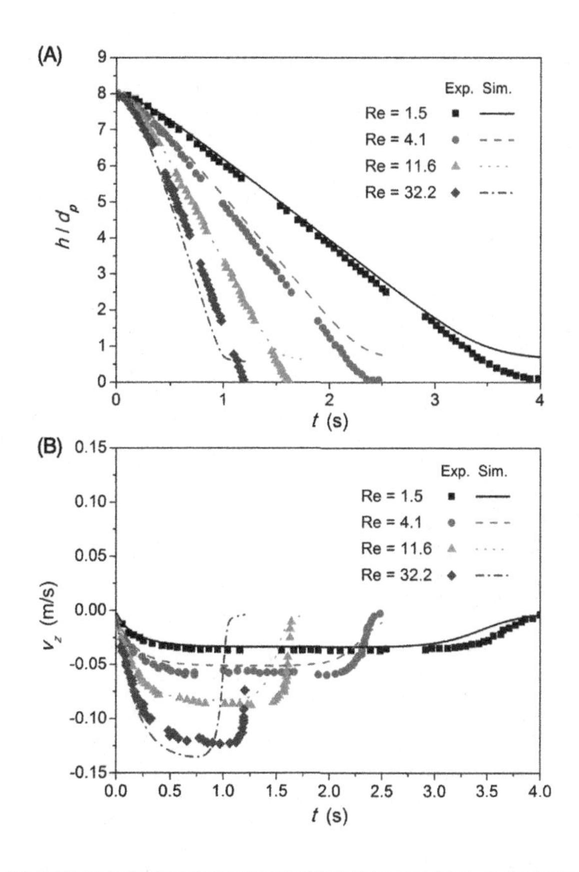

FIGURE 3.9

Height (A) and vertical velocity (B) for the one particle sedimentation in comparison with experimental measurements for the model validation.

simulate moderate scale test beds in combination with prediction of basic behaviors of the two-phase flow. The LBM based coupling schemes, although developing fast, have still many limitations which inhibit its application in the simulation for industrial two-phase flows. For example, the LBM based method is severely limited by its symmetry of lattice and the orders of expansion, its perfect recovery of real system based on a fundamentally perfect analysis is still sometimes in doubt.

In the following chapters, we will demonstrate sample applications of using coupled methods for study of a wide range of gas-particle flows.

PART

Applications

2

CHAPTER

Application in gas-particle flows

4

4.1 Homogeneous turbulence

4.1.1 Governing equations

A homogeneous and isotropic flow is a common hypothesis for a fully developed turbulence. The pseudo-spectral method is used here to simulate turbulence for its high resolution. The governing equations are the nondimensional incompressible Navier–Stokes equations:

$$\frac{\partial \boldsymbol{u}}{\partial t} = \boldsymbol{u} \times \boldsymbol{\omega} - \nabla P + \frac{1}{Re_\lambda}\boldsymbol{u} - \boldsymbol{F_b} \tag{4.1}$$

$$\nabla \cdot \boldsymbol{u} = 0 \tag{4.2}$$

where $\boldsymbol{u}$, $\boldsymbol{\omega}$ $(= \nabla \times \boldsymbol{u})$, P $(= \frac{p}{\rho_g} + \frac{u^2}{2})$, $\boldsymbol{F_b}$ $(= \sum_{i=1}^{N} \boldsymbol{f}_p^i)$ are the fluid velocity, fluid vorticity, fluid total pressure, total feedback force from particle to fluid in a control volume, respectively. N, $\boldsymbol{v}_p^i$, $\boldsymbol{f}_p^i$ $(= \frac{\rho_p^B(\boldsymbol{u}-\boldsymbol{v}_p^i)f}{St})$, ρ_p^B $(= \frac{\rho_p V_p}{\rho_g V_g})$, ρ_p, V_p, ρ_g, V_g, f $(= 1+0.15Re_p^{0.687})$, St $(= \frac{\epsilon^{1/2}\tau_p}{\nu^{1/2}})$ are the instantaneous total number of particle within the control volume, the ith particle's velocity, the nondimensional drag force of particle, the local bulk density, the particle density and volume, the fluid density and volume, the factor of Stokes drag coefficient, and the Stokes number, respectively. The Stokes number is based on the initial Kolmogorov time scale $\tau_\eta(0) = (\nu/\epsilon)^{1/2}$, where ν is the kinematic viscosity and ϵ is the dissipation rate of the turbulence kinetic energy. The initial Taylor microscale Reynolds number $Re_\lambda(0) = \langle u' \rangle \lambda / \nu$ is set according to the permit of numerical grids for spatial resolution of the smallest scales of flow, which is monitored by the value of $k_{max}\eta \geq 1$ [272]. The number of numerical grids is extremely restricted by simulation of the interparticle collision in a deterministic way due to the limit of computer performance and capacity. Hence, as a demonstrative research, the Reynolds number is set relatively at a low value. The other parameters used for the present study are displayed in Table 4.1.

The Navier–Stokes equations (4.1)–(4.2) are simulated on a triply periodic cube of side $L = 2\pi$. Moin et al. [234] suggested a mesh spacing of 1.5η for the Fourier spectral scheme to capture a wave of 3η wavelength. Thus the fluid domain is uniformly discretized into N^3 $(N = 64)$ grid points, which defines the wavenumber components in the Fourier space as $k_j = \pm n_j(2\pi/L)$, where for $n_j = 0, 1, \cdots, \frac{N}{2}$

Gas-Particle and Granular Flow Systems. https://doi.org/10.1016/B978-0-12-816398-6.00013-4
Copyright © 2020 Elsevier Inc. All rights reserved.

Table 4.1 Parameters for the fluid and particle.

Parameters	Values
Fluid viscosity ν	0.027
Fluctuation velocity $\langle u' \rangle$	1.558
Dissipation ϵ	2.25
Maximum wavenumber k_{max}	22
Integral time scale T_f	4.033
Integral length scale l_e	0.419
Taylor length scale λ	0.487
Number of particles N_p	64^3
Restitution coefficient e	0.95
Kolmogorov length scale η	0.054
Kolmogorov time τ_η	0.109
Eddy turn-over time τ_e	0.210
Simulation time $23.8\tau_e$	5
Simulation time step Δt	0.001
Reynolds number Re_λ	28.1
Friction coefficient γ	0.3

for $j = 1, 2, 3$. About 1/3 part of the energy in higher wave numbers is truncated at each step to reduce aliasing errors. The initial fluid velocity is set according to [273] and the initial energy spectrum is set according to [274].

The fluid turbulence is freely decaying, homogeneous and isotropic for the following specific reasons: (1) The homogeneous isotropic turbulence is naturally decaying and physically reasonable. (2) The effect of the interparticle collision on the coagulation process is of particularly interest. Since the behavior of the particle motion is strongly coupled to the fluid motion, the particle kinetic energy is then decaying accompanied with the decaying fluid turbulence, resulting in an effect of acceleration or augmentation of the process of particle coagulation. (3) The interparticle collision frequency has a straightforward dependence on the mean particle velocity. The decaying behavior of the particle velocity leads to a stretching segment of this dependence, which is also of great interest in the simulation results. However, the unsteady turbulence may also lead to additional troubles in statistics of numerical results. Fortunately, the statistics of numerical results is carried out within a small time interval (about a quarter of one eddy turn-over time), which is on one hand large enough to make the statistics meaningful and on the other hand sufficient small to ignore the decaying of turbulence.

Since the particle motion is coupled to the fluid motion, some assumptions should be considered to simulate the particle dispersion in the gas-particle flow. When the size of particle is relatively smaller than the Kolmogorov length scale η of turbulence, the particles can be treated as a mass point. However, for particle-particle interactions, the particles are rigid spheres. For dense particles, the drag force is of leading order, and the gravity force, the Saffman lift force, the Basset force and etc., are neglected. The equation of motion is

$$\frac{\boldsymbol{v}_p}{dt} = \frac{f}{\tau_p}(\boldsymbol{u} - \boldsymbol{v}_p) + \frac{\sum \boldsymbol{F}_c}{m_p} \tag{4.3}$$

which can be nondimensionalized and rewritten as (see in Chapter 3)

$$\frac{\boldsymbol{v}_p}{dt} = \frac{f}{St}(\boldsymbol{u} - \boldsymbol{v}_p) + \sum_i \boldsymbol{a}_i \tag{4.4}$$

where $\boldsymbol{a}_i$ is the acceleration caused by the collision force from all surrounding particles ('i'). Note that the characteristic time scale used for the above dimensionless form of the motion equation is based on the initial Kolmogorov time scale $\tau_\eta(0)$ of turbulence. Moreover, the free motion of particle and the interparticle collision process are decoupled here. The interparticle collision force or impulse is carried out at each time step in the procedure of collision detection.

The particle rotation is treated only as a single-phase motion since only the drag force is considered. The particle rotation is only induced by the interparticle collision and friction. The fluid induced rotation of particle is not considered and vice versa. Thus, the rotational motion of particle is simply govern by

$$\frac{d\boldsymbol{\omega}_p}{dt} = \frac{\boldsymbol{r} \times \boldsymbol{J}}{I_p \Delta t} \tag{4.5}$$

All the variables in Eq. (4.5), including the angular velocity $\boldsymbol{\omega}_p$ and the moment of inertia I_p, are nondimensional.

Initially, the particles are uniformly distributed at the center of each cubic computational cell, and the initial velocity of the particle is equal to the fluid velocity while the particle rotational velocity is set to zero. The initial distribution of particles with velocities in respect of fluid provides a foundation stone for sound comparisons between the statistical results of particle motion with and without considering interparticle collisions. However, a randomly homogeneous distribution or an already 'thermalized' distribution of particles is better, providing the simulation time is sufficient long and the effect of initial distribution is negligible. For simplicity, the present study used the former regular distribution. Moreover, to eliminate the initial effect on the statistics of the particle motion, the statistics of the particle results are computed from the 2nd or 3rd period of eddy turn over time.

Governing Eqs. (4.1) and (4.2) of fluid is solved using the pseudospectral method. A brief introduction to the method is given here. Due to the periodic boundary conditions, the velocity fields as well as the total pressure P, the nonlinear term $\boldsymbol{u} \times \boldsymbol{\omega}$ and the counterforce of particle $-\boldsymbol{F}_b$ are expressed by the truncated Fourier series, e.g.

$$\boldsymbol{u}_j(\boldsymbol{x}, t) = \sum_k \hat{\boldsymbol{u}}_j(\boldsymbol{k}, t) \exp(i\boldsymbol{k} \cdot \boldsymbol{x}) \tag{4.6}$$

where $\boldsymbol{u}_j(\boldsymbol{x},t)$ is the jth component of velocity in physical space and $\hat{\boldsymbol{u}}_j(\boldsymbol{k},t)$ is the Fourier coefficient of u_j at wavenumber $\boldsymbol{k}$. Substituting the Fourier serious representations as depicted above into Eqs. (4.1) and (4.2), and applying the orthogonality property of $\exp(i\boldsymbol{k}\cdot\boldsymbol{x})$ yields ordinary differential equations for $\hat{\boldsymbol{u}}_j(\boldsymbol{k},t)$.

In the pseudospectral method, evaluation of the nonlinear term is efficient since computation of this term in a physical space is less costly than by a spectral method. The ordinary differential equations for the Fourier coefficients are time advanced using the second-order explicit Adams–Bashforth scheme for the nonlinear and the counterforce terms and the Crank–Nicolson method for the rest terms.

The hard sphere collision model is used here to deal with the particle-particle collision.

The direct numerical simulation of the particle-laden flow in a deterministic way is obviously a challenge to computer performance. As interpreted in the above section, the detection of the interparticle collision consumes about $O(N_p)$ computation, and the total particle number in the present work is proportional to the size of numerical grids for the flow field discretization. Hence, for simulation of a large Reynolds number in the framework of treating interparticle collision deterministically, a great computer capacity is needed, which requires massive parallel computing. As a primary research, the present work is restricted to low Reynolds numbers.

In this section, interests are focused on the particle behaviors influenced by the particle-particle collision dynamics with various different Stokes numbers and mass loadings. For comparison and for the sake of saving computational cost, we carried out the simulation with the same number of particles but with various Stokes numbers and with three types of particle sizes. Three cases of different particle sizes – $d_p = 0.05\delta_m$, $d_p = 0.1\delta_m$ and $d_p = 0.2\delta_m$ – are chosen and five typical Stokes numbers $St = 0.01, 0.1, 1, 10, 100$ are used for simulations. The parameters are all listed in Table 4.2, including the particle size and the corresponding mass loading. It is found that the case of $St = 0.01$ and 100 can just be viewed as two limiting cases and from $St = 0.01$ to 100 the gas-solid flow is obviously transited from dilute to dense flow.

As pointed out by [275], the computer limitation is one of the main difficulties for direct simulation of gas-particle flows in the Lagrangian approach, especially when the interparticle collision is simulated deterministically. Hence, the case with a large amount of particles corresponding to high mass loading is hard to carry out. Alternatively, the Lagrangian method always computes a fraction of real particles, and each computational particle represents several real particles. Following the same approach of consideration, the preset study treated the case of higher mass loading via increase in particle density rather than in particle number so as to save the computational capacity. In other words, to vary the mass loading by varying the particle density, keeping the particle size $d_p/\eta < 1$ and maintaining the particle numbers.

Finally, this section is focused on the effect of interparticle collision on the particle behaviors only. The ratio d_p/l_e is limited to be less than 0.1 and the turbulence intensity is suppressed by dispersed particles. Hence, the modification of turbulence is not discussed.

Table 4.2 Simulation conditions of three cases.

d_p/δ_m	St	$\dot{m}$
Case1: 0.05	0.01	1.47×10^{-3}
	0.1	1.47×10^{-2}
	1	1.47×10^{-1}
	10	1.47
	100	1.47×10^{1}
Case2: 0.1	0.01	2.94×10^{-3}
	0.1	2.94×10^{-2}
	1	2.94×10^{-1}
	10	2.94
	100	2.94×10^{1}
Case3: 0.2	0.01	5.89×10^{-3}
	0.1	5.89×10^{-2}
	1	5.89×10^{-1}
	10	5.89
	100	5.89×10^{1}

4.1.2 Collision rates and statistics

4.1.2.1 Particle-particle collision rate R_c

Since the transition from dilute to intermediate dense flow may be directly characterized by the prevailing interparticle collision, we may start from the collision rate R_c, which is defined as the number of interparticle collisions occurring per unit volume per unit time. Based on the kinetic theory, the collision rate of any sampled particle is proportional to the square particle diameter d_p^2, the particle number density n_p and the mean relative velocity $\langle v_p \rangle_r$ between particles. Hence the collision rate of particles in a unit volume is proportional to a combined expression $n_p \cdot (n_p d_p^2 \langle v_p \rangle_r)$. Supposing the particulate ensemble is based on an analogy as the gas molecular ensemble in collision processes, and the particle relative velocity distribution follows the Maxwell's law of velocity distribution, the interparticle collision rate is linearly proportional to $n_p^2 d_p^2 \langle v_p \rangle$.

Fig. 4.1 shows the numerical results corresponding to the particle size $d_p = 0.05\delta_m$, $d_p = 0.1\delta_m$, and $d_p = 0.2\delta_m$ respectively, where δ_m is the mesh grid size. As shown in Fig. 4.1, when $St > 1$ the collision rate R_c is linearly proportional to the integral variables of $n_p^2 d_p^2 \langle v_p \rangle$ despite the particle sizes. The stretching for the $n_p^2 d_p^2 \langle v_p \rangle$ is due to the free decaying gas turbulence, resulting in a reduction in the mean particle velocity $\langle v_p \rangle$ (Fig. 4.1A). To eliminate the influence of the initial simulation condition of the regular distribution of particles, the statistical variables in Fig. 4.1 are all computed after two periods of eddy turn-over time. It is found that for St greater than unity, R_c is directly proportional to the particle velocity, square particle concentration and square particle size. Moreover, the particle behaviors of

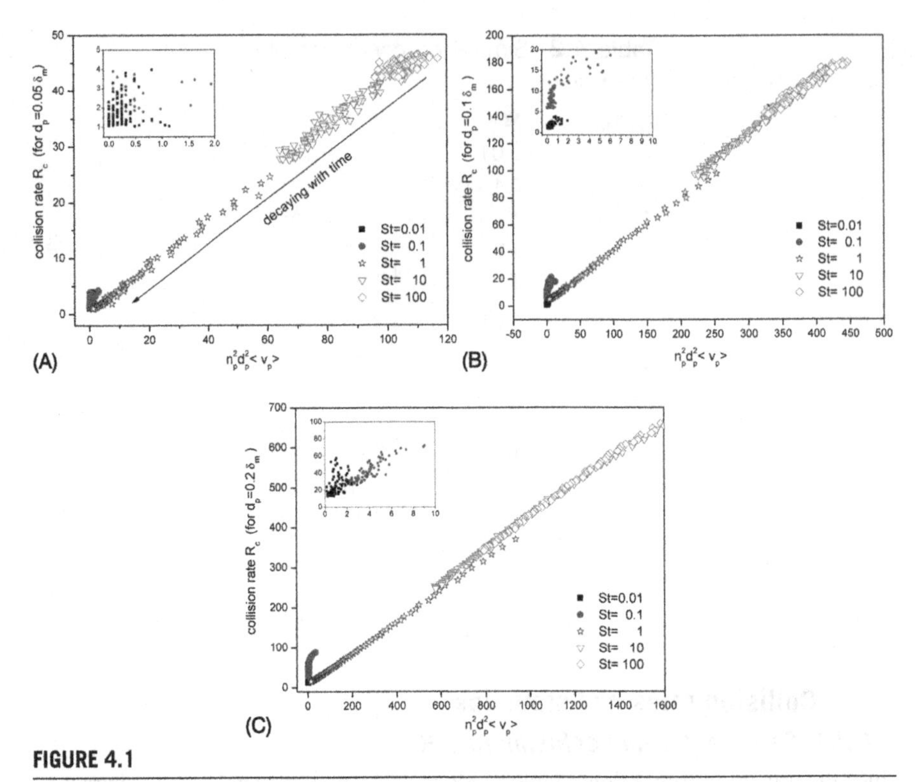

FIGURE 4.1

The particle-particle collision rates at different Stokes numbers (variables are measured in a nondimensional unit): (A) $d_p = 0.05\delta_m$, (B) $d_p = 0.1\delta_m$, (C) $d_p = 0.2\delta_m$.

different particle sizes and loadings bear somewhat resemblances with each other based on the analogy of the kinetic theory. Thus, the transition for the gas-solid flow to the intermediate dense or dense flow can be directly characterized by the particle concentration or particle mass loading.

Contrarily, for St less than unity, the collision rate R_c seems to be independent of the integral variable $n_p^2 d_p^2 \langle v_p \rangle$, appearing in the insets of Fig. 4.1 with a random distribution. For $St = 0.01$ the dependence phenomenon between R_c and $n_p^2 d_p^2 \langle v_p \rangle$ disappears and the particles can be viewed as fluid trackers. In this way, the particle concentration, or mass loading, are less important and the effect of turbulent transport seems to be overwhelming. Thus, the demarcation for the dilute and the dense flows is not appropriate to be characterized by the collision rates, particle concentration or loading. In other words, the characteristics of behavior of the gas phase and solid phase are not clearly distinguished from each other.

4.1.2.2 Mean free path

In fact, the collision rate in the above section for St greater than unity is a direct extension of the concept of molecule collision rate in the kinetic theory. For validation,

Table 4.3 Comparison between the theoretical and numerical slope coefficients R_c

Particle diameter	$d_p = 0.05\delta_m$	$d_p = 0.1\delta_m$	$d_p = 0.2\delta_m$	Theoretical
Slope coefficient	2.388	2.394	2.386	2.221

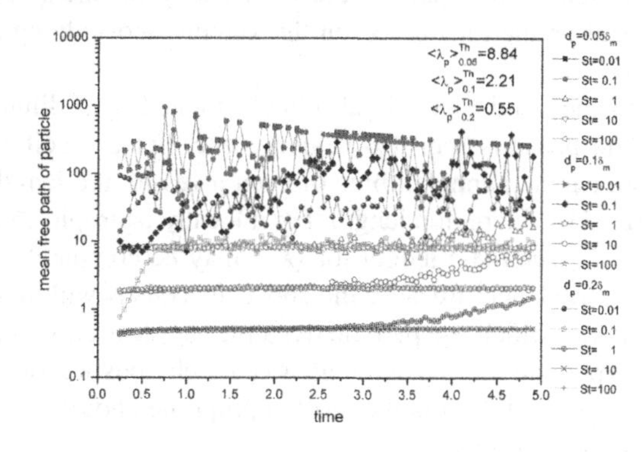

FIGURE 4.2

Time history of the mean free path of particles.

the slope coefficients of the numerical results of Fig. 4.1 with St greater than unity are calculated using the method of least square fit. The results are displayed in Table 4.3 with a comparison to their common theoretical counterpart of $\frac{\sqrt{2\pi}}{2}$. It is found that the accordance is satisfactory and the discrepancies between the numerical and theoretical results are very small. Now, the mean free path of the particle can be derived from R_c. Still based on the kinetic theory, the particle mean free path is a function of the particle size and concentration, since

$$\langle \lambda_p \rangle = \langle \lambda_p \rangle \tau_c = \langle \lambda_p \rangle / \langle Z_p \rangle = 1/(\sqrt{2}\pi d_p^2 n_p) \tag{4.7}$$

where $Z_p = \frac{1}{\tau_c} = \sqrt{2}\pi d_p^2 n_p \langle v_p \rangle = 2\frac{R_c}{n_p}$ is the mean collision rate for any individual particle. The mean free path λ_p is independent of particle dynamic variables, i.e. particle velocity, and it is an integral parameter of particle size and concentration. Therefore, it is a preferable parameter of particle characteristics in transition to dense flow, i.e. for $St > 1$:

$$\begin{aligned} &\langle \lambda_p \rangle / L_c \le 1, \text{dense flow} \\ &\langle \lambda_p \rangle / L_c > 1, \text{dilute flow} \end{aligned} \tag{4.8}$$

where L_c is a characteristic length of the gas-solid turbulent flow.

Fig. 4.2 shows the numerical results of time histories of the mean free path of particle calculated by $\langle \lambda_p \rangle = \langle v_p \rangle / \langle Z_p \rangle$ for all three types of particle diameters and all Stokes numbers. It is found that though the fluid turbulence is freely decaying with time and so is the kinetic energy of particles, the mean free path of a heavy particle for $St > 1$ is almost time independent. Moreover, comparing to the theoretical values defined by particle size and number density in Eq. (4.7), e.g. $\langle\lambda_p\rangle_{0.05}^{Th} = 8.84$, $\langle\lambda_p\rangle_{0.1}^{Th} = 2.21$ and $\langle\lambda_p\rangle_{0.2}^{Th} = 0.55$, it is found that discrepancies are very small. Hence it is appropriate to characterize the flow regime of dense or intermediate dense for $St > 1$ by $\langle\lambda_p\rangle/L_c \leq 1$. However, it is also found that when the turbulence is fully decaying and the coagulation of particle cluster is heavy, the mean free paths are biased from theoretical predictions based on the kinetic theory though the particle is very heavy ($St > 1$).

For $\langle\lambda_p\rangle \to \infty$ the solid phase is absolutely treated as a dilute flow, and for $\langle\lambda_p\rangle \to d_p$ the solid phase is transited to dense one. For a specific gas-solid flow in a limited region, e.g. a channel, a pipe or even a vessel, the length scale may be directly characterized by the region length. For a locally isotropic and homogeneous turbulent flow, the characteristic length for $\langle\lambda_p\rangle$ may be comparable to some characteristic length of turbulence. Because the above analysis is within the condition of St greater than unity, in which the particle dynamics are less correlated with the gas turbulence, and the particle motion and collision are obviously influenced by the region range, it is suggested that it is the most appropriate choice for L_c equal to the characteristic region length scale.

4.1.2.3 Flatness factor $\mathscr{F}$

In Eq. (4.7), the mean free path depends on the particle number density and size. Moreover, the effect of the inter-particle collision will, to different extents, change the distribution of particle number density. Thus it is reasonable to study more on the number density. The most natural way to quantify the distribution of number density is via probability density function (PDF) of the number density. The frequency of extreme events is reflected in the size of the tails of the PDF and thus in the high order moments [276]. As the particle size d_p is set according to the mesh size δ_m for discretizing the flow domain, the instantaneous and local particle number density is simply obtained by dividing the particle numbers in a local numerical cell by the cell volume. Hence the flatness of number density distribution is calculated, which is defined as $\mathscr{F} = \frac{\langle n_p'^4\rangle}{\langle n_p'^2\rangle^2}$, where n_p' is the number density fluctuation. Fig. 4.3A–C shows the flatnesses at different St for particle size $d_p = 0.05\delta_m$, $d_p = 0.1\delta_m$, and $d_p = 0.2\delta_m$ respectively. It is found that regardless of particle size, the flatness at $St < 1$ are far different from that at $St \geq 1$. The increase in $\mathscr{F}$ for $St < 1$ indicates the differentiation of condensation and dilution of the particle number density at different regions and the trend seems to be augmented as time increases. The lower the particle Stokes number, the rapider the differentiation processing.

However, for $St > 1$ the flatness $\mathscr{F}$ seems to converge asymptotically to 3, which is the flatness of the Gaussian distribution. In other words, the differentiation of the particle number density seems to be suppressed and the particle preferentially collection rarely appears.

Based on the numerical results from Fig. 4.1 to Fig. 4.3, it is indicated that the study on the particle-particle collision effects for $St < 1$ is a more complicated and difficult problem than that for $St > 1$, since it is deeply coupled with the gas turbu-

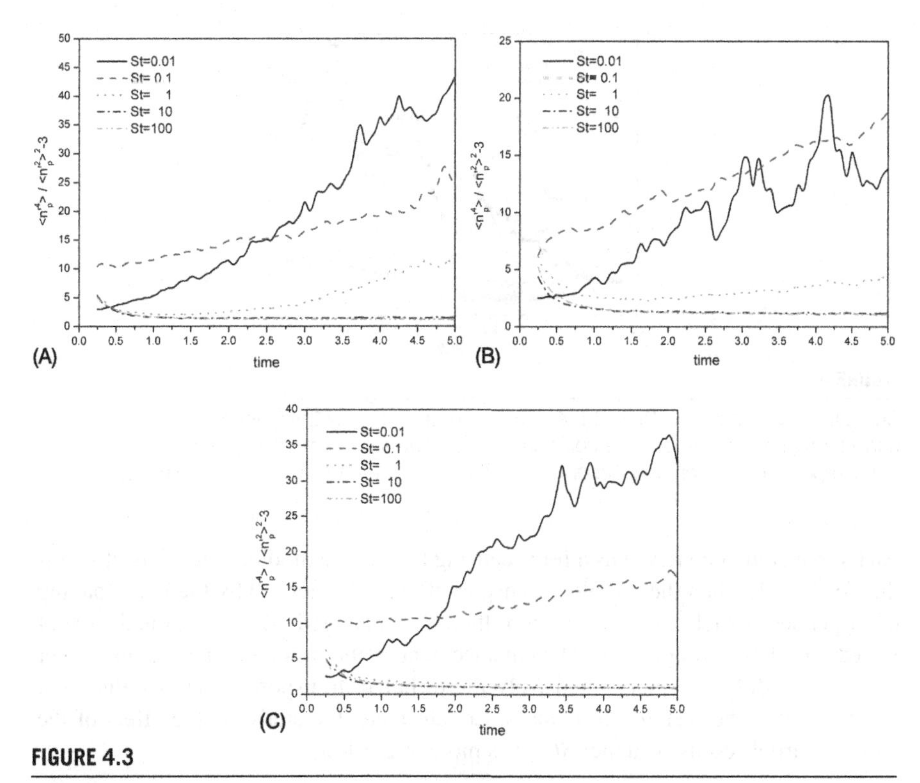

FIGURE 4.3

Time evolution of the flatness of particle number density distribution function at different Stokes numbers: (A) $d_p = 0.05\delta_m$, (B) $d_p = 0.1\delta_m$, (C) $d_p = 0.2\delta_m$.

lence. For example, from Fig. 4.3, it is observed that there are differences between the time evolutions of $St = 0.1$ and 0.01. Specifically, it is observed that for $St = 0.1$, at the beginning of particle statistics after one eddy turn-over time, the flatness has already reached a comparatively large values. It seems that the particle clustering has already formed, but evolves at a comparatively lower speed. On the contrary, for $St = 0.01$, after one eddy turn-over time, the particle condensation and dilution phenomena of the particle number density seem not to be very evident, and the distribution of particle still seems to be homogeneous. However, the particle clustering evolves in time very rapidly and with comparatively large fluctuation due to intensive coupling with the fluid turbulence.

Moreover, it is also observed that at the same Stokes number but different particle sizes, the results are still different. For example, for $St = 1$, the evolution of flatness is clearly increasing for the $d_p = 0.05$, relatively neutral for $d_p = 0.1$ and clearly decaying at $d_p = 0.2$. It seems that the Stokes number is not enough to discriminate the behavior of particles. Referring to Table 4.2, the cases of different particle sizes at the same Stokes numbers are still different in mass loading $\hat{m}$. The $\hat{m}$ influences the particle feedback force upon the carrier phase Eq. (4.1), and the particle-fluid feed-

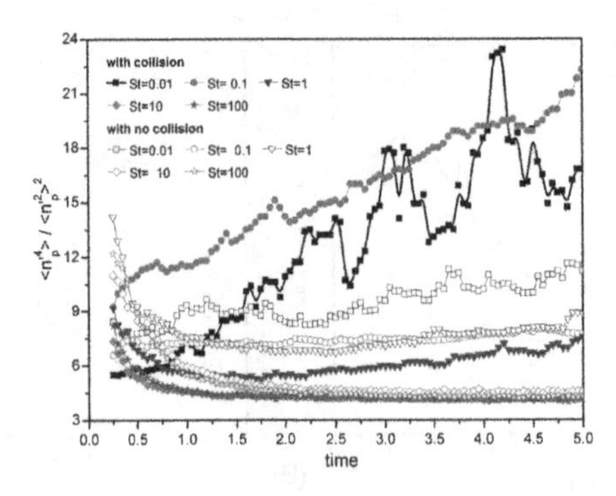

FIGURE 4.4

Comparison of flatness of PDF of the particle number density between with and without considering the particle-particle collision effects. (The solid symbol is for considering the particle-particle collision and the open symbol is not; the particle size is $d_p = 0.1\delta_m$.)

back force could be viewed as a force causing turbulence modulation. Thus, it seems that for $St = 1$, when the turbulent transport effect is influenced by the mass loading of suspended particles, the evolution of flatness is changed. Moreover, the degree of increase in flatness depends on the enhancement of the mass loading, i.e. the larger the mass loading, the more intensively the turbulent transport effect is influenced, and the higher the increase in flatness. Conclusively, the study on the effect of the particle-particle collision under $St < 1$ is more interesting.

4.1.2.4 Comparison with the no-collision case

The present section is motivated by the interests of studying the effect of the particle-particle collision when $St < 1$. The flatness factor is calculated under the same simulation conditions but without considering the particle-particle collision model for comparison. In general, take $d_p = 0.1\delta_m$ for example, Fig. 4.4 illustrates the comparison of time evolutions of $\mathcal{F}$ at different Stokes numbers between the results with and without considering the particle-particle collision. It is interesting to find that $\mathcal{F}$ for $St < 1$ without considering the particle-particle collision (open denotation in Fig. 4.4) increases much more slowly and is much less in magnitude than that with the particle-particle collision (solid denotation in Fig. 4.4). It is said that the particle-particle collision effects under $St < 1$ will intensify the differentiation and make the occurrence of condensed particle collection more frequently. It is against our previous intuitive guess, since the collision model is inherently repulsive and without any effect of attraction in the particle-particle collision.

Moreover, it is found that for $St > 1$ there seems to be an inverse trend for the effect of the particle-particle collision, which is to say that the particle-particle collision effect will make the occurrence of particle collection or coagulation less frequently. Referring to the two competitive effects – the turbulent transport effect and the accumulation effect – in the gas-solid flow, now another important effect of the particle-particle collision would be taken into account in turbulent gas-solid flows, since it influences the particle motion selectively on different Stokes numbers.

It is considered that there must be some physical reasons which underlie the selectively clustering effect of particles with Stokes numbers upper or lower than unit when the interparticle collision is considered. We tried to interpret the physical mechanisms. Firstly, for the case of $St < 1$, the particle collection and clustering are true though the particles are very light and they can follow the streamlines almost completely. Secondly, due to both the contributions of turbulent transport effect and the accumulation effect to the collision rate, it is observed from Fig. 4.1 that the ratio of the particle-particle collision rate R_c to the integral variable $n_p^2 d_p^2 \langle v_p \rangle$ is comparatively larger than that of $St > 1$. It is considered that the above two effects are related, i.e.

- there are already some regions occupied by existing clusters;
- the occurrence of the particle-particle collision will retard or even prevent moving particles from passing through the regions occupied by the particle clusters, causing the cluster to grow up.

In fact there might be another fact of turbulent transport effect which may blow away the existing clusters. Thus, the increase of $\mathscr{F}$ in Fig. 4.4 is partially due to the decaying of turbulence, weakening the turbulent transport effect as time elapses. Moreover, if no particle-particle collision is considered, the particles can penetrate freely into existing clusters and pass through them. The clusters fail to capture new particles and hence the agglomeration is less intensive.

In other aspects, for $St > 1$, the particle trajectories resemble the ballistic trajectories. The particle motion is seldom correlated with the fluid turbulence and the distribution is homogeneous. The particle motion with large inertia responses to collision slowly and usually overlaps with successive collisions. The great impacting impulse of large particles would make the particle agglomeration loose and cause the particle clusters to be more dispersed. Thus, the collision only contributes to make the concentration distribution even more homogeneous.

To say conclusively, it is found that for $St > 1$, similar to the kinetic theory, the particle-particle collision rate is proportional to particle mean velocity, mean square of particle number concentration and mean square of particle size. However, for $St < 1$ it doesn't hold. Due to the precondition of validity of analogy to kinetic theory and in order to determine the relative importance of the particle-particle collision and characterize the regime of dilute and dense flow, a reasonable criterion of the ratio of the particle mean free path to the characteristic scale of the gas-solid flow is suggested. The evaluation of the flatness of the particle distribution function reflects that the particle-particle collision has selective effects on the particle number density distributions. The most interesting thing is that the effect the of particle-particle collision will intensify the particle differentiation into both condensation and dilution for the Stokes number less than unity. The physical interpretation is that it is easy for light particle clusters to capture new elements by the particle-particle collision, which would cause the particle clusters to grow up.

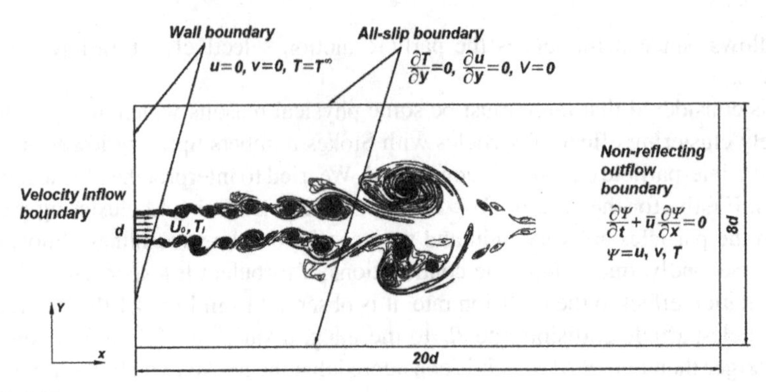

FIGURE 4.5

The sketch of flow configuration and boundary conditions.

4.2 Planar jets

In this section, at first, DNS of two-dimensional planar jets with heat transfer and then DNS of the three-dimensional particle-laden flow are shown to study the effects of two-way coupling.

4.2.1 2D case with the heat transfer

The flow configuration is sketched in Fig. 4.5. The jet is injected through the inlet with a mean velocity U_0 and a width d. In the past studies of the direct numerical simulation of planar jet or mixing layers, the turbulent inflow condition is always given by the direct forcing method [277]. However, in order to observe the evolving process of coherent structure more clearly, it is preferred to not introduce additional turbulence at the inlet. Some types of inflow profiles are studied. At first, the top-hat profile is used as the inflow velocity. Then, a hyperbolic tangent profile is used to study the effect of inflow boundary layer on jet evolution [278]:

$$u(y) = \frac{1}{2}\left(1 + \tanh\left(\frac{0.5 - |y|}{2\theta}\right)\right) \tag{4.9}$$

where θ is the inflow momentum thickness (Table 4.4).

The thermal jet at $T = T_i$ ($> T_\infty$) is cooled by the ambient fluids at temperature $T = T_\infty$. The nonreflecting boundary condition is utilized for the outlet condition [279]. The side walls are set all-slipping boundaries [280]. The entire domain of flow is $20d \times 8d$, discretized by 2000×800 Cartesian mesh grids. As the Kolmogorov length scale for planar jets at $Re = 4000$ is estimated as $\eta_{min} = 0.008d$ under the isotropic assumption [281], it should be larger than $0.008d$ at $Re = 3000$ in this section. Notice that the finest mesh spacing in this study is $h = 0.01d$. It is of the same order as the Kolmogorov length scale, namely $\eta_{min} = O(h)$. It is fine enough to meet the requirement of spatial resolution for the direct numerical simulation [234].

Table 4.4 Parameters used in the numerical simulation.

Characteristic diameter d^* (mm)	10
Scales of the flow domain	$20d \times 8d$
Grid numbers, $N_x \times N_y$	2000×800
Spatial resolution δ (μm)	10
Reynolds number, Re	3000
Prandtl number, Pr	0.1, 1, 10
Inflow temperature, T_i, (°C)	35
Ambient temperature, T_∞, (°C)	25
Dimensionless inflow velocity, U_0	1
Dimensionless time step, Δt	0.001
Total simulation time, T_s	20

Additionally, the dimensionless time step for time integration is $\Delta t = 0.001$, and in total 20000 time steps are simulated. The corresponding real parameters are listed in Table 4.4.

For example, the numerical results of vortex street at $Re = 3000$ is shown in Fig. 4.5. It is well known that the self-preserving zone for a plane free jet begins about $x = 7.3d$. Fig. 4.6A shows the comparison of mean profiles in the self-preserving zone with the early experiments carried out by [282] and [283]. In Fig. 4.6A, $\bar{u}$ is nondimensionalized by $\bar{U}_c$ at the central line, and y is divided by half width b of the jet. The present mean profiles are in good agreement with experimental measurements, although some slight discrepancies still exist. Additionally, to verify grid independence for the simulated results, Fig. 4.6B shows the comparison of mean velocities and fluctuated velocities at $t = 5$ and $x = 5d$ between the cases with 2000×800 grids and 1000×400 grids. By this comparison, one is convinced that to use 2000×800 grind is fine enough to obtain reliable conclusions.

As is well known, the values of Prandtl numbers are far different from each other for different fluids. For example, for air under standard conditions $Pr = 0.7 \sim O(1)$, for water under 10°C, $Pr = 9.4 \sim O(10)$. For engine oil under 60°C, $Pr = 1050 \sim O(10^3)$; and for liquid metal (e.g. Hg under 20°C, $Pr = 0.025 \sim O(10^{-2})$ (where '$O(\cdot)$' indicates the order of magnitude). Thus, it is necessary to use a wide range of Prandtl numbers to represent different properties of various fluids. Thus, three typical values of Prandtl numbers are investigated here (Table 4.4).

4.2.1.1 Vortex structure and temperature fields

This section illustrates the evolution of vortex structure and temperature fields. For example, the snapshots of vortex structure (Fig. 4.7), temperature fields with $Pr = 0.1$ (Fig. 4.8) and $Pr = 10$ (Fig. 4.9) with the top-hat velocity inflow profiles are shown.

At the dimensionless time $t = 1$, the rolling vortex structure is formed and developed symmetrically with opposite signs around $x = d$ caused by the Kelvin–Helmholtz (K-H) instability in the shear layers (Fig. 4.7A). Then, thermal fluids

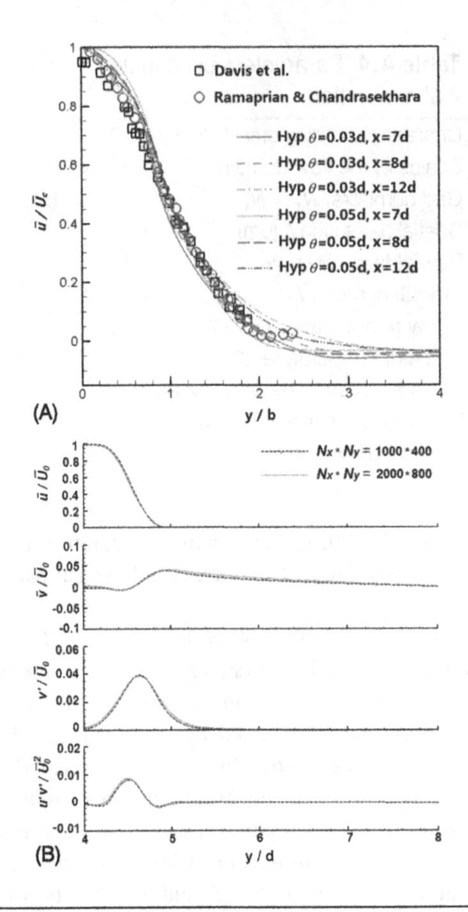

FIGURE 4.6

(A) Comparison of the mean profiles of axial velocity with the experiments. (B) Comparison of the mean and fluctuated velocities with different grid numbers (2000 × 800 and 1000 × 400) at $t = 5$ and $x = 5d$ with $\theta = 0.05d$.

are driven by the vortex, and the temperature field is developed accompanied with the rolling-up process (Fig. 4.8A, and Fig. 4.9A). With continuing formation of new small vortices by the K-H instability, the new vortices are changing into secondary vortices of the primary vortex formed in advance, and moving around it ($t = 5$, Fig. 4.7B). Meanwhile, the temperature fields are developed simultaneously. As illustrated in Fig. 4.8B and Fig. 4.9C, thermal fluids are dominantly driven by the developed primary vortex, as well as the ongoing developing new vortices. Moreover, with greater Pr (Fig. 4.9B), the effect of heat conduction is negligible. As a result, the thermal fluids are almost concentrated within the vortex cores, whereas the region between adjacent hot vortices are still cool (Fig. 4.9B). On the contrary, with $Pr \ll 1$, the effect of thermal conduction is not negligible. Then, the temperature in those intermediate regions between secondary vortices is increased due to thermal conduction from thermal secondary vortices (Fig. 4.8B).

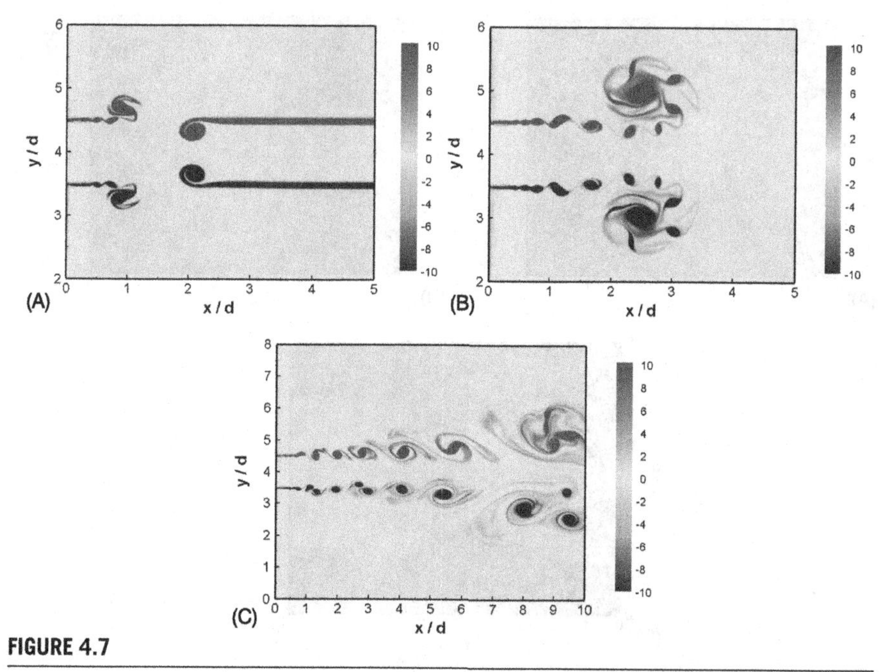

FIGURE 4.7

The snapshots of vortex at dimensionless time $t = 1$ (A), 3 (B) and 9 (C), respectively.

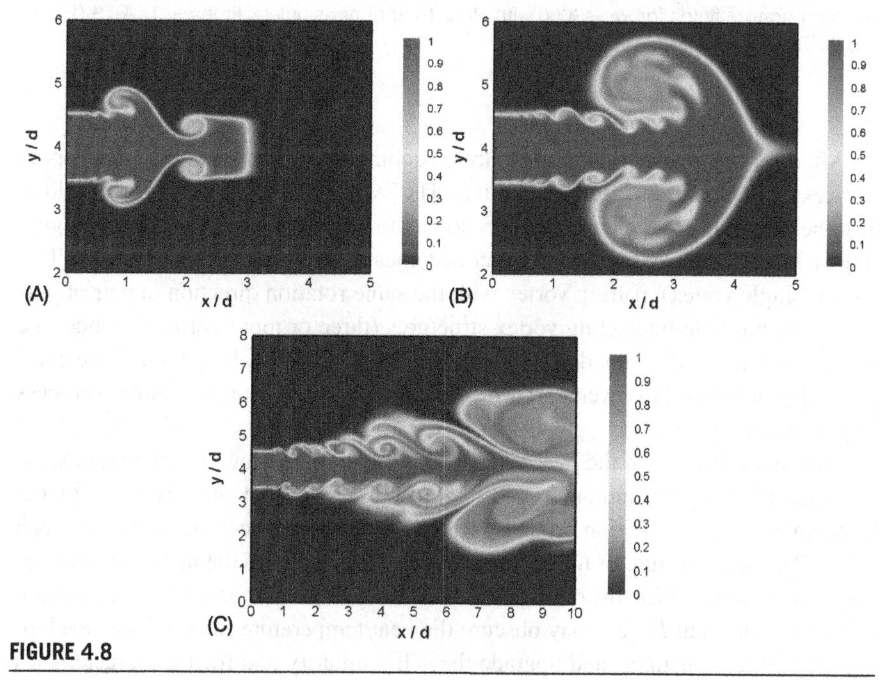

FIGURE 4.8

The temperature fields for $Re = 3000$ with $Pr = 0.1$ at dimensionless time $t = 1$ (A), 3 (B) and 9 (C), respectively.

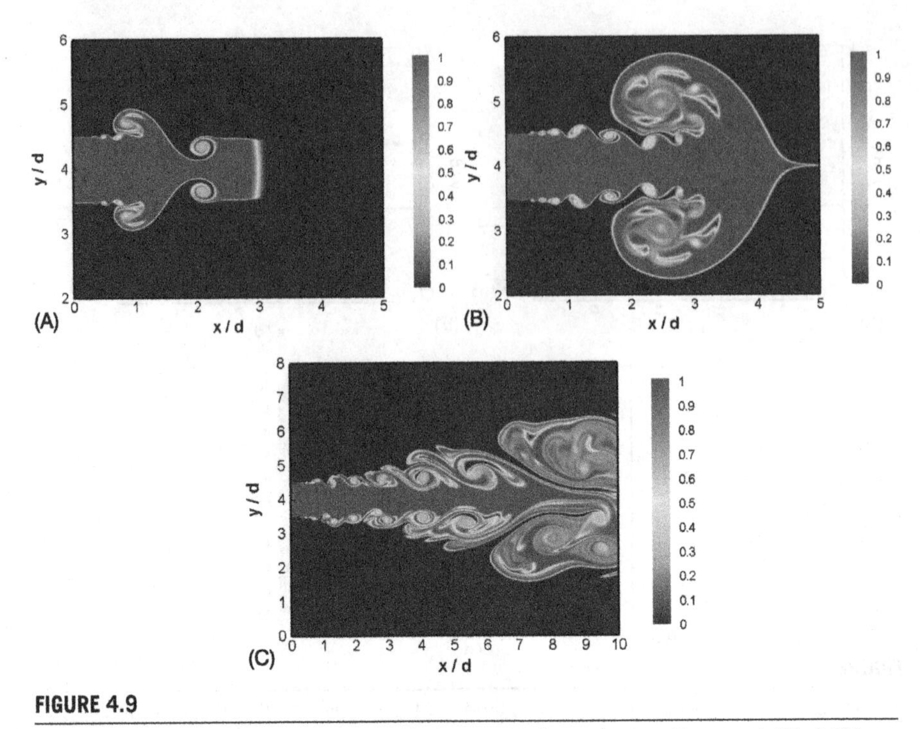

FIGURE 4.9

The temperature fields for $Re = 3000$ with $Pr = 10$ at dimensionless time $t = 1$ (A), 3 (B) and 9 (C), respectively.

After that, in Fig. 4.7, the vortex street is composed of continuously developing vortices of opposite direction of rotation. The symmetric feature of vortex field is disturbed, and an asymmetric mode becomes to appear in the downstream region (Fig. 4.7C). Therefore, the vortex street becomes complicated, composed of rolling vortex (single vortex), pairing vortex with the same rotation direction (a pair of vortices), and multiple-interacting vortex structures (three or more vortices). Under the asymmetric mode of vortex development, the temperature fields become more complicated than before. However it is strongly dominated by the vortex evolution process (Fig. 4.8C and Fig. 4.9C).

The characteristics of the heat/temperature distribution around each vortex seem to be fractal and self-similar. In other words, it seems that the structures of the heat/temperature distribution around different scales of vortices are similar to each other. The finer structure of the secondary vortex is similar to the larger structure of the primary vortex. Moreover, it is inferred that the important process of the thermal conduction in small Pr jets may obscure the heat/temperature interface by developing much finer structures, and upgrade the self-similarity and fractal characteristics of the temperature interface.

4.2.1.2 Fractal characteristics of the heat transfer interface

For validation of the above inference, it is needed to show the fractal feature of the heat transfer interface. In general, to measure the length $L(s)$ of a fractal interface by scale s one should follow the expression

$$L(s) \propto s^{-\beta} \tag{4.10}$$

where β is the fractal dimension of the interface, i.e. $\beta = -\frac{d\log(L)}{d\log(s)}$, and the Napierian base for the logarithm operation is used. Based on Eq. (4.10), the fractal dimension of temperature fields is calculated by using the box-counting method. The method consists of two procedures described below:

- Given a set of small scales s (from about δ to 10δ, where δ is the finest mesh scale), count the number of box elements $L(s)$ which cover and only cover the heat transfer interface. In the region occupied by hot fluids, it has $T^\star = 1$, and in the region with ambient cool fluids, it has $T^\star = 0$. Then, the interface of the heat transfer is identified by $0 < T^\star < 1$. To avoid numerical noises, an improved criteria $\epsilon < T^\star < 1 - \epsilon$ is used, where $\epsilon = 0.05$ is a small value. It means the heat transfer interface is characterized by large temperature gradients.
- Assuming $L(s) = Cs^{-\beta}$, where C is a coefficient, or equivalently, $\log(L) = \log C - \beta \log s = \alpha - \beta \log s$, the values of α and β can be obtained by the linear regression technique. The magnitude of the slope coefficient of the regression lines is the fractal dimension of the heat transfer interface, i.e. $\beta = -\frac{d\log(L)}{d\log(s)}$, and the intercept α indicates the coefficient $C = e^a$.

For example, Fig. 4.10A shows the relationship between $\log(L_T)$ and $\log(s)$ for different Pr flows at $t = 3$, 9 and 15 respectively. In Fig. 4.10A, it is found that: (1) A perfect linearity between $\log(L_T)$ and $\log(s)$ exists, no matter what Pr number is, indicating the evident fractal characteristics of the heat transfer interface. (2) At different time, the linear regression lines have different intercepts on the $\log(L_T)$-axis, indicating the temporal variation of coefficient C. (3) The regression lines for different Pr numbers almost coincide with each other, and vary almost simultaneously in time. It means that the fractal dimensions of the heat transfer interface for different Pr number flows are almost the same. Keep in mind that different Pr flows have the same evolution process of vortex, the simultaneous variation of regression lines indicates the dominant role of vortex evolution on the heat transfer.

Moreover, Fig. 4.10B shows the variation of intercepts α for different Pr flows. It is found that they increase in time almost simultaneously, too. The only reason for the simultaneous increase is the simultaneously evolving process of vortex. The temporal increase of intercept shows that the heat transfer interface is increased as vortex evolves.

More importantly, a more detailed calculation shows the differences of fractal dimension between different Pr number flows, i.e. the effect of the thermal conduction on heat transfer interfaces. Fig. 4.10C shows the variation of fractal dimensions for different Pr number flows. One can see that the fractal dimension fluctuates with

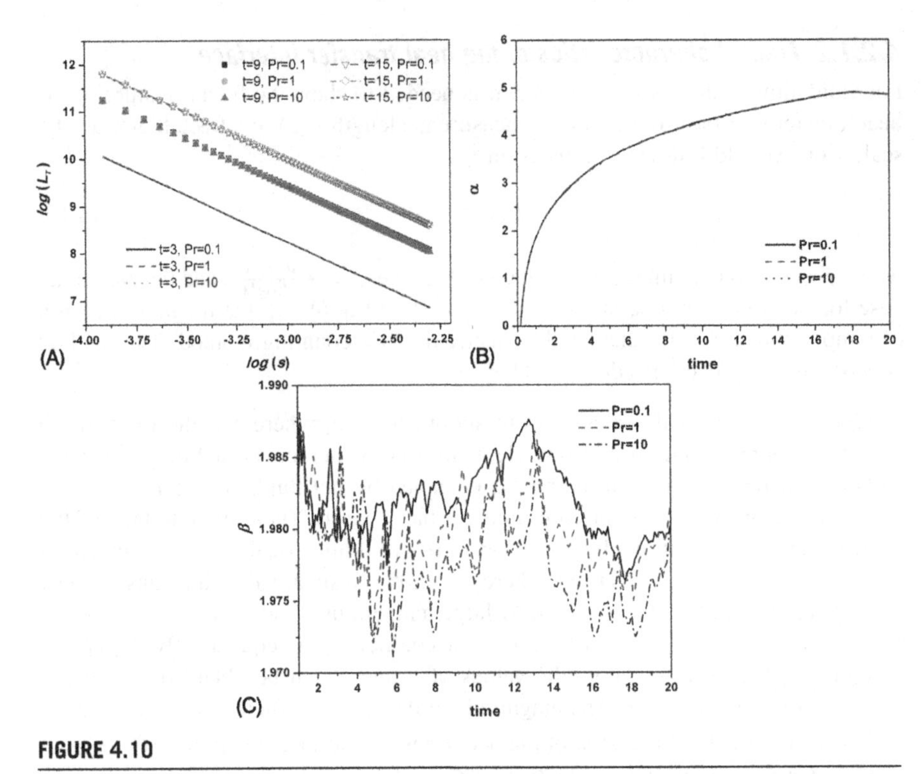

FIGURE 4.10

(A) The relationship between the heat transfer interface $L_T(s)$ and the measure scale s for different Pr flows at $t = 3$, 9 and 15 respectively. (B) The time variation of intercepts α for different Pr flows. (C) The time variation of the fractal dimension β for different Pr flows.

time in a small range (between $1.97 \sim 1.99$). In this way, the effects of the thermal conduction are indicated, i.e. lower Pr number flows have relatively larger fractal dimensions and larger Pr flows have relatively smaller fractal dimensions. It means that the effect of the thermal conduction can change, although not very evidently, the fractal dimension of the heat transfer interface. In low Pr flows, it has $\nu < a$, which indicates that the thermal diffusion is greater than the momentum diffusion, and subsequently the 'thermal mixing layer' is thicker than the 'momentum mixing layer'. Under this condition, the development of the heat transfer interface should be rapider than the momentum interface. The comparison of snapshots of temperature fields in Figs. 4.7–4.9 confirms the conclusion that the rapider development of the thermal diffusion can develop finer structures of the heat transfer interface and upgrade the fractal characteristics.

4.2.1.3 Effect of the inflow momentum thickness θ

The inflow momentum thickness θ can affect the jet evolution, and subsequently, affect the heat transfer interface. Thus, the former conclusions should be confirmed under different momentum thicknesses for validation.

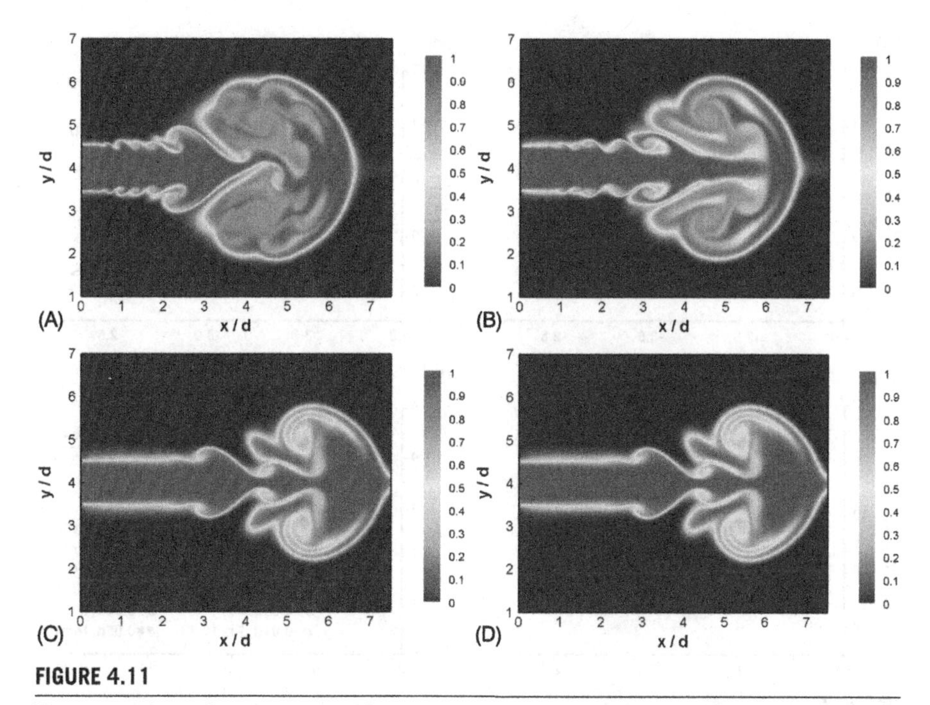

FIGURE 4.11

The comparison of temperature with $Pr = 0.1$ and $t = 5$ for different inflow velocity profiles: The top-hat profile and the hyperbolic tangent profile (A). The hyperbolic tangent profiles under various inflow momentum thicknesses – $\theta = 0.01d$ (B), $\theta = 0.03d$ (C), and $\theta = 0.05d$ (D), respectively.

For this reason, Fig. 4.11 illustrates the comparison of temperature fields for $Pr = 0.1$ between the case using the top-hat profile and hyperbolic tangent profile of inflow velocity (Fig. 4.11A) and the cases using the hyperbolic tangent profiles under various inflow momentum thicknesses θ, namely, $\theta = 0.01d$ (Fig. 4.11B), $\theta = 0.03d$ (Fig. 4.11C) and $\theta = 0.05$ (Fig. 4.11D). It is seen that the K-H instability is not as strong as before, and the jet evolution is not as sensitive to the K-H instability as before, too. The jet evolves from symmetric mode to asymmetric mode more slowly with a thicker inflow boundary. Thus, the core region of the jet becomes lengthened with a large θ. As a result, the small structures of the temperature interface could become reduced as θ increases.

Then, Fig. 4.12A–C shows the fractal heat transfer interface of the jets using the hyperbolic tangent profile of inflow velocity under various inflow momentum thicknesses θ. Referring to Fig. 4.10A, it is confirmed that $\log(L_T)$ is linearly proportional to $\log(s)$ for different θ and Pr numbers. More importantly, for all the cases, the regression lines for different Pr are no longer in coincidence. It means the intercepts α for different Pt flows are not the same. To say specifically, it is much clearer than in the former case that a lower Pr number jet indicates a relatively larger intercept α, or a longer heat transfer interface, and vice versa. In addition, Fig. 4.12D shows

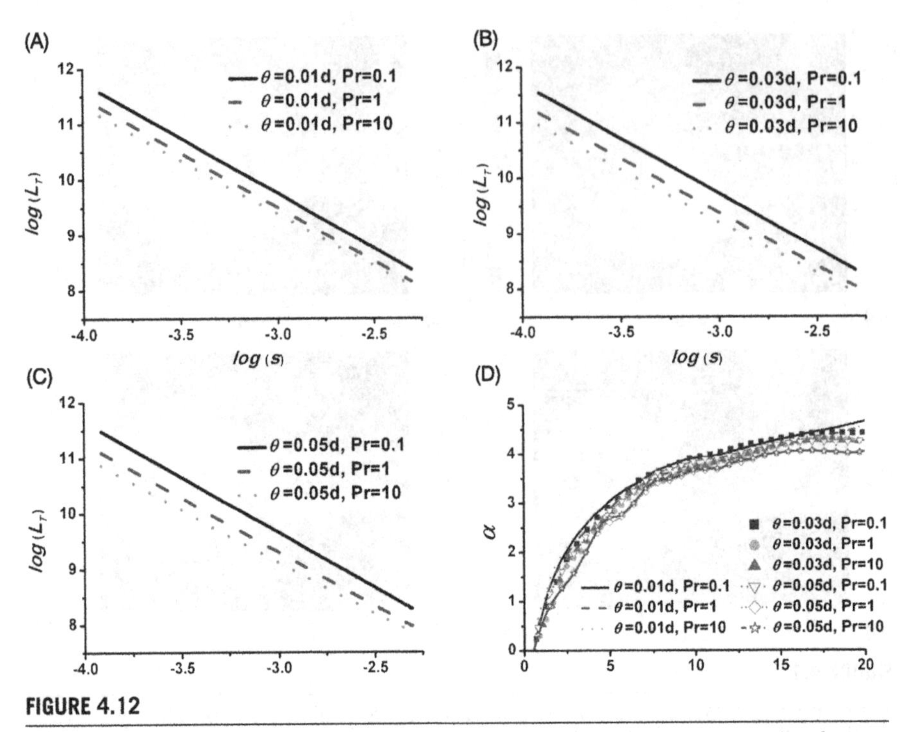

FIGURE 4.12

The relationship between the heat transfer interface $L_T(s)$ and the measure scale s for different Pr flows at $t=9$ with: (A) $\theta=0.01d$, (B) $\theta=0.03d$, (C) $\theta=0.05d$, respectively. (D) The time variation of intercept α for different Pr and θ.

the temporal variation of intercepts α. In general, this confirms again that a larger Pr number indicates a relatively smaller intercept and subsequently a reduced heat transfer interface all the time, and vice versa.

In addition, Fig. 4.13 presents comparison of temporal variations of the fractal dimension of the heat transfer interface for different Pr flows under different inflow momentum thicknesses θ. Under these conditions, the discrepancy of fractal dimensions of the heat transfer interface for different Pr flows becomes much clearer than before (Fig. 4.10C). Thus, it is concluded that not only the intercepts do not coincide, but also the fractal dimensions become significantly different from each other. More specifically, a lower Pr flow has a much larger fractal dimension of the heat transfer interface, indicating the great role of the thermal conduction in developing finer structures of the heat transfer interface, and vice versa. Comparing Fig. 4.13 with Fig. 4.12, one can see that the change in the fractal dimension β (a power law) of the heat transfer interface is more significant than that in the coefficient C (a linear relation).

Moreover, it seems that the larger the inflow momentum thickness is, the greater the discrepancies between the fractal dimensions of different Pr flows are. It is reasonable. Remember that the heat transfer interface originates from the shearing layer

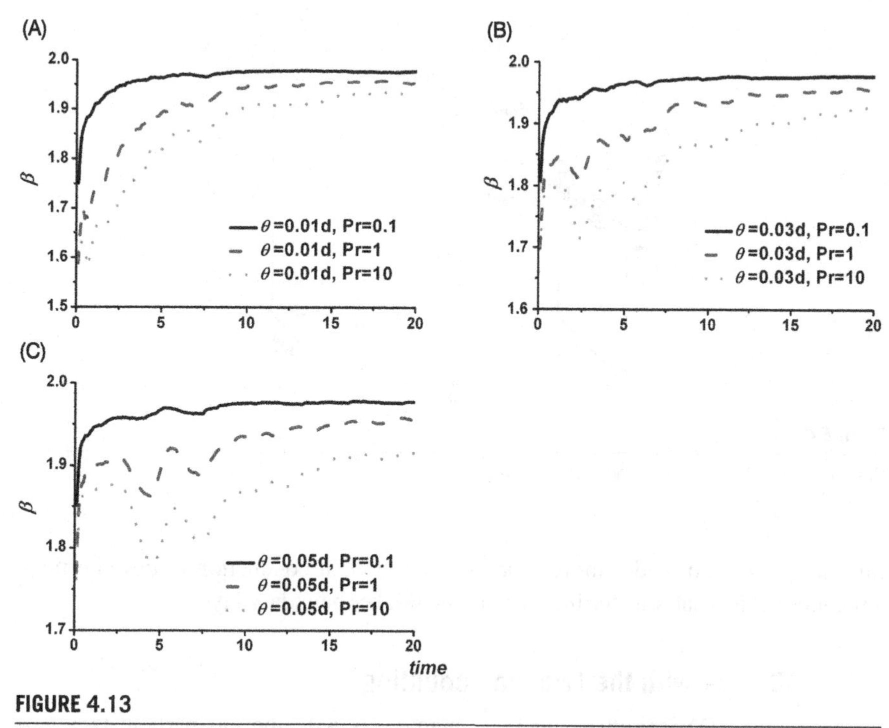

FIGURE 4.13

The time variation of the fractal dimension β for different Pr and θ.

of the jet. A larger inflow momentum thickness would indicate a thicker inflow boundary, as well as a thicker shear layer. Thus, it has more space for the thermal conduction to play its important role in developing finer structures of the heat transfer in the thicker shear layer than that in the thin layer. This is the reason for that the fractal dimension β and the coefficient C in the top-hat-profiled jet flows are almost equal to each other in different Pt flows.

To say conclusively, the heat transfer interfaces are fractal which is mainly determined by the process of vortex evolution. With the top-hat inflow profile, the coefficients of the heat transfer interface coincide with each other perfectly for different Pt flows and the fractal dimensions of the interface fluctuate with very small discrepancies. It shows the dominant role of vortex evolution on the heat transfer characteristics. However, the effect of the thermal conduction cannot be neglected for lower Pr flows, since it develops the finer structure of the heat transfer interface, and subsequently increases the fractal dimension of it.

With the hyperbolic tangent profiles, the heat transfer interfaces are still fractal. But the temporal variations of the coefficient and the fractal dimension of the heat transfer interface no longer agree with each other for different Pr flows. Alternatively, a lower Pr number indicates a larger coefficient and a larger fractal dimension, and vice versa. This trend seems to be enhanced by the increase of the inflow boundary

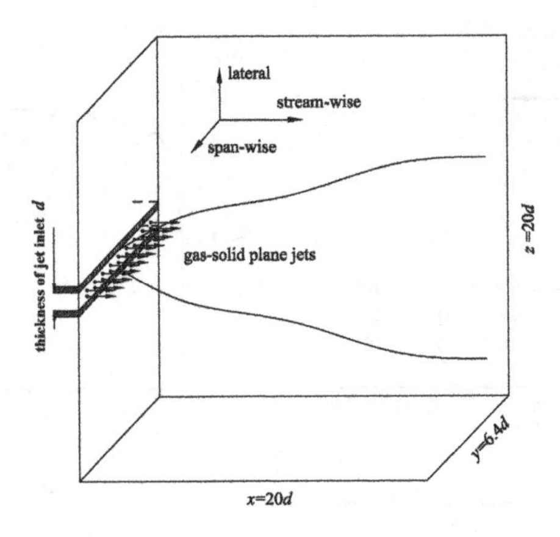

FIGURE 4.14

Sketch of the three-dimensional planar jet.

thickness since it provides more space for the thermal conduction to develop finer structures of the heat transfer interface in the thicker boundary layer.

4.2.2 3D case with the two-way coupling

In this section, DNS simulation of 3D planar jets with the gas-particle two-way coupling is presented. The finite volume method and the fractional-step projection technique are used [284]. The scales of the flow domain are $20d \times 6.4d \times 20d$ (d is the thickness of jet inlet, Fig. 4.14) in the streamwise (x), spanwise (y) and lateral (z) directions, respectively. In the x and y directions, the uniform staggered grids are used. In the lateral direction, the uniform staggered grids are set in the region $-2.25d < y < 2.25d$, and the grids are stretched outside. A total number of $300 \times 64 \times 256$ grid points, in the x, y and z directions respectively, are used. The central differences scheme is used for the spatial discretization to ensure the second-order precision of solution in space, and a low-storage, third-order Runge–Kutta scheme is applied for time integration [285]. The Poisson equation for pressure is solved using a direct fast elliptic method [233]. The particle motion equation is solved considering the leading role of the particle drag only.

The two-way coupling method uses the point-force approximation to describe the momentum coupling between fluid and particle. Based on the conservation of momentum, the integral particle-to-fluid effects in any control volume is the sum of forces from particles to fluid. Assuming there are N_p particles within the control volume, the integral particle-fluid force $\boldsymbol{F}_b$ can be formulated as

$$\boldsymbol{F}_b = -\sum_{i=1}^{N_p} \frac{m_p \boldsymbol{f}_D}{m_g St}(\boldsymbol{u}_g - \boldsymbol{v}_p) \tag{4.11}$$

Table 4.5 Parameters used in this section.

Scales of the flow domain (mm)	20 × 6.4 × 20
Thickness of the plane jet inlet, d (mm)	1
Number of the grids	300 × 64 × 256
Inflow velocity (m/s)	10
Reynolds number, Re	3000
Particle mass loadings, $\dot{m}$ (kg/kg),	0.1; 0.3; 1
Particle Stokes number, St	0.5; 1; 5;
Particle diameter, μm	1.2; 1.7; 3.9
Particle density, ρ_p (kg/kg^3)	2000
Fluid density, ρ_g (kg/kg^3)	1
Fluid viscosity, μ Pa·s	1.67×10^{-5}
Simulation time step, Δt (μs)	0.4
Number of simulation step	2×10^4

where m_g and m_p are the mass of gas and particle, respectively. In addition, the particle-fluid force in each control volume is interpolated to the Eulerian grid points of fluid via the linear interpolation scheme.

The particles are issued into the flow from the plane jet inlet (Fig. 4.14). The Reynolds number, based on the inflow mean velocity of fluid U_0 and the thickness of the jet inlet d, is $Re = U_0 \times d/\nu = 3000$. Initially, the inflow velocity profile of the jet follows the equation

$$u = \frac{U_0}{2}(1 + \tanh\frac{z}{2\theta_0}), v = w = 0 \quad (4.12)$$

where $\theta_0 = 0.05d$ is the initial momentum thickness of the jet. The Neumann condition is applied at the outlet boundary in the streamwise and lateral directions. The periodic boundary condition is used in the spanwise direction. The time step is $\Delta t = 0.4$ μs, and 20000 time steps are simulated. The parameters are listed in Table 4.5.

The mass loading is defined as the ratio of mass fluxes of particle to fluid at the inlet of the jet. Three different values of mass loadings, i.e. $Z_m = 0.1, 0.3, 1$, are simulated with a fixed Stokes number. Moreover, the results of the one-way coupling and two-way coupling are compared. Additionally, due to the dispersed phase is dilute, the interparticle collision is neglected.

4.2.2.1 3D fluid vortex structure

Fig. 4.15 shows the fluid vortex structures for different mass loadings at $t = 0.256$ ms. It is observed that the particle motion changes the fluid vortex structure evidently even with a relatively low mass loading ($Z_m = 0.1$). To say specifically, for the one-way coupling (Fig. 4.15A), there exists clearly a pair of large scale symmetrical spanwise vortices at the onset of the jet. However, for the two-way coupling, the pair of the large spanwise vortices is destroyed (Fig. 4.15B–D) by the laden particles.

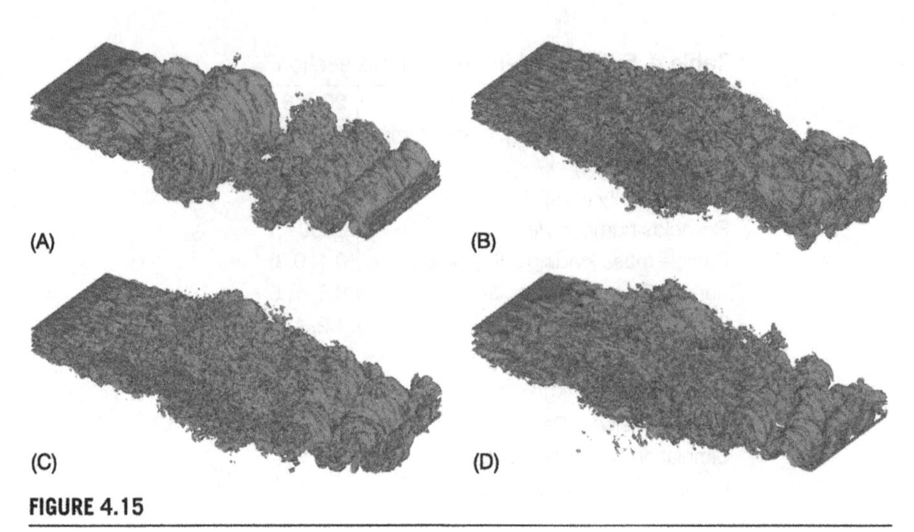

FIGURE 4.15

Fluid vortex structure under the effect of the two-way coupling with different mass loadings at $t = 25.6$: (A) one-way coupling, (B) two-way coupling with $Z_m = 0.1$, (C) $Z_m = 0.3$, (D) $Z_m = 1.0$.

Moreover, as mass loading increases, the large vortices are increasingly broken, with their scales decreased and the distribution of them becoming stochastic. In addition, the downstream vortex evolution becomes more rapid than before. In other words, the axial lengths of the initially undisturbed jet core in the two-way coupling cases are shorter than those in the one-way case. In two-way cases especially when the loadings are heavy, the lengths of the undeveloped initial segments are so short that they disappear almost immediately after the jets are issued into the flow domain.

Not only the pair of large scales spanwise vortices but also the small vortices in the flow are greatly changed under the effect of the particle feedback. This observation is in accordance with the results of a previous study [286]. The interaction between the laden particles and the large vortices are strong, which accelerates the transition from large coherent structures to small vortices. Moreover, the addition of particles attenuates the effect of the spanwise large scale vortices for the jet evolution, leading to a nearly homogeneous distribution of small vortices.

4.2.2.2 Fluid velocity profiles

Fig. 4.16 illustrates the profiles of mean axial velocities at $x = 4d$ and $x = 12d$ under the two-way coupling effect. It is found that the influences on the mean streamwise velocity by the laden particles are different at different axial locations. For $x = 4d$, the laden particles slow down the streamwise velocity in the center of the jet for all mass loadings, whereas they do few influences on the rest lateral regions. In contrast, for $x = 12d$, the laden particles augment the mean streamwise velocity in the center of the jet, whereas attenuate the velocity in the lateral regions. Moreover, the degree of modification increases with the mass loading.

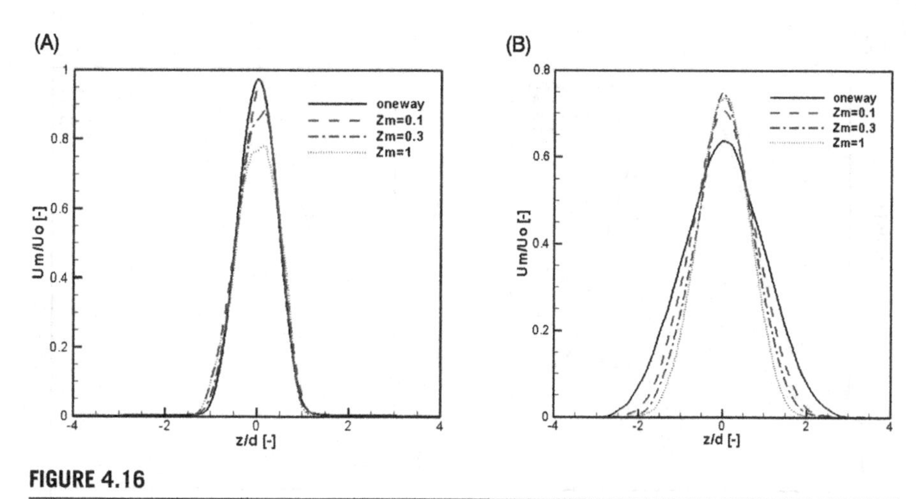

FIGURE 4.16

Effect of the two-way coupling on the mean velocity profiles of fluid for different mass loadings. (A) U_m at $x = 4d$; (B) U_m at $x = 12d$.

The different modifications of velocity profiles in axial locations are due to the transfer of momentum between the gas and solid phase. In the upstream region ($x = 4d$), particles are concentrated near the axis region with its velocity less than fluid, and the momentum is transferred from fluid to particle. In contrast, in the downstream region, the fluid flow is decayed more rapidly than particle because of the lower rigidity of fluid movement than that of particle. For this reason, the fluid velocity is augmented in the center of the jet ($x = 12d$). As a result, the fluid velocities in the other regions are attenuated.

Fig. 4.17 shows the *r.m.s* of fluctuation velocity profiles at $x = 12d$ with different mass loadings. It is found that the degree of modification of *r.m.s* velocity profiles by the two-way coupling increases with mass loadings, and it depends on the directions of fluctuation. The laden particle augments the *r.m.s* streamwise velocity in the jet center and attenuates it in the other regions (Fig. 4.17A). However, the *r.m.s* values are attenuated by the laden particles in all regions for all mass loadings in the lateral and spanwise directions (Fig. 4.17B and C). These results indicate the suppression of fluid fluctuation by laden particles in almost all the regions.

It is noticed that the particles are all very small (for which the point-force approximation is reasonable), and the particle diameter d_p is far smaller than the integral length scale of turbulence l_e. Thus, it is appropriate to find that the above results of Fig. 4.17 are in accordance with the well-known qualitative conclusions [33].

Since the fluid fluctuation in the spanwise direction is periodic, it just shows here the correlation of fluctuations between the streamwise and lateral directions. It is clearly illustrated by Fig. 4.17D that the laden particles suppress the Reynolds stress tensor in the whole region whatever the mass loadings are, which is in accordance with the experimental results obtained by [287]. More specifically, the degree of attenuation of the Reynolds stress depends on the magnitude of mass loading pro-

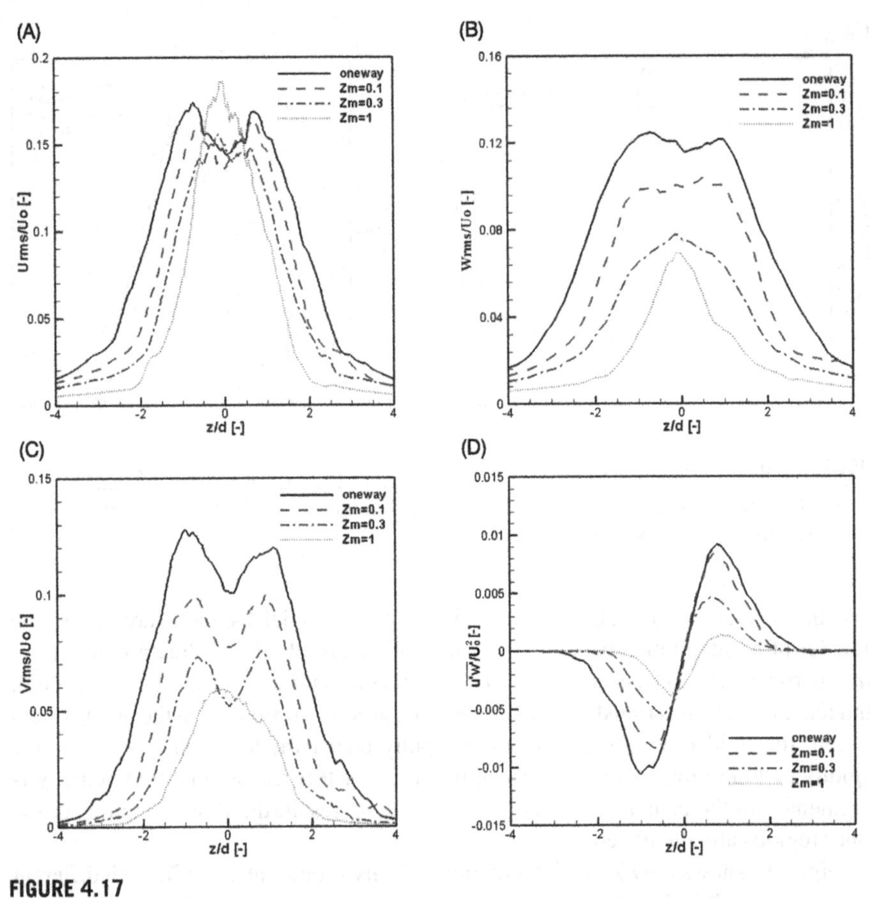

FIGURE 4.17

Effect of the two-way coupling on the mean velocity profiles of fluid for different mass loadings: (A) Streamwise *r.m.s* velocity. (B) Lateral *r.m.s* velocity. (C) Spanwise *r.m.s* velocity, and (D) the Reynolds stress $\frac{\overline{u'w'}}{U_0^2}$ at $x/d = 12$ under different mass loadings.

portionally. A heavy mass loading can cause a high suppression of the Reynolds stress.

4.3 Swirling jets

4.3.1 Vortex breakdown

The simulated dimensions of the swirling flows are $30d \times 10d \times 10d$ in streamwise, lateral and spanwise directions, respectively (Fig. 4.18A). For the spatial discretization, $384 \times 128 \times 128$ structured grids are used, which could resolve the scales of turbulence as fine as about $0.075d$. The dimensionless variables and their corresponding real scales are used (Table 4.6).

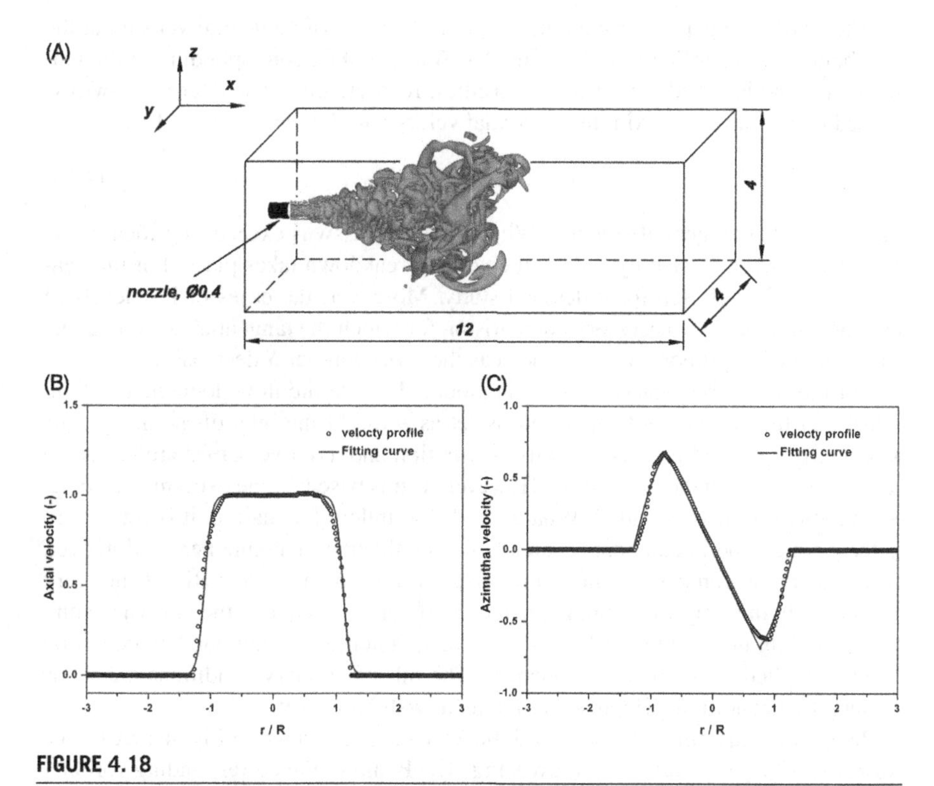

FIGURE 4.18

Sketch of the simulation scales (A) with a view of the simulated swirling jets; azimuthal (B) and axial (C) velocity profiles for $S = 1.42$. The origin of the coordinate is the center of the jet inlet, r is the distance from the origin in the radial direction, and $R = d/2$ indicates the radius of the jet inlet.

Table 4.6 Typical real values of variables used in the numerical simulation.

Scales of the flow domain (mm)	12 × 4 × 4
Scale of jet diameter d (mm)	0.4
Grids number $N_x \times N_y \times N_z$	384 × 128 × 128
Mean inlet axial velocity U_0 (m/s)	1.52
Fluid density ρ_f (kg·m^{-3})	1000
Fluid viscosity (Pa·s)	1.005×10^{-3}
Reynolds number Re	606
Swirl number S	1.42
Time step Δt (μs)	2.5
Simulation time T_s (ms)	50

The swirl is imparted through the initial distribution of tangential velocity at the inlet boundary according to [288] (Fig. 4.18B and 4.18C, corresponding to the tangential (azimuthal) and axial velocity profiles, respectively). The degree of swirl is defined by the ratio of maximum tangential velocity to the mean axial velocity:

$$S = 2 \cdot \max\{U_t\}/U_0 \tag{4.13}$$

where U_t is the tangential velocity. When the level of swirl exceeds a critical value (e.g., $S = 1.42$ suggested by [288]), the vortex breakdown takes place. For this reason, $S = 1.42$ is chosen for a detailed study. Moreover, the cases of low levels of swirl are simulated for necessary comparison, for which the tangential velocity at the inlet boundary is proportionally reduced as the swirl number S decreases.

For the inflow boundary, a nozzle is mounted inside the flow domain. The fluid velocity at the boundary of the nozzle is set as zero. At the inlet of the nozzle, the particularly specified profiles of swirl distribution and axial velocities are set as the velocity inflow boundary condition. However, as it is based on the experimental data, no analytical form is provided. What regards the inflow fluctuation, it is set to zero as there is no experimental data available for the fluctuation components. Moreover, as the Reynolds number is kept at a relatively low value, the fluctuation is not very important. In this way, the evolution behavior of vortices is due to the intrinsic nature of the swirling flow. At the end the flow domain, a nonreflecting boundary condition [279] is applied as the outflow boundary. The other boundary conditions are set as the Dirichlet conditions which act as a nonslip wall boundary.

Jeong and Hussain [289] proposed the λ_2 vortex, as shown in Fig. 4.19A by the isosurface of $\lambda_2 = 1$ vortex. Moreover, Fig. 4.19B shows the corresponding vorticity $|\omega| = |\nabla \times \boldsymbol{u}|$ viewed in the lateral and spanwise directions (Fig. 4.18A), respectively.

As the recirculation zone is characterized by the stagnant or recirculating fluid, a region with zero or nearly zero vorticities enclosed by large vorticities in the central region of the jet is considered as the recirculation zone. As shown by Fig. 4.18A and B, an expansion of vortex is formed near the nozzle exit, establishing a clear recirculation zone with nearly zero vorticities. It is the so-called 'bubble' type vortex breakdown. Before the bubble, the jet is rotated but greatly confined by the nozzle, forming possible small vortices on the wall of the nozzle. After the bubble, a sharp contraction of the vortex follows. The vortices after the bubble are helical, which are initiated from the enclosure of the bubble, and evolve downstream regularly. It is identified as the helical type of vortex breakdown. The behavior of the helical vortex breakdown is mainly characterized by the downstream movement of helical waves. It will be demonstrated in the following section that the bubble and helical types of vortex breakdown are significant spatio-temporal behaviors. For example, it will be show later that they are related to the temporal coherent oscillation effects of scalars.

For quantitative comparison between simulation and experiments, Fig. 4.20 illustrates the mean azimuthal and axial velocities of the simulated results. Due to the axisymmetry, the top half for axial velocity and the bottom half for azimuthal velocity are illustrated together. The dashed lines indicate the minus values of velocities. Referring to the experimental results (Fig. 4.20B from Fig. 7 of Ref. [288]), the

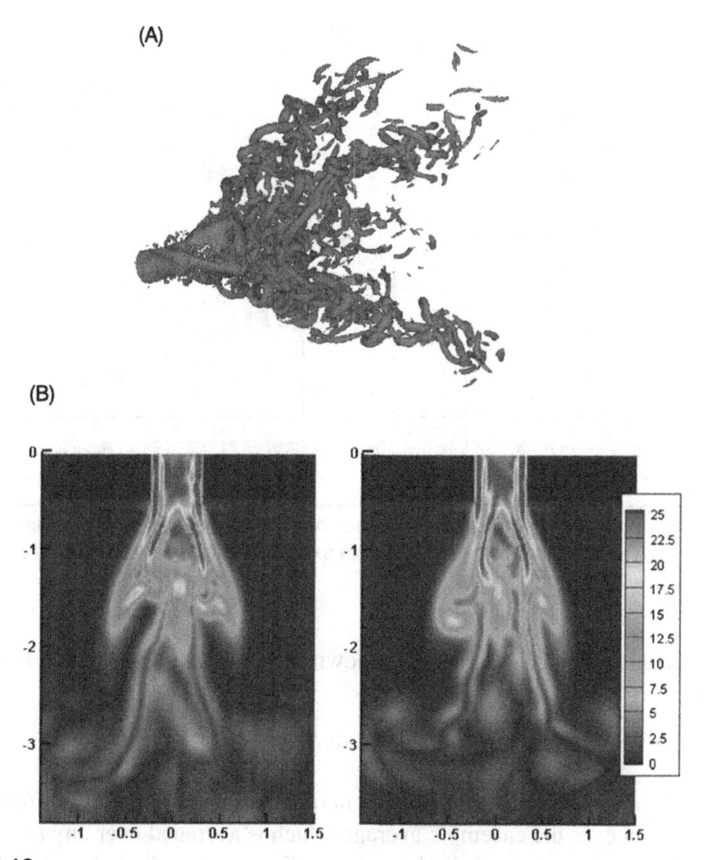

FIGURE 4.19

Bubble vortex breakdown: (A) the three-dimensional isosurface of λ_2 vortex; (B) the cross-sectional visualization of the corresponding vorticity on the median planes through the jet axis.

present simulation is successful in capturing the basic characteristics of the swirling jets, although some differences still exist in the recirculation zone. Thus, the results of the present simulation are mainly reliable. Moreover, Fig. 4.20A shows the location of the sampling point for the following analysis. It is within the center of the recirculation zone, and it indicates the center of the bubble vortex breakdown.

4.3.2 Coherent oscillation

4.3.2.1 Auto-correlation

To quantify the effects of coherent structures, the temporal correlation of any scalar, such as velocities u, v, w, pressure p, and vortices λ_2, on a sampling point at the jet axis with a distance of about $x = 2.5d$ from the exit of the jet are computed. The sampling point is in the right center of the recirculation zone (Fig. 4.20A), which is

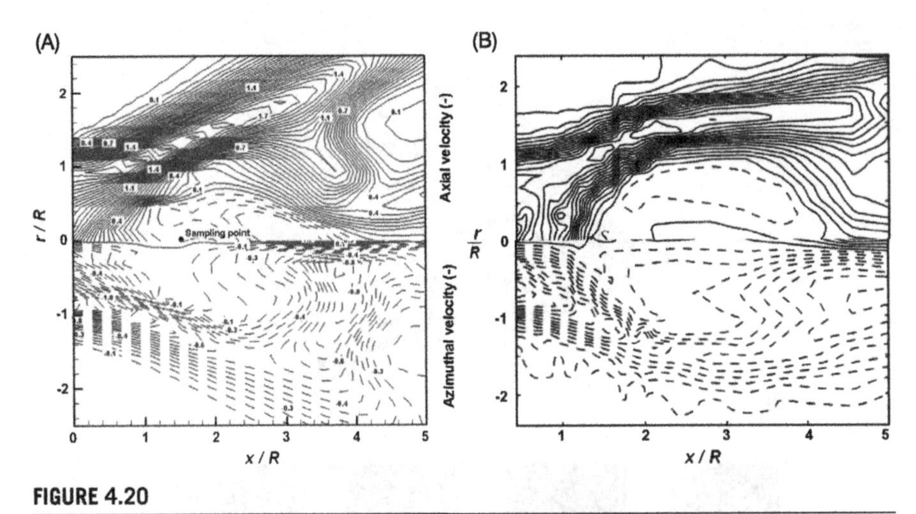

FIGURE 4.20

(A) The axial and azimuthal velocities over the vortex breakdown region and location of the sampling point. (B) The experimental results on axial and azimuthal velocities from Fig. 7 of [288].

closely related to the bubble vortex breakdown. The time auto-correlation function is defined as

$$R_{\phi\phi}(t,t') = \langle \phi(t)\phi(t') \rangle \tag{4.14}$$

where ϕ indicates the fluctuating component of any scalar. t and t' are different time points, and $\langle \cdot \rangle$ means the ensemble average which is averaged over any pair of time points (t, t') based on the ergodic hypothesis. For numerical implementation, it is calculated as follows:

$$R_{\phi\phi}(t,t') = R_{\phi\phi}(\tau)\big|_{(t'-t=\tau=n\Delta t)} = \frac{1}{N_T - n}\sum_{i=1}^{N_T-n} \phi(t_i)\phi(t'_j)\big|_{j-i=n} \tag{4.15}$$

where N_T is the total number of time steps and Δt is the time step. The auto-correlation function is to quantify the relationship of scalars at different time. It is more usual to use the auto-correlation coefficient $\rho_{\phi\phi}(t,t')$ to characterize the relationship of fluctuation of scalars through normalization of the correlation function:

$$\rho_{\phi\phi}(t,t') = \frac{R_{\phi\phi}(t,t')}{\langle \phi(t)\phi(t) \rangle^{1/2} \langle \phi(t')\phi(t') \rangle^{1/2}} \tag{4.16}$$

Firstly, to show the characteristics of correlation of flow quantities of vortex breakdown, it is needed to compare the results with and without the vortex breakdown. Thus, $S = 0.33$ (low swirl), $S = 1.35$ (a critical level of swirl for which the bubble vortex breakdown is about to occur) and $S = 1.42$ (a high level of swirl with bubble vortex breakdown) are chosen for a detailed study. Let ϕ be u', v', w', p' and

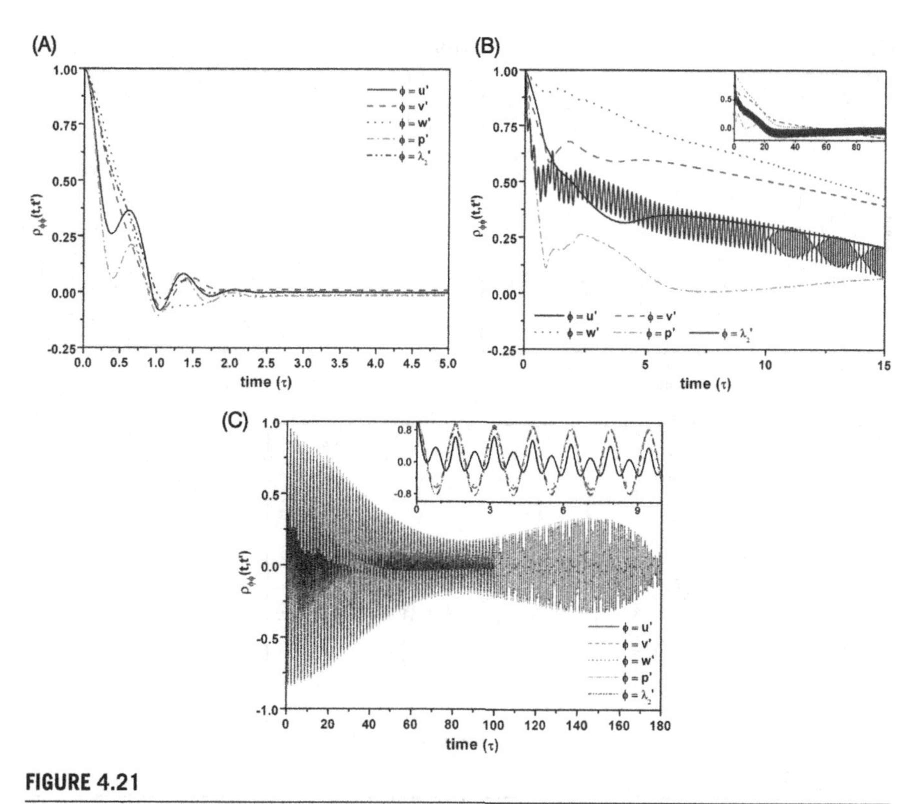

FIGURE 4.21

Auto-correlation functions of the flow scalars for $S = 0.33$ (A), 1.35 (B) and 1.42 (C), respectively at $Re = 606$.

λ_2', respectively, the auto-correlation coefficients under the above levels of swirl are illustrated in Fig. 4.21.

It is observed that for $S = 0.33$ (Fig. 4.21A), the correlation coefficient decreases rapidly down to zero. It means that for low swirling jet there is no long time correlation for the flow quantities. For the critical level of swirl ($S = 1.35$, Fig. 4.21B), it is observed that the correlation coefficient decreases less rapidly than before and a moderately long time correlation exists. From the inset of Fig. 4.21B, it is seen that it is not until $\tau = 20$–50 that the coefficients reach zero. Moreover, for the vortex core λ_2', it is found to be oscillated while decreasing, although the frequency of oscillation for a long time interval is not maintained. For the supercritical level of swirl ($S = 1.42$, Fig. 4.21C), it is founded that the flow quantities, including velocity components, pressure and vortices, are all oscillated during decrease. The correlation of u' oscillates with a frequency of double times as large as the others, although the oscillation amplitudes are not equal. For the rest, u', v', w', p' and λ_2', the correlation functions vary almost in the same frequency. Moreover, it is evident that there exists a common long time correlation of scalars. The amplitudes of the correlation

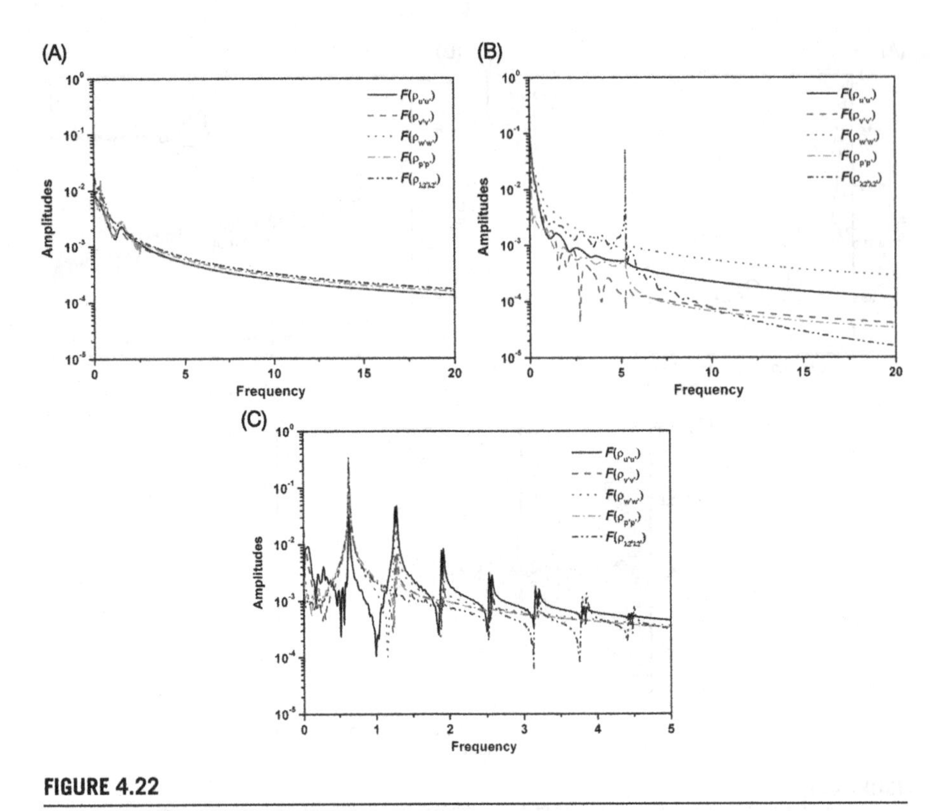

FIGURE 4.22

Fourier transformations of the auto-correlation functions for $S = 0.33$ (A), 1.35 (B) and 1.42 (C) corresponding to Fig. 4.21A–C, respectively.

function decay gradually and linearly first and then increase slightly, dragging a long spindly type of tail as the correlation time s increases further. It is reasonable that the time correlation decreases at first as time increases since the scalar fluctuation is inherently stochastic. However, the persistence of a long time correlation of scalars' variation indicates the effects of coherent motion of regular oscillation.

Recall the definition of coherent motion given by [25] – the correlation of variables over a range of long time larger than the smallest scales of flow is an evidence of coherent oscillating motion. Thus, the regular oscillation and a long time correlation of the scalars are closely related to the evolvement of the coherent structure. For this reason, the correlation functions of scalars u', v', w', and p' are possibly driven by the oscillation of the vortex core λ_2', and consequently by a perfectly simultaneous pace of oscillation. In addition, it is noticed that under the critical state, the vortex core oscillates firstly. It seems to be true that the coherent structure dominates the temporal variation of any scalar in the flow, making most of them correlated in resonance.

In order to specify the oscillation periodicities, Fig. 4.22 shows the Fourier transformations of the correlation functions corresponding to Fig. 4.21, respectively. For

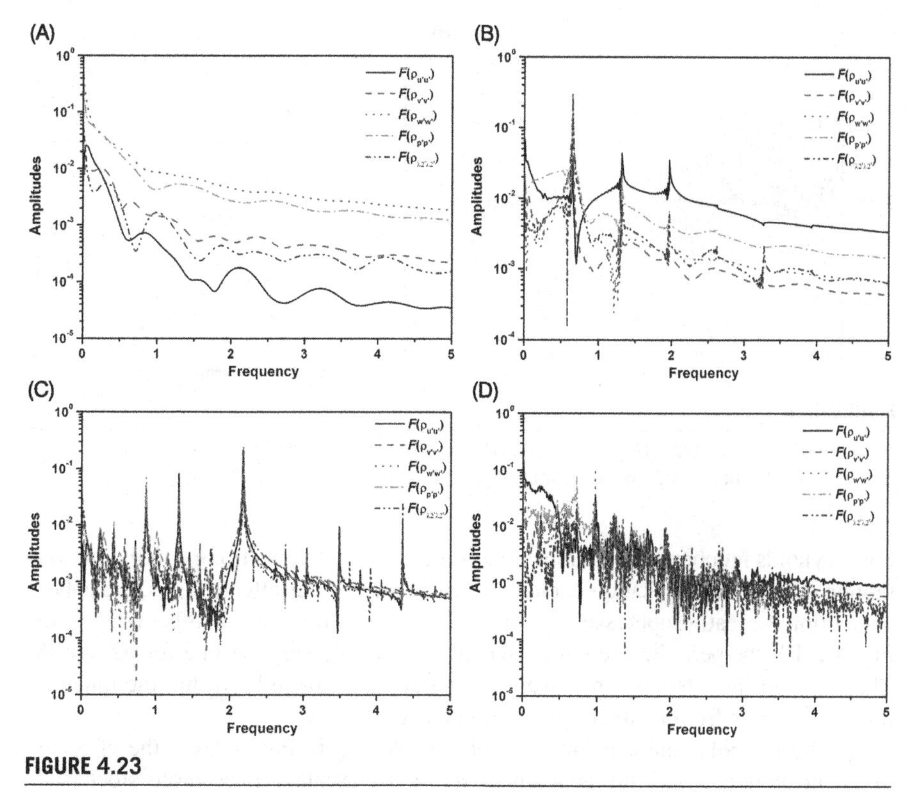

FIGURE 4.23

Fourier transformations of the auto-correlation functions at $S = 1.42$ for $Re = 486$ (A), 546 (B), 726 (C) and 786 (D), respectively.

Fig. 4.22A ($S = 0.33$), there is almost no evident peak of frequency. For Fig. 4.22B ($S = 1.35$), only the vortex core (λ_2') has an evident peak around $f = 5$. However, from Fig. 4.22C ($S = 1.42$), it is clearly observed that all of the correlation functions have a common dimensionless fundamental frequency of $f = 0.634$, which corresponds to the periodicity of $T = 1/f = 1.577$. The evolution of vortices is appropriately assumed to be composed of processing waves with various frequencies and periods. Then T is considered as the period of the fundamental wave with respect to the processing of the helical vortex breakdown (as observed in Fig. 4.19, it shows the snapshot of three-dimensional processing waves of helical vortices). Moreover, the harmonics of oscillation are almost regularly spaced and coincide with each other. It indicates the resonance of flow variables with close relation to the vortex breakdown, and also to the processing waves of helical vortices.

Secondly, it is necessary to study the coherent motion of oscillation under different Reynolds numbers. For this reason, the Fourier transformations of the correlation coefficients are calculated for $S = 1.42$ at $Re = 486$, 546, 726 and 786, respectively, and illustrated in Fig. 4.23 correspondingly. From Fig. 4.23A ($Re = 486$), it is seen that the periodic oscillation is not evident, although it seems to be about to occur. At

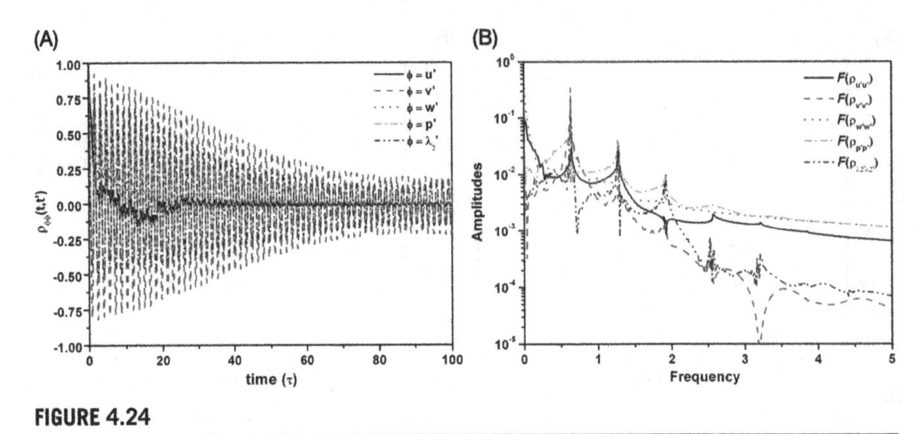

FIGURE 4.24

Auto-correlation functions (A) and Fourier transformations of them (B) for $Re = 606$ and $S = 1.42$ on another sampling point with an offset distance of $2.5d$ in the lateral direction.

low Reynolds numbers, the viscous effects are relatively very strong and the inertial motion of coherent oscillation cannot overcome the viscous effects. Thus, the coherent motion is greatly suppressed. For a relatively larger Reynolds number ($Re = 546$, Fig. 4.23B), the periodic oscillation seems to be very clear, just like for $Re = 606$. Thus, though $Re = 606$ is a relatively low value, it seems to be within the range of transition region from laminar flow to turbulence.

As the Reynolds number increase further ($Re = 726$, Fig. 4.23C), the effect of turbulent motion causes the coherent oscillation of scalars to be more stochastic. As a result, more components of oscillation with different frequencies and locally maximal amplitudes appear, leading the synchronous oscillation of scalars to being disturbed. However, the common frequency for them still exists, and the scalars are still oscillated simultaneously. Finally, for $Re = 786$ (Fig. 4.23D), the Fourier spectrums seem to be composed of a large amount of harmonics with peak amplitudes. In this way, the regular paced coherent motion of oscillation is entirely covered, and consequently, the coherent motion seems to be stochastic.

Lastly, it is necessary to mention that the above depicted coherent motion of scalars in resonance can be detected in many points not exclusively the points at the jet axis. For example, Fig. 4.24 shows the time correlation at another point with an offset distance of $2.5d$ in the lateral direction from the sampling point. It is seen from Fig. 4.24A and the corresponding Fourier spectrum from Fig. 4.24B that the time correlations of the scalars are still in resonance. Thus, the coherent motion of oscillation is an important feature of the region of vortex breakdown in a highly swirling jet, although it only occurs at relatively low level of turbulence transiting from laminar flow to turbulence.

4.3.2.2 Cross-correlation

Remember the aforementioned definition of coherent motion by [25] – the coherent motion is mainly characterized by two aspects: one is that at least one fundamental

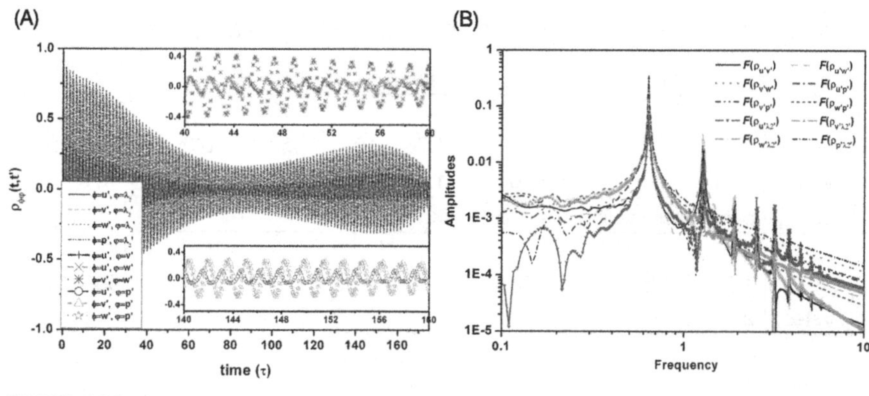

FIGURE 4.25

Cross-correlation functions (A) and corresponding Fourier transformations (B) for $Re = 606$ and $S = 1.42$.

flow variable significantly correlates to itself, and another one is its correlation to another variables. Thus, in order to complete demonstration of the effect of the coherent oscillating motion, it is necessary to study the relationship between different scalars, i.e. to calculate the cross-correlation functions $R_{\phi\varphi}(t,t') = \langle \phi(t)\varphi(t') \rangle$ and the cross-correlation coefficient $\rho_{\phi\varphi}(t,t') = \frac{R_{\phi\varphi}(t,t')}{\langle \phi(t)\phi(t) \rangle^{1/2} \langle \varphi(t')\varphi(t') \rangle^{1/2}}$.

Fig. 4.25A illustrates the cross-correlation coefficients between different scalars. It is found that the cross-correlation functions are oscillated in perfectly regular periodicities, although the amplitudes of oscillation are different from one to the other. As a whole, the cross-correlation functions of all combinations of scalars follow the similar profiles, that is, a mild and approximately linear decrease followed by long and expanded tails. The local details for some cross-correlation functions are shown by the insets of Fig. 4.25A. They indicate the periodic oscillations of the correlation functions and the existence of the consistent fundamental periodicities.

Through the Fourier transformation $F(\rho_{\phi\varphi}(t,t'))$ of the cross-correlation coefficients, it is observed from Fig. 4.25B that all the cross-correlation functions have a common fundamental frequency. Moreover, the fundamental frequency is of the same value as that of the auto-correlation. For the harmonic frequencies, they are also of overlapped values. Thus, it is evident that the variations of scalars are closely related to each other. As is well known, the coherent structure is closely related to the correlation of velocities. Now, based on Fig. 4.25, it seems to be true that the coherent structure is not only closely related to the correlation of velocity and vortices, but also to the cross-correlation of all possible scalars. In other words, all the cross-relationships of scalars are dominated by the evolution of coherent structures, that is, the coherent motion of oscillations. In addition, it is necessary to mention that the coherent motion of oscillation is stable in time (just like a traveling wave). Thus, the coherent motion of oscillation can persist for ever. The convergence at the ending of

Table 4.7 Parameters of the particle phase for case 1 and 2.

Case 1: keep the number flow rates $\dot{C}_p = 10$/step					
St	0.01	0.1	1	10	100
d_p (d)	1.76×10^{-4}	5.56×10^{-4}	1.76×10^{-3}	5.56×10^{-3}	1.76×10^{-2}
m_l (kg/kg)	1.41×10^{-5}	4.45×10^{-4}	1.41×10^{-2}	4.45×10^{-1}	1.42

Case 2: keep the mass loadings $m_l = 0.134$ (kg/kg)				
St	0.5	1	5	10
d_p (d)	1.24×10^{-3}	1.76×10^{-3}	3.93×10^{-3}	5.56×10^{-3}
$\dot{C}_p$ (/step)	268	95	8	3

the correlation functions is due to the finite simulation results. It is practically never converged to zero.

4.3.3 Particle-vortex interaction

In this section, the characteristics of particle-vortex interactions will be shown and analyzed. Assuming that: (1) The fluid is incompressible and Newtonian. (2) The particle phase is dilute, which means the particle volume fractions are very low and the particle-particle interaction doesn't play a leading role compared to the fluid-particle interaction. (3) The particle-particle collision can be neglected in this case. (4) The volume of particle is very small and the wake effects induced by the finite volume of particles are neglected. The coupling of momentum between particle and fluid is through the point-force description, and the particle phase is assumed as mass point. The drag force is of the leading order which is considered here.

The swirling jet is issued according to the experimental setup by [288]. To reduce the computational capacity, a scale-down experimental setup, about 0.01 of the practical scales, is simulated. Fig. 4.18A shows the simulation setup in dimensionless scales. The boundary conditions and inlet velocity profiles are sketched in Fig. 4.18 too. The nonreflecting boundary condition [279] is applied here for the outflow boundary. Otherwise, the velocities on the walls of the container are set zero due to the nonslip wall boundary. A moderate Reynolds number $Re = 3000$ is used here. The dimensionless time step is $\Delta t = 0.005$ and a total time of simulation is 100. Other parameters used here are the same as in Table 4.6.

In gas-solid flows, the Stokes number St and the mass loading m_l are two major parameters which could affect the particle behavior and turbulence modulation dominantly. The mass loading m_l is defined as the ratio of the mass flow rate of particle to that of fluid at the inlet of jet. As the Stokes number and mass loading are correlated with each other, thus, the present study is divided into two parts: one is for the effects of Stokes numbers and mass loadings with the number flow rate fixed (Table 4.7, where the number flow rate $\dot{C}_p$ is defined as the number of particles issued into the flow per unit simulation step), and the other one is for the pure effects of Stokes numbers with the mass loadings fixed (Table 4.7). For case 1, the turbulence modulation

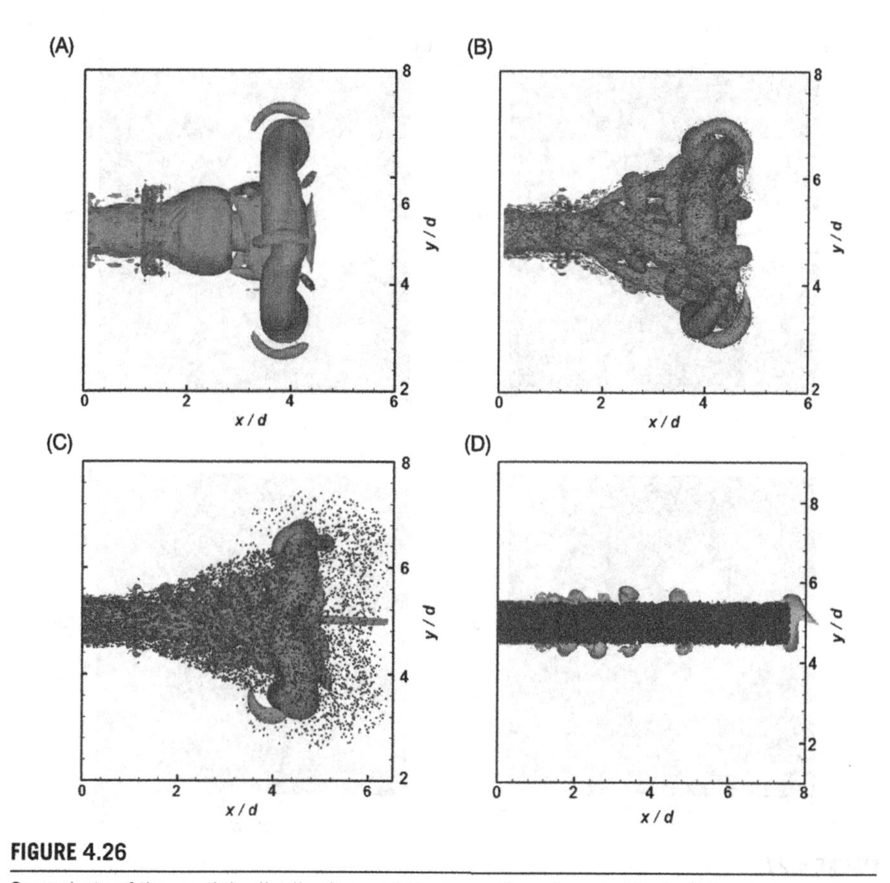

FIGURE 4.26

Snapshots of the particle distribution and the λ_2 vortices for case 1 at $t=3$:
(A) no particles; (B) $St=0.01$; (C) $St=1$; (D) $St=100$.

and the particle-vortex interaction are influenced by both the particle response characteristics and the particle loadings. For case 2, it is influenced solely by the particle response characteristics.

4.3.3.1 Instantaneous characteristics

Fig. 4.26 and Fig. 4.27 are the results of λ_2 vortices for Case 1 ($Re=606$) and Case 2 ($Re=3000$) respectively.

As seen in Fig. 4.26, when no particles are injected (Fig. 4.26A), the fluid vortices are composed of a smooth bubble and a vortex ring. The bubble is followed immediately by the vortex ring at the downstream enclosure of the bubble. When small particles are injected (Fig. 4.26B is for $St=0.01$, and the result is similar as for $St=0.1$), it is seen that the bubble is disturbed by the particles, changing from a smooth bubble feature into a group of twisted vortex cores. On the other hand, the small particles are attached perfectly surrounding the cores of vortices. For interme-

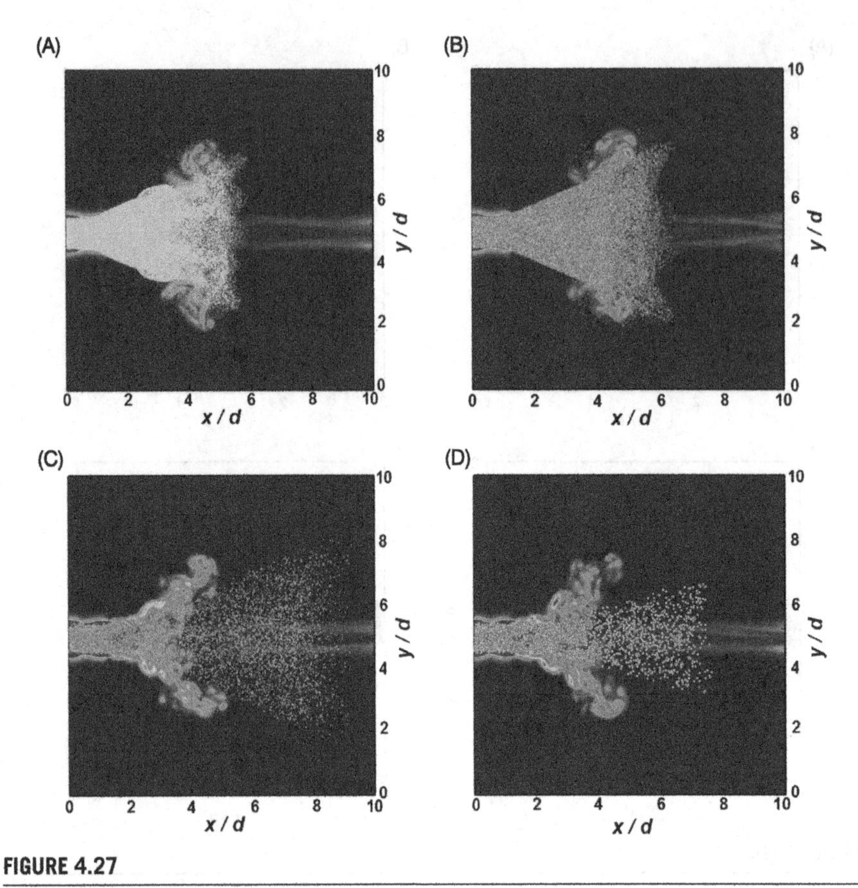

FIGURE 4.27

Snapshots of the particle distribution and the vortices for case 2 at $t=3$: (A) $St=0.5$; (B) $St=1$; (C) $St=5$; (D) $St=10$.

diate particles (Fig. 4.26C, $St=1$), it is observed that the vortices are changed more greatly than the former, e.g. the degree of the radial expansion of vortices is slightly suppressed. Moreover, due to the effect of inertia, the particles are dispersed within a large nearby region around the vortices cores. However, the degree of modulation of vortices by addition of the very heavy particles (Fig. 4.26D is for $St=100$, and the result is similar for $St=10$, too) is of a sharp contrast to the formers. The structure of bubble vortices disappears, and the vortex rings are almost breakdown and isolated during the downstream evolution. The particles are not dispersed in the radial direction. They form a particulate jet of great flow rigidity.

As Fig. 4.26 shows the mutual effects of particles of different response characteristics (Stokes numbers) and different mass loadings on the modulation of vortices, it is necessary to show the sole effect of the particle response characteristic on the modulation of vortices. Thus, Fig. 4.27 is necessary.

As seen in Fig. 4.27, there is a more or less difference between the results of the particle dispersion characteristics for different Stokes numbers. Small particles are dispersed around almost the whole region of vortices, whereas large particles are dispersed less in the radial direction but within a long axial range and with smaller included angles. Hence, the modulation of vortices should be different for different dispersion characteristics.

However, it is also found that the modifications of vortices in Fig. 4.27 between the results of different Stokes numbers are not as evident as that in Fig. 4.26 between the results of both different Stokes numbers and mass loadings, even for the large particles ($St = 10$). According to Table 4.7, it is easy to find that it is due to the significant effect of mass loadings. For Fig. 4.26D the mass loading is 1.42, whereas for Fig. 4.27D the mass loading is 0.134. With heavy loadings, the particle-vortices interactions are intensive. On the contrary, with light loadings, the particle-vortices interactions are comparatively mild.

4.3.3.2 $\mathscr{F}_{\lambda_2}(\boldsymbol{k}, t)$ in the spectrum space

To show the characteristics of modulation of vortices completely, this section will show the spectrum representation of λ_2 vortices cores by the three-dimensional fast Fourier transformation:

$$\mathscr{F}_{\lambda_2}(\boldsymbol{k}, t) = \int \lambda_2(\boldsymbol{r}, t)\exp(-i\boldsymbol{k}\cdot\boldsymbol{r})d\boldsymbol{r} \tag{4.17}$$

where $\boldsymbol{r} = x\boldsymbol{e}_1 + y\boldsymbol{e}_2 + z\boldsymbol{e}_3$, ($\boldsymbol{e}_i$, ($i = 1, 2, 3$) are the unit vectors in the physical space, and x, y, z correspond to the coordinates in the directions of $\boldsymbol{e}_i$, respectively) and $\boldsymbol{k} = k_x\boldsymbol{\xi}_1 + k_y\boldsymbol{\xi}_2 + k_z\boldsymbol{\xi}_3$ ($\boldsymbol{\xi}_i$, ($i = 1, 2, 3$) are the unit vectors in the spectrum space, and k_x, k_y, k_z correspond to the coordinates in the directions of $\boldsymbol{\xi}_i$, respectively). Due to the vortices are symmetrical on the cross-section plane (i.e. the y–z plane in Fig. 4.18), $\mathscr{F}_{\lambda_2}(\boldsymbol{k}, t)$ is appropriate to be decomposed into a 'parallel' component $\mathscr{F}_{\lambda_2,\parallel}$ and a 'perpendicular' component $\mathscr{F}_{\lambda_2,\perp}$, where:

$$\begin{cases} \mathscr{F}_{\lambda_2,\perp} = \int \lambda_2(\boldsymbol{r}, t)\exp(-i\boldsymbol{k}\cdot\boldsymbol{r})\big|_{k_x=\text{const}}d\boldsymbol{r} \\ \mathscr{F}_{\lambda_2,\parallel} = \int \lambda_2(\boldsymbol{r}, t)\exp(-i\boldsymbol{k}\cdot\boldsymbol{r})\big|_{\sqrt{k_y^2+k_z^2}=\text{const}}d\boldsymbol{r} \end{cases} \tag{4.18}$$

For example, Fig. 4.28 and Fig. 4.29 show the typical configurations of $\mathscr{F}_{\lambda_2,\perp}$ when $k_x = 1$ for case 1 and case 2 respectively.

4.3.3.3 $\mathscr{F}_{\lambda_2,\perp}(\boldsymbol{k}, t)$ for case 1

For case 1, it is seen from Fig. 4.28A that the spectrum representations are perfectly smooth. It is due to the bubble vortex breakdown for low Re number ($Re = 606$) is smooth (Fig. 4.26A). Moreover, the peaks in Fig. 4.26A on the k_y–k_z plane appear regularly when $r_\perp = \sqrt{k_y^2 + k_z^2}$ is constant, i.e. the peaks occur mainly around the rings with $r_\perp = k_1, k_2, k_3$ (Fig. 4.28A) respectively. Remember that $\exp(-i\boldsymbol{k}\cdot\boldsymbol{r})$

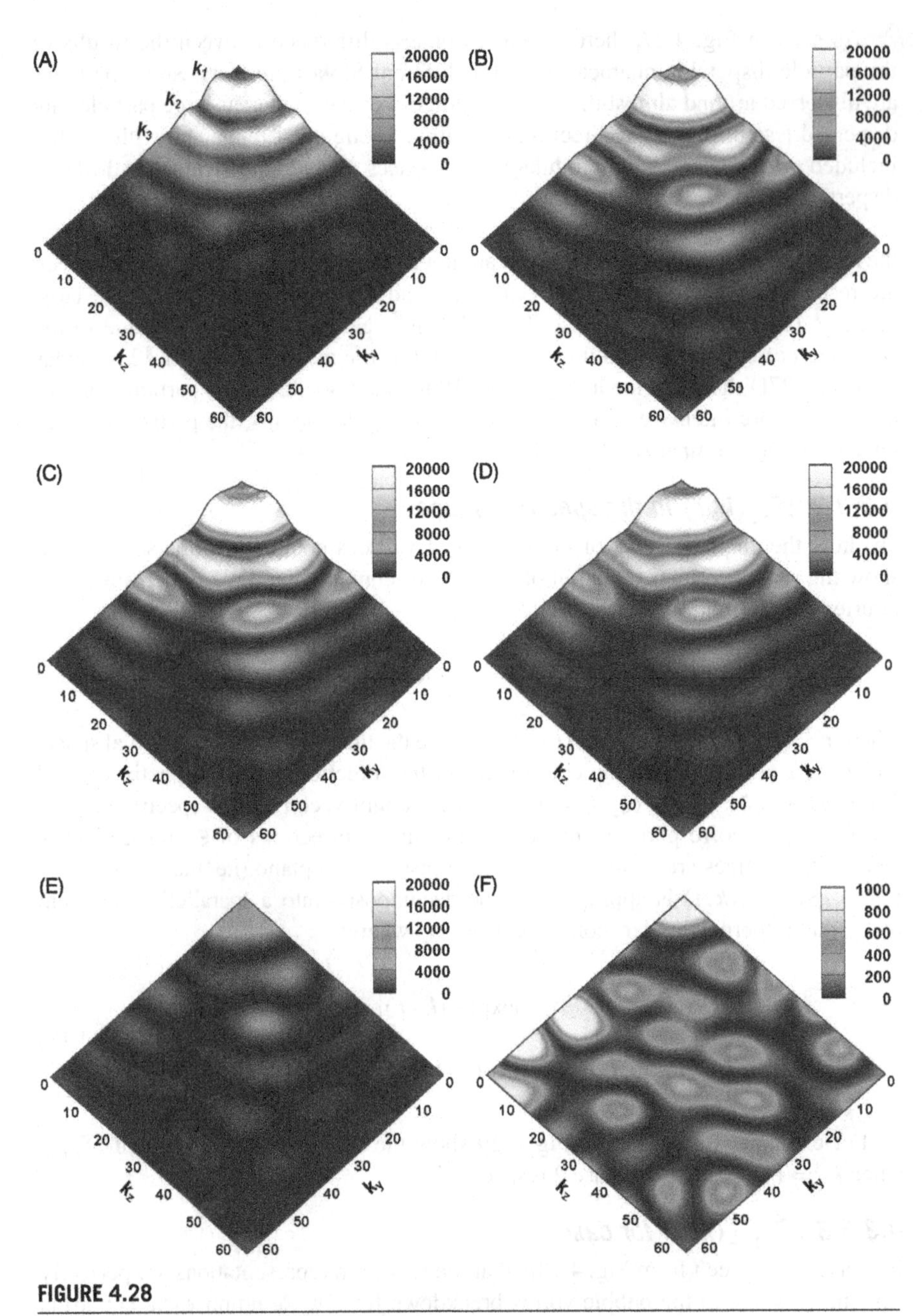

FIGURE 4.28

FFT of λ_2 for case 1 on the k_y–k_z plane at $t = 100$ (k_1, k_2 and k_3 in (A) indicate the waves in the k_y–k_z plane with the characteristic wave numbers): (A) no particles; (B) $St = 0.01$; (C) $St = 0.1$; (D) $St = 1$; (E) $St = 10$; (F) $St = 100$.

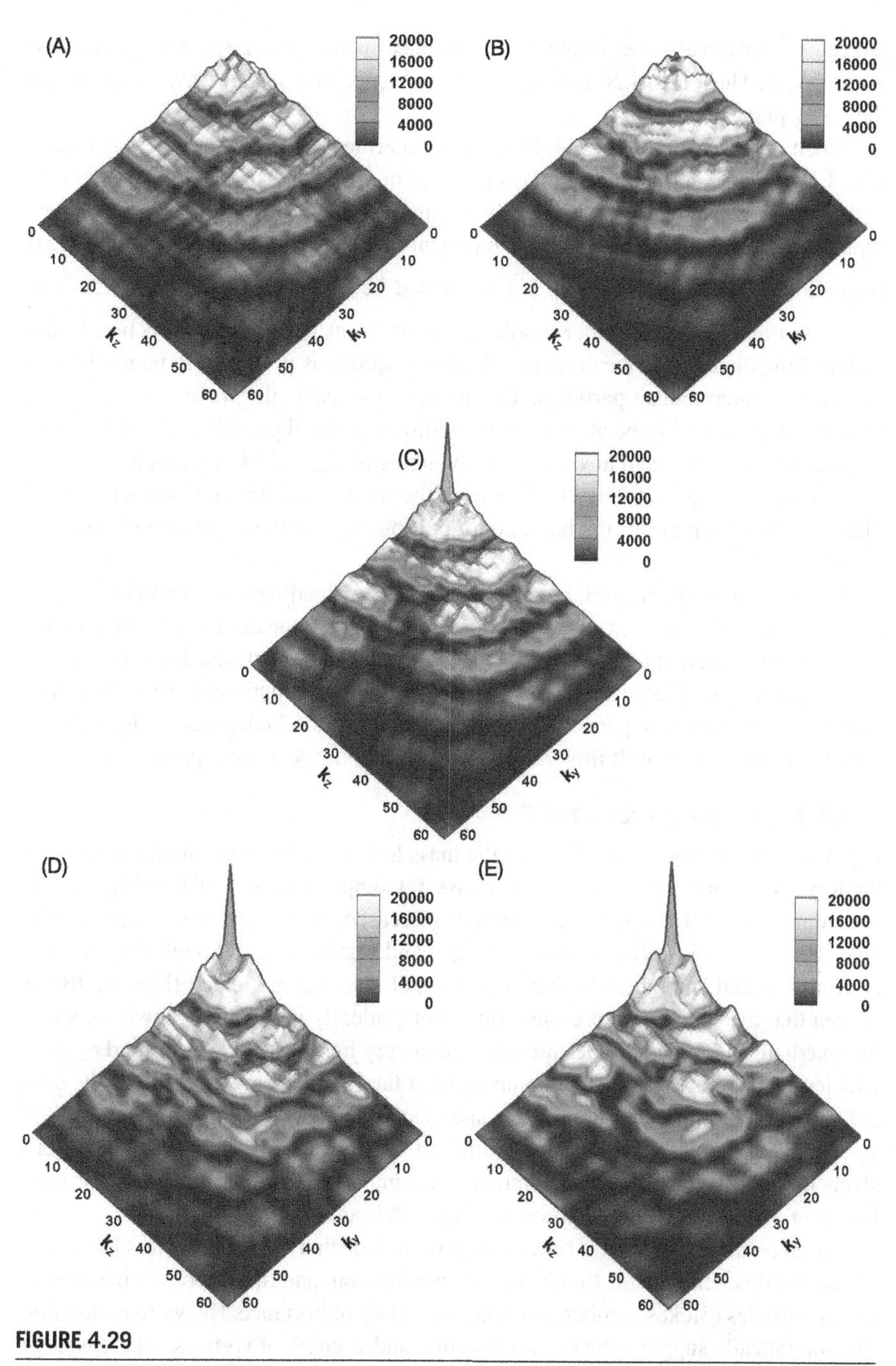

FIGURE 4.29

FFT of λ_2 for case 2 on the k_y–k_z plane at $t = 100$: (A) no particles; (B) $St = 0.5$; (C) $St = 1$; (D) $St = 5$; (E) $St = 10$.

indicates a spherical wave in the physical space, where $\lambda = 2\pi/k$ ($k = |\boldsymbol{k}|$) is the wave length. Thus, Fig. 4.26A indicates the existence of a spatial wave projected on the k_y–k_z plane.

When small particles (Fig. 4.26B, C) or intermediate particle (Fig. 4.26D) are added, it is observed that at least the k_3-ring is interrupted. The k_3-ring corresponds to the small wave length, or equivalently the small scale vortices. Supposing the interrupted location is (k'_y, k'_z), the wave component corresponding to $\exp[-i(k'_y y + k'_z z)]$ is modulated. It is seen that (k'_y, k'_z) is located between $(k_3, 0)$ and $\frac{\sqrt{2}}{2}(k_3, k_3)$ or $\frac{\sqrt{2}}{2}(k_3, k_3)$ and $(0, k_3)$. In other words, the wave front between the direction to the wall and the direction to the corner of the cross-section is proponed to be modulated by small or intermediate particles. To say more precisely, the wave fronts in these regions are proponed to be attenuated by addition of small particles. Thus, this conclusion interprets why and how the smooth bubble in Fig. 4.26A changes into several twisted cores in Fig. 4.26B and C. However, the wave of the largest scale is not modulated much, which means the basic configuration, such as the expansion of vortices, is still maintained.

Quite on the contrary, for large particles and heavy loadings, as seen in Fig. 4.28E and F, all rings of $r_\perp = k_1, k_2, k_3$ are suppressed greatly. It means all the wave components are intensively modulated, even for the largest scale vortices. Thus, the downstream expansion of large scale vortices is dramatically suppressed. They lose their energy by transferring to particles or small scale vortices. The legend of Fig. 4.28F is adjusted to make it clear. It illustrates almost a random spectrum representation.

4.3.3.4 $\mathscr{F}_{\lambda_2,\perp}(\boldsymbol{k}, t)$ for case 2

Fig. 4.29 corresponds to case 2 where the mass loading is kept low and the same, and the Reynolds number is higher than for case 1. Comparing Fig. 4.29A to Fig. 4.28A, it is found that with the Reynolds number increased, the spectrum becomes fluctuated. However, it is still composed of clear and regular rings, though they are not as complete and smooth as before. When small particles are added (Fig. 4.29B), it is seen that the spectrum representation is not evidently modulated. However, when intermediate and large particles are injected, a very high peak occurs around $r_\perp = 0$, which corresponds to a constant component of the Fourier representation of the spatial wave. More important, the regular and clear appearance of rings of peaks on the k_y–k_z plane becomes a random distribution of peaks. It means the vortices are intensively modulated from a regular spherical wave into a random distribution of vortices. However, the peaks are not suppressed (Fig. 4.29E and F).

Based on these results, it is reasonable to conclude that the particle mass loading is the main influencing factor to the vortices modulation, and the response characteristics of particles (Stokes number) is of the secondary importance. Heavy mass loading can dramatically suppress the spatial structure and strength of vortices, whereas large particles can change mildly the structure of vortices, especially from regular spherical waves into random spatial waves.

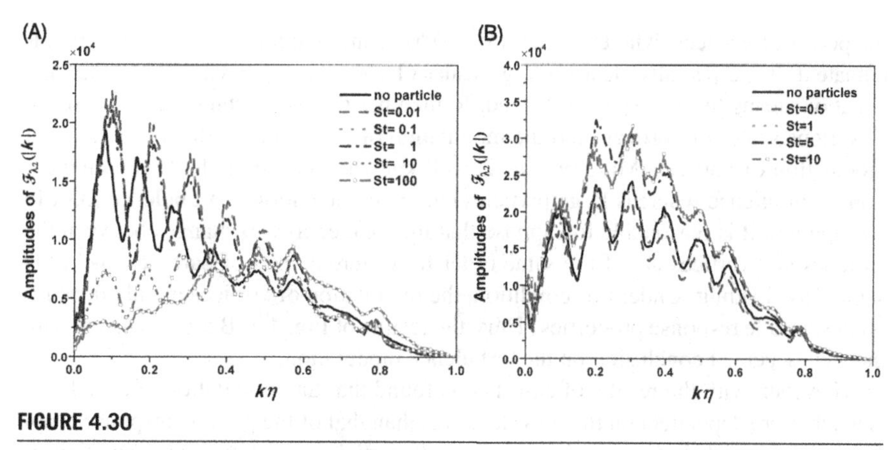

FIGURE 4.30

FFT of λ_2 for case 1 (A) and case 2 (B).

4.3.3.5 $\mathscr{F}_{\lambda_2}(k, t)$ for case 1

This section shows the representation of vortices in the full spectrum space $\mathscr{F}_{\lambda_2}(k, t) = \mathscr{F}_{\lambda_2}(|\boldsymbol{k}|, t)$. For comparison, the wave number $k = (k_x^2 + k_y^2 + k_z^2)^{1/2}$ in Fig. 4.30A and B is nondimensionalized by η, which is the reciprocal of the maximum wave number k_{max}.

As shown in Fig. 4.30A for case 1, the peaks are observed more clearly than before. For $St \leq 1$, it seems to be true that: (1) the amplitudes of $\mathscr{F}_{\lambda_2}(\boldsymbol{k}, t)$ are augmented by the particles, i.e. small particles modulate the vortices by strengthening the power; (2) smaller particles have more augmentation effects on the amplitudes than larger particles; (3) compared to that without particles, the peaks occur for larger wave numbers k, which indicates the addition of small particles makes the scales of vortices smaller than that without particles. It is reasonable since the addition of particles can break the smooth large scale weak vortices into small scale strong vortices cores when they pass through them.

Meanwhile, for $St > 1$, it seems to be true that: (1) The large scale vortices, which is appropriately identified here by the wave numbers with $k\eta < 0.5$, are suppressed greatly, especially for the heaviest mass loading ($St = 100$). (2) On the contrary, for the range of large wave number k with $k\eta > 0.5$, which indicates the range of small scales vortices, the amplitudes are augmented due to addition of the large particles. This phenomenon may be interpreted as: (1) Firstly, due to the heavy loading, the particle-vortices interaction should be very intensive. Thus, the large particles suck a lot of energy from the large scale vortices when they pass through them. (2) Secondly, large particles are fairly inertial, and their motion is not easy to be changed. Thus, the energy is fed back to fluid when they induce a lot of small scale vortices. As a result, the power of the smallest vortices is augmented.

4.3.3.6 $\mathscr{F}_{\lambda_2}(k, t)$ for case 2

As seen in Fig. 4.30B, it is interesting to find that the results seem to be quite contrary to those for case 1. Under the same mass loading, large particles ($St = 5, 10$) augment the power of vortices, whereas small ($St = 0.5$) or intermediate ($St = 1$) particles attenuate it. It is especially clear for large scales of vortices ($0.1 < k\eta < 0.5$). Thus, the situation seems to be very complicated. Remember that the turbulence is a cascade of various scales of vortices, and the modulation of turbulence is the direct result of modulation of vortices. Moreover, as is well known, small particles cause attenuation of turbulence whereas large particles cause augmentation of turbulence [33]. In our opinion, it is reasonable to suppose that the total energy exchanges between the particles and vortices are of the same order for different sizes of particles under the same mass loading. Under this condition, the modulation of vortices mainly depends on the particle response properties. Thus, the results of Fig. 4.30B are consistent with the well-accepted conclusion on the turbulence modulation.

However, with the results of case 1, it is found that the modulation of turbulence is much more dependent on the mass loadings than that of the particle response properties. With much light mass loadings, small particles can cause augmentation of vortices within the range of large scale vortices by breaking them and reducing their scales. More important, with heavy mass loadings, large particles can cause attenuation of large scales of vortices, suppressing them and transfer their energy to small scales of vortices.

Besides, it is necessary to mention that, due to the limitation of the point-force approximation, the wake effects of finite volume particles cannot be considered here. Thus, the comparison is only between the effects of particle mass loadings and particle dynamical response properties on the modulation of vortices and turbulence. However, the point-force method captures the essential characteristics of the mechanism of modulation based on the assumption that the particle-fluid momentum exchange plays a leading role in the particle-fluid interaction.

4.3.3.7 The energy spectrum

In order to validate the above analysis, the energy spectrum of fluid is calculated as shown in Fig. 4.31A and B for case 1 and 2 respectively. The energy spectrum $E(k)$ is defined here as

$$E(k) = \iint \hat{u}_i^\star(k)\hat{u}_i(k)dA(k) \tag{4.19}$$

where $\hat{u}_i(k)$ is the spectrum of fluctuation velocity u_i, and $A(k)$ is a spherical shell with the radius k.

As is mentioned above, small particles break large vortices into small vortices when they pass through them. Thus, there exists a shift of spectrum from small wave number k to large wave number $k + \Delta k$. However, the turbulence fluctuation energy is slightly reduced (Fig. 4.31A, $St \leq 1$). More important, as the mass loadings are heavy, the turbulence fluctuation is not augmented but attenuated (Fig. 4.31A, $St > 1$). Moreover, for $St = 100$, it is observed that the level of $E(k)$ for $k\eta \approx 0.001$ (large scales of vortices) are almost equivalent to that for $0.05 \leq k\eta \leq 0.5$ (relatively small scales of vortices). In addition, when $0.05 \leq k\eta \leq 0.5$, $E(k)$ for $St = 100$ is larger than that for $St = 10$, whereas it is smaller than that of $St = 10$ for $k\eta \approx 0.001$. (Note

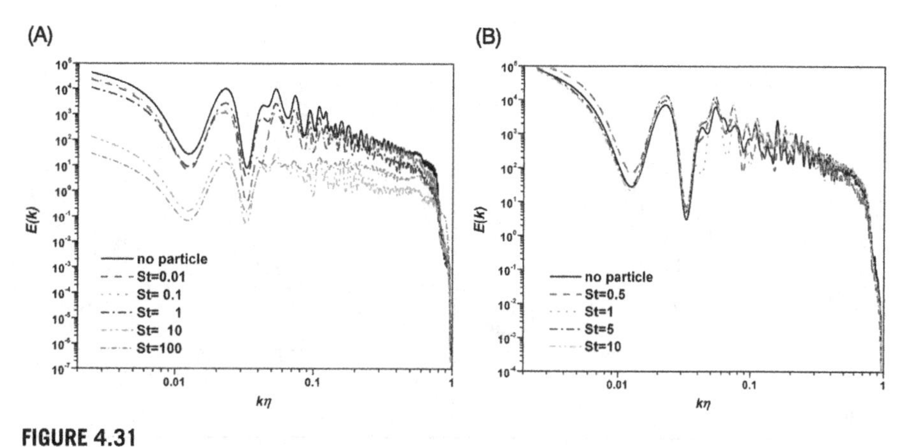

FIGURE 4.31

The energy spectrum $E(k)$ for case 1 (A) and case 2 (B).

that it is a log-log plot.) It validates the previous analysis of transfer of the turbulent kinetic energy from large scales of vortices ($k\eta \approx 0.001$) to relatively smaller scales of vortices.

In contrast, for Fig. 4.31B, it is seen that the energy spectrum is not much modulated when the mass loadings are kept the same. Hence, by comparing Fig. 4.31B to Fig. 4.31A, it validates the conclusion that the modulation of turbulence is more dependent on the mass loadings than that of particle properties.

Besides, it is noticed that the energy spectrum is oscillated, especially for small wave numbers. Recall the results of Fig. 4.28 and Fig. 4.29 – there are some clearly separated rings of small wave numbers in the spectrum presentation of vortices, which correspond to the spatial waves of large scale vortices. Then, the spatial waves of large scale vortices can consequentially affect the characteristics of the turbulence fluctuation and energy spectrum. Thus, the oscillation in energy spectrum reflects the nonnegligible effect of large scale coherent vortices on turbulence fluctuations. As is well known, swirling flows are typical anisotropic turbulent flows. Hence, it is considered that Fig. 4.31 reflects a remarkable difference for the anisotropic turbulence to the isotropic turbulence due to the existence of large scale coherent vortices [33].

In addition, Fig. 4.32 illustrates the characteristics of modulation of the dissipation rate of turbulent kinetic energy in Fig. 4.32A and B, corresponding to case 1 and 2 respectively. The dissipation rate ϵ_t is indicated by

$$\epsilon_t = \int k^2 E(k) dk \tag{4.20}$$

From Fig. 4.32A, it is found that for small particles and light mass loadings, the peaks occur around wave numbers close to the case of no particles. However, for intermediate particles and moderate mass loading ($St = 1$ or $St = 10$), the peak occurs around smaller wave numbers, which indicates the dissipation for relatively large scale vortices is the most intensive. In contrast, although it is still of the leading order in energy

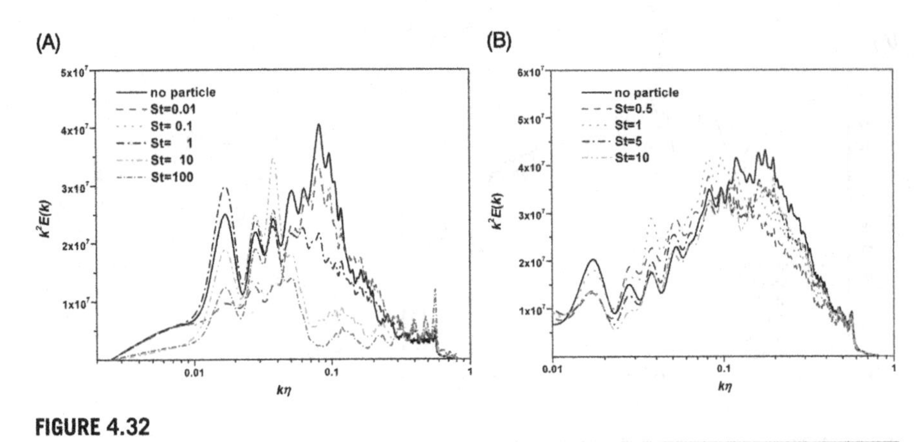

FIGURE 4.32

The energy dissipation $k^2E(k)$ for case 1 (A) and case 2 (B).

dissipation for large scales of vortices for very large particle and heavy mass loading ($St = 100$), there is a secondary peak around the very large wave numbers, indicating the augmented highly dissipative characteristics of very small scale vortices. It is reasonable since the turbulent kinetic energy is transferred from large scales of vortices to small scales of vortices by the large particles. The augmentation of kinetic energy in small scales of vortices can result in consequential augmentation of dissipation of kinetic energy to keep the balance of energy.

Finally, it is found again that the distributions of energy dissipation in the spectrums for different sizes of particles are not modulated much since the mass loadings are kept the same (Fig. 4.32B). It again verifies the relative importance of the effects of mass loading and particle response property on the turbulence modulation.

4.3.4 Four-way coupling

In this section, the governing equations of the three-dimensional, time-dependent and incompressible viscous nondimensional Navier–Stokes equations are applied to solve the swirling jet of the Newtonian fluid in a rectangular container (Fig. 4.18). The scales of the numerical setup are only 1% of the practical scales of the experimental setup [288] so as to meet the limitation of computational capacity for DNS. The boundary conditions are the same as in the former section and the parameters are listed in Table 4.8.

For the particle phase, the drag force is considered here. For the particle-particle collision, some fundamental assumptions are followed: (1) The particles are rigid spheres and with uniform densities and diameters. (2) The particle-particle collision follows the conservation of momentum. (3) The loss of kinetic energy due to the inelastic collision and damping friction is taken into account by two empirical parameters, i.e. the restitution coefficient e and the friction coefficient γ respectively. (4) Each particle-particle collision is solved deterministically without introducing any

Table 4.8 Typical real values of variables used in the numerical simulation.

Scales of the flow domain (mm)	$8 \times 4 \times 4$
Scale of jet diameter d (mm)	0.4
Grids number $N_x \times N_y \times N_z$	$256 \times 256 \times 256$
Mean inlet axial velocity U_0 (m/s)	125
Fluid density ρ_g (kg·m^{-3})	1.29
Fluid viscosity (Pa·s)	1.85×10^{-5}
Reynolds number Re	3486.5
Swirl number S	1.42
Time step Δt (µs)	0.01
Simulation time T_s (ms)	0.2
Restitution coefficient, e	0.95
Friction coefficient, γ	0.3
Total number of particles,	2×10^5
Diameters of particle, d_p (µm)	0.25; 0.56; 0.79; 1.77; 2.5; 7.93
Stokes numbers, St	0.1; 0.5; 1; 5; 10; 100
Particle mass loadings, m_l	0.25; 0.56; 0.79; 1.77; 2.5; 7.93

statistical treatment and the particle-particle collision is solved by the hard sphere model. The particles are injected through the inlet of the jet homogeneously distributed and with equal number flow rates, i.e. ten particles per simulation step. Six Stokes numbers are simulated with the particle size decided by the Stokes number, as well as the mass loadings (Table 4.8). As for different Stokes numbers, the mass loadings are changed greatly ($\sim St^{3/2}$), to keep the same mass loadings for comparison requires a great number of particles when St is very small, which is far beyond the computational capacity. Thus, the number flow rate of particles is kept the same, and the cases with and without collision are compared.

This section attempts to explore the characteristics of the particle-particle collision in swirling jets. Since the particle-particle collision is correlated with the anisotropic characteristics of turbulence, some aspects are considered, i.e. the relationship between the PDFs of the particle-particle collision and the turbulence statistical characteristics, such as turbulent kinetic energy (TKE) k_t, dissipation rate (TDR) ϵ, the intensity of turbulence $\langle u'_i \rangle$ and correlation of fluid fluctuations (Reynolds stress tensor $\langle u'_i u'_j \rangle$). From this point of view, it is appropriately to assume that the PDFs of the particle-particle collision can be expressed as $f(k_t, \epsilon, \langle u'_i \rangle, \langle u'_i u'_j \rangle)$. The particle-particle collision is complicated due to multiple and nonlinear superimposition of these influencing effects. For simplicity, it assumes that these effects are decoupled and separately affect the particle-particle collision behavior, namely:

$$\hat{f}(k_t, \epsilon, \langle u'_i \rangle, \langle u'_i u'_j \rangle) \sim \hat{f}_k(k_t) \hat{f}_e(\epsilon) \hat{f}_u(\langle u'_i \rangle) \hat{f}_{uu}(\langle u'_i u'_j \rangle) \qquad (4.21)$$

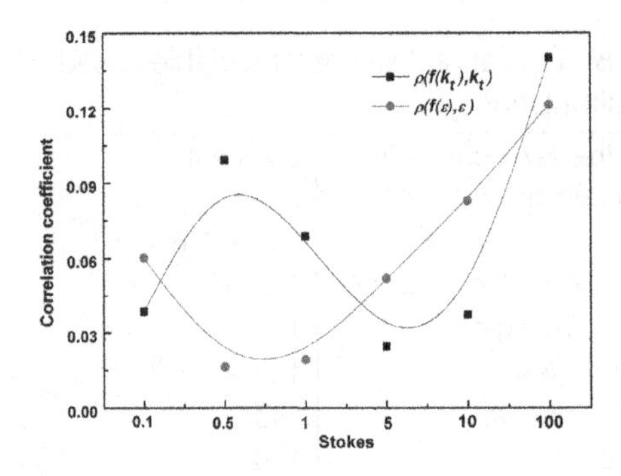

FIGURE 4.33

The coefficients of correlation between the PDFs of the particle-particle collision and the turbulent kinetic energy and dissipation rate.

This section aims to demonstrate the dependency feature of $\hat{f}_k(k_t)$, $\hat{f}_e(\epsilon)$, and $\hat{f}_{uu}(\langle u'_i u'_j \rangle)$, respectively.

4.3.4.1 Dependency on TKE and TDR

The turbulence kinetic energy and the dissipation rate are defined as $\hat{k}_t = \langle \frac{1}{2} u'_i u'_i \rangle$ and $\hat{\epsilon} = \nu \langle \frac{\partial u'_i}{\partial x_k} \frac{\partial u'_i}{\partial x_k} \rangle$, respectively. For the present study, the expressions are reduced by omitting the coefficients, i.e.

$$\hat{k}_t = u'_i u'_i \tag{4.22}$$

$$\hat{\epsilon} = \langle \frac{\partial u'_i}{\partial x_k} \frac{\partial u'_i}{\partial x_k} \rangle \tag{4.23}$$

The correlation coefficient is defined as $\varrho(f(\varphi), \varphi) = \frac{Cov(f(\varphi),\varphi)}{\langle f(\varphi)' \rangle \langle \varphi' \rangle}$, where $Cov(f(\varphi), \varphi)$ is the covariance of the probability density function f and any characteristic variable φ of turbulence. $\langle \varphi' \rangle$ means the $r.m.s$ value of φ.

Fig. 4.33 illustrates the correlations between the PDF of collision and TKE or TDR. The correlation between the PDF and TDR has a similar trend to the PDF itself, i.e. very small and very large particles have great values of correlation coefficients and moderate particles have relatively small values. For $St \leq 1$ and $St \geq 1$ the particle-particle collision and the TDR are relatively well correlated, and large TDR has a relatively higher probability of the particle-particle collision compared to the intermediate Stokes numbers. However, the correlation between the PDF and TKE seems to be more complicated, since the correlation coefficient for $St = 0.1$ is of a relatively small value whereas for $St = 0.5$ it is larger than that for $St = 0.1$ and $St = 1$. As pointed by [35], for intermediate Stokes numbers, the behavior is complicated by two effects, i.e. the preferential concentration due to the centrifugal effect and particles being less strongly correlated with each other. However the scalings of these two effects are different. This is the reason why the observed behavior is so complicated. To say specifically, for $St = 0.5$ which is close to unity and belongs to

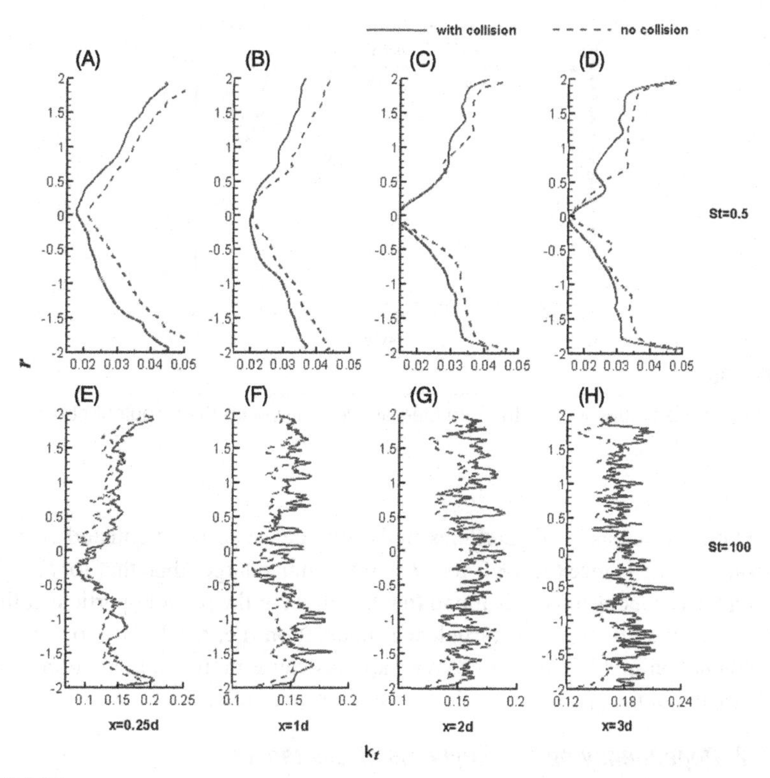

FIGURE 4.34

The radial distribution of the turbulent kinetic energy for $St = 0.5$ (A)–(D) and $St = 100$ (E)–(F) at four axial locations of $x = 0.25d$, $1d$, $2d$, $3d$, respectively.

the moderate Stokes numbers, it is prone to be driven by vortices and more intensively correlated with the kinetics of turbulence than that of $St = 1$. For $St = 0.1$, the motion of particle is obviously well correlated with the turbulence. But it is also well correlated with each other, resulting in a decrease of the mean relative velocity of particles. Thus, it has a relatively smaller value of probability of the particle-particle collision in the local region, causing a low correlation between the particle-particle collision and the turbulent kinetic energy. Conclusively, the correlation coefficient for particles with $St = 0.5$ is an optimum equilibrium between the turbulence transport effect and the correlation effect of relative motion of particles.

Moreover, it is necessary to show the effects of the particle-particle collision on modulation of turbulence when high correlation between them exists for $St = 0.5$ and $St = 100$. Fig. 4.34 shows the comparisons of TKE between the results with and without considering particle-particle collisions at four typical axial locations, namely $x = 0.25$, 1, 2 and 3d respectively. For $St = 0.5$, the TKE seems to be attenuated by the particle-particle collision effect whereas it seems to be slightly augmented for $St = 100$. As is well known, the addition of large particles can cause a turbulence

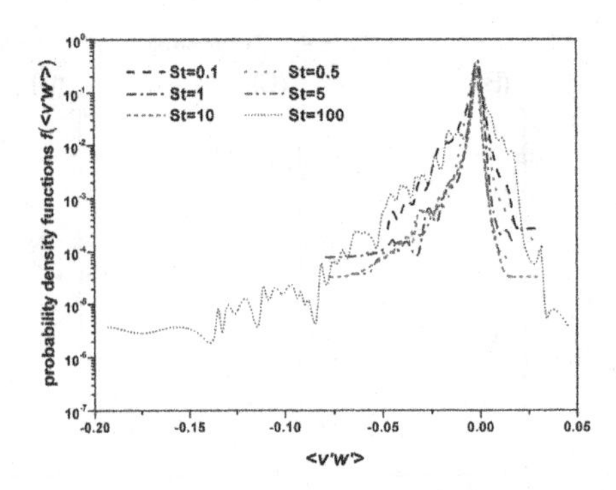

FIGURE 4.35

Distribution density functions of the particle-particle collision on the turbulent correlated fluctuation of $\langle u'_i u'_j \rangle$.

augmentation whereas small particles make turbulence to be attenuated (it is also seen from Fig. 4.34 since the TKE for $St = 100$ is much larger than that for $St = 0.5$). Based on these conclusions, it is found that the effect of the particle-particle collision can intensify the effect of the turbulence modulation, i.e. the light particles cause more attenuation in TKE and the heavy particles cause more augmentation in TKE than that without the particle-particle collision, respectively.

4.3.4.2 Dependency on the Reynolds stress tensor

It is well known that the Reynolds stress tensor $\langle u'_i u'_j \rangle$ can be divided into normal components $\langle u'_i u'_i \rangle$ and shearing components $(\langle u'_i u'_j \rangle_{i \neq j})$. As depicted above, the TKE is related to the self-correlation of turbulence fluctuations. Thus, the effects of the normal components have been investigated in the above section. This section just studies the correlation coefficient between the PDF of collision and the shearing components.

For example, Fig. 4.35 illustrates the PDF profiles of the collision probability on the shearing component $\langle u'_i u'_j \rangle$. It shows that the profiles of $St = 0.1$ and $St = 100$ are wider than the others, especially for $St = 100$. Moreover, it is clear that these profiles have longer tails in the negative direction than in the positive direction. The correlation coefficients between the PDFs of collision and the Reynolds stress tensors are calculated as shown in Table 4.9. It is clearly seen that the correlation coefficients between the collision PDFs and the normal components of correlated fluctuations of turbulence are positive, whereas the collision PDFs are negatively correlated to the shearing components of the Reynolds stress tensor. It means that the normal fluctuation of turbulence as well as the TKE is beneficial to the particle-particle collision whereas the shearing fluctuation of turbulence is adverse to the particle-particle collision. It is reasonable since the 'normal fluctuation' can induce an increasing level of the mean relative velocity between particles and the 'shearing fluctuation' can lead pairs of particles to be separated in the parallel directions. This result could be useful

Table 4.9 Correlation coefficients $\varrho(f(\varphi), \varphi)$.

St	$\varphi = \langle u' \rangle$	$\varphi = \langle v' \rangle$	$\varphi = \langle w' \rangle$	$\varphi = \langle u'v' \rangle$	$\varphi = \langle u'w' \rangle$	$\varphi = \langle v'w' \rangle$
0.1	0.02006	0.02491	0.02466	−0.01895	−0.00987	−0.02152
0.5	0.0813	0.0962	0.0957	−0.0154	−0.00741	−0.0153
1	0.0619	0.078	0.0778	−0.00484	−0.00168	−0.0126
5	0.0257	0.0233	0.0225	−0.0125	−0.0113	−0.0165
10	0.0379	0.022	0.0219	−0.0204	−0.0203	−0.0255
100	0.0458	0.063	0.0639	−0.182	−0.286	−0.019

for predicting and estimating the intensity of the particle-particle collision frequency in shearing flows.

4.4 Bubbling fluidized bed

4.4.1 3D bubbling fluidized testing bed

4.4.1.1 Computation of ϵ_{3D} and configuration

The void fraction exerts its influence on the evolution of the dense gas-solid flow considerably. Thus it is necessary to compute the void fraction of each cell accurately. The phenomenon of particle crossing a border of neighboring computational cells is of common and frequent appearance in simulation. Hence the segmentation of a sphere must be taken into consideration to account for it. In this section, a numerical technique is used to calculate the volume of parts of a sphere. When one particle move into neighboring cells, the particle is subdivided into small elements and its volume fraction in each cell is obtained by summing up the elements it contains. For example, let the volume of each subdivided element be $(\frac{1}{40})^3$ of the volume of the cubic covering the sphere, the relative error contributing to void fraction ϵ_{3D} is less than 0.25% (with $d_p = 883\mu$ and meshing size $\Delta_m = 2$ mm).

Table 4.10 shows a summary of numerical and experimental conditions. The particle material is quartz sand, and the fluidization is shot by a high-speed camera (MotionXtra HG-100K). The dimensions of the bed are actually $100 \times 300 \times 600$ mm^3 (Fig. 4.36), which are reduced to $20 \times 80 \times 500$ mm^3 to save computational cost. Since the transverse section of the bed for simulation is very small, the boundary is set to be uniform velocity inlet at bottom and outflow condition at top. In experiment, the gas is injected into the bed through distributed pipes at bottom. The time step should be about 20 (μs). Here, due to numerical stability consideration and after practical testing, $\Delta t_p = 4 \times 10^{-6}$ for solid phase and $\Delta t_g = 2 \times 10^{-5}$ for gas phase are chosen.

4.4.1.2 Force behavior analysis

In the dense gas-solid flow, the particle-particle and the gas-particle interactions are two types of the most important interactions which play leading roles in bubble and

Table 4.10 Numerical and experimental conditions.

Parameters	In experiment	In simulation
Cell sizes $\Delta_x \times \Delta_y \times \Delta_z$ (mm)	–	2 × 2 × 2
Cell numbers $N_x \times N_y \times N_z$	–	10 × 40 × 250
Dimensions of fluidized bed (mm)	100 × 300 × 600	20 × 80 × 500
Tube radius R_s (mm)	10	10
Particle diameter d_p (μm)	883	883
Particle density ρ_p (kg·m^{-3})	2650	2650
Particle number n_p	–	178200
Stiffness factor k_n, k_t (N·m^{-1})	–	(1000, 250)
Poisson ratio ν	–	0.25
Restitution coefficient e	–	0.9
Friction coefficient μ	–	0.3
Superficial gas velocity U_0 (m^{-1})	1.16	1.16
Gas density ρ_g (kg·m^{-3})	1.22	1.22
Gas viscosity μ_g (Pa·s)	1.80×10^{-5}	1.80×10^{-5}
Time step (Δt_g; Δt_p) (s)	–	2×10^{-5}; 4×10^{-6}

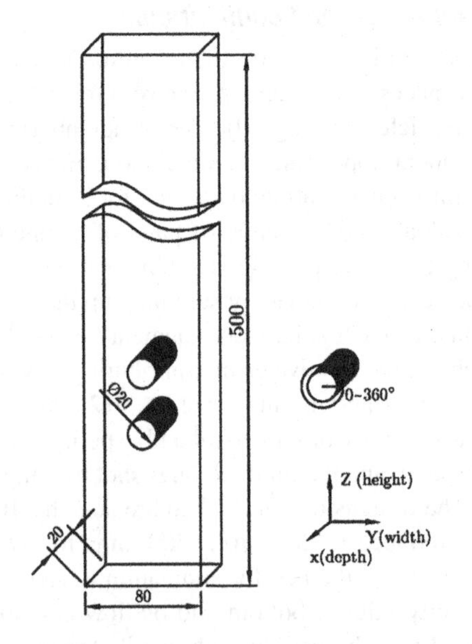

FIGURE 4.36

Sketch map of the fluidized bed with immersed tubes.

particle evolution. Taking mean value of all particles, Fig. 4.37A and B present the evolution of the mean gas-particle interaction force $\langle \boldsymbol{F}_{g-p} \rangle$ (only drag force and pressure gradient force under consideration) and the particle-particle contact force $\langle \boldsymbol{F}_{p-p} \rangle$ (divided by particle's gravity force to get dimensionless values), where the

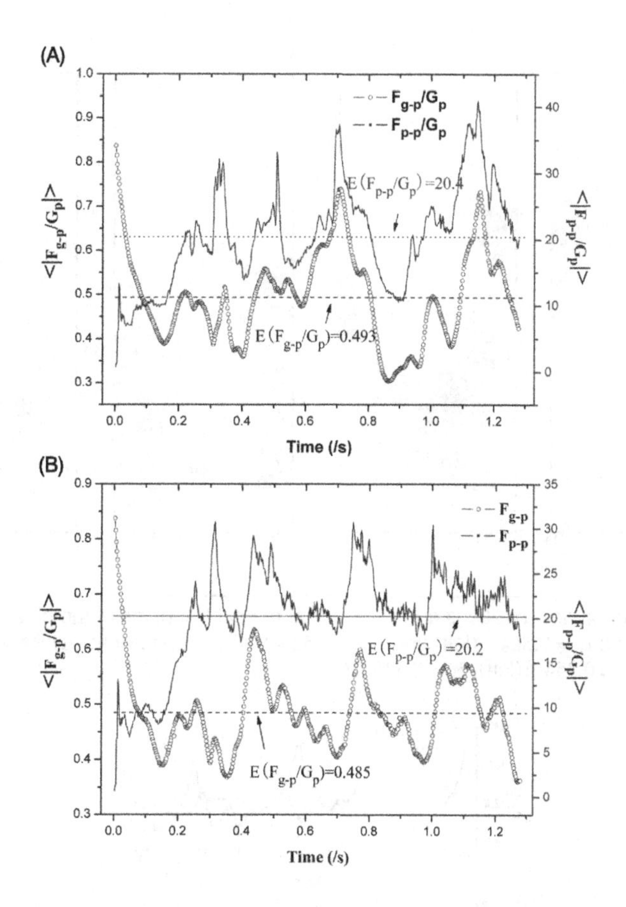

FIGURE 4.37

Comparison of F_{g-p} and F_{p-p} for results with and without LES: (A) results with LES; (B) results without LES.

operator $\langle\cdot\rangle$ means ensemble average. Fig. 4.37A is obtained when LES is used while for (b) it is not used. The comparison between $\langle \boldsymbol{F}_{g-p}\rangle$ and $\langle \boldsymbol{F}_{p-p}\rangle$ indicates that the particle-particle interaction is approximately two orders of magnitude as large as the gas-particle interaction. Furthermore, although differ in magnitude, both Fig. 4.37A and B show that the evolution of $\langle \boldsymbol{F}_{g-p}\rangle$ and $\langle \boldsymbol{F}_{p-p}\rangle$ seem to follow the same pace to increase and decrease, appearing with a periodic variation.

It is interestingly that $\langle \boldsymbol{F}_{g-p}\rangle$ and $\langle \boldsymbol{F}_{p-p}\rangle$ reach their minimum and maximum values. Four time points at $t_1 = 0.864$ s, $t_2 = 1.152$ s in Fig. 4.37A and $t_3 = 0.354$ s, $t_4 = 0.436$ s in Fig. 4.37B are chosen to visualize the particle velocities and positions, which are displayed in Fig. 4.38A and B respectively. It is found that at t_1 and t_3 (the left parts of Fig. 4.38A and B) the particles reach a top height and conglomerate at two sides of the bed – the state when gas bubbles coalesce into the largest one before eruption. Quite on the contrary, at t_2 and t_4 (the right parts of Fig. 4.38A

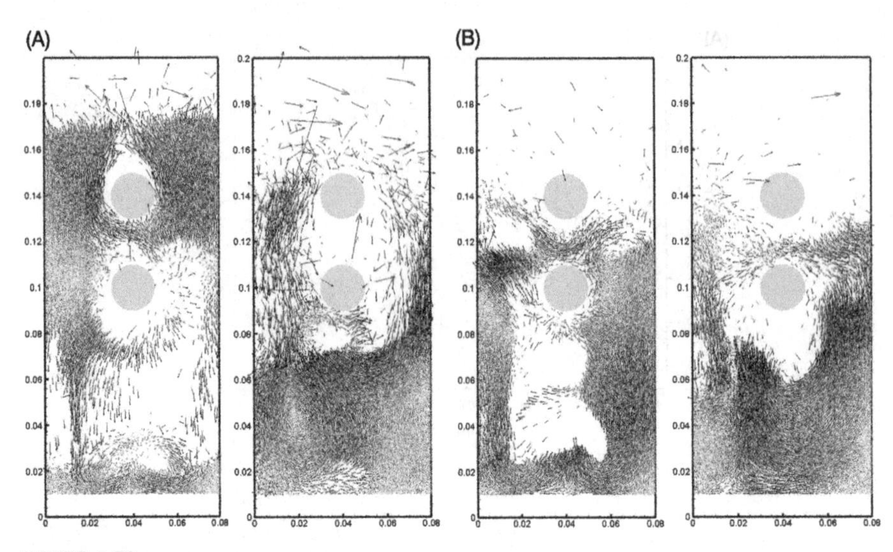

FIGURE 4.38

Visualization of particle positions and velocities at the central plane of the bed. (A) Results with LES at time $t_1 = 0.864$ s (left) and $t_2 = 1.152$ s (right) respectively. (B) Results without LES at time $t_3 = 0.354$ s (left), $t_4 = 0.436$ s (right) respectively.

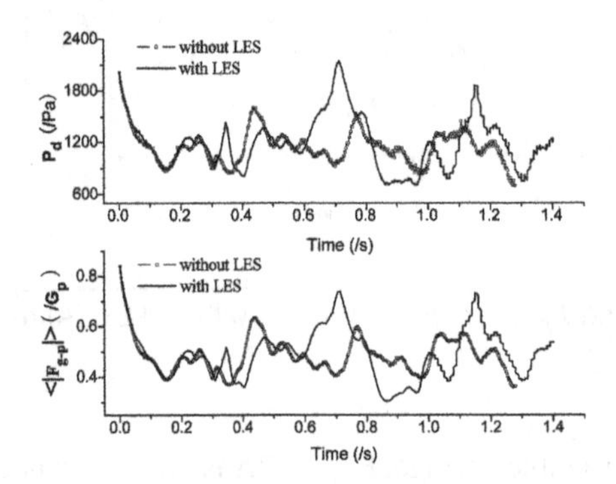

FIGURE 4.39

Comparison of F_{g-p} and P_d for results with and without LES.

and B) the particles show an impetuous back-mixing image and form a large 'eddy' of particulate flow at the bottom of the bed before new gas bubbles emerge. In other words, the bubble's motion and the inter-phase interaction in the dense gas-solid flow seem to bear a strong correlation.

Since the value of $\langle \boldsymbol{F}_{g-p} \rangle$ is composed of the pressure gradient force and the drag force –other forces, like the Magnus lift force etc., are neglected. In order to know more about the gas-particle interaction, the mean pressure drop P_d at a height of 0.2 m of the bed is shown in Fig. 4.39 together with $\langle \boldsymbol{F}_{g-p} \rangle$. It is interesting that no matter if LES is used or not, P_d and $\langle \boldsymbol{F}_{g-p} \rangle$ seem to follow a perfect similar pace of evolution. Hence it is found that $\langle \boldsymbol{F}_{g-p} \rangle$ is to a great extent determined by P_d rather than by the drag force in the dense gas-solid flow though it may not be true for specific condition, such as inside bubbles where the pressure gradient can be neglected.

It is considered that, the variation of $\langle \boldsymbol{F}_{g-p} \rangle$ is mainly dominated by the variation of P_d, and the variation of P_d is to a great degree determined by whether or not the 'channel' for gas flow is blocked. Since the back-mixing particles considerably block off the gas flow, P_d increases and so do $\langle \boldsymbol{F}_{g-p} \rangle$ and $\langle \boldsymbol{F}_{p-p} \rangle$ and consequentially new bubbles emerge at the bottom of the bed. On the other hand, when small bubbles grow, coalesce into each other, and erupt finally, the P_d turns to decrease and so do $\langle \boldsymbol{F}_{g-p} \rangle$ and $\langle \boldsymbol{F}_{p-p} \rangle$.

4.4.2 Pulsed fluidization

This section aims to investigate the effects of pulsed fluidization on the characteristics of fundamental inter-phase and intra-phase interactions in bubbling fluidized beds. Also the effect of oscillation of the gas flow on the particle fluctuation motion is studied. A pulsed gas inflow is used with four frequencies of oscillation ranging from 0 to 20 Hz and $U_0 = U_s$ in which the inactive phase occur only at discrete time point. Moreover, due to the inherent advantage for the Lagrangian description of the particle motion and the particle-particle interaction, a DEM-LES coupling simulation is used here for investigation and comparison of the fundamental interactions.

A 2D bubbling fluidized bed with several immersed tubes is simulated. The immersed tubes are fixed cylindrical surfaces set in bed horizontally for investigation of the influence of pulsed fluidization on the particle-surface interaction. The fluidized bed is 300 mm in width and 800 mm in height, with 5 staggered arranged immersed tubes (case 1) and 6 inline arranged immersed tubes (case 2), respectively. The diameter of tube is 20 mm and the particle material is sand with its diameter $d_p = 883$ μm, and density $\rho_p = 2650$ kg/m^3. The configurations of immersed tubes for case 1 and case 2 are sketched in Fig. 4.40. The coordinates of immersed tubes in fluidized bed as well as other parameters used in the present simulation are listed in Table 4.11. The gas flow is discharged from the bottom of the bed and its outlet is at the top. The inflow of gas from the bottom consists of a constant component U_0 and a pulse component $U_0 \sin(2\pi f t)$. Thus, the function of pulsed gas inflow $U(t)$ is

$$U(t) = U_0(1 + \sin(2\pi f t)) \tag{4.24}$$

where the oscillation frequency f is 5, 10, 15 and 20 Hz, respectively.

For numerical implementation, initially the particles are regularly arranged at the bottom of the bed, with zero velocities and a height of about 220 mm, and the initial gas velocity is U_0. Then as time advances, the inflow velocities of gas at the inlet are uniformly varied according to Eq. (4.24). The outflow boundary follows the Neumann condition, i.e. $\frac{\partial u}{\partial n} = 0$.

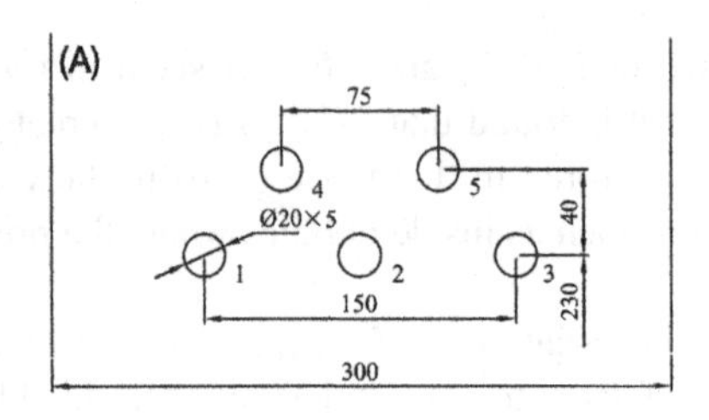

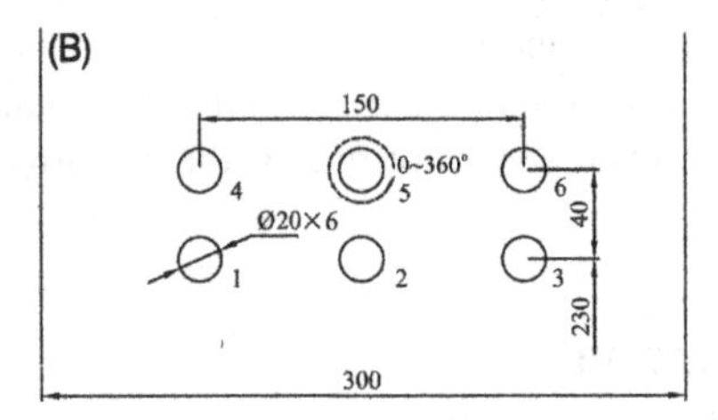

FIGURE 4.40

Configurations of the immersed tubes. (A) Case 1: the staggered arrangement. (B) Case 2: the inline arrangement.

Table 4.11 Parameters used in simulation and experiment.

Bed scale (mm)	**300 x 800**	**Oscillation frequency (Hz)**	**5, 10, 15, 20**
Case 1: Locations of five stagger immersed tubes (mm)	(75, 230), (150, 230), (225, 230), (112.5, 270), (187.5, 270)	Superficial gas velocity (m/s)	$U_0 = 1.8$
		Gas density (kg/m³)	$\rho_g = 1.29$
		Gas viscosity (Pa·s)	$\mu_g = 1.85 \times 10^{-5}$
		Particle numbers	80343
		Particle diameter (m)	$d_p = 883 \times 10^{-6}$
		Particle density (kg/m³)	$\rho_p = 2650$
Case 2: Locations of six inline immersed tubes (mm)	(75, 230), (150, 230), (225, 230), (75, 270), (150, 270), (225, 270)	Tube diameters (mm)	$D_p = 20$
		Restitution coefficient	$e = 0.9$
		Friction coefficient	$\gamma = 0.3$
		Collision stiffness	$K_n = 1000$
		Time step (s)	$\Delta t = 10^{-5}$
		Simulation time (s)	3.0
		Mesh grid size (mm)	$\Delta x = \Delta y = 2$

With regard to time integration, there are at least three types of stability requirements: (1) CFL condition for numerical solution of fluid motion equations; (2) integration of the particle motion equation to obtain accurate particle motion trajectories; (3) treatment with the particle-particle or particle-surface interactions. Among these three types of stability requirement, the last one is far stricter than the former two types. For simulating the particle-particle collision, a fairly fine time step is required on account of the calculation of particle deformation. Thus, to obtain stable and re-

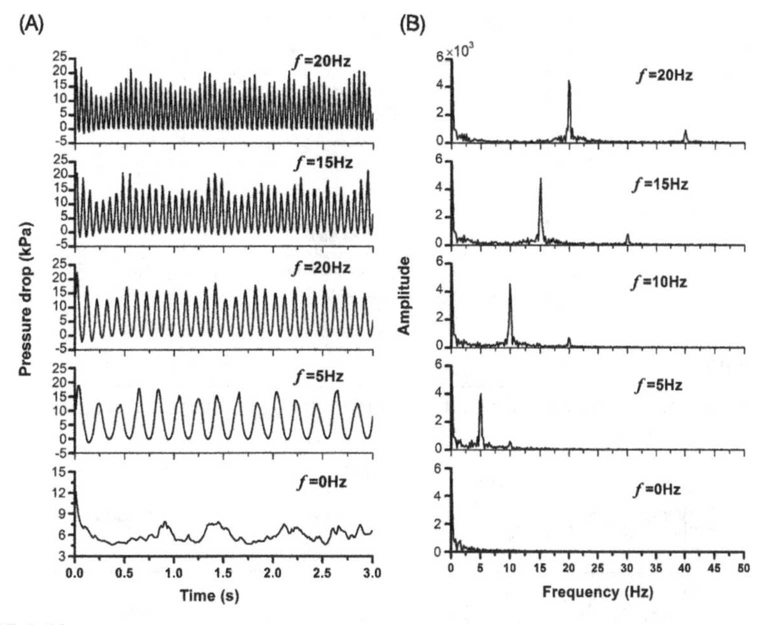

FIGURE 4.41

Pressure drop and its spectrum of all pulsating frequencies for case 1.

liable results, $\delta t = 10^{-5}$ s is used for time advancement. Meanwhile, the other two types of requirements are also satisfied.

4.4.2.1 Pressure drop

Firstly, the straightforward characteristic of pulsed fluidization is the pulsed variation of bed pressure drop. Take case 1 for example – Fig. 4.41 shows the time history of fluctuation of pressure drop for frequencies of oscillation of $f = 0$, 5, 10, 15, 20 Hz and their Fourier spectrums. It is evident that for pulsed fluidization in spite of the arrangement mode of the immersed tube bank, the pressure drop fluctuates at a dominating fundamental mono frequency. On the contrary, for nonpulsed fluidization ($f = 0$) the pressure drop has no fundamental frequency and the time evolution of it is irregular.

Moreover, in Fig. 4.41, the amplitudes of the pulse of pressure drop are very large since the amplitude of the pulse velocity $U(t)$ ranges from 0 to $2U_0$. When $U = 0$ the pressure drop reaches zero, and when $U = 2U_0$ the pressure drop reaches the maximum value of about 1.5×10^4 Pa. In Fig. 4.42 the relationship between the time-averaged pressure drop and the pulse frequency is illustrated. It presents a gradual increase of the average pressure drop when the pulse frequency increases.

4.4.2.2 Force behavior analysis

Secondly, it is interesting to explore the behavioral characteristics of particles, especially the fundamental inter- or intra-phase interactions in the pulsed fluidized bed.

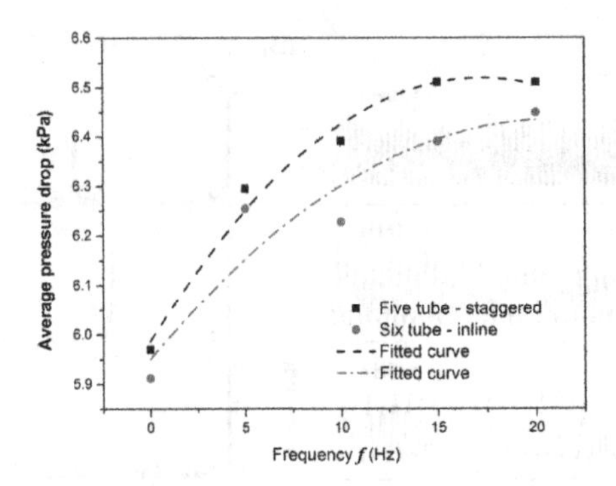

FIGURE 4.42

Relationship between the time-averaged pressure drop and the pulsating frequency.

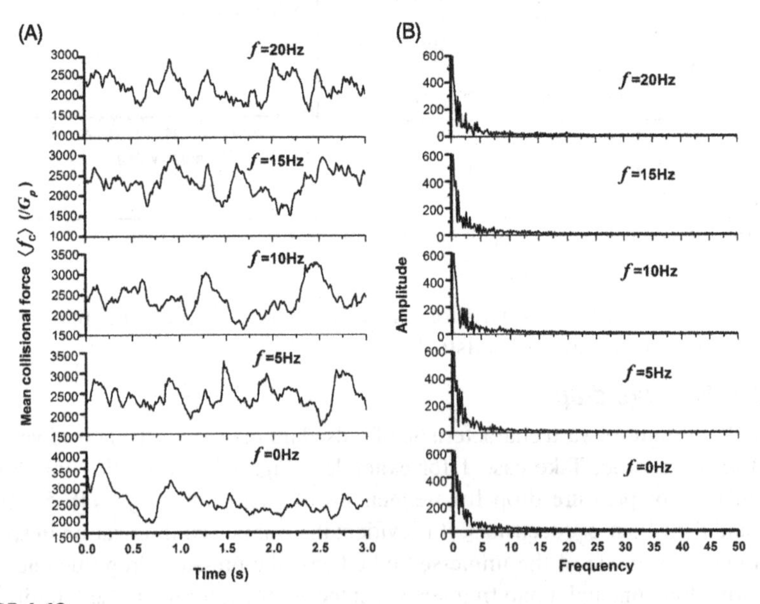

FIGURE 4.43

Time history of mean collisional force $\langle f_c \rangle$ and its spectrum for case 2.

The aim of the present section is to investigate the characteristics of several fundamental interaction forces, such as mean collisional force $\langle \boldsymbol{f}_c \rangle$, mean drag force $\langle \boldsymbol{f}_D \rangle$ and mean pressure gradient force $\langle \boldsymbol{f}_P \rangle$.

The mean collisional force $\langle \boldsymbol{f}_c \rangle$ is an average value of all kinds of collisional forces of all particles, including the particle-particle collisional force, the particle-wall collisional force and the particle-tube collisional force. Take case 2 for example – Fig. 4.43 shows the time history of the mean collisional force and its Fourier transformations. For clarity and comparison, $\langle \boldsymbol{f}_c \rangle$ is nondimensionalized by the particle

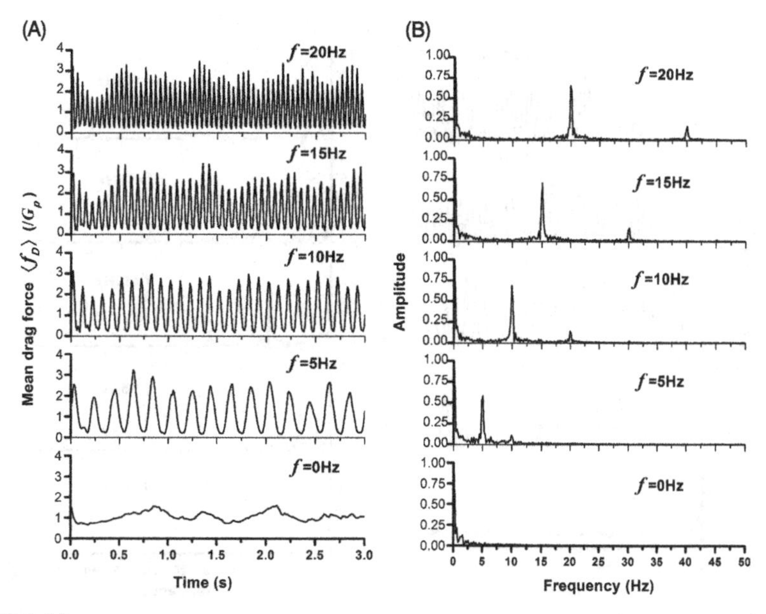

FIGURE 4.44

Time history of the mean drag force $\langle f_D \rangle$ and its spectrum for case 1.

gravity G_p. It is evident that $\langle \boldsymbol{f}_c \rangle$ is fluctuating irregularly with large amplitude and frequency. The fluctuation has no apparent period and has no evident dependence on the pulse frequency $\boldsymbol{f}$ and the mode of arrangement of the tube bank.

The collisional force is a kind of intra-solid phase interaction, which is not directly influenced by the mode of gas pulsation. Thus, it has no evident dependence on the frequency of pulsed fluidization. Moreover, the collisional force is an interaction which may be viewed as random forces acting on any individual particle. Compared to any particle cluster, the collisional force can be regard as an inner interaction between any pair of colliding particles within the cluster. As the motion of particle cluster is a type of bulk behavior which is dominantly influenced by gas flow or bubble, the collisional force seems to be of second importance compared to the hydrodynamic force.

Among the many components of the gas-solid interactions, the drag force is shown to be of the leading order, provided the particle density is far larger than the fluid density [235]. Though in the dense gas-solid flow, the drag force is relevant to the local concentration of a particle cluster, the order of magnitude for the drag force are still of the first importance. In numerical simulation, the drag force $\langle \boldsymbol{f}_D \rangle$ is a function of voidage ϵ_g, inter-phase velocity discrepancy $|\boldsymbol{u}_g - \boldsymbol{u}_p|$, etc. Thus, it is prone to be influenced by the mode of the gas pulsation. Still take case 1 for example – Fig. 4.44 shows the time history of the mean drag force and its spectrum. In contrast to $\langle \boldsymbol{f}_c \rangle$, $\langle \boldsymbol{f}_D \rangle$ is dominatingly influenced by the pulsed fluidization, showing a forced oscillation with its forced frequency exactly the same as that of pulsed

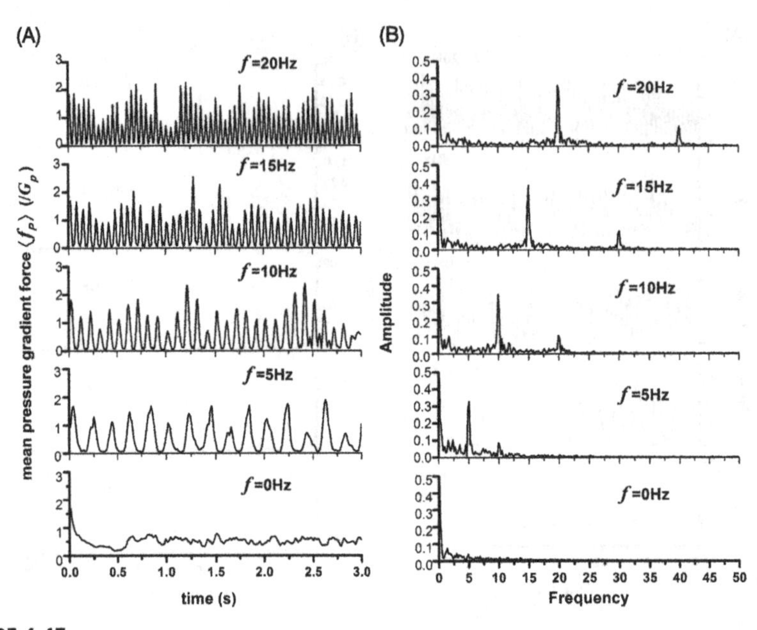

FIGURE 4.45

Time history of the mean pressure gradient force $\langle f_P \rangle$ and its spectrum for case 2.

velocity inflow. The time history of $\langle f_D \rangle$ is similar to the pressure drop, and so is the Fourier transformation of it.

From Fig. 4.44, it is observed that the frequency of oscillation of $\langle f_D \rangle$ is almost exclusively controlled by the pulse frequency of gas velocity. Moreover the peak values of $\langle f_D \rangle$ for the pulsed fluidization always exceed the maximum value of $\langle f_D \rangle$ for the nonpulsed fluidization. This is particularly beneficial for control or enhancement of the inter-phase momentum exchange. As a result, this has potential feasibility for controlling inter-phase mixing, heat transfer, product dissipation etc., in combustion industry, disregarding the mode of arrangement of tube bank.

In the same way, the time history of the mean pressure gradient force $\langle f_P \rangle$ for case 2 is shown in Fig. 4.45. Similarly, the variation of $\langle f_P \rangle$ is again in perfect accordance with the variation of pressure drop and the pulsed inflow of gas.

Since the hydrodynamic force in the present study is composed of the drag force and the pressure gradient force (other forces, like the Magnus force, virtual mass force, the Saffman lift force etc. are neglected in the present study for their low magnitudes and difficulty to calculate), the simultaneous time variation of components of the hydrodynamic force with the pulsating gas velocity reveals the feasibility for controlling the fundamental interactions in the pulsed fluidized bed. Since the behavioral characteristics of fundamental interactions in a fluidized bed are still far from being well understood by researchers, the possibility of controlling the fundamental interphase interactions even in a limited scope provides an alternative way for specific utilizations in industrial applications, e.g. enhancement of heat transfer, etc.

As is mentioned above, the drag force and the pressure gradient force are fluid-to-particle forces which belong to the type of inter-phase interactions, while the collisional force can be treated as the type of intra-phase interaction if particles are treated as a single solid phase when uniform particle properties are used. By comparing Fig. 4.44 and Fig. 4.45 to Fig. 4.43, with reference to Fig. 4.41, it is found that there exist second peaks of frequency in the spectrum analysis of Fig. 4.41, Fig. 4.44 and Fig. 4.45. The secondary frequencies are about double frequencies of the first fundamental frequencies respectively. Remember that a double-frequency or multiple-frequency property (up to 'n'-times of the fundamental frequency) is one of the basic properties of the forced oscillation. The occurrence of a multiple-frequency is due to existence of fraction factors and damping factors. As a result, it is concluded that in the pulsed fluidization, both the variation of fluid motion and fluid-to-particle interactions are forced oscillated, whereas the particle-particle interactions are not affected and oscillated.

4.4.2.3 Collision on immersed tubes

The fluidized bed is equipped with immersed tubes, and it is necessary to know the time dependence or the pulse frequency dependence of collisions on tubes. To say specifically, it is necessary to explore the influence of the mode of fluidization on the behavior of the particle collision on immersed tubes, especially the circumferential distribution of collision on the outer surface of tubes. As DEM simulation is powerful in dealing with any specific calculation of the particle-particle, particle-wall (particle-tube surface) collision, all particle-tube collisions are traced easily. Therefore, the circumferential distribution of collision is analyzed through counting and accumulating at each time step the number of collisions occurred in every circumferential degree (Fig. 4.40) on the outer surface of immersed tubes.

Let's choose the 2nd tube and the 4th tube for case 1 and the 2nd tube and the 5th tube for case 2 as representative tubes for the case study. Referring to Fig. 4.40 for the mode of arrangements of immersed tubes, the 2nd tube in both cases is located in the center of the lower row of tubes and the 4th tube for case 1 and the 5th tube for case 2 are located in or near the center of the upper row of tubes. Fig. 4.46 shows respectively the distribution of the total collision numbers on outer surface of the above representative tubes. Figs. 4.46A and 4.46B are for the 2nd tube for case 1 and case 2, respectively, while Figs. 4.46C and 4.46D are for the 4th tube for case 1 and the 5th tube for case 2, respectively.

It is observed from Fig. 4.46A that when $f = 0$ Hz, the circumferential distribution of the particle-tube collision has two peaks with their peak values around lower and upper center (90° and 270°) while the first peak is greatly suppressed and the second peak is not obviously influenced when $f > 0$ Hz (pulsed fluidization). However, Fig. 4.46B shows no apparent change upon the pulsed fluidization compared to the nonpulsed fluidization on the circumferential distribution of collision. It shows the relation between the mode of arrangement and the circumferential distribution of the particle-tube collisions under the pulsed fluidization.

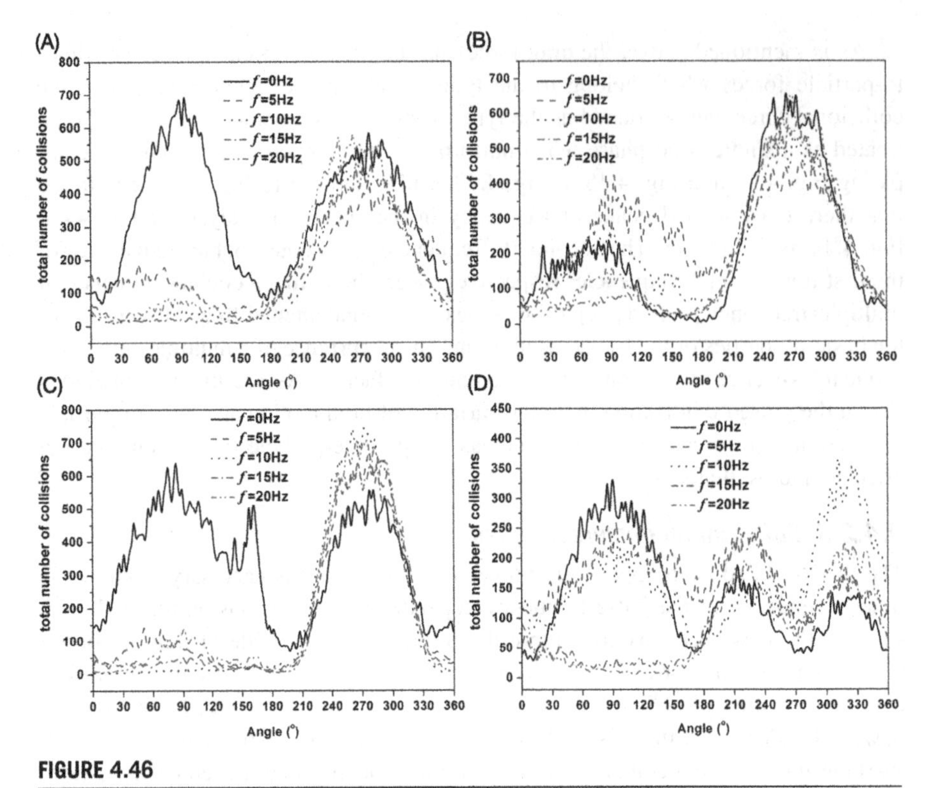

FIGURE 4.46

Circumferential distribution of the particle-tube collision on representative tubes: (A) on the 2nd tube for case 1; (B) on the 2nd tube for case 2; (C) on the 4th tube for case 1; (D) on the 5th tube for case 2.

From Fig. 4.46C, it is again observed that the circumferential distribution is similar to Fig. 4.46A, i.e. the collision on the upper half tube surface is greatly suppressed while on the lower half is slightly enhanced. It seems that the change of circumferential distribution is somewhat uniform for case 1. From Fig. 4.46D, the circumferential distribution is also changed, and seems to be more complicated. For example, for relatively high frequency ($f = 15$ and 20 Hz) the collision on the upper half surface (0–180°) is also suppressed while for relatively low frequency ($f = 5$ and 10 Hz) the collision is not apparently influenced. For the lower half (180–360°) surface, the circumferential distribution of collision has two small peaks with their peak values appearing around (270°±45°). It is due to the inline arrangement of immersed tubes and the special location of tube 5 with axisymmetric characteristics. In this way, the distribution of collision on it is symmetrically influenced by the two bilateral upstream tubes of tube 1 and tube 3. More important, the frequent direct impacts around the central bottom surface (270°) of tube 5 are greatly attenuated by tube 2, which is placed right upstream of tube 5 and acts as a shield of it, protecting it from direct normal impact at the bottom center. Meanwhile, the distribution is not greatly influenced by the pulsed fluidization since the main influencing factor is still the arrangement of tubes.

4.4.2.4 Particle phase fluctuation

The particle phase fluctuation is influenced by the pulsed gas flow, though discrepancies in the mean collisional force $\langle f_c \rangle$ under the pulsed and nonpulsed fluidized beds are not apparent. The particle statistics is carried out based on the Eulerian approach of fluid statistics. In every simulation time t, the particle concentration $N_p^E(x, y, t)$ is discretized by a grid mesh (cell dimension $d_x = d_y = 2$ mm, superscript 'E' means the Eulerian quantity), and velocities of particles v_p^L (superscript 'L' means Lagrangian quantity) in each cell is summed up and divided by the number of particles N_p^E in the cell to obtain an averaged instantaneous particle velocity of the cell $v_p^E(x, y, t)$. Then the particle statistics is carried out on the analogy of the fluid statistics in Eulerian approach. To investigate the intensity of the particle fluctuation, it is appropriate to define the particle fluctuating intensity as $I_x = \frac{u_x^{r.m.s}}{\bar{U}}$, where $\bar{U} = \sqrt{\bar{v}_x^2 + \bar{v}_y^2}$ is the time averaged velocity of particles. For example, Figs. 4.47A and B show comparison of I_x and I_y, respectively, between the nonpulsed fluidization (the left subfigure) and the pulsed fluidization (the right subfigure) for case 1 with the pulse frequency $f = 20$ Hz. From the left subfigures of Figs. 4.47A and B, it is observed that the horizontal and vertical fluctuating motion of particle of the nonpulsed fluidization is intensive, ranging from the immediate downstream behind the immersed tubes to the outlet of the fluidized bed. However, the horizontal (right subfigure of Fig. 4.47A) and vertical (right subfigure of Fig. 4.47B) fluctuating motion of particle in the pulsed fluidized bed is suppressed in the immediate downstream region of immersed tubes. Until relatively far downstream of the immersed tube the suppressed fluctuating motion of particle is released or recovered.

Thus, it is interesting to find that the pulsed fluidization seems to suppress the particle fluctuation both in horizontal and vertical direction. To confirm this phenomenon, Figs. 4.48A and B show the time history of the fluid velocity at different heights of the bed, i.e. height (H for short in Figs. 4.48) equal to 0.2, 0.4, 0.6 and 0.8 m. From Fig. 4.48A, it is observed that though the variation of the fluid velocity is irregular, the Fourier spectrum of it shows the low fundamental frequencies in variation. Quite on the contrary, from Fig. 4.48B it is observed that the fluctuations of the fluid velocities below the height of $H = 0.6$ m have large high components of frequency (around 20 Hz) in their Fourier spectrums. However at the height of $H = 0.8$ m which is far downstream from the immersed tubes the high frequency components of the fluid velocity disappear. The results in Fig. 4.48 are in accordance with those in Fig. 4.47, and it is find that the high frequency component of the fluctuation in the fluid velocity might lead to suppression in the particle fluctuating motion. Moreover, it is found that the high frequency components in Fig. 4.48B (about $f = 20$ Hz) is the frequency of pulsed inflow, which means that the pulsed fluidization might lead to suppression of the particle fluctuation motion. Additionally, the disappearance of high frequency components of the fluid velocity in Fig. 4.48 at

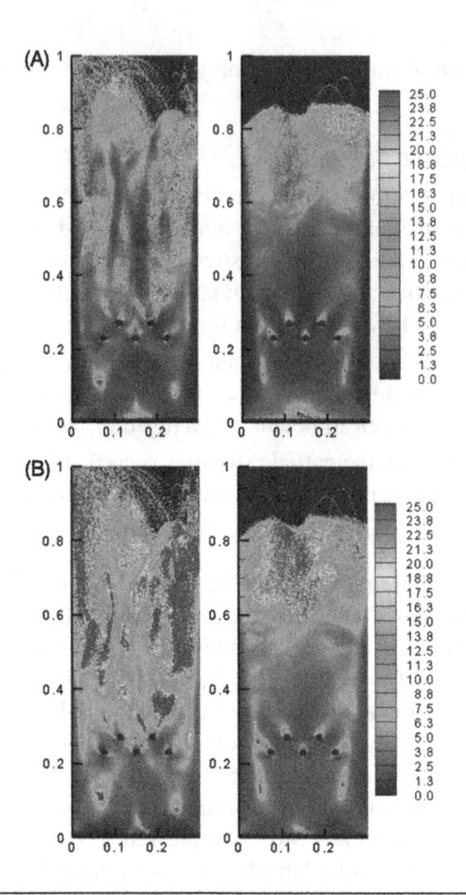

FIGURE 4.47

Visualization and comparison of the particle fluctuating intensity I_x (A) and I_y (B) for case 1. (A) I_x: the left subfigure is for nonpulsed and the right one is for pulsed. (B) I_y: the left subfigure is for nonpulsed and the right one is for pulsed.

the height of 0.8 m is due to the attenuation of the pulsed gas flow, resulting in release or recovery of an intensive particle fluctuating motion.

4.5 Spouted bed

As a large particle is always encountered in a spouted bed, an experiment work of spouted bed [290] is used here as the referential case for the model validation. The lower part of the spouted bed is shown in Fig. 4.49A. Table 4.12 lists the bed dimension scales and parameters used in the current simulation in comparison of the parameters used in the experiment.

Two main categories of simulation are performed here, i.e. the two-dimensional cases (the gas phase is two-dimensional and named as "2D case") and the three-

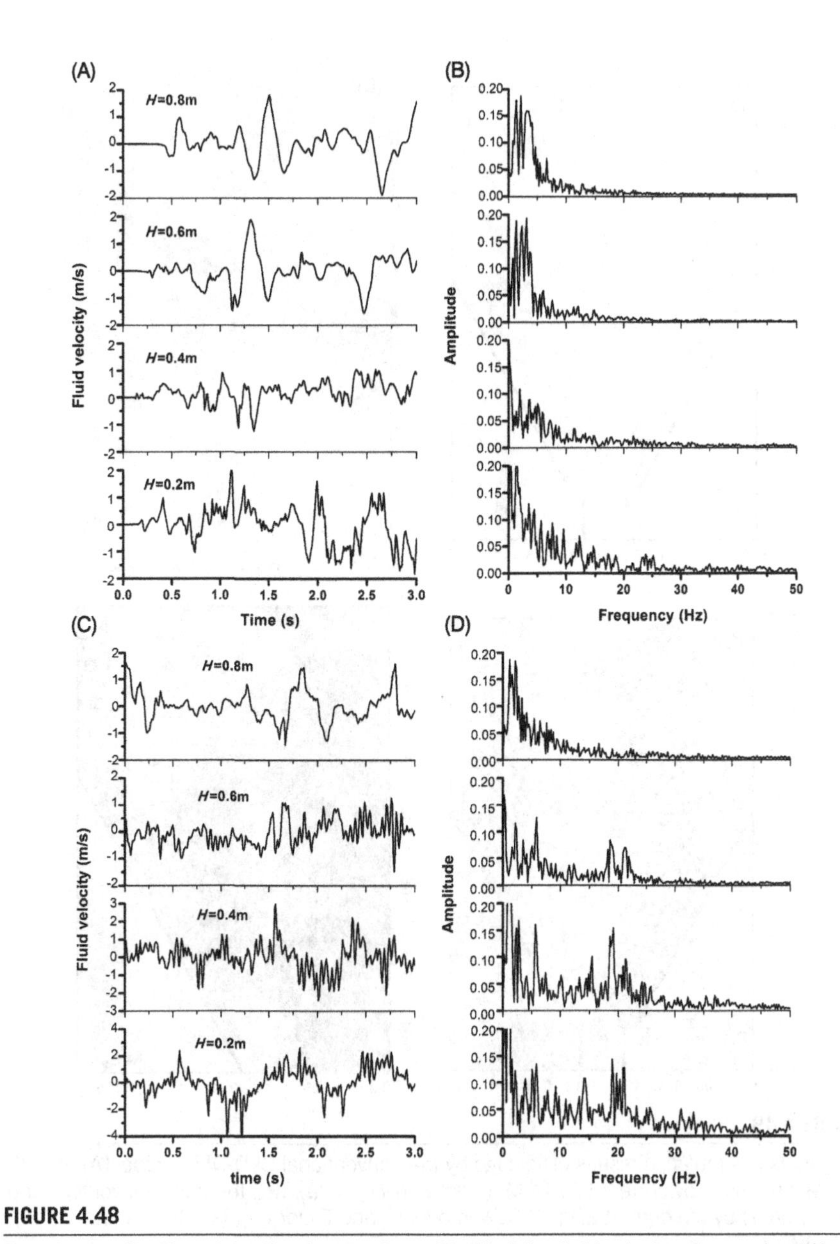

FIGURE 4.48

Time history (A, C) and the Fourier spectrum (B, D) of the fluid velocity of nonpulsed (A, B) and pulsed (C, D) fluidization at different heights of $H = 0.2$, 0.4, 0.6 and 0.8 m for case 1 with $f = 20$ Hz.

dimensional cases (the gas phase is three-dimensional with the depth of bed $D_B = d_p$, and named as "3D case"). In addition, a demonstrative three-dimensional case of $D_B = 3d_p$ is also performed here only to show the capability of the current model for

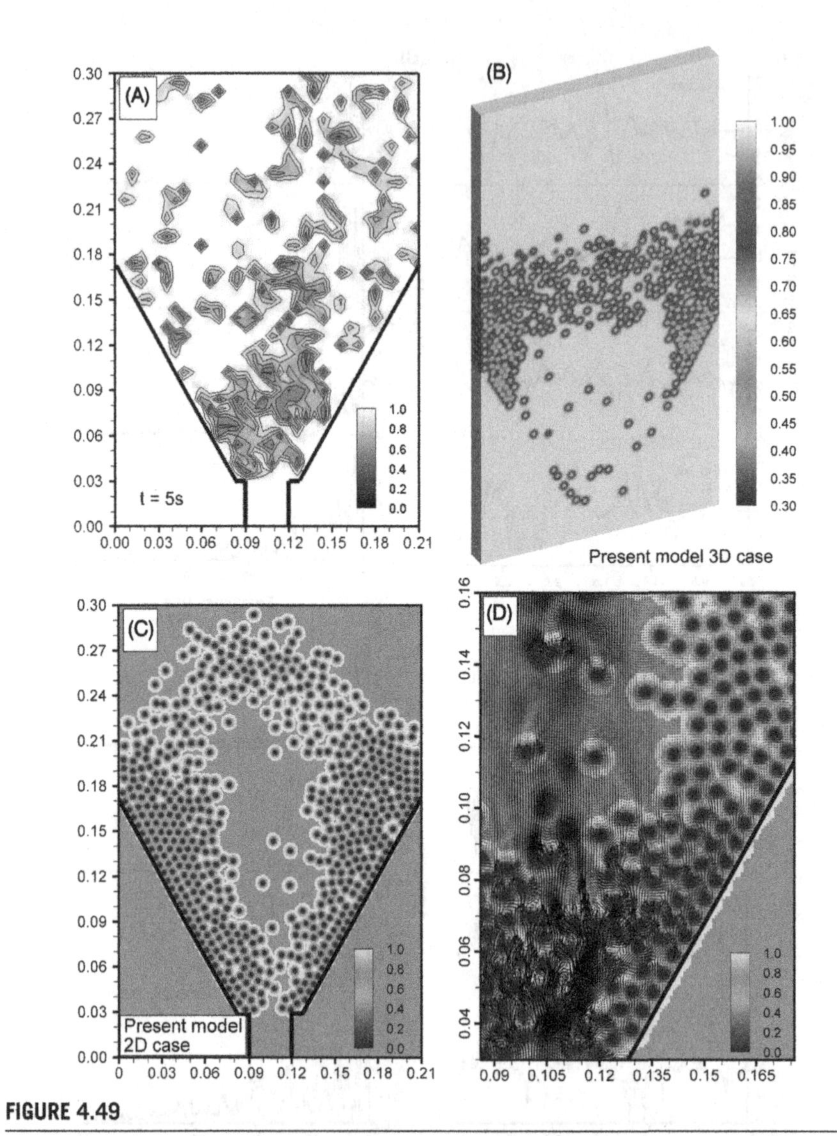

FIGURE 4.49

Snapshots of simulation results of voidage by the conventional CFD-DEM model (A), the 3D simulation results by current LES-DEM model with $D_B = 3d_p$ (B), the transient voidage and velocity fields by the current 2D LES-DEM model (C and D) for $U_{g,s} = 4.39$ m/s and $H_b = 13.4$ cm.

a 3D simulation of a deeper bed. As it is meant to be a two-dimensional spouted bed of very small thickness (36 mm ($D_B = 6d_p$)) in experiment, the bed depth is reduced to 6 mm ($D_B = d_p$ for one particle diameter in depth only) in current 2D and 3D simulations to save the computational cost. Note that there is no technical difficulty to extend the simulation from two-dimensional to three-dimensional system, except the

Table 4.12 Parameters used in simulation and experiment.

Parameters	In experiment	In simulation
Bed width W_B (mm)	210	210
Bed total height H_B^T (mm)	1200	710
Bed depth D_B (mm)	36	6 (2D with LES) 6 (3D with or without LES) 18 (3D demonstration only)
Included angle of wedge φ (°)	60	60
Width at wedge bottom W_b (mm)	45	45
Slot width below wedge W_s (mm)	30	30
Particle diameter d_p (mm)	6	6
Particle density ρ_p (kg·m^{-3})	2518	2518
Static material height H_b (cm)	8, 9.2, 10.4, 11.4, 12.4, 13.3	8, 9.2, 10.4, 11.4, 12.4, 13.3
Stiffness factor k_n (N·m^{-1})	–	1.0 × 10^3
Poisson ratio ν	–	0.3
Restitution coefficient e	–	0.9
Friction coefficient μ	–	0.3
Gas density ρ_g (kg·m^{-3})	1.22	1.22
Gas viscosity μ_g (Pa·s)	1.85 × 10^{-5}	1.85 × 10^{-5}
Superficial gas velocity $U_{g,s}$ (m^{-1})	3.20, 3.72, 4.39	3.20, 3.72, 4.39
Mesh scale for gas phase δ (mm)	–	for 2D case: 0.75, 1 for 3D case: 1
Ratio of δ/d_p	–	for 2D case: $\frac{1}{8}$, $\frac{1}{6}$ for 3D case: $\frac{1}{6}$
Time step Δt (s)	–	10^{-5}
Simulation time T_s (s)	–	5

computational time. Therefore, the 3D case of $D_B = 3d_p$ is a case for demonstration only.

The DEM simulation of the particle phase is always three-dimensional but the friction from the front and rear walls are neglected. This configuration may be regarded as a bed of one-layer thickness extracted from the median plane of the real bed. Moreover, after numerical tests, it is noticed that there is no need to simulate a bed of a practical height, since the particle cannot be conveyed to such a high level. Thus, the bed height in simulation is reduced to 710 mm only to save the computational capacity, too.

The gas is incompressible air issued from the bottom of silo underneath the spouted bed. The Dirichlet condition is set as the inlet boundary. A fixed parabolic profile is used to mimic the velocity distribution in a steady pipe flow as below:

$$u_g(x) = \begin{cases} U_{gc,0}(1 - (\frac{x-x_c}{r_i})^2), & \text{if } |x - x_c| \leq r_i \\ 0, & \text{otherwise} \end{cases} \tag{4.25}$$

where r_i and $U_{gc,0}$ are the half-width of the silo inlet and the maximum velocity at the center x_c, respectively. The maximum inflow velocity $U_{gc,0}$ is set according to the superficial gas velocity $U_{g,s}$ in the experiment of [290] (Table 4.12) to keep the same mass flow flux. For the convenience of comparison, the superficial gas velocity $U_{g,s}$ is used to name each case (Table 4.12). The outlet at the top is set as the Neumann boundary with $\frac{\partial \phi}{\partial n} = 0$, where $\boldsymbol{n}$ is the normal direction of the outflow area.

To show the effect of the mesh size on the virtual void fraction distribution, two mesh grids of $\delta_1 = 0.75$ mm, and $\delta_2 = 1.0$ mm are used here for the 2D gas phase solution. They are all smaller than the particle diameter with the ratios of $\frac{\delta_1}{d_p} = \frac{1}{8}$ and $\frac{\delta_2}{d_p} = \frac{1}{6}$, respectively. However, for the 3D case, only $\frac{\delta_2}{d_p} = \frac{1}{6}$ is used here for saving the computational costs.

4.5.1 CFD-DEM vs. SVFM-based fine LES-DEM

As a first step, Fig. 4.49 shows the comparison of flow patterns (visualized by the void distribution of the conventional CFD-DEM simulation (Fig. 4.49A) and the current simulation (the full 3D case in Fig. 4.49B and the 2D case in Fig. 4.49C and D). The snapshots of the 2D case correspond to the transient particle dispersion, void fraction, and transient velocity field at ($t = 5$ s) for $U_{g,s} = 4.39$ m/s and $H_b = 13.4$ cm based on the mesh scale of $\delta_1 = \frac{1}{8}d_p$. In contrast, the snapshot is on a scale of $\delta_2 = \frac{1}{6}d_p$ for the 3D case (with $D_B = 3d_p$ in Fig. 4.49B) and on an even coarser scale of $\delta_c = d_p$ in the conventional CFD-DEM simulation (Fig. 4.49A). Although $\delta_c = d_p$ is the finest mesh scale which could be used in the conventional CFD-DEM model, the results obtained by the convectional method are clearly worse than the current 2D and 3D cases. By using subparticle scales, the current model captured the clearer features of the spouting of large particles, the void dispersion, and the gas phase flow successfully. They are evidently different from the conventional CFD-DEM methods by providing more details on the gas turbulence in subparticle scales. As the most important task of present work is on the development of the subparticle LES-DEM model, attention is paid to the validation and demonstration of the model in the following work.

4.5.1.1 Comparison with experimental results

The results of the flow pattern simulation are compared to the experiment of [290]. For example, the results in Fig. 4.50 are obtained when $U_{g,s} = 4.39$ and $H_b = 12.4$ according to the conditions in the experiment [290]. As the spouted flow pattern is cyclic, this section just shows the particle motion in one period in Fig. 4.50. But the time interval $\Delta t \approx 0.05$ s is set the same as that in [290]. In details, Fig. 4.50B shows the void distribution of the 3D case at $\delta_2 = \frac{1}{6}d_p$, and Fig. 4.50C shows the particle dispersion of the 2D case at $\delta_1 = \frac{1}{8}d_p$.

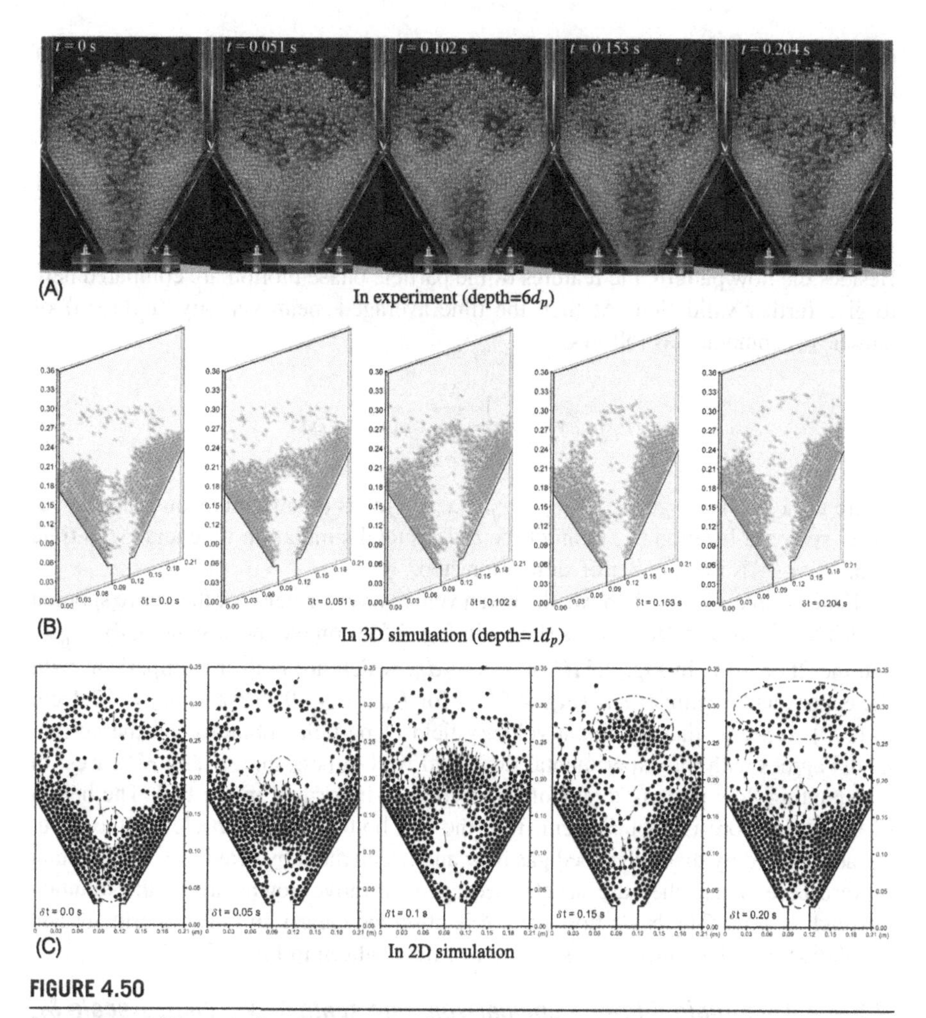

FIGURE 4.50

Comparison of the spouting flow pattern of the current 3D simulation with $D_B = d_p$ (B), the current 2D simulation (C) and the experimental results (A) of [290] for $U_{g,s} = 4.39$ m/s and $H_b = 12.4$ cm.

The spouted flow pattern of particles can be seen clearly in Fig. 4.50B and C. Both the 2D can 3D simulations indicate that the particles in the bottom center (e.g. enclosed by red dashed curves in Fig. 4.50C) are exposed to a high speed gas flow issued from the bottom silo. They are conveyed upward through the spout, upcasting and falling back in the fountain, and collected and transferred downward in the annulus. After a detailed inspection, one sees that the particles are more widely dispersed in the 2D simulation than in the 3D case. In particular, compared to the 2D case (Fig. 4.50C), they are more dilute in the spout and denser in the annulus in the 3D case (Fig. 4.50B). However, the general resemblance between the flow patterns of 2D and 3D simulation confirms that the simulation captures the basic feature of the spouting flow patterns and agrees well with the experimental observations phenomenologically.

4.5.2 Particle phase behavior

4.5.2.1 Mean velocity field comparison: 2D, 3D vs. Exp

Besides the flow pattern, the features of the particle phase motion are compared here to give further validation. At first, the time averaged mean velocity $\langle \boldsymbol{v}_p(x,y)\rangle$ of particles is computed as follows:

$$\langle \boldsymbol{v}_p(x,y)\rangle = \frac{1}{N_s}\sum_{i=1}^{N_s} \boldsymbol{v}_p(x,y,i\Delta t) \tag{4.26}$$

where $\boldsymbol{v}_p(x,y,t) = v_{p,x}(x,y,t)\boldsymbol{e}_x + v_{p,y}(x,y,t)\boldsymbol{e}_y$ is the instantaneous particle velocity vector at location (x,y) and time t. T_s is total simulation time and Δt is time step. $N_s = \frac{T_s}{\Delta t}$ is the number of simulation step.

For example, Fig. 4.51 shows the mean velocity vector fields of the entire spouted field. The 3D case of $D_B = d_p$ is shown in Fig. 4.51B on the mesh scale of $\delta_2 = \frac{1}{6}d_p$ and the 2D case is in Fig. 4.51C on $\delta_1 = \frac{1}{8}d_p$, which are both in comparison with the experimental counterpart (Fig. 4.51A) for $U_{g,s} = 4.39$ m/s and $H_b = 12.4$ cm in [290]. It seems that the mean velocity field of particles obtained by the current model agrees with the experimental flow field well in both the 2D and 3D cases. In the spouting flow pattern, a pair of large vortices is formed in the bed. The height of the vortex core is about 15 cm from the bed bottom. The particles in the spout are accelerated by the high speed gas flow and decelerated by gravity in the fountain upward core. Then, they are accelerated again to move downward in the fountain descending zone. The basic flow pattern is observed clearly in both experiment and simulation, which confirms the similarity of flow pattern in Fig. 4.50.

4.5.2.2 Longitudinal profile comparison: fine scale δ_1 vs. coarse scale δ_2, and LES vs. no LES

Secondly, Fig. 4.52 shows a comparison of the longitudinal time-averaged mean velocity on the bed axis, where the simulation results are obtained both from the mesh scale of $\delta_1 = \frac{1}{8}d_p$ (Fig. 4.52A) and from $\delta_2 = \frac{1}{6}d_p$ (Fig. 4.52B and C). In Fig. 4.52A and B for the 2D case on different mesh scales, the velocity profiles are appropriately divided into four stages. It is seen that: (1) The particle velocity in simulation agrees well with that in experiment at the initial region ($\Delta H \le 2.5$ cm in stage 1) during acceleration in spout. (2) However, $\langle V_{yc}\rangle$ in simulation begins to deviate from that in experiment after about $\Delta H = 2.5$ cm in the acceleration process in the end of stage 1. In particular, in experiment, the particle velocity can be still gradually accelerated in stage 2 although the acceleration is lower than that in stage 1. However, in simulation, such a process cannot be observed in stage 2 or it cannot be maintained as high as that in experiment. (3) As a result, although the decrease of velocity in current

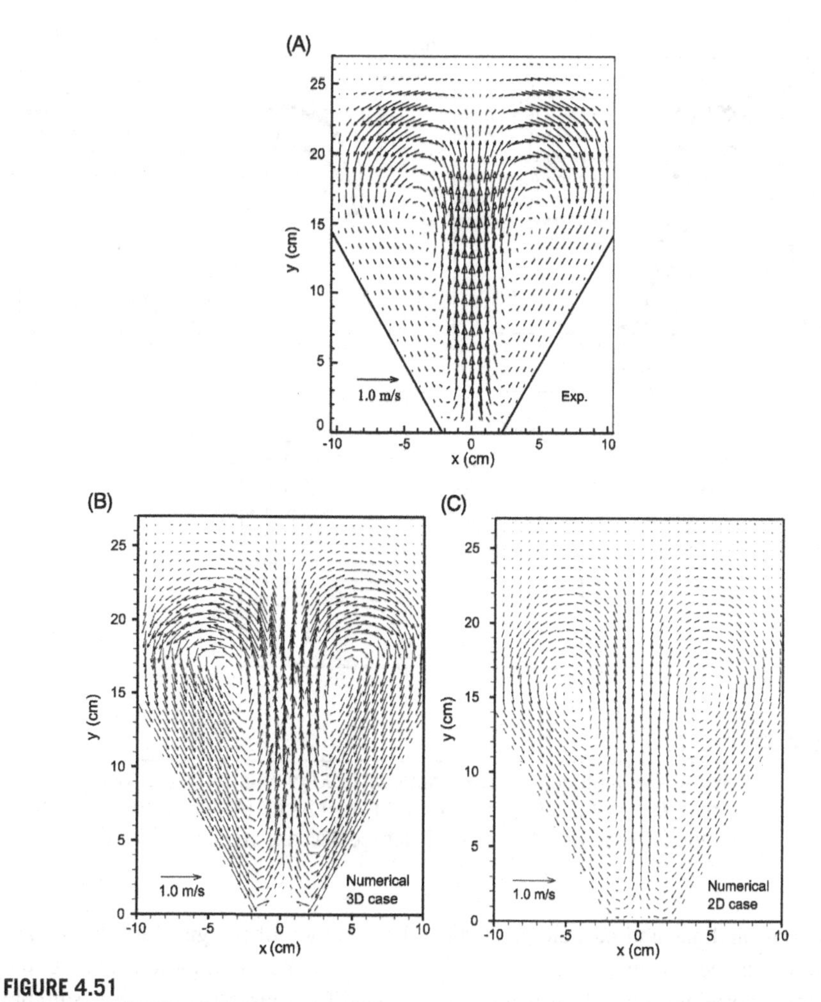

FIGURE 4.51

Comparison between the numerical ((B) for the 3D case of $D_B = d_p$ and (C) for the 2D case) and experimental (A) [290] time-averaged velocity vector fields of the entire spouted field when at $U_{g,s} = 4.39$ m/s and $H_b = 12.4$ cm.

simulation in stage 3 and 4 are similar to that in experiment, it may occur earlier or on lower heights (smaller ΔH) as that in experiment.

In our opinion, the reason for the major difference in stage 2 may lie in the following items: (a) The difference in bed dimension in the thickness direction. In experiment, it is a pseudo-2 dimensional apparatus 6 times thicker than the present 2D or 3D case. (b) The difference in the velocity profile of the silo inlet. As the velocity profile of gas in the silo inlet is lacking in experiment, it is assumed that it follows the parabolic velocity profiles in fully developed pipe flows. (c) The accuracy of the drag force and the pressure gradient force. Besides the setup configurations, operating

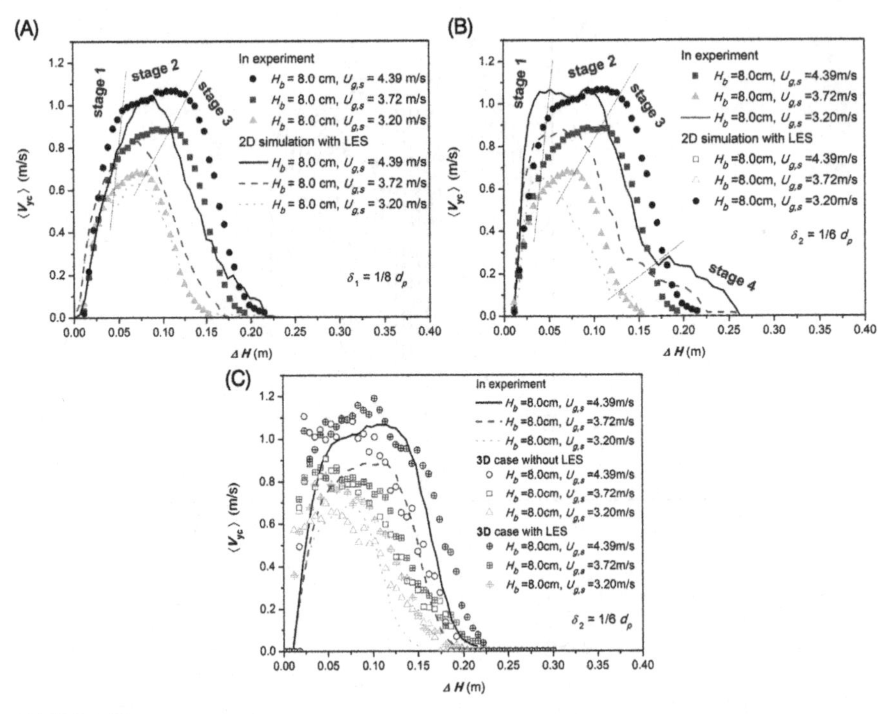

FIGURE 4.52

Comparison between the numerical and experimental [290] time-averaged velocity vector in the central line for $U_{g,s} = 4.39$, 3.72, 3.20 m/s and $H_b = 8$ cm: (A) for the 2D case with $\delta_1 = \frac{1}{8}d_p$; (B) for the 2D case with $\delta_2 = \frac{1}{6}d_p$; (C) for the 3D case with or without LES.

conditions, and gas and particle properties, the accuracy of the gas-particle interaction forces in the current simulation may also influence the simulation results. As the drag force correlation by [200] is regarded as good for a large particle, it is suitable to regard the drag force as precise enough here. However, the pressure gradient force, or equivalently, the pressure drop in bed, should be checked in detail.

Although the $\langle V_{yc} \rangle$ in experiment and simulation are not rigorously the same, the comparison of them in Fig. 4.51A and B still give some good indications on the model validation. It seems that, in general, the current virtual void fraction model is suitable for reasonable prediction of the motion of large particles in spouted beds on either the fine or coarse scales.

In addition, if the additional SGS stress tensor (Eq. (3.3)) is not included in Eq. (3.2), the present subparticle LES-DEM model is a subparticle scale CFD-DEM model. Therefore, Fig. 4.52C shows the comparison of $\langle V_{yc} \rangle$ obtained by the LES simulation and without LES simulation to compare the subparticle LES-DEM and the subparticle CFD-DEM model. It is seen that, except the case of $U_{g,s} = 3.2$ m/s, the results obtained by the LES-DEM are better than those without LES, since the

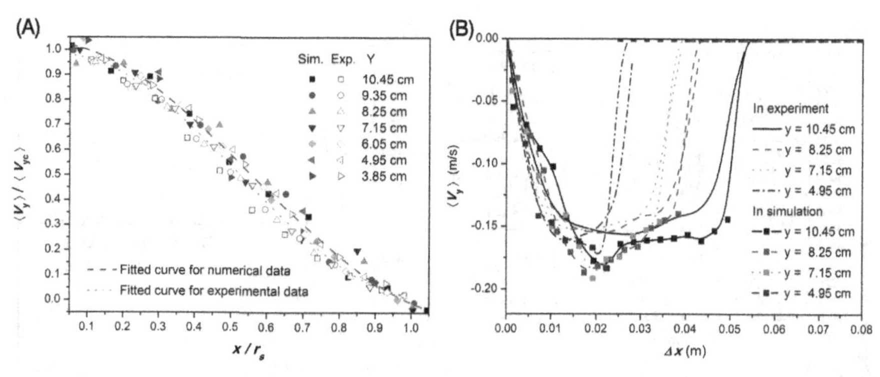

FIGURE 4.53

Comparison of the lateral distribution of the time-averaged particle velocity between simulation and experiment [290] on different heights for $U_{g,s} = 4.39$ m/s and $H_b = 12.4$ cm: (A) data in the spout on $\delta_1 = \frac{1}{6}d_p$; (B) data in the annulus on $\delta_2 = \frac{1}{8}d_p$.

LES-DEM results are generally nearer the experimental results than those without LES. This justifies the rationality of the current subparticle LES model.

4.5.2.3 Lateral profile of the particle mean velocity

On the other hand, the lateral profiles of the time-averaged particle velocity of the 2D case in the spout and annulus are shown in Fig. 4.53A and B, respectively. In Fig. 4.53A, the lateral distribution of $\langle V_y(x) \rangle$ is normalized by dividing $\langle V_{yc} \rangle$, and the lateral coordinate x is divided by the spout radius r_s (the radius of $\langle V_y(x) \rangle = 0$ in the spout), to show the feature of self-similarity. It was found that the self-similarity of a third-degree polynomial function for the lateral profile in the spout is also true in the current simulation, which agrees very well with the experimental observation in [290].

On the other hand, Fig. 4.53B shows the downward vertical velocity of particles in the annulus. It is seen that although the absolute peak values of the time-averaged velocity in the annulus are somewhat larger than those in the experiment, the general variation of the lateral profile of the particle mean velocity in the simulation are reasonably consistent with that in the experiment. Therefore, it is appropriate to conclude that the current model is also valid for predicting the lateral distribution of the motion of large particles.

4.5.3 Gas phase behavior

4.5.3.1 Pressure drop: 2D, 3D vs. Exp

The above section shows the validation and basic features of the particle phase, whereas the gas phase motion has still not been well illustrated. Therefore, the gas phase is explored in this section for a further validation. At first, the pressure drop of the bed is essential for demonstrating the validity of the current model. For that

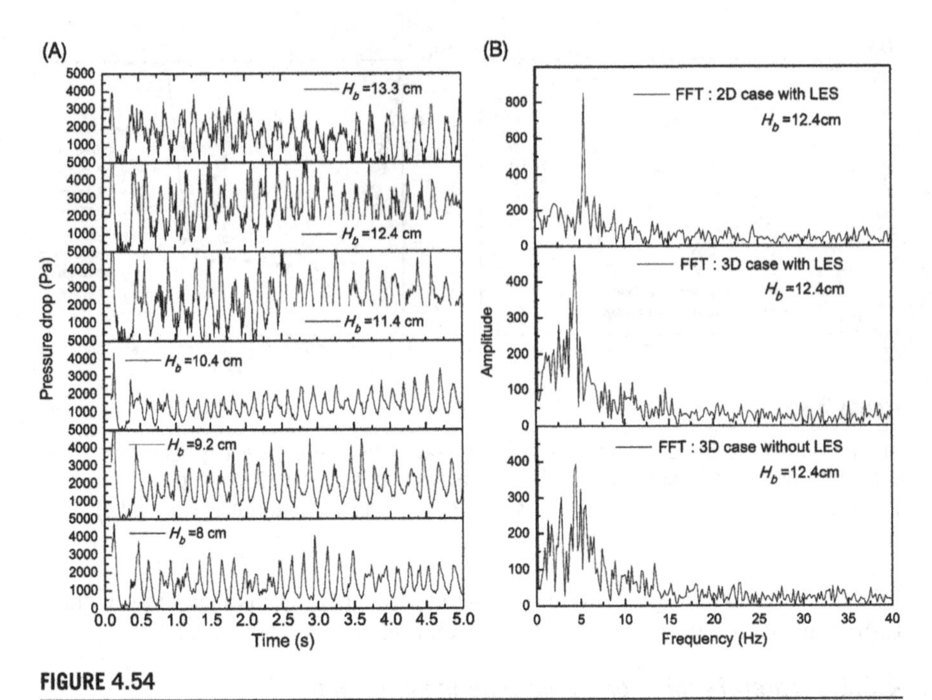

FIGURE 4.54

(A) The pressure drop of the spouted bed for different H_b at $U_{g,s} = 4.39$ m/s; and (B) the fast Fourier transformation of the pressure drop for the 2D case and the 3D cases with or without LES at $H_b = 12.4$ cm.

purpose, Fig. 4.54A shows the pressure drop of the spouted beds of different static material heights H_b at $U_{g,s} = 4.39$ m/s. It is clearly seen that the pressure drop is fluctuating periodically at a stable steady pace for all kinds of bed heights. Thus, there should be a fundamental frequency for this kind of variation. Taking $H_b = 12.4$ cm for example, Fig. 4.54B shows the Fast Fourier Transformation of the bed pressure drop of the 2D case and the 3D cases with or without LES. It proves the existence of the fundamental frequency around $F_{max} = 5.499$ Hz (in the 2D case) and $F_{max} \approx 4.436$ Hz (in the 3D case), which are very close to $F_{max} = 4.88$ Hz measured in the experiment [290].

In addition, Fig. 4.55 shows the simulation data on the relationship between the fundamental frequency F_{max} and the static material height H_b in comparison with the experimental data reported in [290], [291], and [292]. It includes the 2D simulations on $\delta_1 = \frac{1}{8}d_p$ and $\delta_2 = \frac{1}{6}d_p$ and the 3D simulations with or without LES. For each case, the numerical data obtained from the current model confirm that F_{max} is decreasing linearly with the increase of H_b. Compared to the gradient of F_{max} with respect to H_b in the experiment of [290], the simulation results obtained from the finer mesh scale of $\delta_1 = \frac{1}{8}d_p$ are better than those obtained from the coarser mesh scale $\delta_2 = \frac{1}{6}d_p$. However, the numerical data on $\delta_2 = \frac{1}{6}d_p$ are closer to those in [291]. For the 3D cases, the gradient of the dependence of F_{max} on H_b in the LES-DEM simu-

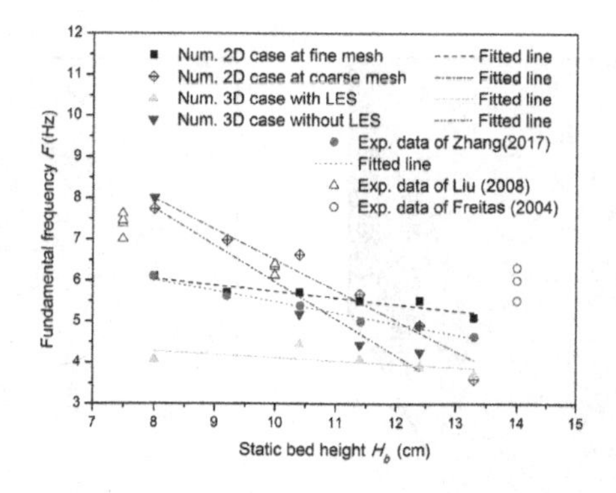

FIGURE 4.55

Comparison of the fundamental frequencies of the pressure drop (including the 2D case and the 3D case with or without LES) with literature for different static bed heights H_b at $U_{g,s} = 4.39$ m/s.

lation is closer to the experimental results of [290] than those without LES. However, the absolute values of F_{max} in LES-DEM are somewhat lower than those in the experiment, and the scatting data without LES are nearer the experimental data than those with LES. Therefore, although with some discrepancies, the 3D simulations still show generally acceptable consistency with the experimental data in different extents.

In conclusion, the current 2D LES-DEM simulation on different mesh scales and the 3D simulation with or without LES are all reliable. They demonstrate the validity of the current model for predicting the basic features of the gas phase, too. As the simulation conditions cannot be exactly the same with the experimental conditions, some differences still exist between experiment and simulation. But the simulation data are still generally in good accord with the experimental results. Therefore, together with the demonstrative data on the particle phase comparison, it is certain that the current model is viable for gas-large-particle flow simulation on the basis of subparticle scales with $\delta < d_p$.

4.5.3.2 Velocity of gas phase: 2D

Because of the inherent advantage of the current LES-DEM simulation on fine scales, it is necessary to show more features of the gas phase motion in the spouted bed.

For example, defining

$$\langle \varphi_g(x, y) \rangle = \frac{1}{N_s} \sum_{i=1}^{N_s} \varphi_g(x, y, i\Delta t) \tag{4.27}$$

$$\langle \varphi_g'(x, y) \rangle = \left(\frac{1}{N_s} \sum_{i=1}^{N_s} \left(\varphi_g(x, y, i\Delta t) - \langle \varphi_g(x, y) \rangle \right)^2 \right)^{1/2} \tag{4.28}$$

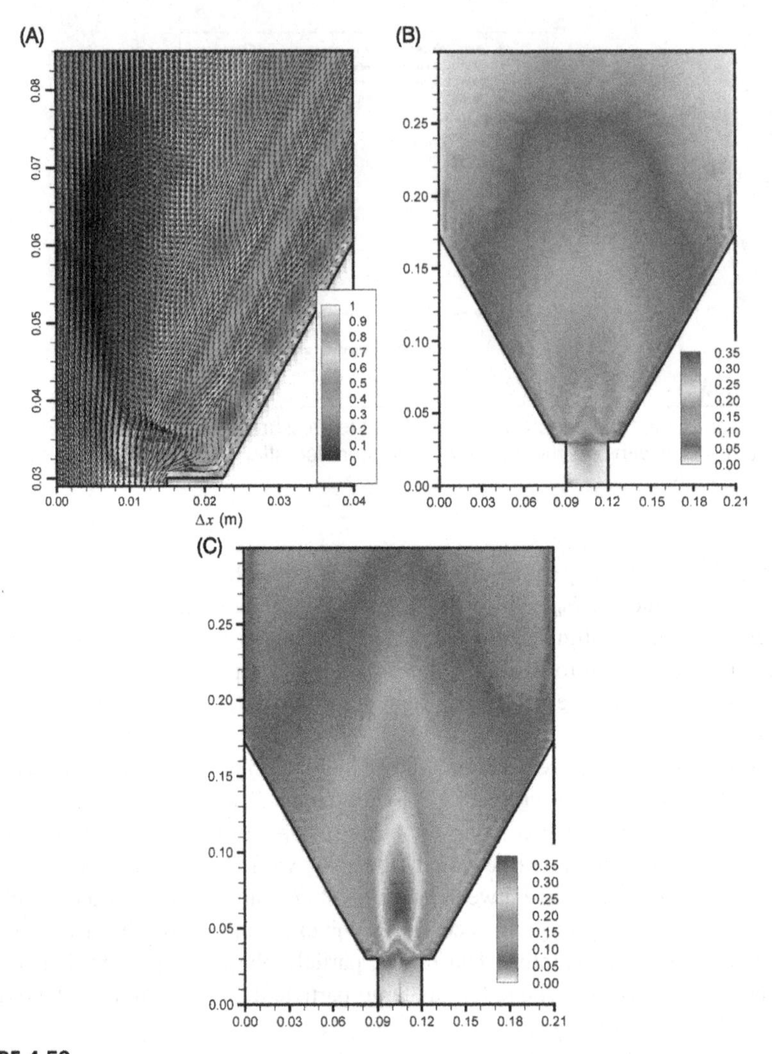

FIGURE 4.56

(A) The time-averaged gas phase velocity for $H_b = 13.4$ cm and $U_{g,s} = 4.39$ m/s. (B) The root of the mean square of horizontal fluctuating velocity $u_{g,x}/U_{gc,0}$, and the vertical fluctuating velocity $u_{g,y}/U_{gc,0}$.

it is possible to show the time-averaged and root of mean square values of gas velocity and the void fraction by letting $\varphi_g = u_{g,x}, u_{g,y}$ and ϵ_g, respectively.

Taking the 2D case for example, Fig. 4.56A shows the time-averaged velocity field of the gas phase $\langle u_{g,x}\rangle \boldsymbol{e}_x + \langle u_{g,y}\rangle \boldsymbol{e}_y$ for the case of $H_b = 13.4$ cm and $U_{g,s} = 4.39$ m/s. Fig. 4.56B and C show the root of mean square of the horizontal velocity $\langle u'_{g,x}\rangle$ and the vertical velocity $\langle u'_{g,y}\rangle$ (normalized by $U_{gc,0}$) of the gas

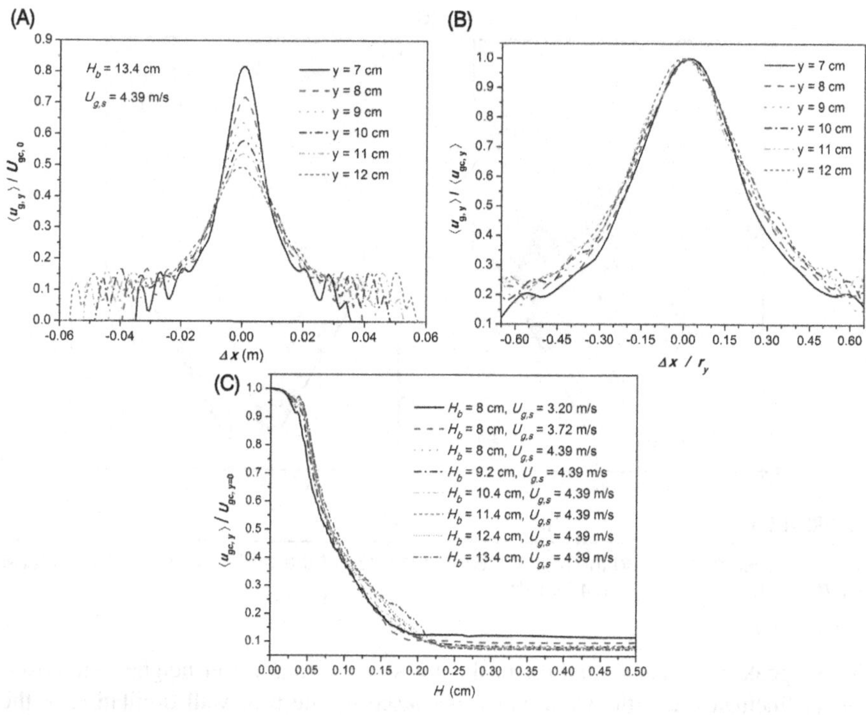

FIGURE 4.57

The time-averaged (A) and normalized time-averaged gas phase vertical velocity (B) for $H_b = 13.4$ cm and $U_{g,s} = 4.39$ m/s on the heights from $y = 7$ to 12 cm. (C) The time-averaged gas phase vertical velocity on the central axis of the bed for different H_b and $U_{g,s}$.

phase, respectively. The color in Fig. 4.56A illustrates the time-averaged voidage distribution. There exist clear stripes along the lateral wall because the particle is large. The particle size (6 mm) is comparable to the inlet size (30 mm) within one order of magnitude. The near wall effect [293,294] of the void fraction oscillation must exist for such condition. The colored stripes of the void fraction indicate the time-averaged arrangement of large particles on and near the lateral wall. It is also seen that the gas velocity in the central spout is high, whereas it becomes lower in the interfacial region between the spout and the annulus, even lower in the annulus, and of particularly lowest values in the near wall region. Similarly, in Fig. 4.56B and C, the horizontal and vertical fluctuating velocity is higher in the spout especially near the inlet. The highest $\langle u'_{g,x}\rangle$ region near the inlet is relatively shorter than the region of highest $\langle u'_{g,y}\rangle$, and the highest level of $\langle u'_{g,x}\rangle$ (about $0.2U_{gc,0}$) is lower than that of $\langle u'_{g,y}\rangle$ (about $0.35U_{gc,0}$).

For more details, Fig. 4.57A extracts the mean $\langle u_{g,y}\rangle$ on the heights of $y =$ 7–12 cm from the data of the case $H_b = 13.4$ cm and $U_{g,s} = 4.39$ m/s. The $\langle u_{g,y}\rangle$

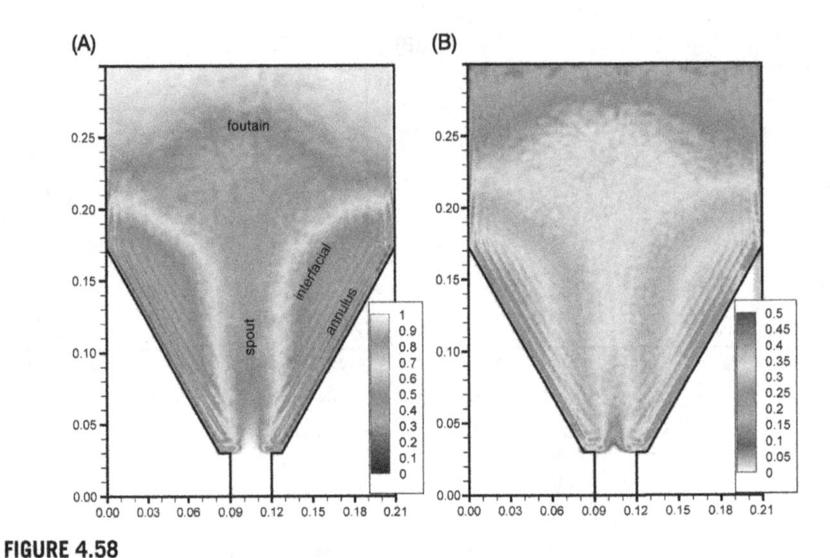

FIGURE 4.58

The time-averaged (A) and the root of mean square (B) of the void fraction of the gas phase for $H_b = 13.4$ cm and $U_{g,s} = 4.39$ m/s.

have a peak in the center and the peak values decrease at larger heights. Moreover, $\langle u_{g,y} \rangle$ fluctuates near the wall, which is caused by the near wall oscillation of the void fraction [293,294] and the time-averaged regular arrangement of large particles near walls. It confirms the color contour of voidage in Fig. 4.56A.

The width of the data dispersion in the x direction becomes larger at higher y. Therefore, the $\langle u_{g,y} \rangle$ is normalized by the highest value of it in the center $\langle u_{gc,y} \rangle$, and dividing x by the half-width r_y of the bed at height y. The normalized data are shown in Fig. 4.56B. It is found that the normalized $\langle u_{g,y} \rangle$ have similar characteristics on different heights, especially in the spout. As the larger particles are spouted by the gas, the similarity of the gas phase velocity of $\langle u_{g,y} \rangle$ in Fig. 4.57B should be the main reason for the existence of the self-similarity characteristics in the particle velocity in Fig. 4.53A.

In a similar manner, the gas velocity $\langle u_{gc,y} \rangle$ on the central axis of the bed is normalized by $\langle u_{gc,y=0} \rangle$ at the inlet center, and $\langle u_{gc,y} \rangle / \langle u_{gc,y=0} \rangle$ of different cases in Fig. 4.57C is shown. It is seen that the $\langle u_{gc,y} \rangle / \langle u_{gc,y=0} \rangle$ has, although not exactly, also a pseudo-similar feature among different cases. It indicates that the pseudo-similar feature may be mainly dominated by the pseudo-similar flow pattern in the spout. In other words, if the particle concentration is overall sufficiently dilute in the spout, the gas phase may dominate the flow pattern in the spout. Therefore, $\langle u_{gc,y} \rangle$ may have pseudo-similar characteristics, just like a jet flow in bed.

4.5.3.3 Void fraction of the gas phase: 2D

Let $\varphi = \epsilon$ in Eqs. (4.26)–(4.27), then the fields of time-averaged $\langle \epsilon_g \rangle$ and root of mean square $\langle \epsilon_g' \rangle$ of the void fraction are shown in Fig. 4.58A and B, respectively. In

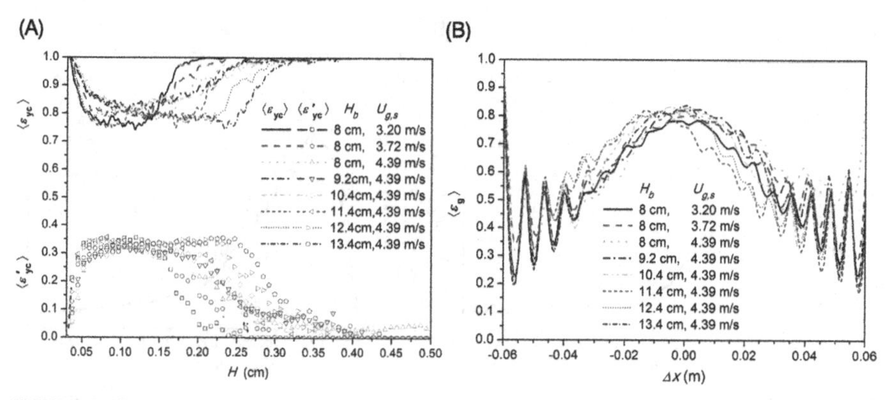

FIGURE 4.59

(A) The time-averaged $\langle \epsilon_g \rangle$ and R.M.S value $\langle \epsilon_g' \rangle$ of the void fraction of the gas phase on the central axis of the bed for different H_b and $U_{g,s}$. (B) The time-averaged void fraction $\langle \epsilon_g \rangle$ on $y = 13$ cm for different H_b and $U_{g,s}$.

Fig. 4.58A, it is clear that the overall bed can be divided into four regions, the spout, the fountain, the annulus, and the interfacial between the three. Differently from the conventional three-region division of the spouted bed, it is appropriate to separate the annulus region into an annulus (consisting of dense particles and low $\langle \epsilon_g \rangle$) and an interfacial region (intermediate dense and having medium $\langle \epsilon_g \rangle$). As indicated by Fig. 4.58, it usually has $\langle \epsilon_g \rangle \geq 0.7$ in the spout and the fountain, $\langle \epsilon_g \rangle \leq 0.4$ in the annulus, and $0.4 \leq \langle \epsilon_g \rangle \leq 0.7$ in the interfacial region between them. On the other hand, $\langle \epsilon_g' \rangle$ is the largest in the interfacial region (above 0.3), and intermediate in the fountain (around $\langle \epsilon_g' \rangle \approx 0.3$). Otherwise, $\langle \epsilon_g' \rangle$ is always less than 0.3.

For quantitative illustration, Fig. 4.59A extracts $\langle \epsilon_g \rangle$ and $\langle \epsilon_g' \rangle$ on the central vertical axis for $H_b = 13.4$ cm and $U_{g,s} = 4.39$ m/s. $\langle \epsilon_g \rangle$ is above 0.75 in the central axis, and $\langle \epsilon_g' \rangle$ is below 0.4 for all cases. In particular, on the central axis in the spout, it has about $0.75 \leq \langle \epsilon_g \rangle \leq 0.85$, and $0.25 \leq \langle \epsilon_g' \rangle \leq 0.35$, within the range of 7.5 cm $\leq y \leq 15$ cm. On the other hand, the lateral distribution of $\langle \epsilon_g \rangle$ on height $y = 13$ cm for different cases are shown in Fig. 4.59B. $\langle \epsilon_g \rangle$ is always decreasing and fluctuating near the wall, and the fluctuating of voidage near the wall is just similar to that observed in the packed beds of large particles [293,294].

In conclusion, the flow behaviors of large particles in the spout are relatively dilute two-phase jet flows with $\langle \epsilon_g \rangle \approx 0.7$–$0.8$, whereas it is like a relatively dense packed bed in the annulus and with $\langle \epsilon_g \rangle$ fluctuating around 0.2–0.5. More importantly, the above results of the gas phase behavior demonstrate the ability of the current model for capturing the gas phase behaviors on the fine meshes of $\delta < d_p$.

4.5.3.4 The effect of the parameter τ: 3D

The gas velocities and voidages in the 3D case are used here to analyze the effect of the parameter τ. For example, Fig. 4.60A and B show the normalized mean gas velocity $\langle u_{gc,y} \rangle$ on the central axis of the bed and the root of mean square of $\langle u_{gc,y}' \rangle$

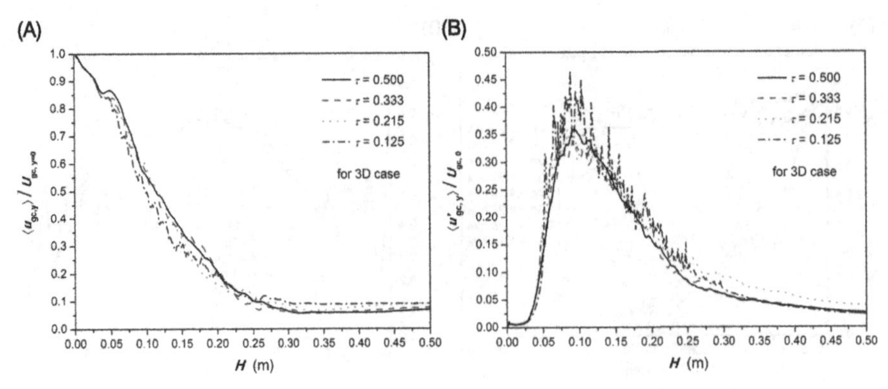

FIGURE 4.60

The time-averaged (A) and the root of mean square (B) of the gas phase vertical velocity on the central line under different τ for $H_b = 13.4$ cm and $U_{g,s} = 4.39$ m/s.

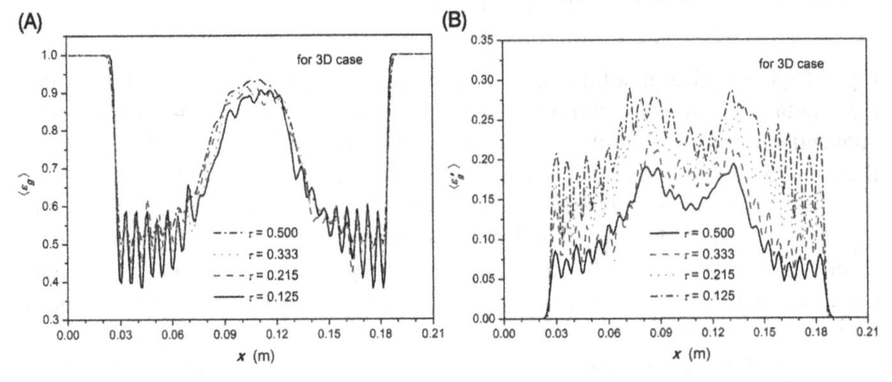

FIGURE 4.61

The time-averaged (A) and the root of mean square (B) of the gas phase voidage on $y = 13$ cm under different τ for $H_b = 13.4$ cm and $U_{g,s} = 4.39$ m/s.

(normalized by $\langle u_{gc,0} \rangle$), respectively. It is clearly seen that the change of parameter τ from 0.125 to 0.5 doesn't change the distribution of the mean gas velocity a lot (Fig. 4.60A). However, it can change the degree of fluctuation of the gas velocity, especially for lower values of τ. A lower τ indicates the closer sharp distribution of voidage near the particle surface. Thus, using the smooth distribution of void function of $\phi(\zeta, \tau)$ may suppress the fluctuation of the gas velocity slightly, especially when τ is large. To this regard, using a small τ is benefitable, provided the subparticle scale LES-DEM model is feasible.

In addition, Fig. 4.61A and B show the mean values and the root of mean squares of the gas voidages under different τ, respectively. It also indicates that: (1) The variation of the mean gas voidages does not change much at different τ (Fig. 4.61A). A smaller τ may generate a sharp oscillation of $\langle \epsilon_g \rangle$ near the bed wall, which is closer to the real condition. On the contrary, a large τ may attenuate the sharp oscillation of the void distribution and result in a smoothed variation. A smoothed voidage is preferable for numerical stability in computational performance, and unpreferable for building a real sharp gas-particle interface. (2) A large τ can also suppress the fluctuation of voidages of gas (Fig. 4.61B), which is similar to that indicated by Fig. 4.60B. Therefore, to say concludingly, a small value of τ is preferable for performing the current subparticle LES-DEM model. A large τ may filter out much details, especially on the gas-particle/wall interface. But, if $\tau = 0$, the subparticle scale details are all diminished and the current model is degenerated into the conventional CFD-DEM model, which gives much coarser prediction of the gas-particle interaction (as demonstrated in Fig. 4.49A).

In addition, the magnitude of the fluid velocity in the cells fully occupied by particles is determined by the voidage in that cell, which is influenced by the parameter τ. When τ changes, the mean value of the fluid velocity and the mean voidage do not change much. But the degree of fluctuation of the fluid velocity or oscillation of the gas voidage changes greatly. Therefore, the physical meaning of the subparticle scale fluid velocity in the cells fully occupied by particles may be regarded as the modulated subparticle turbulent fluctuation of the gas velocities. To this regard, the SGS model given by the LES determines the subparticle scale stress tensor, and the smoothing parameter τ determines the modulated gas turbulence by the smoothed void fraction. The former scale is finer than the latter. Therefore, the results obtained by the current model with or without LES are not quite different from each other.

4.5.4 Additional remark

In this section, a virtual void fraction model is proposed to be incorporated in the LES-DEM model to simulate the gas-large-particle flow on the basis of subparticle scales $\delta < d_p$, which is named as the subparticle LES-DEM coupled method. This approach is quite different from the conventional CFD-DEM methods basically for two points: (1) The solution level for the gas phase motion. The current LES-DEM is finer than the conventional CFD-DEM. (2) The inter-phase coupling levels. The conventional CFD-DEM approach is based on the $\Delta \geq d_p$ level. In contrast, in the current fine CFD-DEM model, the fluid-to-particle drag is based on the coarse graining from $\delta < d_p$ to d_p, whereas the particle-to-fluid feedback is based on fine graining from d_p to $\delta < d_p$ through the conservative virtual distribution function. Thus, the virtual distribution function is the key issue in this scheme.

The criteria for generating the conservative function of virtual distribution are clearly provided, and an example of virtual distribution function based on the solution of the diffusion equation is given. Incorporated into the subparticle LES-DEM method, the capability and accuracy of the current model, including both 2D and 3D cases, by exploring the longitudinal and lateral distributions of the particle mean velocity and the pressure drop in comparison with the reported data on the gas-particle flows in a spouted bed [290] have been demonstrated:

- The current model has been validated after the complete comparison on the gas phase and particle phase, e.g. voidage, gas phase velocity, particle velocity, pressure drop, with experiments. Moreover, the 2D and 3D applications can both give finer (subparticle) details on the gas-particle flow than the conventional CFD-DEM model.
- The particle phase flow pattern, the velocity fields, and the similarity characteristics of the lateral distribution of the particle velocity in simulation are quite like or close to that observed in experiment. In addition, the similarity of the lateral distribution of the gas phase velocity on different heights is found, which is considered as the dominant mechanism for the similarity of the lateral distribution of the particle velocity in the spout.
- Although some differences still exist in the longitudinal distribution of particles which mainly lie in the stage 2 to keep enough long stage for accelerating the particle motion in the spout, the particle motion velocity in the stages 1, 3, and 4 are still quite similar to that in experiment.
- The bed pressure drop is found to be periodically fluctuating with the fundamental frequency F_{max} quite close to that in experiment. The linear relationship between F_{max} and the bed height H_b is still found to be close to that in experiment, including the absolute values and the gradients. In particular, the results on the finer grid $\delta_1 = \frac{1}{8}d_p$ are better than those on the coarse grid $\delta_1 = \frac{1}{6}d_p$.
- The systematic comparison has demonstrated the validity of the current work for predicting the gas and particle phase flow behaviors in the gas-large-particle flows. The gas phase velocity and voidage, which has not been fully explored in the experiment, are illustrated, which demonstrates the advantages of the current model for capturing finer information of the gas phase.
- The parameter τ of the current model can be viewed as a smooth-degree parameter, which may smooth or filter out the subparticle scale sharply varied distributions and fluctuations. Therefore, a small nonzero value of τ is preferable, since a zero τ may degenerate the current model into the conventional CFD-DEM model.

4.6 Summary

This chapter has demonstrated the application of coupled methods with DNS/LES-DEM schemes to simulate and analyze the two-phase flow behaviors. The DNS-DEM method is particularly useful for detailed computation and mechanical study of the particle-turbulence interaction, although the key issue of this method is still in the algorithm for accurate computation of the particle-fluid interfacial interaction force through integral of stress tensor on the interface. The conventional CFD-DEM, as well as the coarse-scale LES-DEM method, is suitable for engineering application, supposing the particle size effect is negligible compared to the scale of resolution required in industry. The SVFM-based LES-DEM method solves the gas-large-particle system through an intermediate spatial scale, which may meet requirements for both the fundamental research and engineering application. In particular, it is suitable for the spouted bed, the packed bed, and the pebble bed in industry. The key issue of this method may be in the degree of smoothing which may affect the drag computation.

Using these kinds of coupled methods, the interaction mechanisms in many kinds of gas-particle two-phase flows can be analyzed. However, the coupled approaches are still under development. One may choose appropriate scheme depending on the particular situation encountered solving problems in engineering.

CHAPTER

Application in granular flows

5

5.1 Some functions

5.1.1 Evaluation functions of mixing degree

5.1.1.1 Dimensionless concentration difference

For the convenience of mixing study, e.g. in 2D circular tumbler, the particles are divided horizontally into two separate and equal parts. At the beginning, the west part is marked by red and the east part is blue. Then, the mixing between the two parts of particles in the tumbler advances when the tumbler rotates. To quantify the degree of mixing microscopically, a measurement function is defined based on the local difference of particle concentrations ($c_a - c_b$) between the red (denoted by the subscript 'a') and blue (denoted by the subscript 'b') parts of the particles. For comparison, the difference of particle concentration is nondimensionalized by the corresponding local concentration of the mixture, which is actually the sum of c_a and c_b. Thus, the measurement function is expressed as:

$$\xi(c_\alpha, c_\beta) = \begin{cases} \frac{c_\alpha(\boldsymbol{r}) - c_\beta(\boldsymbol{r})}{c_\alpha(\boldsymbol{r}) + c_\beta(\boldsymbol{r})}, \text{ if } (c_\alpha(\boldsymbol{r}) \cdot c_b(\boldsymbol{r}) > 0) \\ 0, \text{ otherwise} \end{cases} \tag{5.1}$$

where $c(\boldsymbol{r})$ is the local concentration of particles at $\boldsymbol{r}$, and $\boldsymbol{r}$ is a position vector. For the region solely occupied by the red or blue particles, i.e. the two parts of particles thoroughly separated, it has $\xi = 1$ or $\xi = -1$. For the region of well mixed, it has $\xi = 0$. As the tumbler rotates, the dynamic characteristics of mixing are observed through the variation of the measurement function. Moreover, the detailed configuration of the mixing degree on the interface of the two parts can also be evaluated by ξ.

5.1.1.2 Radial distribution function (RDF)

For the convenience of mixing analysis in the macroscopic and statistical way, another function is introduced which is the so-called radial distribution function (RDF). The radial distribution function is defined as:

$$f_{\alpha\beta}(r) = \frac{1}{\Omega_\alpha} \int_{\Omega_\alpha} \frac{1}{\Omega_\beta} \int_{\Omega_\beta} |\boldsymbol{x}_\alpha - \boldsymbol{x}_\beta|_{|\boldsymbol{x}_\alpha - \boldsymbol{x}_\beta| = r} d\boldsymbol{x}_\alpha d\boldsymbol{x}_\beta, \tag{5.2}$$

Gas-Particle and Granular Flow Systems. https://doi.org/10.1016/B978-0-12-816398-6.00014-6
Copyright © 2020 Elsevier Inc. All rights reserved.

where α, β denote the red and blue parts of particles, respectively, and Ω_α and Ω_β are the area of regions occupied by the 'α' or 'β' particles respectively. The $|\boldsymbol{x}_\alpha - \boldsymbol{x}_\beta| = r$ represents the radial distance between the 'α' particle at $\boldsymbol{x}_\alpha$ and the 'β' particle at $\boldsymbol{x}_\beta$. Thus, Eq. (5.2) indicates the averaged distance for all of 'β' particles to any of the 'α' particle. In other words, the radial distribution function means the averaged probability to find a 'β' particle with a distance r from any given 'α' particle. It is easy to observe from Eq. (5.2) that the radial distribution function is symmetrical for the 'α' and 'β' parts of particles, i.e. $f_{\alpha\beta}(r) = f_{\beta\alpha}(r)$. Moreover, we call $f_{\alpha\alpha}(r)$ and $f_{\beta\beta}(r)$ as the auto-radial distribution function (*a*-RDF) when $\alpha = \beta$ in Eq. (5.2), and $f_{\alpha\beta}(r)$ as the cross-radial distribution function (*c*-RDF). For numerical implementation, Eq. (5.2) is expressed in the discrete form

$$f_{\alpha\beta}(r) = \frac{1}{N_\alpha}\sum_{\alpha}^{N_\alpha}\frac{1}{N_\beta}\sum_{\beta}^{N_\beta}|\boldsymbol{x}_\alpha - \boldsymbol{x}_\beta|_{|\boldsymbol{x}_\alpha - \boldsymbol{x}_\beta| = r}, \tag{5.3}$$

5.1.1.3 Mixing information entropy via RDF

With the radial distribution, which is regarded as a macroscopic statistical mixing evaluation, it is possible to define the Shannon information entropy as

$$S_M = -\int f_{\alpha\beta}(r)\log_2 f_{\alpha\beta}(r)dr \tag{5.4}$$

The unit of the Shannon information is decibel (dB) [295]. The information entropy may have greater values when the radial distribution of particles corresponds to more uncertainties of particle distribution in the system.

It is necessary to mention that the Shannon information entropy based on a radial distribution function provides a macroscopic view of the overall level of mixing degree. The benefit of the radial distribution function is that it is not only related to the local information of concentration of particles, but is also related to the long distance relation of particles.

5.1.1.4 Mixing information entropy via local concentration

The Shannon information entropy [297] can also be defined based on the local concentration of particles:

$$M_E = \int_r -c_A(\boldsymbol{r})\log_2(c_A(\boldsymbol{r})) - c_B(\boldsymbol{r})\log_2(c_B(\boldsymbol{r})) \tag{5.5}$$

where $c_A(\boldsymbol{r})$ and $c_B(\boldsymbol{r})$ are the concentrations of two types of particles at $\boldsymbol{r}$. In the system composed of two types of particles, the concentration at r can be computed as $c_A(r) = N_A(r)/(N_A(r) + N_B(r))$. When $c_A = 0$ or $c_A = 1$, it leads to $c_A(r)\log_2(c_A(r)) = 0$. Otherwise, $c_A(r)\log_2(c_A(r)) \neq 0$. By integrating over the whole area of the drum, M_E can be used to quantify the overall degree of particle mixing in the drum.

5.1.1.5 Improved information entropy via coordination number fraction (CnIE)

The entropy can also be presented based on the coordination number fraction [298]. The coordination number of a particle is the number of particles in contact with the particle. The contact is defined as a state when the distance between the surfaces of two particles is less than a small value, e.g. $1\%d$, where d is diameter of particle. For a mono-system, if the core particle j is type-'A', its concentration function is defined as

$$\begin{cases} c'_{a,j} = \dfrac{N'_{a,j}+1}{N'_{a,j}+N'_{b,j}+1} \\ c'_{b,j} = \dfrac{N'_{b,j}+1}{N'_{a,j}+N'_{b,j}+1} \end{cases} \tag{5.6}$$

When the core particle is type-'B', its concentration function is defined as

$$\begin{cases} c'_{a,j} = \dfrac{N'_{a,j}}{N'_{a,j}+N'_{b,j}+1} \\ c'_{b,j} = \dfrac{N'_{b,j}}{N'_{a,j}+N'_{b,j}+1} \end{cases} \tag{5.7}$$

where $N'_{a,j}$ and $N'_{b,j}$ are the type-'A' and type-'B' particles' contact numbers with the core particle, respectively. $c'_{a,j}$ and $c'_{b,j}$ are the concentrations of type-'A' and type-'B' particles, respectively.

Then, the information entropy based on coordination number fraction (named 'CnIE' for short) is defined as

$$S'_j = -c'_{a,j}\log_2(c'_{a,j}) - c'_{b,j}\log_2 c'_{b,j} \tag{5.8}$$

$$S_c = \frac{1}{S'_0}\sum_j \frac{n'_{a,j}+n'_{b,j}}{N_A+N_B} S'_j \tag{5.9}$$

where S'_0 denotes the information entropy of completely mixed system. The subscript 'c' of S_c denotes that the information entropy definition is based on the coordination number fraction.

5.1.1.6 General mixing information entropy for $N_a : N_b \neq 1:1$ and multiple particle system (GMIE)

The information entropy based on local concentration is the most widely used evaluation function to assess the degree of mixing. The relation between the mathematical definition of information entropy and physical mixing should guarantee that: the greater information entropy should correspond to the better particle mixing degree, and vice versa. However, for a binary system, when the number of particles of one kind is not equal to that of another kind, i.e. $N_a : N_b \neq 1:1$, the information entropy may not meet the relation between mathematical quantifications and physical states. Therefore, the improved general mixing information entropy is proposed below.

At first, let the tumbler be divided by Cartesian grids into several cubic cells. In one cubic cell, the local concentration of particle is defined as

$$\begin{cases} c_{a,j} = \dfrac{n_{a,j}}{n_{a,j} + n_{b,j}} \\ c_{b,j} = \dfrac{n_{b,j}}{n_{a,j} + n_{b,j}} \end{cases} \tag{5.10}$$

where j denotes the index of the cubic cell. 'a' and 'b' denote the predefined two kinds of particles, respectively. n_a and n_b are the number of 'a' and 'b' particles in the cell j, respectively. Obviously, $c_{a,j} + c_{b,j} = 1$. Conventionally, the information entropy in cell j is defined as [299]: $S_j = -c_{a,j}\log_2 c_{a,j} - c_{b,j}\log_2 c_{b,j}$. Then, the total entropy of the system is

$$S = \begin{cases} \sum_j S_j \\ \frac{1}{S_0}\sum_j \frac{N_{cell,j}}{N} S_j \end{cases} \tag{5.11}$$

where N and N_{cell} are the total number of particles and the number of particles in the jth cell, respectively. S_0 denotes the entropy of the completely mixed system, i.e. the maximum entropy value.

For such definition of information entropy we have: when $c_{a,j} \in (0, \frac{1}{2})$, S_j monotonously increases with $c_{a,j}$; and when $c_{a,j} \in (\frac{1}{2}, 1)$, S_j monotonously decreases with $c_{a,j}$. Then S_j achieves the maximum value 1 at $c_{a,j} = \frac{1}{2}$. Meanwhile, when $c_{a,j} = 0$ or $c_{a,j} = 1$, S_j obtains the minimum value 0. Therefore, for $N_a : N_b = 1:1$, the information entropy will reach the maximum when the degree of mixing is the highest. However, when $N_a : N_b \neq 1:1$, the biggest value of information entropy does not match the best mixed state.

An example is given here to illustrate this issue: For a system with $N_a = 200$, $N_b = 100$, let all the cells contain the same (or very close) distributions of particles 'a' and 'b'. For instance, for a cell (denoted as "Cell-1") with ($N_a = 4$, $N_b = 2$),

$$S_1 = -\frac{1}{S_0}\left[\frac{6}{300}\left(\frac{2}{6}\log_2\frac{2}{6} + \frac{4}{6}\log_2\frac{4}{6}\right)\right] = \frac{0.018366}{S_0} \tag{5.12}$$

For comparison, for a cell with ($N_a = 3$, $N_b = 3$) (denoted as "Cell-2"),

$$S_2 = -\frac{1}{S_0}\left[\frac{6}{300}\left(\frac{3}{6}\log_2\frac{3}{6} + \frac{3}{6}\log_2\frac{3}{6}\right)\right] = \frac{0.02}{S_0} \tag{5.13}$$

It has $S_1 < S_2$. However, as $N_a = 200$, $N_b = 100$, the overall mixing degree under the cell distributions similar to "Cell-1" is better than that close to "Cell-2". From the viewpoint of overall mixing degree, the best mixing condition should correspond to the state where any local ratio of particle numbers $n_{a,j} : n_{b,j}$ is equal to the total number ratio $N_a : N_b$, i.e. $n_{a,j} : n_{b,j} = N_a : N_b, \forall j$. Otherwise, supposing the distributions in the majority of cells are similar to "Cell-2", almost a half of type-'a' particles ($N_a^R = N_a - N_b = 100$) are totally unmixed. In this way, the physical condition and the mathematical expression do not show a good coincidence. Therefore, when $N_a : N_b \neq 1:1$, especially when $N_a \gg N_b$ or vice versa, the information entropy evaluation of mixing levels should be improved to meet the following condition:

$$f_a(c_{a,j}(n)) = \begin{cases} 0, (c_{a,j}(n) = 0) \\ 1, (c_{a,j}(n) = 1) \\ \frac{1}{2}, (c_{a,j}(n) = P_a = \frac{N_a}{N_a + N_b}) \end{cases} \tag{5.14}$$

Then, a linear piecewise function is proposed here to give the expression in the two intervals of $(0, P_a)$ and $(P_a, 1)$:

$$\hat{f}_a(c_{a,j}) = \begin{cases} \frac{1}{2P_a} c_{a,j}, (c_{a,j} \in (0, P_a)) \\ \frac{1}{2-2P_a} c_{a,j} + \frac{1-2P_a}{2-2P_a}, (c_{a,j} \in (P_a, 1)) \end{cases} \tag{5.15}$$

In this way, the improved expression of local concentration $\hat{f}_b(c_{b,j})$ can be used for better estimation of the real mixing degree under the extended ranges of $N_a : N_b$, where the consequently improved definition of information entropy is

$$S = -\frac{1}{S_0} \left[\sum_{j=1}^{l} \frac{N_{cell}}{N} [\hat{f}_a(c_{a,j}) \log_2 \hat{f}_a(c_{a,j}) + \hat{f}_b(c_{b,j}) \log_2 \hat{f}_b(c_{b,j})] \right] \tag{5.16}$$

It can be proved that: (1) $\forall c_{a,j} + c_{b,j} = 1$, $\hat{f}_a(c_{a,j}) = 1 - \hat{f}_b(c_{b,j})$. (2) The information entropy in a single cell can be represented as

$$S_j = -\hat{f}_a(c_{a,j}) \log_2 \hat{f}_a(c_{a,j}) - (1 - \hat{f}_a(c_{a,j})) \log_2 [1 - \hat{f}_a(c_{a,j})] \tag{5.17}$$

where S_j is a composite function with respect to $c_{a,j}$. When $c_{a,j} \in (0, P_a)$, $\hat{f}_a(c_{a,j}) \in (0, \frac{1}{2})$ and is monotonously increasing; when $c_{a,j} \in (P_a, 1)$, $\hat{f}_a(c_{a,j}) \in (\frac{1}{2}, 1)$ and it is monotonously decreasing. Thus, S_j obtains the maximum value 1 at $c_{a,j} = P_a$ and the minimum value 0 at $c_{a,j} = 0$ or $c_{a,j} = 1$.

To sum up, the improved information entropy has both the basic feature of the information entropy function and the consistency with the real degree of mixing at all ranges of $N_a : N_b$. In addition, this improved definition can also be extended for m ($m > 2$) kinds of particles. The local concentration should satisfy the following conditions:

$$f_i(c_{i,j}(n)) = \begin{cases} 0, (c_{i,j}(n) = 0) \\ 1, (c_{i,j}(n) = 1) \\ \frac{1}{m}, (c_{i,j}(n) = P_i = \frac{N_i}{\sum_i N_i}) \end{cases} \tag{5.18}$$

where $i = 1, 2, 3, \cdots, m$ denote the kinds of particles; $P_i = \frac{N_i}{N}$ denotes the ratio of the number of type-'i' particles to the total number of particles. The formula of general mixing multiple particle mixing information entropy (named GMIE for short) is written as follows:

$$S = -\frac{1}{S_0}\left[\sum_{j=1}^{n}\sum_{i=1}^{m}\frac{N_{cell}}{N}[\hat{f}_i(c_{i,j})\log_2\hat{f}_i(c_{i,j})]\right]$$
$$S_0 = \sum_{j=1}^{n}\sum_{i=1}^{m}\frac{1}{n}[\hat{f}_i(P_i)\log_2\hat{f}_i(P_i)] = \sum_{i=1}^{m}[\hat{f}_i(P_i)\log_2\hat{f}_i(P_i)] = \log_2\frac{1}{m} = -\log_2 m \tag{5.19}$$

where

$$\hat{f}_i(c_{i,j}) = \begin{cases} \frac{1}{mP_i}c_{i,j}, (c_{i,j} \in (0, P_i)) \\ \frac{(m-1)}{m(1-P_i)}c_{i,j} + \frac{1-mP_i}{m(1-P_i)}, (c_{i,j} \in (P_i, 1)) \end{cases} \tag{5.20}$$

5.1.1.7 Lacey mixing index (LMI)

The Lacey index is based on local concentration [300]. Let M_l denote the Lacey index, defined as

$$M_l = \frac{S^2 - S_0^2}{S_R^2 - S_0^2} \tag{5.21}$$

where S_0^2, S_R^2 and S^2, respectively, present the variances of fully demixed system, random mixed system and the fraction of type-'B' particle (or type-'A' particle). They are expressed as

$$\begin{cases} S^2 = \frac{1}{N}\sum_{i=1}^{N}(x_i - x_m) \\ S_0^2 = x_m(1 - x_m) \\ S_R^2 = \frac{x_m(1 - x_m)}{n} \end{cases} \tag{5.22}$$

where x_i is the local concentration of type-'A' particles in cell i. x_m is the overall proportion of type-'B' particle in the system, n is the average number in a cell and N is the number of cells with particles. x_i, x_m, and n can be calculated by

$$
\begin{cases}
x_i = \dfrac{N_{a,i}}{N_{a,i} + N_{b,i}} \\
x_m = \dfrac{N_a}{N_a + N_b} \\
n = \dfrac{1}{N} \sum_{i=1}^{N} (N_{a,i} + N_{b,i})
\end{cases}
\tag{5.23}
$$

$N_{a,i}$ and $N_{b,i}$ are defined as the numbers of type-'A' particles and type-'B' particles in cell i, respectively. N_a and N_b are total numbers of type-'A' particle and type-'B' particle respectively.

5.1.1.8 Improved mixing index for multiple particle systems (MPMI)

According to the definition of information entropy based on coordination number fraction, a new mixing index is proposed here, which is a kind of Lacey index based on the coordination number fraction rather than on local concentration. The definition of the sample is the same as that of information entropy based on coordination number fraction. A sample includes a core particle and the particles in contact with it. The core particle can be a particle of any kind.

To evaluate the system with multiple kinds of particles, the new mixing index is defined as follows:

$$
\begin{cases}
M_c = \sum_k \dfrac{N_k}{\sum N_k} M_{c,k} \\
M_{c,k} = \dfrac{S_k'^2 - S_{k,0}'^2}{S_{k,R}'^2 - S_{k,0}'^2}
\end{cases}
\tag{5.24}
$$

where subscript 'c' means the index is based on coordination number fraction and 'k' means the type of particle. The index can evaluate a system composed of multiple kinds of particles through the weighted average index of all kinds of particles. 'k' can be 2, 3 or a larger number.

The definitions of variances $S_k'^2$, $S_{k,0}'^2$ and $S_{k,R}'^2$ are similar as those in M_l:

$$
\begin{cases}
S_k'^2 = \dfrac{1}{N_k'} \sum_{j=1}^{N_k'} (x_{k,j}' - x_{k,m}') \\
S_{k,0}'^2 = x_{k,m}'(1 - x_{k,m}') \\
S_{k,R}'^2 = \dfrac{x_{k,m}'(1 - x_{k,m}')}{n_k'}
\end{cases}
\tag{5.25}
$$

where $x_{k,j}'$ is the concentration of type-'k' particles in sample i. $x_{k,m}'$ is the overall proportion of type-'k' particles in the system, n is the average sample size and N_k' is the number of samples where a type-'k' particle is the core particle. In order to distinguish two kinds of Lacey indices, these variances are marked by superscript '$\prime$' and subscript 'k'. Corresponding to the sample based on coordination number fraction, $x'_{k,j}$, $x'_{k,m}$ and N'_k can be defined as follows:

$$\begin{cases} x'_{k,j} = \dfrac{N'_{k,j}+1}{\sum_k N'_{k,j}+1} \\ x'_{k,m} = \dfrac{N_k}{\sum_k N_k} \\ n'_k = \dfrac{1}{N'_k}\sum\limits_{j=1}^{N'_k}(\sum\limits_k N'_{k,j}+1) \\ N'_k = N_k \end{cases} \tag{5.26}$$

where $N'_{k,j}$ is the number of type-'k' particles in sample j. N_k is the total number of type-'k' particles.

The PSMI method [301] looks similar to the new mixing index. Nevertheless, the two ideas are different for the following reasons: (1) In PSMI, any particle is regarded as the core particle. The concentration of type-'A' particles or type-'B' particles in the sample is calculated and used to replace the local concentration in the conventional Lacey method. Then, the variances are calculated based on the concentration to obtain the PSMI. (2) However, in the currently proposed 'MPMI' index, a type-'k' particle is regarded as the core particle. The concentration of the type-'k' ($k = 1, 2, \ldots$) particles in the sample is calculated and used to replace the local concentration in the conventional Lacey method. Then, the index of type-'k' particles based on the concentration is calculated. Finally, the weighted average of the indexes for all kinds of particles is calculated to obtain the new 'MPMI' index (Eq. (5.24)).

The benefit of a current new 'MPMI' index is that it can evaluate mixing state of a system with three or more kinds of particles, what, I think, PSMI cannot. The reason is that the PSMI method chooses a kind of particle first and only then calculates its variance. As a result, it can only evaluate a system with two kinds of particles. However, the current MPMI index can evaluate multiple particle systems.

5.1.2 Evaluation functions of heat transfer degree

5.1.2.1 Weighted temperature (WT) and temperature discrepancy function (TDF)

In the heat transfer evaluation, the concentration weighted temperature (WT) is

$$\Theta(T_a, T_b; n_a, n_b) = \begin{cases} \frac{1}{n_a(r)+n_b(r)}\left(\sum_{j=1}^{n_a} T_{a,j} + \sum_{j=1}^{n_b} T_{b,j}\right), \text{ when } n_a + n_b > 0 \\ 0, \text{ when } n_a = n_b = 0 \end{cases} \tag{5.27}$$

where $T_{a,j}$ is the temperature of the jth particle. $\Theta(T_a, T_b; n_a, n_b)$ contains four independent variables, and it does not depend solely on the thermal conduction characteristics, as both mixing and conduction can cause variation of $\Theta(T_a, T_b; n_a, n_b)$. Eliminating the mixing effect on $\Theta(T_a, T_b; n_a, n_b)$ requires the evaluation of the degree of discrepancy of $\Theta(T_a, T_b; n_a, n_b)$ from $\Theta(T_a^0, T_b^0; n_a, n_b)$ or the so-called temperature discrepancy function (TDF) $\Pi(T_a, T_b; T_a^0, T_b^0; n_a, n_b)$, where the superscript '0' denotes the initial variables at $t = 0$.

$$\Pi = \begin{cases} \frac{1}{n_a+n_b}\left(n_a T_{a,j}^0 - \sum_{j=1}^{n_a} T_{a,j} + \sum_{j=1}^{n_b} T_{b,j} - n_b T_{b,j}^0\right), & \text{when } n_a + n_b > 0 \\ 0, & \text{when } n_a = n_b = 0 \end{cases} \tag{5.28}$$

The derivation of Eq. (5.28) can be explained as follows. (1) Consider the ideal case with granular mixing only, i.e., the particles are adiabatic, and the particle-particle heat conduction does not occur. The weighted local difference of temperature is

$$\Theta(T_a^0, T_b^0) = \frac{1}{n_a + n_b}\left(n_a T_{a,j}^0 - n_b T_{b,j}^0\right) \tag{5.29}$$

It indicates that the difference in particle temperature is purely caused by particle mixing. (2) Then, consider the practical case with heat conduction. The concentration weighted local temperature difference is

$$\Theta(T_a, T_b) = \frac{1}{n_a + n_b}\left(\sum_{j=1}^{n_a} T_{a,j} - \sum_{j=1}^{n_b} T_{b,j}\right) \tag{5.30}$$

It indicates that the difference in particle temperature is caused by both mixing and conduction. (3) Subtract Eq. (5.30) from Eq. (5.29) to obtain Eq. (5.28). Taking type 'a' for example, a difference can be found between Eqs. (5.30) and (5.29). It is clear that $n_a T_{a,j}^0 = \sum_{j=1}^{n_a} T_{a,j}^0$ and $n_a T_{a,j}^0 - \sum_{j=1}^{n_a} T_{a,j} = \sum_{j=1}^{n_a}(T_{a,j}^0 - T_{a,j})$. Herein $\frac{1}{n_a+n_b}\left(\sum_{j=1}^{n_a} T_{a,j}^0 - \sum_{j=1}^{n_a} T_{a,j}\right)$ infers the weighted local total temperature difference in the 'a'-type particle between the cases of pure mixing $\frac{1}{n_a+n_b}\sum_{j=1}^{n_a} T_{a,j}^0$ ($\equiv \frac{n_a T_a^0}{n_a+n_b}$, adiabatic condition without thermal conduction, constant temperature), and time-varied temperature $\frac{1}{n_a+n_b}\sum_{j=1}^{n_a} T_{a,j}$ (caused by both mixing and thermal conduction). Thus, the TDF can be explained as a type of quantification of temperature difference purely caused by the effect of particle-particle heat conduction.

5.1.2.2 Dimensionless temperature probability density function (TPDF)

In general, the distribution of particle temperature can be characterized by the probability density functions of dimensionless temperatures. Without external input and output of particle heat, the particle temperatures are always within the initially given temperature ranges, i.e. $T_\beta < T < T_\alpha$. Using the dimensionless temperature as

$$T^\star = \frac{T - T_\beta}{T_\alpha - T_\beta} \subseteq [0, 1], \tag{5.31}$$

the probability density function of temperature at time τ could be defined as follows:

$$f_\tau(T^\star) = \frac{n(T^\star)}{N_p \delta T^\star} \tag{5.32}$$

where $n(T^\star)$ is the number of particles within $T^\star \in \left[T^\star - \frac{\Delta T^\star}{2}, T^\star + \frac{\Delta T^\star}{2}\right) \subseteq [0, 1]$. $N_p = N_\alpha + N_\beta$ is the total number of particles, and $N_\alpha \approx N_\beta$ are the numbers of particles with the given high and low initial temperatures, respectively.

5.1.2.3 Mean temperature discrepancy (MTD)

At the beginning, the particles with the high and low temperatures are separated completely, with $T^\star_\alpha\big|_{t=0}$ and $T^\star_\beta\big|_{t=0}$, respectively. For the limit of fully developed state of heat conduction, temperature difference should not exist between any two particles, i.e. $T^\star_\alpha\big|_{t=\infty} = T^\star_\beta\big|_{t=\infty} = 0.5 = T^\star_\infty$. At any time, it is possible to define the mean temperature discrepancy between the state at the current time and the fully developed state to evaluate the degree of heat conduction by the following expression:

$$\langle T^\star_\delta \rangle(\tau) = \int_0^1 \left|T^\star - T^\star_\infty\right| f_\tau(T^\star) dT^\star = \int_0^1 |T^\star - 0.5| f_\tau(T^\star) dT^\star \tag{5.33}$$

where the operator $\langle\cdot\rangle$ denotes the mean value. The subscript 'δ' denotes the discrepancy, and 'τ' denotes time. Thus, $\langle T^\star_\delta \rangle(\tau)$ evaluates the instantaneously evolved characteristics of MTD.

Furthermore, the mean rate of variation of MTD can be expressed as

$$\langle \dot{T^\star_\delta} \rangle = \frac{\Delta \langle T^\star_\delta \rangle}{\Delta \tau} = \frac{\langle T^\star_\delta \rangle(\tau) - \langle T^\star_\delta \rangle(0)}{\tau} \tag{5.34}$$

Meanwhile, the variance of MTD can be expressed by

$$\sigma \langle T^\star_\delta \rangle = \int_0^1 (T^\star_\delta - \langle T^\star_\delta \rangle)^2 f_\delta(T^\star) dT^\star \tag{5.35}$$

5.1.2.4 Thermal radial distribution function (TRDF) and thermal information entropy (TIE) S_T

The radial distribution of temperature can be evaluated in a similar manner as RDF in Eq. (5.2), which is the so-called thermal radial distribution function (T-RDF)

$$g_{\alpha\beta}(r) = \frac{1}{\Omega_\alpha} \int_{\Omega_\alpha} \frac{1}{\Omega_\beta} \int_{\Omega_\beta} |T_\alpha(\boldsymbol{x}_\alpha) - T_\beta(\boldsymbol{x}_\beta)|_{|\boldsymbol{x}_\alpha - \boldsymbol{x}_\beta| = r} d\boldsymbol{x}_\alpha d\boldsymbol{x}_\beta. \tag{5.36}$$

Assuming that the temperature field within the tumbler is homogeneous in developed thermal conduction state, T-RDF $g_{ab}(r)$ in all 'r' ranges must be zero. In contrast, nonzero T-RDF may indicate the macroscopic discrepancy of temperature between the red (hot) and green (cool) particles.

Using temperature radial distribution function, similar to Eq. (5.4), with respect to the global heat transfer state, the thermal information entropy (TIE) is defined as

$$S_{T1} = -\int g_{\alpha\beta}(r)\log_2 f_{\alpha\beta}(r)dr \tag{5.37}$$

or

$$S_{T2} = -\int g_{\alpha\beta}(r)\log_2 g_{\alpha\beta}(r)dr \tag{5.38}$$

The state in which the discrepancy of temperature exists in the radial direction may indicate a global level of heat transfer. As g_{ab} and f_{ab} indicate the information of temperature and concentration at a radial distance r, the integral of them by Eq. (5.37) and (5.38) may indicate a macroscopic level evaluation of the degree of heat transfer. In contrast to S_M, a large value of S_T indicates a low level of heat transfer between the hot and cool particles, whereas a small value of S_T indicates the opposite. The advantages of RDF and T-RDF are that they are not only related to the local information on the concentration of particles but also related to the information on the concentration of particles within long distances.

5.2 Circular drum mixers

5.2.1 Flow pattern and mixing evolution

For the microscopic study of mixing evolution, the results of mixing measurement function $\xi(c_\alpha, c_\beta)$ for different rotations, i.e. 0.5, 1, 1.5 and 2 rotations respectively are compared, at a typical rotation speed of $\omega = 1$ (π/s) (Fig. 5.1) in the clockwise direction. Then, the results at a typical rotation (2 rotation) with different rotation speeds, i.e. $\omega = 0.75$, 1, 1.25, 1.5, 1.75 and 2 (π/s) respectively, are also illustrated in Fig. 5.2[1]. As Fig. 5.1 shows, the mixing interface becomes coarser, more hierarchical and cascading as the tumbler rotates. The red and the blue particles evolve into more and more detailed levels of mixing, and eventually evolve down into the discrete particle size. In this way, the striated structure of mixing interface appears. The internal mixing interface in Fig. 5.1 is quite similar to the results reported by [302]. Moreover, it is also observed with a flowing layer upon a static base, separated by a high shear rate region. During the tumbling motion, the particles move into the flowing layer in the left-top high shearing region, and flows rapidly downstream at

[1] For interpretation of the colors in this and the following figures, the reader is referred to the web version of this chapter.

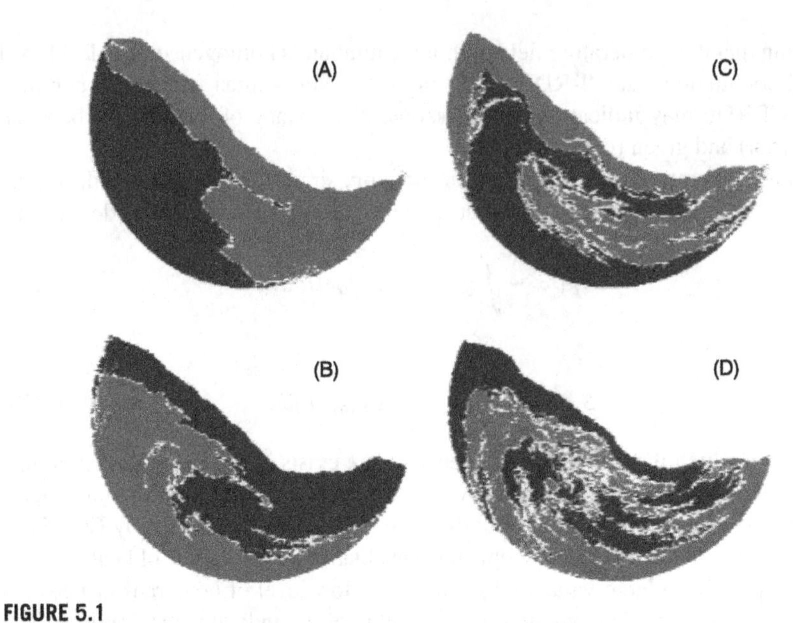

FIGURE 5.1

The measurement functions after (A) 0.5 rotation, (B) 1.0 rotation, (C) 1.5 rotation, and (D) 2 rotation, respectively, for the rotation speed $\omega = 1$ (π/s).

the end of the flowing layer into the static base through the right-bottom region. By the cycling movement of the above process, the mixing interface is stretched into a jagged pattern of curve with fine striations. Finally, a breakdown of the stratification structure into a near homogeneous distribution of the distorted interface is reached. It is called the fully developed state of mixing.

Fig. 5.2 shows more clearly the stratification structure of mixing interface around the shearing layer. Moreover, it is observed that with lower rotation speed, the stratification structure has more hierarchical details (Fig. 5.2A and B) and each of them is very thin, whereas for higher rotation speed, the stratification becomes thicker and with fewer hierarchies (Fig. 5.2E and F). It is obviously that the former has a more positive effect for mixing. It indicates that the lower rotation will help the mixing since it has more time for the stratification structure to develop in details.

5.2.2 Dimension analysis

It is observed from Fig. 5.1 and Fig. 5.2 that the mixing structure has many hierarchical details. For geometrical consideration, it seems to be of fractal characteristics. Moreover, it is reported by [302] that the motion of granular particles in the tumbler is chaotic. Thus, some dimension analyses on the mixing interface are performed here.

For a fractal interface, the length of the curve $R(s)$ is proportional to

$$R(s) \propto s^{-D_H} \tag{5.39}$$

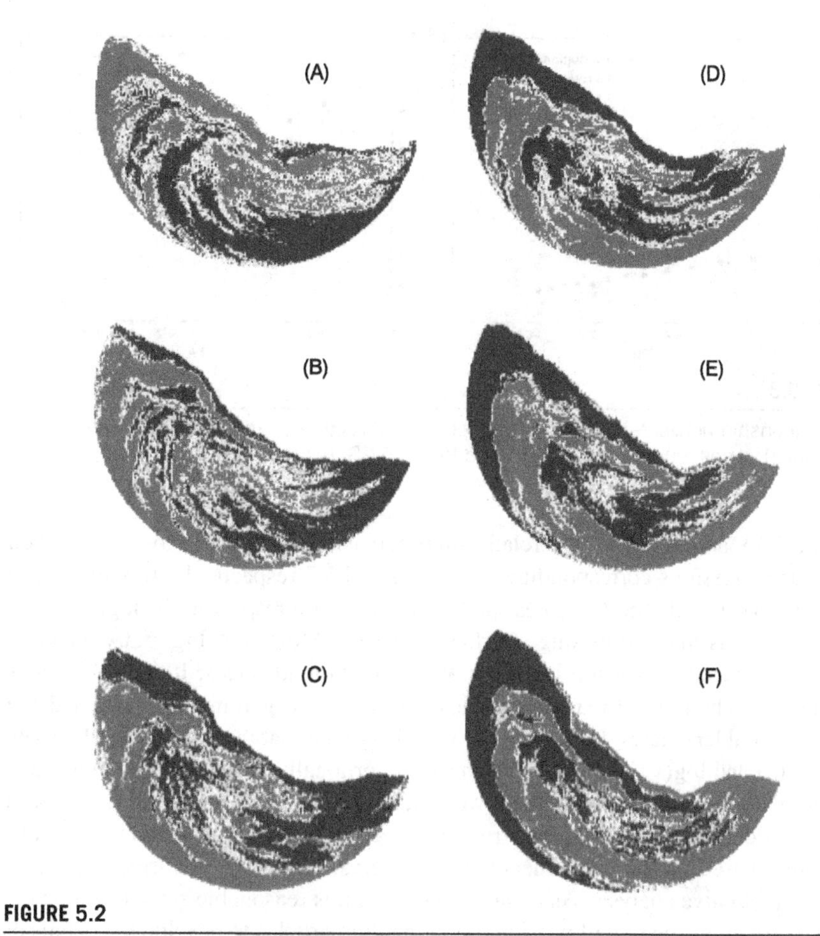

FIGURE 5.2

The measurement functions for the rotation speeds (in π/s) $\omega = 0.75$ (A), 1.0 (B), 1.25 (C), 1.5 (D), 1.75 (E), 2.0 (F), respectively, after 2 rotation.

where D_H is the fractal dimension of the interface. $D_H = -\frac{d\log R(s)}{d\log(s)}$, where the Napierian base is used for the logarithm operation. Based on Eq. (5.39), the fractal dimension of the interface is calculated by using a box-counting method. The procedures are as following:

- Given a set of small scales s, it is needed to count the number of all small box elements $R(s)$ which cover and only cover the mixing interfaces. It is noticed that the mixing interface is identified as $|\xi| < 1$ and $\xi \neq 0$, or equivalently $c_\alpha \cdot c_\beta \neq 0$. It means the interface is characterized by none zero local concentrations of both the red and blue particles.
- Then, the relationship of $\log R(s)$ and $\log(s)$ is obtained by the linear regression. After that, the magnitude of the slope coefficient k_s of the linear regression is just the dimension of the mixing interface, i.e. $D_H = -k_s$.

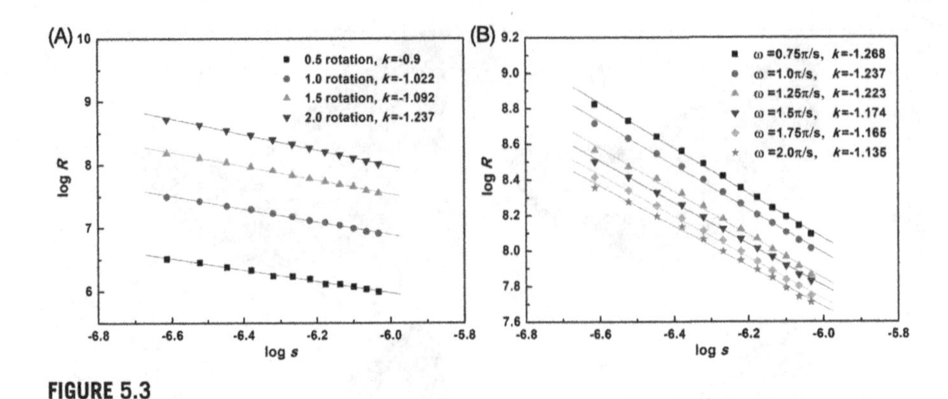

FIGURE 5.3

The relationship between the log$R(s)$ and log(s) (solid scatters) and their linear regressions (solid lines) corresponding to Fig. 5.1A and to Fig. 5.2B, respectively.

- Fig. 5.3A and B shows the relationship between log$R(s)$ and log(s) and their linear regressions corresponding to Fig. 5.1 and 5.2 respectively. It is found that the log$R(s)$ is of a perfect linear proportionality with respect to the log(s). Thus, it is obvious that the mixing interface is fractal. Moreover, Fig. 5.3A indicates that the fractal dimension increases as the rotation advances. It could be easily interpreted by Fig. 5.1 that the stratification structure is getting developed in detail as the tumbler rotates. For Fig. 5.3B, it is also found that the relationship between log$R(s)$ and log(s) is of a perfect linear proportionality. Moreover, it is indicated that, with the same rotation, the lower rotating speed has a larger value of fractal dimension than that of the higher rotating speed. The same conclusion in the above section based on the phenomenological observation is now confirmed based on the quantitative analysis. As mentioned before, it is reasonable since reaching the same rotation slowly will provide more time for particles to develop the details of the mixing stratification structure.
- As Fig. 5.3A shows, the slope coefficient of the linear regression line increases in magnitude when the rotation advances. However, what is the limit state under which the mixing is fully developed? As mentioned above, a breakdown of the stratification structure into a nearly homogeneous distribution will take place when the number of rotations is enough. For this consideration, the dimension analysis is performed at the final simulation time ($t = 5$ s). It is shown in Fig. 5.4. For Fig. 5.4, with the same elapsed time but different rotation speeds, the higher speed experiences more number of rotations than that of the lower speed. Thus, it is found that the mixing state at $t = 5$ s for $\omega = 3.0$ (π/s) (Fig. 5.4A) is very close to the fully developed state (yet does not reach it). Moreover, for a sufficiently high rotation speed, such as $\omega = 2.0$, 2.5 and 3.0 (π/s), there is almost an asymptotic location for the regression lines (Fig. 5.4B). It is called the limit line. To calculate the slope of this limit line, i.e. the limit dimension of the mixing interface, more simulations were conducted for $\omega = 3.0$ (π/s) until it almost reaches the fully de-

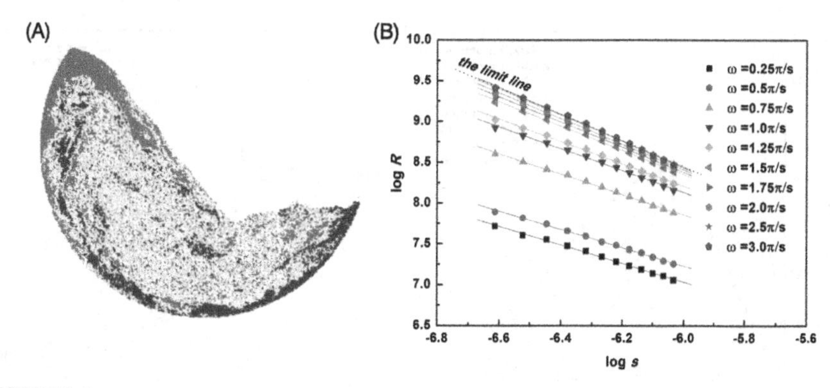

FIGURE 5.4

The relationship between the log$R(s)$ and log(s) (solid scatters) and their linear regressions (solid lines) for various rotation speeds (B) at the final simulation step (A).

veloped state (about 10 s). After linear regression, it is about $D_H = 1.856$. That is appropriately the ultimate dimension of the tumbler mixing interface.

In conclusion, the mixing interfaces of particles of tumbling motion are fractal. The fractal dimension increases as the mixing process advances, until the fully developed state, which is indicated by the limit line, is reached. Under the same number of rotations, the lower rotation speeds is of more positive effect for the mixing level than that with large rotation speeds, since a slow mixing advancement provides more time for particles to develop the details of the hierarchical stratification structure.

5.2.3 Information entropy analysis

5.2.3.1 Radial distribution function

For the macroscopic study of mixing, the auto- and cross-radial distribution functions are introduced (denoted as a-RDF and c-RDF, respectively). The results of a-RDF and c-RDF at different rotations with the rotation speed $\omega = 1.0$ (π/s) are compared in Fig. 5.5A.

From Fig. 5.5A, it is observed that initially there exist evident differences between the a-RDF and c-RDF, whereas the discrepancies are getting eliminated as the rotation proceeds. For example, with 0.5 rotation, the a-RDF has a relatively greater peak value around $r = 90d_p$, and after that it is rapidly decreased, whereas the c-RDF has a lower peak value around $r = 180d_p$, and it goes down more slowly than a-RDF. In contrast, after 2 rotations, the discrepancy between them is almost negligible. The reasons for the discrepancy between a-RDF and c-RDF are as follow: (1) The red and blue parts are preliminarily separated. As a result, it is required to have a longer radial distance at the same probability to find a 'blue' particle around any given 'red' particle with comparison to a 'red' particle. (2) However, with the advancement of rotation and mixing, the red and blue particles are dissolved into each other. Thus, it

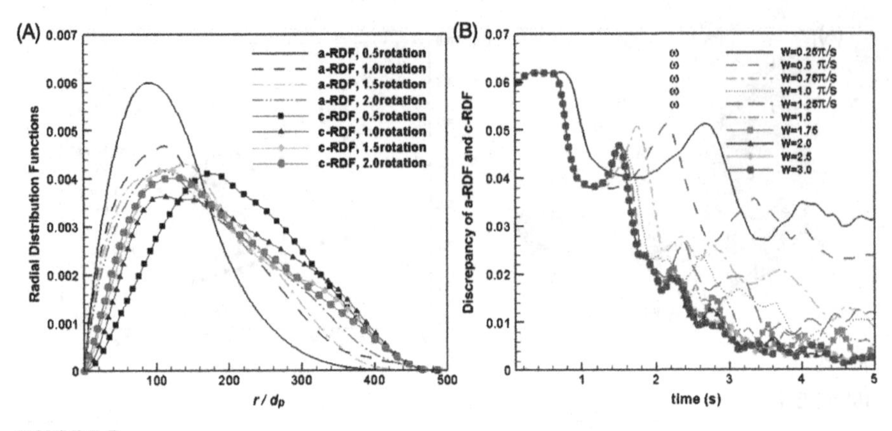

FIGURE 5.5

The auto- and cross-radial distribution functions (A) at different rotations for $\omega = 1\ (\pi/\text{s})$, and the evolution of 2-norms of the discrepancies between the a-RDF and c-RDF for different rotation speeds (B).

has the same probability to find a red and blue particle around the given red particle at a radial distance. For this consideration, it is found that the RDF functions are good macroscopic statistical measurements for mixing quality.

For other rotation speeds, the results and conclusions are similar. In Fig. 5.5B, the evolution of discrepancy between the a-RDF and c-RDF is illustrated for various rotation speeds. The discrepancy in Fig. 5.5B is formulated as the 2-norm of the difference between a-RDF and c-RDF, which is:

$$||a\text{RDF} - c\text{RDF}||_2 = \left(\int (f_{\alpha\alpha}(r) - f_{\alpha\beta}(r))^2 dr \right)^{1/2} \tag{5.40}$$

It is observed that the difference between a-RDF and c-RDF are getting reduced when the mixing advances, although some locally inverse trends occur due to the external effects of tumbling acceleration.

As mentioned above, the radial distribution function $f_{\alpha\beta}(r)$ is regarded as the probability to find a type of particle at the distance r from the given particle. So, what is the total probability? It is straightforward to introduce the Shannon information entropy as it is a good formulation of the total probability. Hence, the Shannon information entropy used here is a good evaluation of mixing which reflects the macroscopic statistical characteristics of mixing. One should remembered that the mixing process is an entropy-increased process in thermodynamics. It is analogously considered that the information entropy is increasing as the mixing of particles proceeds. The larger information entropy corresponds to the more developed state of mixing. Notice that the entropy in thermodynamics is not exactly the Shannon entropy but closely related to the information entropy. Therefore, the Shannon information entropy is regarded as a good macroscopic measurement of the uncertainties, as it plays an equivalent role as the entropy in thermodynamics.

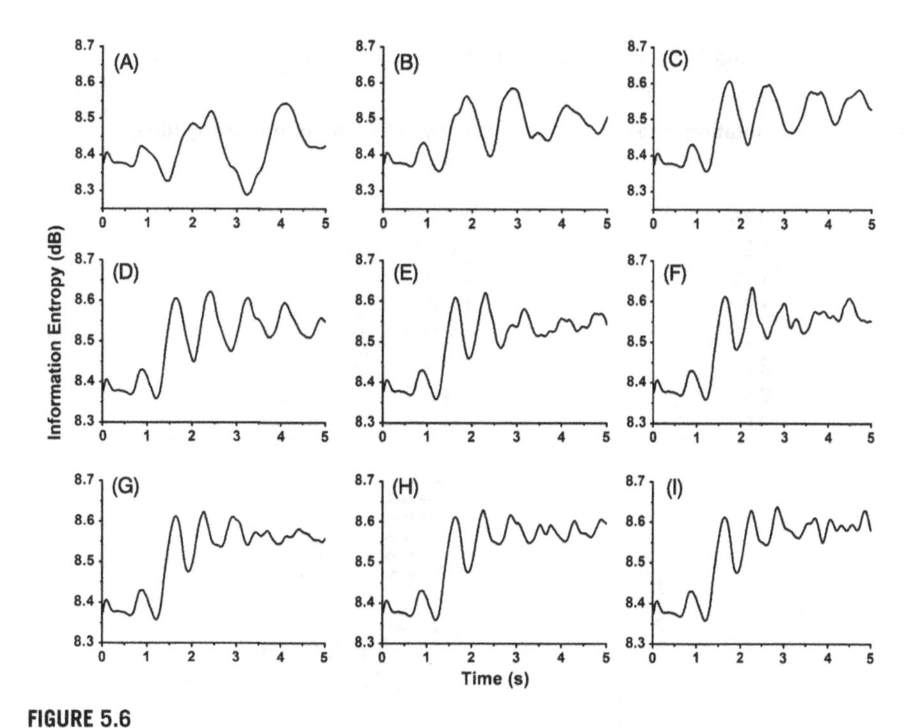

FIGURE 5.6

Evolution of Shannon information entropy for various rotation speeds: (A) $\omega = 0.5$, (B) $\omega = 0.75$, (C) $\omega = 1.0$, (D) $\omega = 1.25$, (E) $\omega = 1.5$, (F) $\omega = 1.75$, (G) $\omega = 2.0$, (H) $\omega = 2.5$, (I) $\omega = 3.0$ (π/s).

In Fig. 5.6A to I, the evolutions of Shannon information entropies under various rotation speeds are illustrated. It is seen from Fig. 5.6 that the entropy is oscillating and increasing with time, and their oscillation amplitudes are decreasing. The increase of the information entropy confirms the characteristics of process of mixing, i.e. an uncertainty-increased process. It is observed that at $t = 1.5$ s the initial oscillating and increasing stage of mixing is nearly finished and it is very close to the eventual level of mixing after that. Thus, the mean values of information entropy are calculated after $t = 1.5$ s. As Table 5.1 shows, the mean values of Shannon entropy after $t = 1.5$ s increase with the rotation speeds. Remember that in Fig. 5.4B, we have already shown that the mixing level is increased with the rotation speed at the same time. Herein it is a validation of the previous conclusions from a macroscopic point of view.

However, the oscillation in entropy indicates more complex dynamical characteristics of the mixing development. When the developed state of mixing is reached or closely approached, the oscillation is attenuated and reduced to be a fluctuating process. The interpretation for the oscillation in entropy should be interesting. It is noticed that, as shown in Fig. 5.5B, the norm of difference between a-RDF and c-RDF is decreased in an oscillating manner. Hence, the oscillation of entropy is

Table 5.1 Mean information entropies for various rotation speeds after $t = 1.5$ s.

Rotating velocity ω/π (rad/s)	Mean Information Entropy (dB)
0.25	8.399
0.5	8.432
0.75	8.497
1.0	8.536
1.25	8.545
1.5	8.542
1.75	8.561
2.0	8.564
2.5	8.572
3.0	8.580

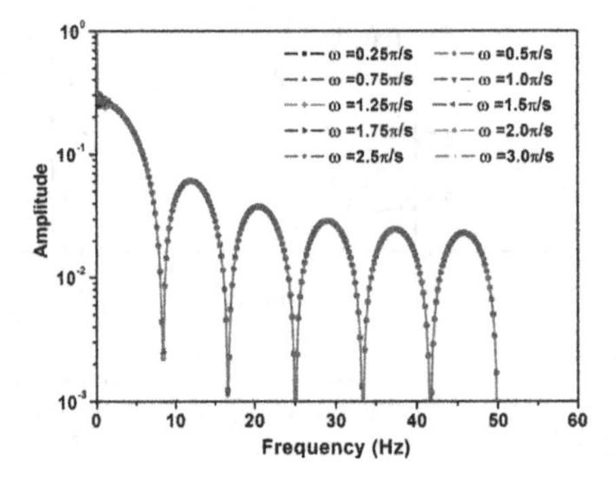

FIGURE 5.7

Frequency spectrums of Fig. 5.6A–I respectively by Fast Fourier Transformation.

closely related to the evolution of radial distribution functions. Based on the physical consideration of mechanism, the main process of mixing in the tumbler is periodic. The particles are brought upwards by the moving wall of the tumbler and flowing downwards in the shearing layer over the static base. Thus, the periodical characteristics of movement of particle can be certainly reflected by the evolutions of RDFs and subsequently the information entropies. Thus, the Fourier Transformations of the evolution of information entropies in Fig. 5.6 are carried out and shown in Fig. 5.7.

From Fig. 5.7, it is found that the frequency spectrums of the information entropies for different rotation speeds are almost exactly identical, and regularly spaced. This result indicates the common characteristics of mixing for different rotation speeds. As the accelerations at the initial stage are the same for all rotation speeds ($\dot{\omega} = \pi$ rad/s^2), and the entropy oscillation is especially large in the initial stage, the only common characteristic is the acceleration of rotation speed. Hence, there exists large entropy oscillation during the process of transition from one state to another. It is the so-called relaxation process. Moreover, the oscillation of entropy is closely related to the properties of external effects, i.e. the rotational acceleration of the tumbler. It is analogous to the description of molecular systems in kinetic theory. The transition process from one state in equilibrium to another is defined as the relaxation process, within which no 'state' variable exists since it is far from equilibrium. For these processes, the information entropy is fairly disordered in variation. Thus, it is reasonable that the oscillation of entropy could occur, and after a new equilibrium is established (a developed state of mixing), the oscillation in large magnitude vanishes or is degraded to the fluctuating process.

To say conclusively, the difference between a-RDF and c-RDF is a well indication of the macroscopic degree of mixing. The mixing process is indeed a process to eliminate the difference between them.

The entropy-based analysis indicates the mixing process of particle is a Shannon information entropy-increased process, during which the uncertainty increases. The final level of mixing is appropriate to be evaluated by the Shannon information entropy. Under this approach, the dynamical characteristics of mixing can be explored. It is found that during the mixing process, the Shannon information entropy is oscillating as well as increasing. The oscillation of entropy indicates the transition procedure from one equilibrium state to another due to the external effects, such as acceleration of the system. Finally, the oscillation is reduced or degraded to the fluctuation when a new equilibrium state is established.

5.2.4 Heat conduction features

The DEM coupled with the thermal conduction model is applied here to explore the intrinsic characteristics of particle mixing and thermal conduction characteristics in rotating tumblers and the effect of macroscopic and microscopic mixing structures on heat transfer. At start, the tumbler is partially filled with spherical granular particles that settle at the bottom. For the boundary condition, the tumbler is considered as a rigid, motion-controlled large particle with finite radius, infinite mass, and frictional adiabatic walls. Then, the tumbler is rotated with gradually increasing rotating angular velocity until it reaches a target speed (the rotating acceleration is $\dot{\omega} = \pi$ rad/s^2). Six typical rotation speeds are simulated, ranging from $\omega = 0.5\pi$ rad/s to $\omega = 3\pi$ rad/s. A sufficiently fine time step of $dt = 5$ μs is used to stabilize the particle-particle collision simulation.

The particles are divided into Part 'a' with high temperature and Part 'b' with low temperature. The tumbler boundary is adiabatic and rotates clockwise. The parameters used in the simulation are listed in Table 5.2.

5.2.4.1 Motion of gyration

To illustrate the cyclic motion characteristics of particles, a radius of gyration R_g is defined here as the instantaneous distance of any particle from the center of the tumbler (Fig. 5.8).

In this section, the time variation of the radius of gyration, the velocity $|u_p|$ and the temperature for $\omega = 1, 2$ and 3 (π/s) are shown in Fig. 5.9A, B and C, respectively. The intermittent pulse variation of R_g is observed for all cases. The flat segment of

Table 5.2 Parameters used in the tumbler.

Radius of gyration, R_0 (mm)	288
Number of particles, (N_a, N_b)	(45177, 45424)
Initial temperate of particles, (T_a, T_b) (°C)	(100, 20)
Particle diameter, d_p (mm)	1.2
Particle density, ρ_p (kg/m^3)	7800
Particle specific heat, C_p (J/(kg·K))	460
Particle thermal conductivity, λ_p (W/(m·K))	46.52
Restitution coefficient, e	0.95
Friction coefficient, λ	0.3
Collision stiffness, k_n (N/m)	10000
Rotating velocity, ω/π (rad/s)	0.5, 1.0, 1.5, 2.0, 2.5, 3.0
Simulation time step, Δt (s)	5.0×10^{-6}
Total simulated time, T_I (s)	25

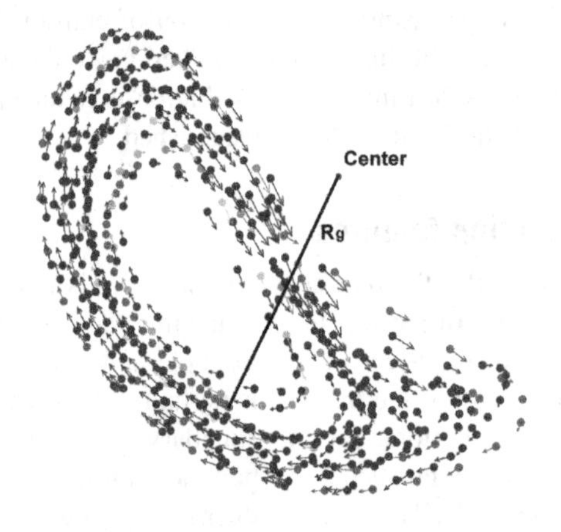

FIGURE 5.8

Typical trajectory of one sampled hot particle, colored by the instantaneous temperatures.

R_g indicates the stable gyration motion in the static base from the right bottom to the left top side of the tumbler (Fig. 5.8) with a nearly equal radius of gyration and low radial fluctuations. The pulse segment of R_g indicates the rapid downhill backward motion of the particles from the left top side to the right bottom in the freely flowing thin layer. Thus, the variation of R_g validates the main feature of the cycled particle trajectory in the rotating tumbler.

On the other hand, the particle velocity $|u_p|$ is varied intermittently in the same pace but in the opposite trend of R_g. In the flat segment of R_g, the particle velocity $|u_p|$, although fluctuated greatly, is of relatively low magnitudes, indicating the motion characteristics in the static base. In contrast, in the pulse segment of R_g, a rapid increase in velocity $|u_p|$ is observed, which is almost simultaneously varied with the

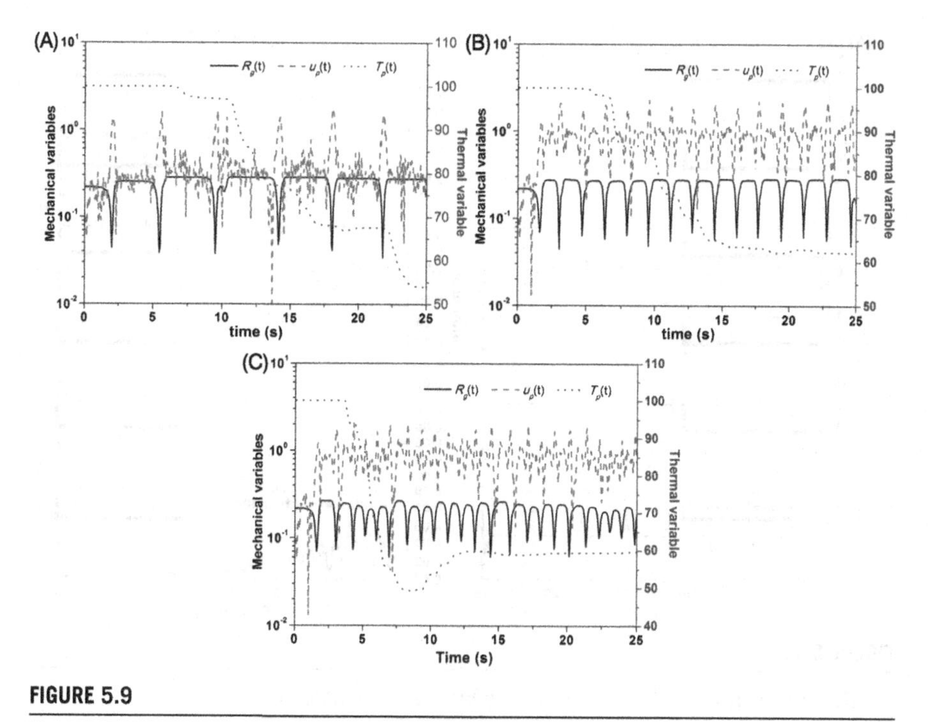

FIGURE 5.9

Time variations of instantaneous radius of gyration, velocity $|u_p|$ and temperature for $\omega = 0.5$ (π/s) (A), $\omega = 1.5$ (π/s) (B) and $\omega = 3$ (π/s) (C), respectively.

pulsation of R_g. It indicates the quick grow in kinetic energy during the downhill motion of particles.

In addition, the cyclic motion and periodic variation of mechanic variables (R_g and $|u_p|$) seem to be rather regular and stable in time, even if the flat segments become shorter for large rotation velocities ($\omega = 3$ π/s).

It is observed and naturally reasonable that the flat segment of R_g is unnecessary on the same level in different cycles. To compare and quantify the intermittent variation on the same level, the time derivative $\frac{dR_g}{dt}$ is calculated and shown in Fig. 5.10A. It is clear that the periodic pulsed motion is present for all the rotation velocities. The periodicity of the pulsed variation decreases as the rotation speed increases. In an equivalent manner, the fundamental spectrum (FFT) of $\frac{dR_g}{dt}$ increases as the rotation speed increases (Fig. 5.10B). But the degree of periodicity increase becomes smaller. Till $\omega = 1.5$–3 (π/s), the fundamental frequencies are almost the same, which indicates the same period of time variation. That is to say the periodicity of the cyclic motion is not linearly dependent on the rotation speed. Consequently, it is not a sensible approach to intensify or maximize particle mixing by increasing the rotation speed. It is appropriate and can be explained by the sliding motion of the bulk of particles, namely the bulk particle-wall sliding. As the rotation speed increases, the velocity difference between the wall and the particles contacting the wall becomes

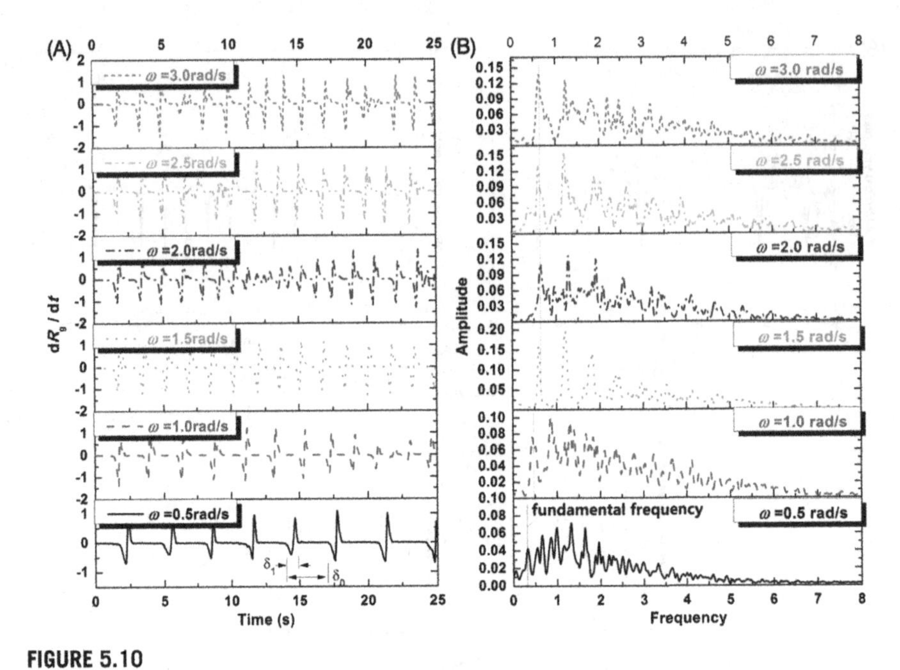

FIGURE 5.10

(A) The time derivative $\frac{dR_g}{dt}$. (B) The fast Fourier transformation (FFT) of $\frac{dR_g}{dt}$.

larger, too. Thus, the augmented bulk-sliding between the bulk of particles and the wall may weaken the trends of increased cyclic motion of particles, which indirectly decrease the ratio of the mixing degree increment to the powder input increment of the tumbler.

However, the above results are for one of the sampled particles. To derive more reliable conclusion, it is necessary to analyze the results of all the sampled particles. Basically, the variation of radius of gyration is pulsed intermittently in time. To quantify the intermittent characteristics, an intermittency factor β is defined by the time ratio of the pulsed segment (δ_1) to δ_0, which is the sum of the pulsed segment (δ_1) and the flat segment (δ_2) in one cycle (as sketched in the bottom inset of Fig. 5.10A, $\delta_0 = \delta_1 + \delta_2$), i.e. $\beta = \frac{\delta_1}{\delta_0}$.

To quantify the intermittency factor β, the flat segment of variation of R_g can be approximated by the condition of $\frac{dR_g}{dt} < \epsilon$, where $\epsilon = 0.05$ is a small threshold value. The probability density functions $p(\beta)$ of the intermittency factor β over all sampled particles are shown in Fig. 5.11. The $p(\beta)$ is defined as

$$p(\beta) = \frac{f(\beta)}{\sum(f(\beta)\Delta\beta)} \tag{5.41}$$

where $f(\beta)$ is the number of particles in $\beta \in [\beta, \beta + \Delta\beta)$, and $\sum f(\beta)$ is the number of all sampled particles.

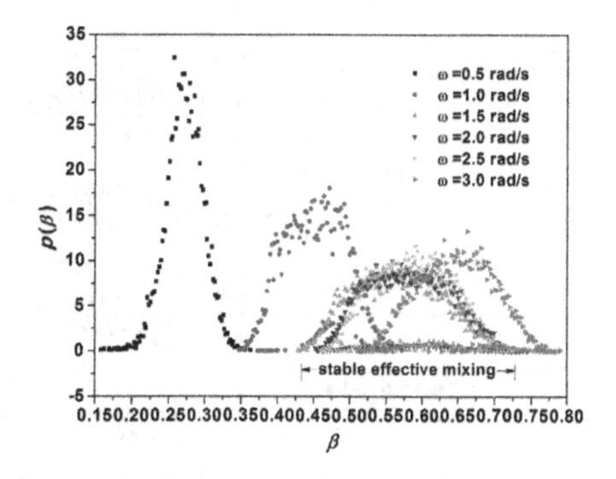

FIGURE 5.11

The probability density functions $p(\beta)$ of the intermittency factor β over all sampled particles.

In Fig. 5.11, it is seen that $p(\beta)$ for different rotation velocities are different in shapes and ranges. For low speeds ($\omega = 0.5\ \pi$/s), $p(\beta)$ occurs in the ranges of small intermittency factors. This indicates that the duration of pulse motion (namely in the freely flowing layer) is far smaller than the duration in the static base. Due to dominating role of flowing shear layer in particle mixing, the mixing efficiency at low speeds of rotation is very low.

When the rotation velocity increases ($\omega = 1.0\ \pi$/s), $p(\beta)$ moves to higher values of β and has a wider range. It means the duration of particle motion in the freely flowing layer (δ_1) becomes relatively longer and the duration in the static base (δ_2) becomes shorter than before. Recall that the periodicity of cyclic motion is decreased from $\omega = 0.5\ \pi$/s to $\omega = 1.0\ \pi$/s. At this stage, the change in rotation velocity augments particle mixing through at least two mechanisms: (1) Decreasing the overall period of cyclic motion to speed up the particle mixing, namely to increase the number of cycles at the same time to make more chance for particles to move in the flowing shear layer. (2) Increasing the ratio of duration in the flowing layer to that in the static base, leading to a larger time fraction for particles staying in the flowing shear layer in a cycle. Thus, within the stage of low rotation speed, to increase rotation speed is an effective way to intensify particle mixing.

However, the conclusions are changed when the rotation speed is sufficiently high ($\omega = 1.5$–$3.0\ \pi$/s). Firstly, the period of cyclic motion is no longer increasing with the rotation speed. Alternatively, the fundamental spectrum of $\frac{dR_g}{dt}$ remains the same, indicating that the main period of cyclic motion is almost independent of the rotation speed at this stage. It is caused by the increased effect of the bulk-particle-wall sliding. In other words, at this stage, the increase in rotation velocity cannot intensify particle mixing as effectively as before.

Secondly, the probability density function of the intermittency factor, except $\omega = 3.0\ \pi$/s, are almost stable in variation of rotation velocities (Fig. 5.11). In other words, the fraction of the pulsed variation of R_g, corresponding to the downhill flowing motion, to the nearly constant variation of R_g in the static base, is almost

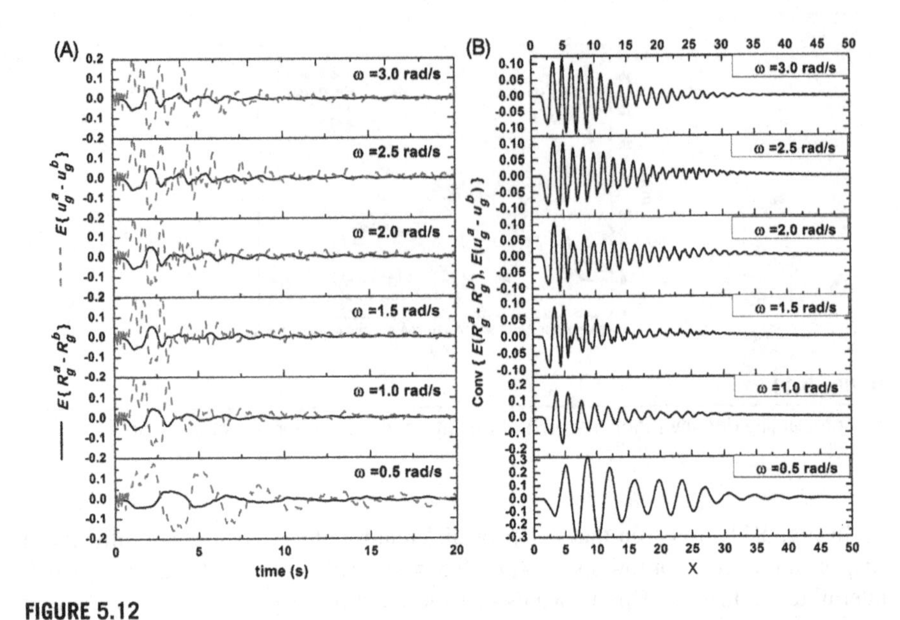

FIGURE 5.12

(A) The mean differences of $(R_{g,h} - R_{g,l})$ and $(u_h - u_l)$ between the hot and cool sampled particles. (B) The convolution of the former two.

independent of the change of rotation velocity, too. At this stage, it is demonstrated that the cyclic motion of particles is mainly determined by the relatively bulk-sliding of the particles from the wall of tumbler. To say more accurately, the cyclic motion of particles depends mainly on the characteristics of bulk-sliding motion. Then, the motion of particles, or the characteristics of particle mixing, is stabilized in variation of rotation velocities. It establishes a dynamically self-maintained state of equilibrium. In this state, the efficiency of mixing reaches a locally maximum point, whereas the efficiency of increase of rotation speed for augmenting particle mixing reaches a locally minimum point.

In addition, compared to $\omega = 1.5$–$2.5\ \pi$/s, a slight difference of $p(\beta)$ at $\omega = 3.0\ \pi$/s is observed in Fig. 5.11. It may be due to the transition of flow regimes from the rolling/cascading to the fully cataracting. In the fully cataracting flow regime, the particles are brought up more higher in the left-top side, and a larger and longer cyclic trajectory is necessary, which may make the intermittency factor much larger.

In above section, the deficiency is that the spectrum of the cyclic motion periods is just for one of the sampled particles. Therefore, it is necessary to analyze the mean periodicity of the cyclic motion for all the hot and cool sampled particles. Fig. 5.12A shows the mean differences of $(R_{g,h} - R_{g,l})$ and $(u_h - u_l)$ between the hot and cool sampled particles:

$$E(R_{g,h} - R_{g,l}) = \frac{1}{N_{Sh}} \sum_{i=1}^{N_{Sk}} R_{g,h,i} - \frac{1}{N_{Sl}} \sum_{j=1}^{N_{Sl}} R_{g,h,j} \tag{5.42}$$

$$E(u_h - u_l) = \frac{1}{N_{Sh}} \sum_{i=1}^{N_{Sk}} |\boldsymbol{u}_{p,i}| - \frac{1}{N_{Sl}} \sum_{j=1}^{N_{Sl}} |\boldsymbol{u}_{p,j}| \tag{5.43}$$

where N_{Sh} and N_{Sl} are the numbers of the sampled hot and cool particles, respectively (Table 5.2).

Fig. 5.12B shows the convolution of the two items, named $Conv\{E(R_{g,h} - R_{g,l}), E(u_h - u_l)\}$, where '$E$' means the expectation over the sampled particles. In general, in Fig. 5.12A, $E(R_{g,h} - R_{g,l})$ and $E(u_h - u_l)$ also evolve in the same pace as observed in Fig. 5.8, but in the opposite directions of variation. Moreover, the variations of $E(R_{g,h} - R_{g,l})$ and $E(u_h - u_l)$ are decaying in time, which is different from the results for the single sampled particle. It is due to the development of mixing which produces a developing uniform mixing state and the difference of the averaged characteristics between the hot and cool parts of particles vanishes.

On the other hand, the decaying periodic variation of $E(R_{g,h} - R_{g,l})$ and $E(u_h - u_l)$ can be more clearly observed by the convolution of them (Fig. 5.12B). The convolution algorithm is defined as

$$Conv(f, g) = (f \star g)(t) = \int_0^{T_s} f(\tau)g(t - \tau)d\tau \tag{5.44}$$

where T_s is the total simulation time, and '$\star$' means the convolution algorithm. With the convolution algorithm, the variation becomes more smoothing than before since the convolution algorithm has features of smoothing and filtering. Moreover, the FFT and convolution algorithms have the relation

$$\mathscr{F}(conv(f, g)) = \mathscr{F}(f) \cdot \mathscr{F}(g) = \hat{f}(\omega_1) \cdot \hat{g}(\omega_1) \tag{5.45}$$

where the operator '$\mathscr{F}$' indicates the FFT algorithm, and '$\hat{\cdot}$' denotes the Fourier spectrum presentation. Taking the FFT operation on $Conv(f, g)$, the product of the Fourier spectrums of $\mathscr{F}(f)$ and $\mathscr{F}(g)$ is obtained. As the convolution operator filters high frequency components, the product of $\mathscr{F}(f)$ and $\mathscr{F}(g)$ is the product of fundamental frequencies of them, respectively, i.e.

$$\widehat{(f \star g)}(\omega_1) \approx \hat{f}(\omega_1) \cdot \hat{g}(\omega_1) \tag{5.46}$$

where ω_1 is the fundamental frequency.

Fig. 5.13 shows $\widehat{(f \star g)}$ when $f = E(R_{g,h} - R_{g,l})$ and $g = E(u_h - u_l)$. In Fig. 5.13, the $\widehat{(f \star g)}$ for all rotation velocities have one peak only, i.e. the fundamental frequency of $\mathscr{F}(Conv(E(R_{g,h} - R_{g,l}), E(u_h - u_l)))$, corresponding to the fundamental frequency of $\mathscr{F}(E(R_{g,h} - R_{g,l}))$, $\mathscr{F}(E(u_h - u_l))$. As $E(R_{g,h} - R_{g,l})$ and $E(u_h - u_l)$ are varied almost simultaneously, the fundamental frequencies for these items are the same, namely ω_1 (compare Fig. 5.13 to Fig. 5.10B for validation). Moreover, the trends remain that: the fundamental frequency ω_1 increases with the rotation velocity ω when the tumbler rotates slowly ($\omega = 0.5$–$1.5\ \pi$/s), whereas ω_1

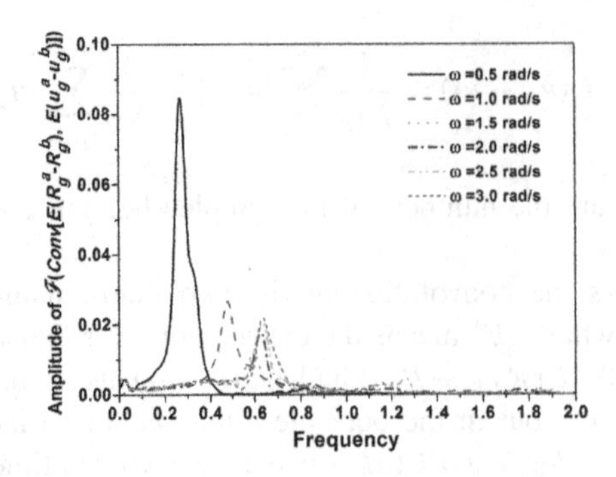

FIGURE 5.13

The Fourier spectrum $\mathcal{F}(Conv(E(R_{g,h} - R_{g,l}), E(u_h - u_l)))$ of the convolution of $E(R_{g,h} - R_{g,l})$ and $E(u_h - u_l)$.

is independent of ω when $\omega = 1.5$–3 π/s. Thus, this consolidates the analysis and conclusions obtained in the above section.

5.2.4.2 After the same evolution time

The thermal conduction between the particles in the drum at the same evolution time as well as the same revolution (rotation) is studied here to show the coupling between mixing and thermal conduction characteristics, since the amount of thermal conduction is dependent on the duration of the particle-particle contact and the degree of mixing in the tumbler.

Fig. 5.14 shows the comparisons of the typical particle mixing state, the distribution of WT $\Theta(T_a, T_b; n_a, n_b)$, and TDF $\Pi(T_a, T_b; T_a^0, T_b^0)$ at $t = 8.25$ s for $\omega = 0.5$ (π/s) and $\omega = 1.5$ (π/s). In the mixing and conduction processes (Figs. 5.14A and B): (1) the granular particles enter the flowing layer on the left-top side; (2) they move downhill to the right-bottom side, becoming stretched; (3) they enter the static base, folded into jagged patterns; and (4) they are driven by the rotary static base with a gradual change in orientation and a slight deformation in mixing interface until they reach the left-top side again. When this process is repeated, the mixing and heat transfer interfaces develop into increasingly finer structures (Figs. 5.14E and F).

Comparing Figs. 5.14A and B, after 8.25 s with only two tumbler revolutions for $\omega = 0.5$ (π/s), the mixing contour appears to be in perfect accordance with the WT distribution contour. The consistency of the mixing/conduction interfaces between the hot (red) and cool (green) particles exists not only in the fine-scale structures and large-scale skeletons of the interfaces but also in the perfectly separated two-color contours. Moreover, TDF $\Pi(T_a, T_b; T_a^0, T_b^0)$ (Fig. 5.14C) shows nonzero values only around the mixing interface, indicating the close dependence of conduction interface on the mixing interface.

In contrast, for $\omega = 1.5$ (π/s) after 5.63 revolutions (Figs. 5.14D–F) with the same time of $t = 8.25$ s, the WT distribution is composed of multiple color contours, with the TDF contours having large values in almost the entire tumbler, showing a

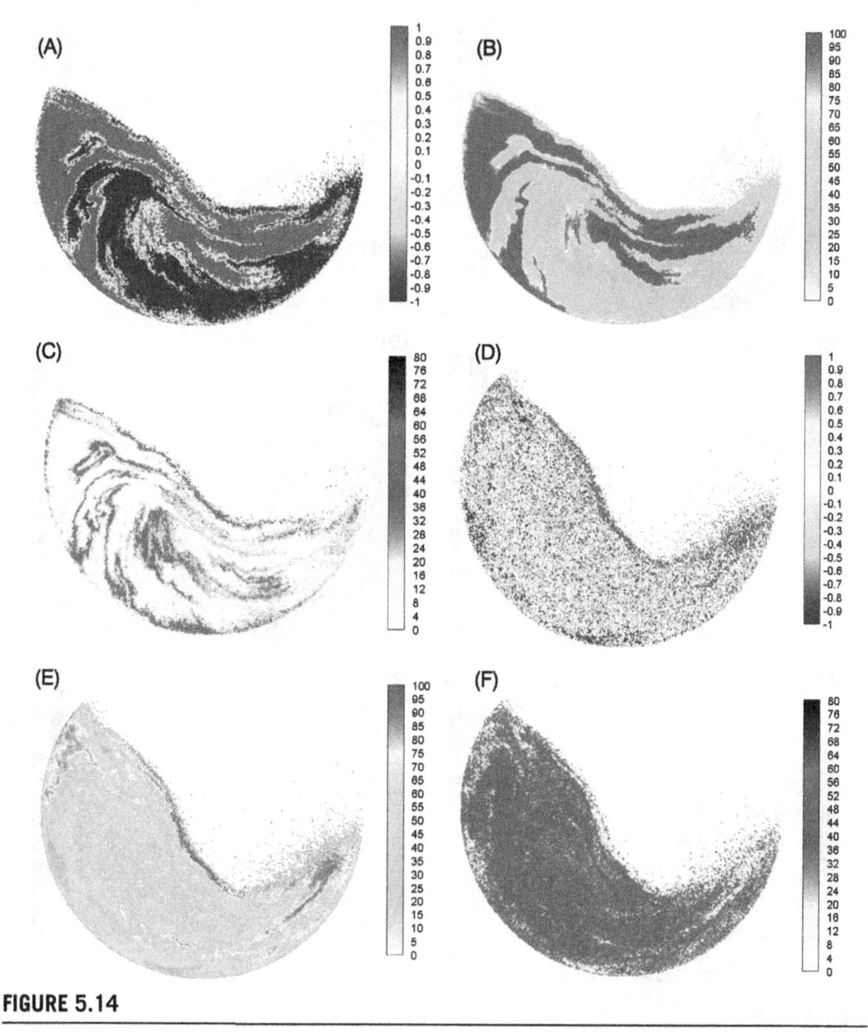

FIGURE 5.14

Comparison of mixing state (A, D), WT distribution $\Theta(T_a, T_b; n_a, n_b)$ (B, E) and TDF $\Pi(T_a, T_b; T_a^0, T_b^0)$ (C, F) at $t = 8.25$ s for $\omega = 0.5$ (π/s) (A, B, C) and $\omega = 1.5$ (π/s) (D, E, F), respectively.

sufficiently developed thermal conduction between the hot and cool particles after several revolutions. The mixing and conduction interface structures still agree with each other. The agreement between mixing and WT contour and the comparison of TDFs in Figs. 5.14C and F indicate the significant effect of mixing on the particle-particle heat transfer under the same temporal evolution.

5.2.4.3 After the same revolution

Fig. 5.15 shows the comparison of snapshots of mixing and thermal conduction after two revolutions for $\omega = 1$ (π/s) and $\omega = 2$ (π/s), respectively. The mixing state for

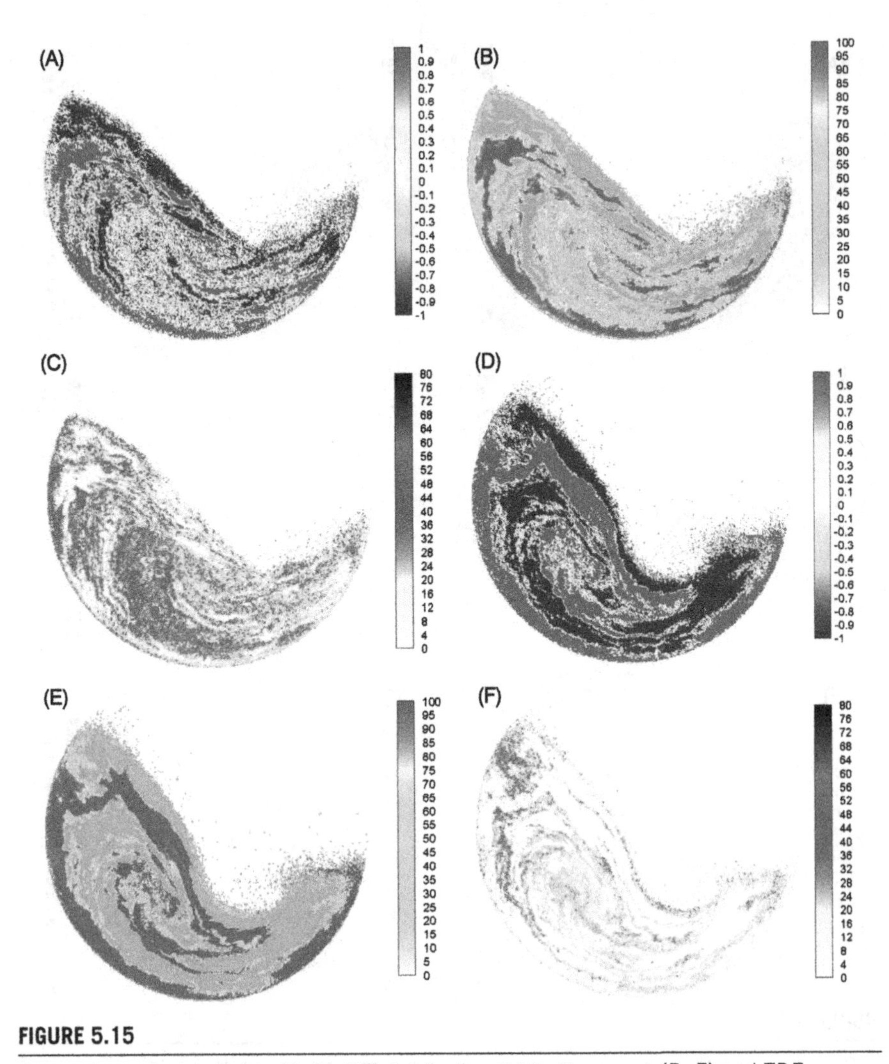

FIGURE 5.15

Comparison of mixing state (A, D), WT distribution $\Theta(T_a, T_b; n_a, n_b)$ (B, E) and TDF $\Pi(T_a, T_b; T_a^0, T_b^0)$ (C, F) at two revolutions for $\omega = 1$ (π/s) (A, B, C) and $\omega = 2$ (π/s) (D, E, F), respectively.

low revolution speed (Fig. 5.15A) appears slightly more developed than that for the high speed (Fig. 5.15D) because the red and blue parts of the particles are more slender in Fig. 5.15A than those in Fig. 5.15D, and the structure of the mixing interface is composed of finer striations. The WT contour shown in Fig. 5.15B has many finer striations and more light-blue/yellow-colored regions, corresponding to intermediate temperatures, than that in Fig. 5.15E. The TDF contours (Figs. 5.15C and F) confirm the results of Figs. 5.15B and E of a more developed particle-particle thermal conduction at $\omega = 1$ (π/s) than at $\omega = 2$ (π/s).

Observing that the same revolution at low speed requires more time than that at high speed when the degrees of mixing do not differ much, the results indicate that duration and mixing degree have common effects, especially the role of duration, on the particle-particle heat transfer in the tumbler.

5.2.4.4 Probability density function of temperature discrepancy functions (TDF)

Figs. 5.14C–F and 5.15C–F show the distribution and differences of TDFs. Therefore, it is necessary to investigate the TDF characteristics quantitatively. Thus, the probability density function (PDF) of TDFs in Figs. 5.16A and B corresponding to Figs. 5.14 and 5.15, respectively, is explored. In general, PDF is formulated as the number of area cells ($\delta N(\Pi)$) with $\Pi \in [\Pi_0, \Pi_0 + \delta\Pi]$ normalized by the total number of area cells ($N = \sum \delta N(\Pi)$) covering all the heat transfer interfaces, i.e.,

$$p(\Pi) = \frac{N(\Pi)}{N\delta N} \tag{5.47}$$

However, to incorporate the total amount of heat transfer interface, the current study uses the nonnormalized PDF instead:

$$p(\Pi) = \frac{N(\Pi)}{\delta N} \tag{5.48}$$

By using nonnormalized PDF, the area below $p(\Pi)$ indicates the absolute amount of heat transfer interface within $[\Pi_0, \Pi_0 + \delta\Pi]$. In this way, the degree and amount of heat transfer under different conditions can be directly compared.

Figs. 5.16A and B show the PDFs of TDF corresponding to Figs. 5.14 and 5.15, respectively. In Fig. 5.16A, the PDF peak of TDF for $\omega = 0.5$ (π/s) is approximately 27°C, whereas peaks for other ω are roughly from 35°C to 37°C. The lesser temperature discrepancy for $\omega = 0.5$ (π/s) than that for other ω indicates a low degree of thermal conduction. Moreover, the peak values for $\omega = 0.5$ (π/s) are significantly less than that for other ω, which indicates a significantly insufficient development of heat transfer. Recalling the fractal dimension of mixing interface in Fig. 5.16A, the nearly consistent log(R)-log(s) for $\omega = 1.0$–3.0 (π/s) leads to the nearly consistent PDFs of TDFs for $\omega = 1.0$–3.0 (π/s). In other words, the degree of mixing of particles dominates the heat transfer characteristics.

In Fig. 5.16B, after the same revolution, the PDF peaks for high rotation speed occur at low TDFs (about 15°C to 20°C), whereas those for low rotation speed occur at high TDFs (about 30°C to 40°C). These results indicate that, compared with low rotation speed, high rotation speed is relatively unfavorable for increasing TDFs with maximum probability because it is adverse to the thermal conduction of particles under the same revolution condition. This is caused by the decreased evolution time for the high rotation speed to reach the same revolution for larger TDF, as the more homogeneous temperature distribution has no sufficient time to develop. Moreover, the absolute PDF peaks at high rotation speed are smaller than those at low rotation speed. This finding indicates that high rotation speed is unfavorable in increasing

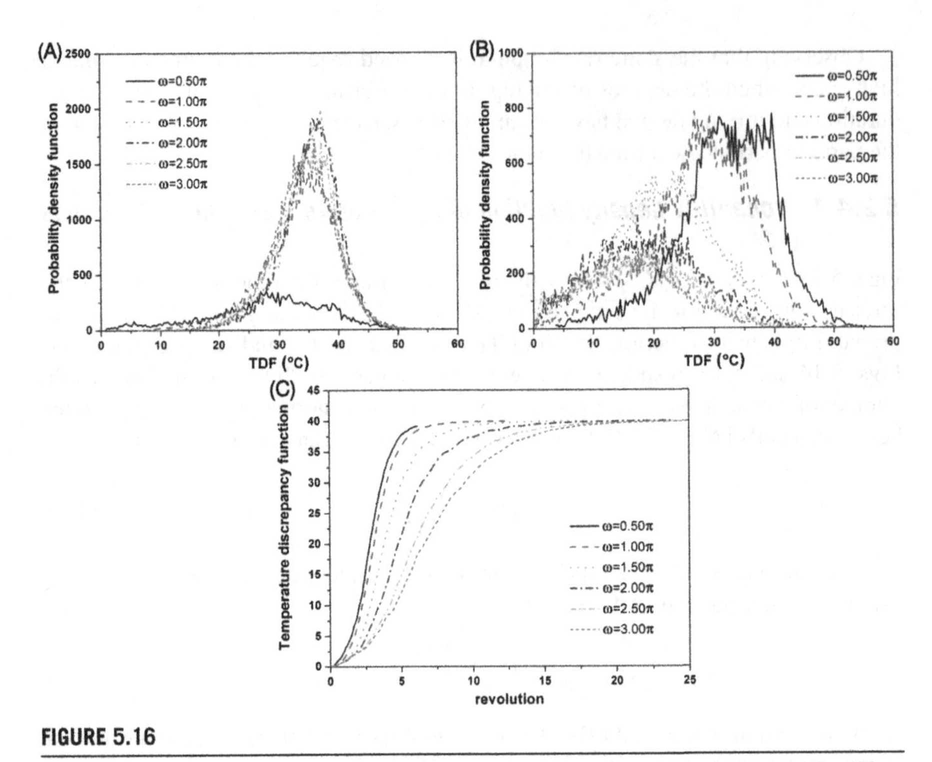

FIGURE 5.16

Probability density functions of TDF corresponding to Fig. 5.14 – (A), and Fig. 5.15 – (B), respectively, and the temporal evolution of mean TDFs over the tumbler – (C).

the PDF peaks because it is also adverse to thermal conduction development. This condition is caused by the relatively lower mixing level on the fine scales of the mixing interface compared with that of the low rotation speed, in which the total 'probability' for the thermal conduction is reduced. Thus, the two factors, mixing levels of the fine scale interface and temporal length, jointly contribute to thermal conduction development.

In addition, Fig. 5.16C shows the temporal evolution of the mean TDFs over the tumbler. Under the same revolution, the TDF for low rotation speed is larger than that for high rotation speed. The thermal conduction at low rotation speed is more sufficiently developed than that at high rotation speed. Thus, because of the common contribution of duration and fine mixing scale to the thermal conduction, using low rotation speed is beneficial to heat transfer in tumblers. In conclusion, using low rotation speed would be more favorable, provided the flow regimes are maintained (at least, the rolling regime), to enable sufficient heat transfer between particles, i.e., to obtain a homogeneous temperature distribution in the tumbler. It is based on the assumption that the total energy input for the rotating the tumbler should be proportional to the number of revolutions so that the same revolution results in the same energy consumption.

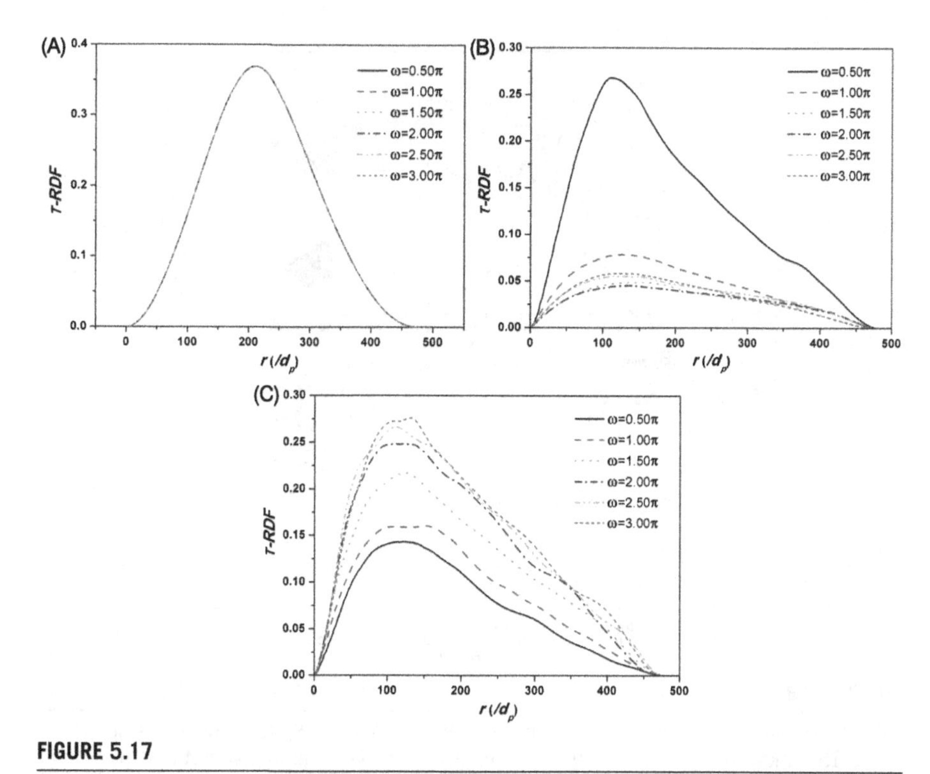

FIGURE 5.17

(A) Initial T-RDF for all speeds. (B) The T-RDFs at $t = 8.25$ s. (C) The T-RDFs after 2 revolutions.

5.2.4.5 Thermal radial distribution function

In this section, the T-RDFs are used (Fig. 5.17) for the macroscopic study of heat transfer. Figs. 5.17B and C show the T-RDFs corresponding to Figs. 5.14 and 5.15, respectively, and Fig. 5.17A shows the initial T-RDF for all speeds. Fig. 5.17A indicates that the T-RDFs for all rotation speeds are the same because they are evolved from the same initial state. However, after the same temporal evolution duration (Fig. 5.17B), the T-RDFs change. (1) The T-RDFs for the relatively high rotation speed $\omega = 1.0$–3.0 (π/s) decrease throughout the whole range, indicating a decrease in temperature difference in the radial direction which is attributed to the mixing and thermal conduction between particles. (2) For low speed $\omega = 0.5$ (π/s), the attenuation in the peak value is less than the former because of a lesser mixing degree effect. (3) The T-RDF peaks shift to a small radial distance from about $r_p = 200d_p$ to $r_p = 100d_p$. The shift of the peaks is mainly caused by the evolution of particle mixing, as the dynamic mixing process changes the relative inter-displacements between the red (hot) and blue/green (cool) particle parts. Alternatively, the attenuation of T-RDF peaks is caused by the heat transfer effect, as the relatively static thermal

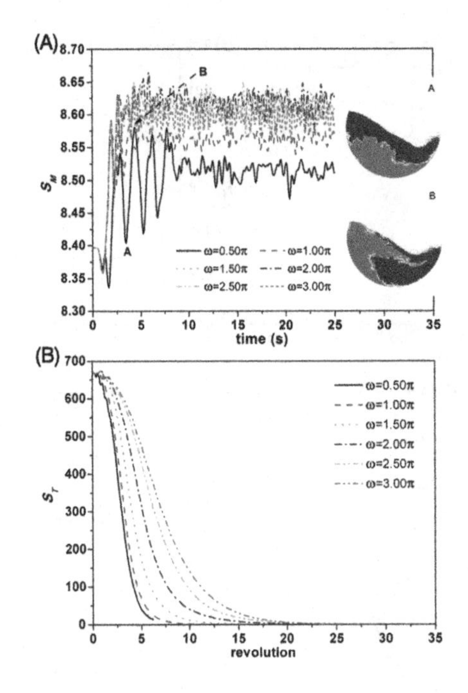

FIGURE 5.18

Information entropy for mixing S_M (A) and thermal conduction S_T (B). The insets of Fig. 5.18A show the corresponding distributions of particles at time points A and B, respectively.

conduction leads to a reduction in temperature difference between the hot and cool particles.

Thus, under the same revolution condition (Fig. 5.17C), the peak values of the T-RDFs decrease because of thermal conduction. Moreover, less decrease of T-RDF may take place at higher speeds or vice versa, since the evolution time is shorter at higher speeds than at lower speeds. Additionally, the shifts in T-RDF peaks are caused by the mixing evolution.

In summary, the change in T-RDF shows the different approaches of static thermal conduction and dynamic mixing, which jointly influence the heat transfer characteristics of granular materials in a particular rotary system. The integration of these approaches is reflected in the following entropy analysis.

5.2.4.6 Entropy analysis

Based on Eqs. (5.4), (5.37), and (5.38), information entropy or its analogical definition can be calculated as shown in Figs. 5.18A and B. In mixing information entropy, S_M increases generally with the increase in rotation speed (Fig. 5.18A and Table 5.3). Using a higher rotation speed can augment the degree of mixing in the tumbler, provided that the flow regimes in the tumbler are maintained. The slight decrease in the main mixing information entropy or the mixing degree for $\omega = 2.0$–3.0 (π/s) is

Table 5.3 Mean information entropies for various rotation speeds after $t = 2.5$ s.

Rotating speed ω/π (rad/s)	Froude number $Fr = \omega^2 R_0/g$	Flow regime	Mean information entropy (dB)
0.5	7.25×10^{-2}	Rolling	8.513
1.0	0.290	Rolling/Cascading	8.563
1.5	0.653	Rolling/Cascading	8.611
2.0	1.160	Cascading/Cataracting	8.618
2.5	1.813	Cataracting	8.612
3.0	2.610	Cataracting	8.595

caused by the transition of the flow regime from the cascading ($Fr = 1.16$, where $Fr = \omega^2 R_0/g$ is the Froude number) to the cataracting regimes ($Fr = 2.61$). The pulsating variation of S_M in the early stage of Fig. 5.18A is caused by the dissymmetrical distribution areas (Ω_a and Ω_b) of the hot and cool particles during the rotation of tumbler. For example, the insets of Fig. 5.18A shows the typical distributions corresponding to the locally lowest and highest points of S_M at $t = 3.45$ and 4.35 s, respectively. It is seen that the former is an up-down distribution whereas the latter one is a left-right distribution, where the sizes of Ω_a and Ω_b for these two types of distribution are not the same. The rotation makes the variation of dissymmetry of Ω_a and Ω_b periodical. However, the periodic fluctuations are rapidly attenuated as the mixing between the hot and cool particles advances.

Moreover, the information entropies corresponding to the radial temperature discrepancies S_T decrease as the rotations increase (Fig. 5.18B). The decrease in S_T is caused by the reduction of the temperature difference, as mixing evolves and subsequently causes reductions in T-RDF and S_T. More importantly, when the rotation speed increases, S_T decreases more slowly, which indicates the less effective evolution of heat transfer among the granular materials with higher rotation speeds and validates the earlier analysis and conclusions on this issue.

Finally, the clear differentiation in S_T with different rotation speeds shows its feasibility through the good indication of heat transfer evaluation. The duration time and mixing levels have been shown to contribute jointly to the thermal conduction characteristics. Evaluating the heat transfer efficiency of the tumbler under the sole contribution of particle mixing is now feasible by using information entropy based on the radial temperature discrepancy. If the information entropy increment $|\Delta S_T|$ is divided by the time duration $|\Delta\tau|$, the mean increase rate of information entropy is obtained as follows:

$$|\bar{\dot{S}}_T| = \frac{S_T(\tau_1) - S_T(\tau_0)}{\tau_1 - \tau_0} = \frac{\Delta S_T}{\Delta\tau} \tag{5.49}$$

Eq. (5.49) may indicate the mean efficiency of heat transfer arising from the sole mixing effect. Fig. 5.19 shows that the maximum increase rate of information entropy for each rotation speed exists, indicating that the maximum operating condition has

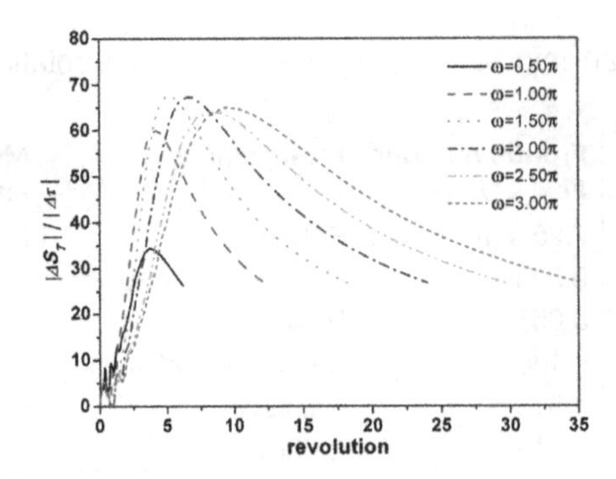

FIGURE 5.19

Mean increase rate of temperature information entropy.

the best efficiency. For example, for $\omega = 1.0$ (π/s), the maximum increase rate occurs at about five revolutions, after which the heat transfer efficiency cannot be maintained to be as high as before by the rotation because the temperature discrepancy is very small (Fig. 5.18B). Moreover, for different rotation speeds, the maximum increase rate occurs at different revolutions. The larger the rotation speed is, the larger the critical point of revolution for maximum increase rate. Before the critical point is reached, low rotation speeds (except $\omega = 0.5$ π/s for the pure rolling regime) always produce higher heat transfer efficiency. The opposite is true after the critical point is reached. Eq. (5.49) can be regarded as a performance plot for the tumbler. It not only shows the evaluation of heat transfer levels, increase rates, and transition points with maximum efficiency but also provides clearly the method for determining the most effective operating conditions for the tumbler. For example, for using a constant rotation velocity, it is best to use $\omega = 1.0$ π/s for less than three revolutions, but to use $\omega = 3.0$ π/s for more than 10 revolutions. Moreover, considering the envelope of the curves, the best efficient way is to use $\omega = 1.0$ π/s before three revolutions and to transit to use $\omega = 1.5$ π/s till 5 revolutions, and then go on to transit to $\omega = 2.0$, 2.5, 3.0 π/s, etc. In other words, this enveloping approach has the highest mean increase rate of entropy S_T, or the highest efficiency of heat conduction, provided the flow regimes remain valid for mixing and conduction.

5.2.4.7 Mean temperature differences and increasing rates

In this section, the averaged variables of all the sampled particles are computed. The mean temperature difference between the hot and cool particles can be calculated as follows:

$$E(H_h^\star - H_l^\star) = \frac{1}{N_{Sh}} \sum_{i=1}^{N_{Sk}} \left(\frac{T_{p,h}^i(t) - T_l^0}{T_h^0 - T_l^0} \right) - \frac{1}{N_{Sl}} \sum_{j=1}^{N_{Sl}} \left(\frac{T_{p,l}^j(t) - T_l^0}{T_h^0 - T_l^0} \right) \tag{5.50}$$

where $T_h^\star$ and $T_l^\star$ are dimensionless temperature nondimensionalized by the initial temperature difference between the hot and cool particles. The global degree of the particle-particle heat conduction can be evaluated by Eq. (5.50).

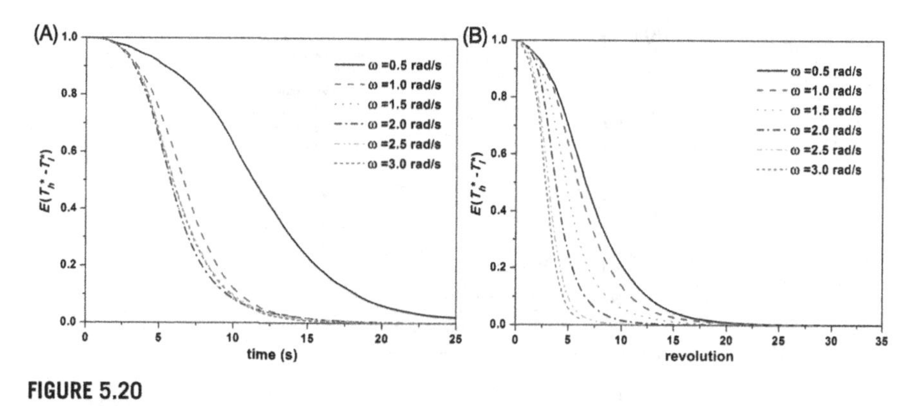

FIGURE 5.20

The variations of $E(H_h^\star - H_l^\star)$ in time (A) and in different revolutions (B).

The variations of $E(H_h^\star - H_l^\star)$ in time as well as in different revolutions are shown in Fig. 5.20A and B, respectively. Comparing Fig. 5.20A and B, it is seen that $E(H_h^\star - H_l^\star)$ is decreasing in time and revolution, and varied in different speeds under different rotation velocities. In general, the speed of variation of $E(H_h^\star - H_l^\star)$ in both time and revolutions increases with the rotation velocity. But slight difference still exists. The least rapid variation of $E(H_h^\star - H_l^\star)$ occurs always at the lowest rotation velocity of $\omega = 0.5$ (π/s) in both Fig. 5.20A and B. The rapidest variation of $E(H_h^\star - H_l^\star)$ in revolutions occurs at $\omega = 3.0$ (π/s). However, the rapidest variation of $E(H_h^\star - H_l^\star)$ in time occurs at about $\omega = 1.5$ (π/s) and $\omega = 2.0$ (π/s) (Fig. 5.20A). Recalling the difference of probability density function $p(\beta)$ of temporal intermittency factor for $\omega = 3.0$ (π/s) in Fig. 5.11, it is caused by the transition of flow regime from the cascading ($Fr = 1.16$) to the cataracting regimes ($Fr = 2.61$) (Table 5.3). In the cataracting regime ($\omega = 2.5$ and 3.0 π/s) with a larger temporal intermittency factor, it makes a reduction of the particle-particle contact time as some particles are discretized from each other during the augmented cataracting motion from the left-top side to the right-bottom side. Thus, the particle-particle heat conduction duration is reduced, and subsequently the variation of $E(H_h^\star - H_l^\star)$ in time becomes slower (Fig. 5.20A).

However, the variation of $E(H_h^\star - H_l^\star)$ in revolutions does not directly depend on time. Intuitively, to reach the same revolution, a large rotation speed needs less time than with a small one, which may cause less reduction of $E(H_h^\star - H_l^\star)$ due to less duration. In contrast, the reduction of $E(H_h^\star - H_l^\star)$ for a large rotation speed is always greater than that for a small one (Fig. 5.20B). It is caused by the more important and dominant role of mixing for the particle-particle heat conduction than the duration of the particle-particle contact. With a large rotation speed, the mixing degree becomes augmented as the intermittency factor becomes greater than with a small one. Thus, the particle-particle heat conduction becomes intensified due to the increased probability of contact between the hot and cool particles.

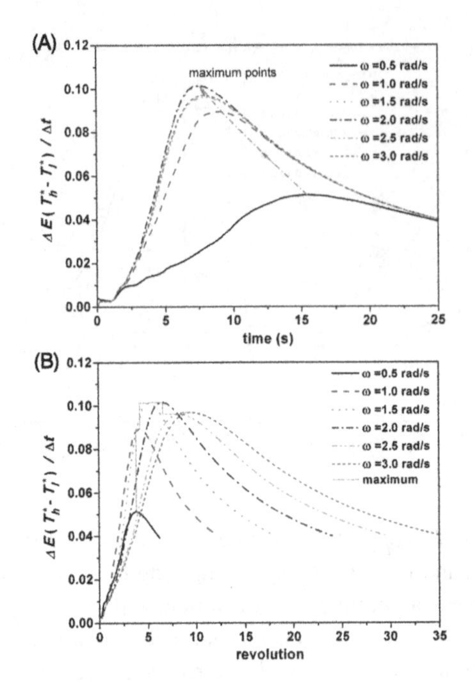

FIGURE 5.21

The time averaged increase rates of $E(H_h^\star - H_l^\star)$ in time (A) and in different revolutions (B).

In addition, the time averaged increase rates of $E(H_h^\star - H_l^\star)$ are shown in Fig. 5.21A and B, corresponding to Fig. 5.20A and B, respectively. It is shown that the time averaged increase rates of $E(H_h^\star - H_l^\star)$ have peak values, what indicates the maximum efficiency of the particle-particle heat conduction under different operation conditions. A shorter operation time always indicates the optimum operation condition for enhancing particle-particle heat conduction. Before the maximum point is arrived, a longer time or greater revolution always increases the efficiency of heat conduction. However, the opposite becomes true after the critical point.

In Fig. 5.21A, as the rotation velocity increases, the occurrence of maximum heat conduction efficiency becomes earlier ($\omega = 0.5$–$2.0\ \pi/\mathrm{s}$), whereas over-increasing of rotation velocity may cause the maximum efficiency to occur later ($\omega = 2.0$–$3.0\ \pi/\mathrm{s}$). The maximum efficiency occurs at about $\omega = 1.5$–$2.0\ \pi/\mathrm{s}$, and it is hard to say which is more efficient. However, in Fig. 5.21B, the heat conduction efficiencies for $\omega = 1.5$–$2.0\ \pi/\mathrm{s}$ are distinguished from each other. With $\omega = 1.5\ \pi/\mathrm{s}$, it needs less revolutions to reach almost the same maximum point of the heat conduction efficiency than that for $\omega = 2.0\ \pi/\mathrm{s}$.

Fig. 5.21A and B depict a full performance plot of operation. In Fig. 5.21A, the time point of maximum heat conduction efficiency for different rotation velocities is indicated. The comparison of time points of maximum heat conduction efficiency indicates the most effective operation velocity of rotation. In Fig. 5.21B, the most

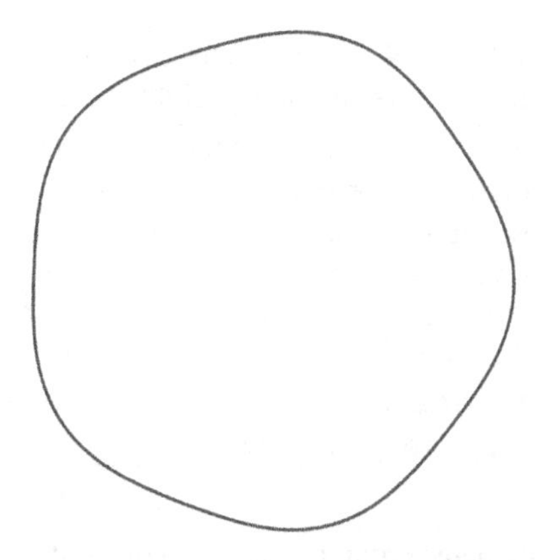

FIGURE 5.22

Sketch of drum with wavy wall ($k_\lambda = 7$).

effective way to reach the same revolution, namely to spend the same order of work, is indicated, which is more distinguishable than the former.

In the rolling/cascading flow regimes, it is deduced by Fig. 5.21B that the most effective way is: (1) To use a lower rotation velocity ($\omega = 1.0\ \pi$/s) at the early stage before the maximum point of this rotation velocity is reached. (2) Then, to use a larger rotation velocity ($\omega = 1.5\ \pi$/s) after the maximum point. (3) When the maximum points of the all rotation speeds are reached ($\omega = 1.5\ \pi$/s), it is better to change to much larger velocities ($\omega = 2.0\ \pi$/s) with the same maximum point of efficiency at a greater number of revolutions so as to slow down the decrease of the heat conduction efficiency. (4) Finally, provided the flow regime is maintained, the changing of operation velocity stops at the largest values, since the decrease rate of the heat conduction efficiency is the slowest then.

5.3 Wavy drum mixers

5.3.1 Wavy wall configuration

In general, the boundary of the tumbler can be modeled by a traveling sine wave on a fixed circle (Fig. 5.22). In the polar coordinate system, let the center of the tumbler be the origin. The governing equation of the motion of the controllable boundary is

$$\rho(t) = \rho_0 + A\sin(k_\lambda \theta + \omega_a t) \tag{5.51}$$

where ρ_0 is a constant corresponding to the radius of gyration of the tumbler, and θ ranges from 0° to 360°. A is the amplitude of the sine wave, or equivalently the amplitude of the wavy boundary. k_λ is the wave number, which is defined as the number of waves within 2π, or equivalently the number of full wave configurations on the tumbler boundary. Thus, k_λ is a dimensionless parameter. ω_a is the angular

Table 5.4 Simulation parameters used in the simulation of this section

Parameter	Value
Radius of gyration ρ_0, (mm)	1000
Number of particles N_p	4000
Restitution coefficient e	0.95
Friction coefficient μ_f	0.3
Rotating speed Ω, (rad/s)	15.7
Particle diameter d_p, (mm)	10
Time step δ_t, (s)	10^{-5}
Mesh size $d_x = d_y$, (mm)	20

frequency. Here, the parameter group (A, k_λ, ω_a) constitutes the controllable parameters of tumbler configuration, as well as the motion characteristics of the tumbler. A and k_λ are related to the amplitudes and phases of the sinusoidal boundary respectively, which determine the static characteristics of the tumbler configuration. ω_a determines the rotating speed, controlling the dynamic characteristics of rotation. In this section, the effects of k_λ, ω_a and A on particles' kinetic energy and motion pattern are studied.

The particles are rigid spheres, and the tumbler is two-dimensional. The motion of particles is restrained within a 2D plane, and the velocity in the third direction is zero. Moreover, a set of Eulerian meshes is used here to reduce the requirement of computational capacity for detecting the particle-particle collision. Only a local region around any particle is searched for collision detection. The parameters used in this simulation are listed all in Table 5.4.

To simulate the particle-wall collision, it is necessary to compute the inner normal vector, which is

$$n = \frac{1}{\sqrt{1+F^2}} \begin{pmatrix} -\cos\theta - F\sin\theta \\ -\sin\theta + F\cos\theta \end{pmatrix} \tag{5.52}$$

where $F = (Ak_\lambda \cos(k_\lambda\theta + \omega_a t)/\rho)$. The rotational linear velocity of the wall at contact point (ρ, ω_a) is

$$u(\rho, \omega_a) = \rho\Omega = (\rho_0 + A\sin(k_\lambda\theta + \omega_a t))\frac{\omega_a}{k_\lambda} \tag{5.53}$$

where

$$\Omega = \frac{\omega_a}{k_\lambda} \tag{5.54}$$

is the rotation velocity of the tumbler (the phase velocity of the sine wave traveling on the circle). With known normal vector and velocity of the wall at the contact point, the particle-wall collision is just the same as the process of the particle-particle collision if the wall of particle is regarded as a particle of $m_p \to \infty$. The motion of tumbler's boundary depends on the amplitude A and the wave number k_λ at a fixed rotation velocity.

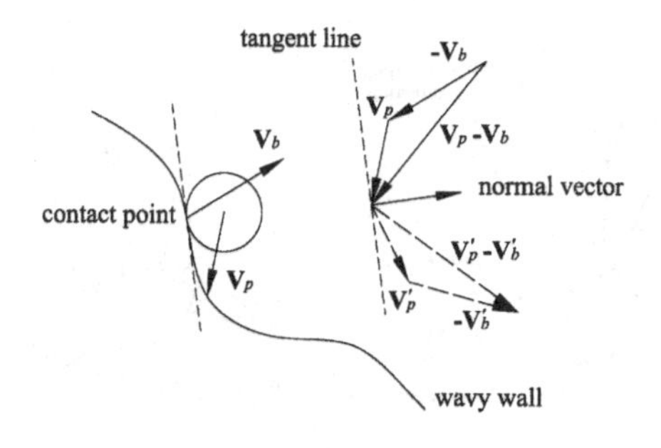

FIGURE 5.23

Sketch of the particle-wavy wall collision.

The numerical procedures for performing the particle-wavy wall collision are the following: (1) The wavy wall is discretized by a set of discretized points, just as a dense cover of the wavy wall. (2) At each time step, an overall sequential search of the discretized points is carried out to compute the normal inner vectors as well as the tangent lines at these points. (3) Then, search for possible collisions of particles with the boundary points. At each boundary point, computation of the distance of any particle from the discretized point to the tangent line is computed to determine whether or not the particle-wall collision takes place. (4) The hard sphere model is used to compute the post-collision variables. Additionally, for numerical stability of temporal advancement of particle motion, the time step interval is required to be sufficiently small so that $\Delta t \cdot \max\{u_i\} < \frac{d_p}{2}$ be met.

As a numerical demonstration, the results of energy analysis of particles and clusters are based on two basic aspects: the effect of rotating velocity Ω (or phase velocity) and the effect of wave numbers on the boundary of the tumbler. For energy analysis, it needs a method to identify the clusters, which are certain coagulations of many particles and have large concentrations of particles in comparison with the sparsely dispersed particles. Since the particle number is very low, a simple but useful method for identifying the particle cluster is used here.

The method is based on the following considerations: in general, a particle cluster is mainly characterized by the local high number density. Thus, it is suitable to assume that there is a threshold value of local number density above which the particle cluster is present. Following this assumption: (1) The maximum particle concentration in discretized mesh cells at each time point is calculated at first. (2) After searching all the cells, those cells, in which the particle number density is greater than a certain threshold value, are extracted. (3) To make sure the majority of the extracted cells are continuous, after some tests, it is found that a half of the maximum can be used as the threshold value.

5.3.2 Analysis and prediction

Before simulation, it is necessary to do some analyses on the characteristics of the particle-wall collision. The motion of particle in the wavy tumbler is stochastic. How-

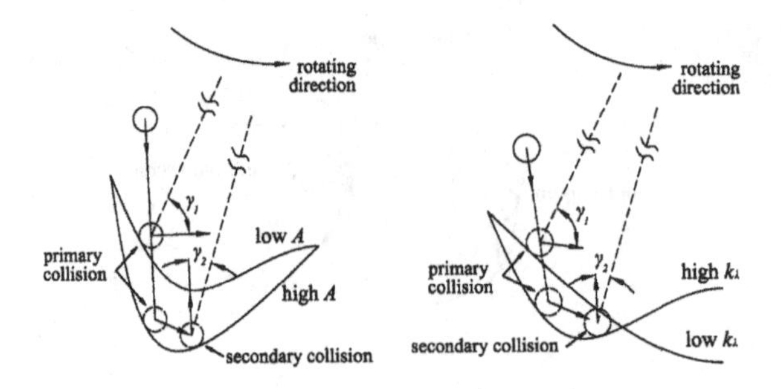

FIGURE 5.24

Sketch of the primary and secondary collision pattern with different amplitudes (A) and different wave numbers (B), respectively.

ever, the motion of the wavy wall is fairly regular referring to Eq. (5.51). Compared to the circular tumbler, the wavy tumbler is more complicated, as the point on the wavy wall is oscillating in the radial direction. The motion of the wavy wall is regarded as certain type of 'organized motion', whereas the motion of particle is regarded as 'stochastic motion'. As shown on Fig. 5.23, the particle velocity is $\boldsymbol{v}_p$, and the wall velocity is $\boldsymbol{v}_b$. Hence the incident velocity of the particle relative to the wavy wall is $(\boldsymbol{v}_p - \boldsymbol{v}_b)$. These three vectors constitute a 'vector triangle'. Moreover, the post-collisional counterparts of them are $(\boldsymbol{v}'_p)$, $(\boldsymbol{v}'_b)$ and $(\boldsymbol{v}'_b - \boldsymbol{v}'_p)$, respectively. As the particle and wall are rigid, and the restitution coefficient is very close to unity, the loss of kinetic energy due to inelastic collision and friction contact is negligible. Thus, it is appropriate to consider the particle-wall collision as a specular reflection process.

It is observed from Fig. 5.23 that the post-collisional velocities form a vector triangle, too. Comparing these two triangles, it is considered that the organized motion and the stochastic motion can be separated into two independent types of motion. $(\boldsymbol{v}_b, \boldsymbol{v}'_b)$ constitute the organized motion, whereas $(\boldsymbol{v}_p - \boldsymbol{v}_p)$ represents the stochastic motion. Both of them follow the law of specular reflecting collision. Consequently, the collision between the moving particle and the moving wavy wall is a superposition of the organized and the stochastic motion. The effect of the stochastic motion of particles on the particle motion pattern is hard to predict, whereas it is necessary to analyze the effect of the organized motion of wavy wall on the particles motion pattern.

Thus, let the incident velocity of particle be $(-\boldsymbol{v}_b)$, and Fig. 5.24A and B sketch the typical particle-wall collision processes for Case (3) and Case (2), respectively. Let's assume the tumbler is rotating in anticlockwise direction, which is positive. In Fig. 5.24A, with the amplitude of boundary varied, it is found that there exist two patterns: (1) for low boundary amplitude, the rebound velocity is in the same direction of rotation, and the angle γ_1 is positive; (2) for high boundary amplitude, the second collision with the wall occurs. The rebound velocity is in the opposite direction of rotation, and the angle γ_2 is negative. The angles γ_1 and γ_2 are defined as $\gamma = \frac{\rho \exp(i\theta) \times (-\boldsymbol{v}'_b)}{|\rho| \cdot |-\boldsymbol{v}'_b|}$, which is the angle corresponding to the moment of rebound

Table 5.5 Control parameters for cases 1–3.

	$\rho_0(d_p)$	$A(d_p)$	k_λ	ω_a (rad/s)	Ω (rad/s)
Case 1:	100	5	8	10.8, 21.6, 32.4, 43.2	1.35, 2.7, 4.05, 5.4
Case 2:	100	5	8, 10, 12, 14	43.2, 54, 64.8, 75.6	5.4
Case 3:	100	3; 5; 7; 9	5	78.5	15.7

velocity with respect to the center of the tumbler. The former pattern is called the primary collision pattern and the latter is called the secondary collision pattern. The difference in the particle-wall collision patterns may influence the motion pattern of particles. Note that the incident and rebound velocity ($-\boldsymbol{v}_b$ and $\boldsymbol{v}_b'$) are thoroughly determined by the motion and configuration characteristics of the tumbler, and they have no relation with particle motion. Thus, suppose every point of the wall belongs to the primary collision pattern, the particles can move predominantly in the same direction of rotation of the tumbler. On the contrary, the particles can move in the opposite direction of rotation of the tumbler, provided every point of the wall belongs to the secondary collision pattern. Thus, the motion pattern of particles may be determined mainly by the intrinsic configuration of the collision patterns of wall boundary. It provides the qualitative guiding information for the design of efficient mixing tumbler.

Similarly, in Fig. 5.24B, it is found that the boundary with low wave number plays the same role as the low amplitude, whereas the boundary with high wave number plays the same role as the large amplitude.

Although Fig. 5.24 is just a sketch of the particle-wall collision process, the characteristics of the collision pattern of the wavy wall may be mixed, i.e. some points are of primary collision pattern and some are of secondary collision pattern. Thus, the balance of the two types of collision patterns is very important for the motion pattern of particles within it. Based on the statistical point of view, it is reasonable to believe that for low boundary amplitude or wave number, the primary particle-wall collision pattern can dominate, whereas for high boundary amplitude or wave number, the secondary particle-wall collision pattern can dominate.

5.3.3 Effects of phase velocity, wave number and amplitude

5.3.3.1 *Case 1: Effect of phase velocity*

Let $k_\lambda = 8 = Const.$, and $A = 2$ m. The effect of rotating velocity (or equivalently the phase velocity ω_a) is analyzed here. To investigate the effect of the rotating velocity Ω of the tumbler, it is only necessary to vary the angular frequency Ω (Eq. (5.54)) in the governing equation of the wavelike boundary of tumbler. In this way, the rotating velocity of the tumbler in Eq. (5.54) is characterized by the phase velocity of sine waves, which only depends on ω_a here. As illustrated in Table 5.5 for case 1, four different operating conditions are simulated with $\omega_a = 10.8, 21.6, 32.4, 43.2$ rad/s respectively, corresponding to the rotating velocities (phase velocity) of $\Omega = 1.35$, 2.7, 4.05 and 5.4 rad/s respectively.

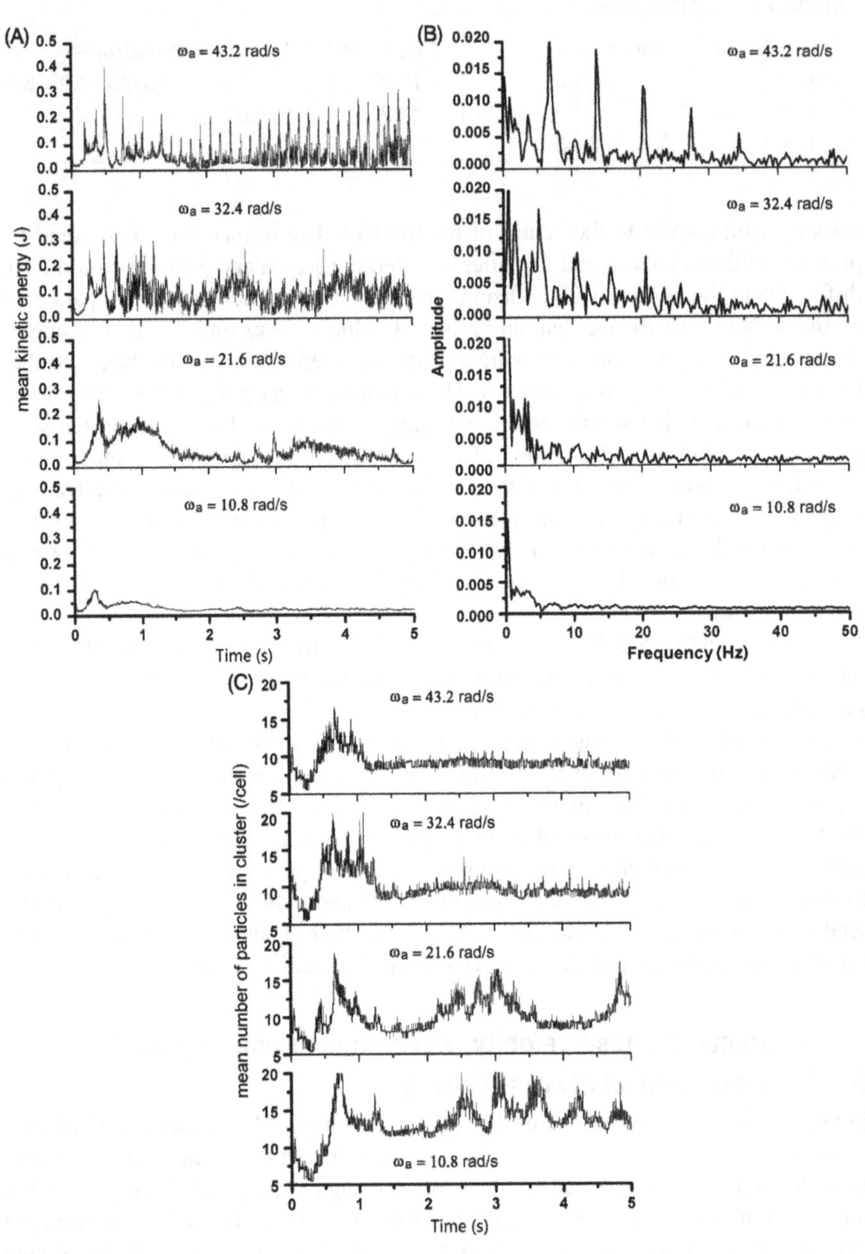

FIGURE 5.25

(A) The time history of the mean kinetic energy of the cluster for Case 1. (B) The frequency analysis of the kinetic energy. (C) Time history of the number of particles in the cluster.

Fig. 5.25A is the time history of the mean particle kinetic energy, including the translational and rotational energy, for all rotating velocities. It is observed that the kinetic energy increases and fluctuates are more intensive as Ω increases. Moreover, it seems to be a regular periodic fluctuation for the highest rotating velocity $\omega_a = 43.2$ rad/s. The periodicity is computed by the Fourier transforms (seen from Fig. 5.25B) of the data in Fig. 5.25A. For $\omega_a = 43.2$ rad/s, the peak values of amplitudes are at $f = 6.8856$, 13.6776, 20.4696, 27.4417, 34.4065 Hz respectively. The fundamental frequency is 6.88 Hz and the corresponding period is

$$T^{\star} = \frac{1}{f} = 0.1452 \text{ s} \tag{5.55}$$

Remembering that the rotating motion of the tumbler's wavelike boundary can be considered as a sinusoid or a sine wave traveling on a circle, the phase periodicity of the wave is

$$T = \frac{2\pi}{\omega_a} = 0.1454 \text{ s} \tag{5.56}$$

The accordance between T and $T^{\star}$ interprets the mechanism of periodic fluctuation of the kinetic energy of particle clusters.

To say specifically, the above interpretation is based on several assumptions. (1) A wavelike boundary, at least not a circle, is needed and then a stationary or oscillated state for clusters of particles which interact with the wavelike boundary in a periodic pattern may take place. (2) The operation mode is needed to be maintained stable, i.e. the number of the particles in the cluster is stable and the kinetic energy transferred from the rotation of wavelike surface to the cluster is stable, too. (3) The response time or time interval of energy transfer from the rotating boundary to the cluster is also constant, which means the energy transfer mode is stable. Any breakdown of each condition may lead to breakdown of the periodic motion. For example, as shown in Fig. 5.25C, compared to the results at $\omega_a = 10.8$ and 21.6 m/s, the results at $\omega_a = 32.4$ and 43.2 rad/s are more stable or temporally independent except the initial segment. It indicates that a stable particle number in clusters ($\omega_a = 32.4$ and 43.2 rad/s) is a contribution to the periodic pattern of motion of clusters, whereas the fluctuation of particle number ($\omega_a = 10.8$ and 21.6 m/s) is a contribution to the breakdown of periodic motion.

5.3.3.2 Case 2: Effect of the wave numbers k_λ ($\omega_a = Const.$)

In this section, let the phase velocity $\omega_a = 43.2$ rad/s (corresponding to the rotating velocity $\Omega = 5.4$ rad/s) be a constant, and the wave numbers (w.n. for short) $k_\lambda = 8$, 10, 12 and 14 on the wavelike boundary of the tumbler are studied respectively.

Fig. 5.26A shows the time history of kinetic energy of the cluster for all wave numbers. It is validated that the energy transfer from the rotating tumbler to the colliding particles depends strongly on the surface configuration of the tumbler. The larger the wave number, the greater the amount of kinetic energy transferred to the

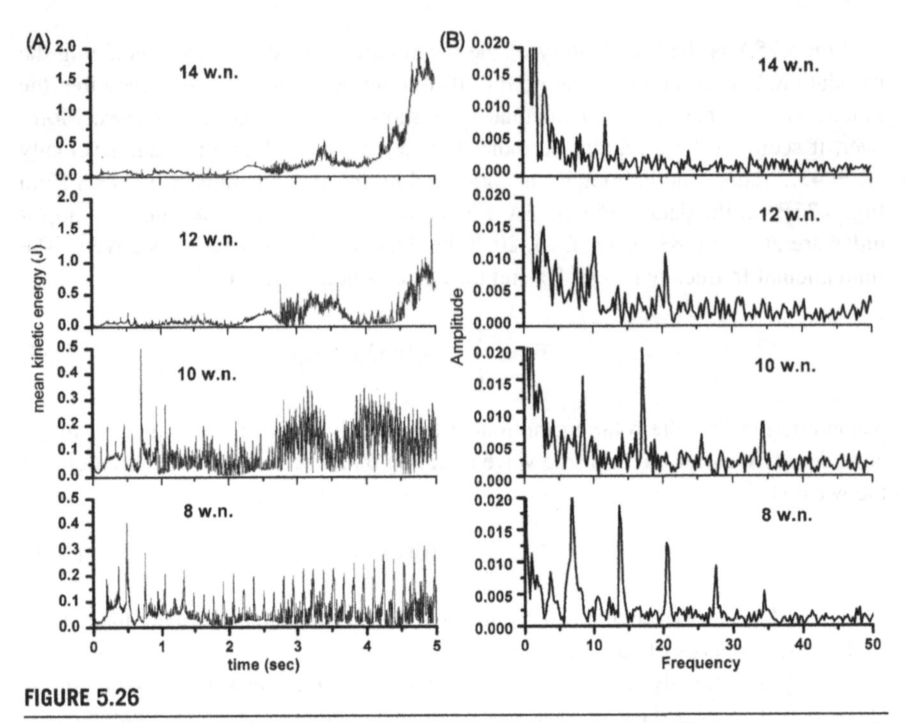

FIGURE 5.26

(A) Time history of the mean kinetic energy of the cluster for Case 2. (B) Frequency analysis of the kinetic energy in (A).

particle cluster. Fig. 5.26B shows corresponding Fourier transforms of the kinetic energy variation. The peak values of amplitudes occur at the frequencies with regular spacing, which indicates some types of periodic variations (Fig. 5.27).

Fig. 5.27A–P are snapshots of the motion of particles for all wave numbers. For $k_\lambda = 8$ w.n. (Fig. 5.27A–D), as depicted in the above section, and similarly for $k_\lambda = 10$ w.n. (Fig. 5.27E–H), the particle cluster seems to be a stationary bulk which stays in a local region or a cluster which oscillates in a local region. The energy transfer from the rotating tumbler balances the energy dissipation in inelastic collisions. The stationary bulk (Fig. 5.27A–D) or the oscillated cluster (Fig. 5.27E–H) is a necessary condition for the pattern of periodic variation of kinetic energy of the cluster. Hence, together with the periodic rotation of the wavelike surface, there exist large peak values of amplitudes at characteristic frequencies in Fig. 5.26B. Moreover, as discussed in the above section, the characteristic periodicities are determined by many factors, e.g. k_λ, ω_a, number of particles in cluster, cluster bulk oscillation, etc.

However, for $k_\lambda = 12$ and 14 w.n., the kinetic energy transferred to the cluster seems to be overwhelming compared to the energy loss in dissipation. The great amount of input of kinetic energy of the cluster breaks the balance. As illustrated in Fig. 5.27I–L ($k_\lambda = 12$) and Fig. 5.27M–P ($k_\lambda = 14$), the cluster is not a stationary bulk or a oscillated cluster in local regions. The cluster is firstly brought upward and

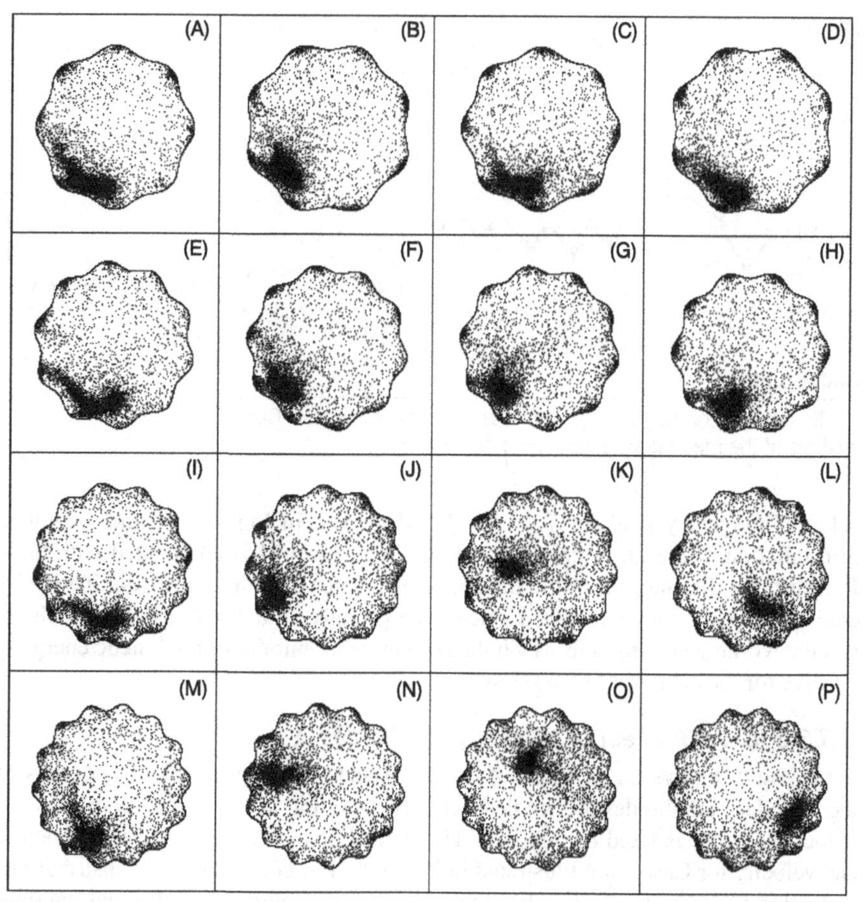

FIGURE 5.27

Snapshots of the particle motion in the tumbler at various wave numbers: (A–D) 8 w.n., when $t = 1.2, 2.0, 2.8, 3.6$ s; (E–H) 10 w.n., when $t = 1.0, 1.8, 2.5, 3.4$ s; (I–L) 12 w.n., when $t = 3.0, 3.7, 4.4, 4.8$ s; (M–P) 14 w.n., when $t = 3.0, 3.7, 4.4, 4.5$ s.

then threw away from the region near the wavelike boundary to the center of tumbler, like in a large cyclic motion. A larger scale factor k_λ corresponds to a larger cycle of these brought-upward and threw-way motions. The large cycle is not directly indicated by temporal variation of the kinetic energy, since the gravitational potential energy is changing what influences variation of the kinetic energy. Thus, in Fig. 5.26B, peak values of the characteristic frequencies disappear when k is large.

Alternatively, the large cycle of brought-upward and threw-way motions of the cluster might be indirectly revealed by the variation of gravitational potential energy. Fig. 5.28 shows the validation of existence of the large cycle motion. As shown in Fig. 5.28A and B, the variation of the mean gravitational potential energy for large k_λ is apparent. The periodicity of large cycle motion is about $T = 2$ s since the peak

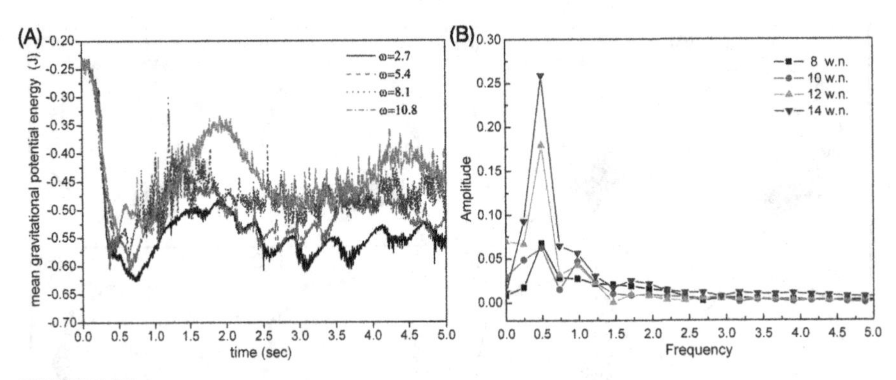

FIGURE 5.28

(A) Time history of the mean gravitational potential energy for Case 2. (B) Frequency analysis of the mean gravitational potential energy.

value of frequency is about $f = 0.5$ Hz. Moreover, in Fig. 5.28A, the amplitude of oscillation of the mean gravitational energy at $k_\lambda = 14$ is augmented with time. It means that the energy input to the cluster is greater than the energy dissipation by collision. Considering industrial application especially for the coal-grinding machine, it is the working condition in which the increment of amount of the kinetic energy is positive for the coal grinding process.

5.3.3.3 Case 3: Effect of boundary amplitudes A

Let $\omega_a = 78.5$ rad/s, $k_\lambda = 5$, and $A = 3$, 5, 7 and $9d_p$ respectively. The effects of the boundary amplitudes are studied here. The rotating speed of the tumbler is anticlockwise, and is fixed as constant. The typical snapshots of particle distribution and velocity for Case 3 are illustrated in Fig. 5.29. For comparison, the distributions and velocities of particles for the smallest ($A = 3d_p$, Fig. 5.29A–C), the intermediate ($A = 5d_p$, Fig. 5.29D–F) and the largest boundary amplitudes of the tumbler ($A = 9d_p$, Fig. 5.29G–I) are shown. The time for each snapshot is determined by the following time-dependent variation of angular momentum of particle, i.e. corresponding to the locations with either maximum or minimum angular momentums. Fig. 5.29A–C illustrates the typical snapshots of particle distribution in the tumbler within a tumbling period. Fig. 5.29D–F and 5.29G–I illustrate the snapshots of particle distribution at the same time of Fig. 5.29A–C, respectively.

In Fig. 5.29, one can observe that, at low boundary amplitudes (Fig. 5.29A–C), the particles are brought upwards by the tumbler and moving with the rotating boundary. Their bulk motions are tumbling and periodic. In contrast, with high boundary amplitudes (Fig. 5.29D–F and Fig. 5.29G–I), the particles are not brought upwards by the tumbler. In Fig. 5.29D–F, the particles just agglomerate at the bottom of the tumbler. It is not like the periodic tumbling bulk motion. Moreover, it seems that, at the largest boundary amplitudes (Fig. 5.29G–I), the particle cluster moves in the opposite direction of the tumbling motion. The particle cluster moves left and stays at the left-bottom side, while the tumbler is rotating in the anticlockwise direction.

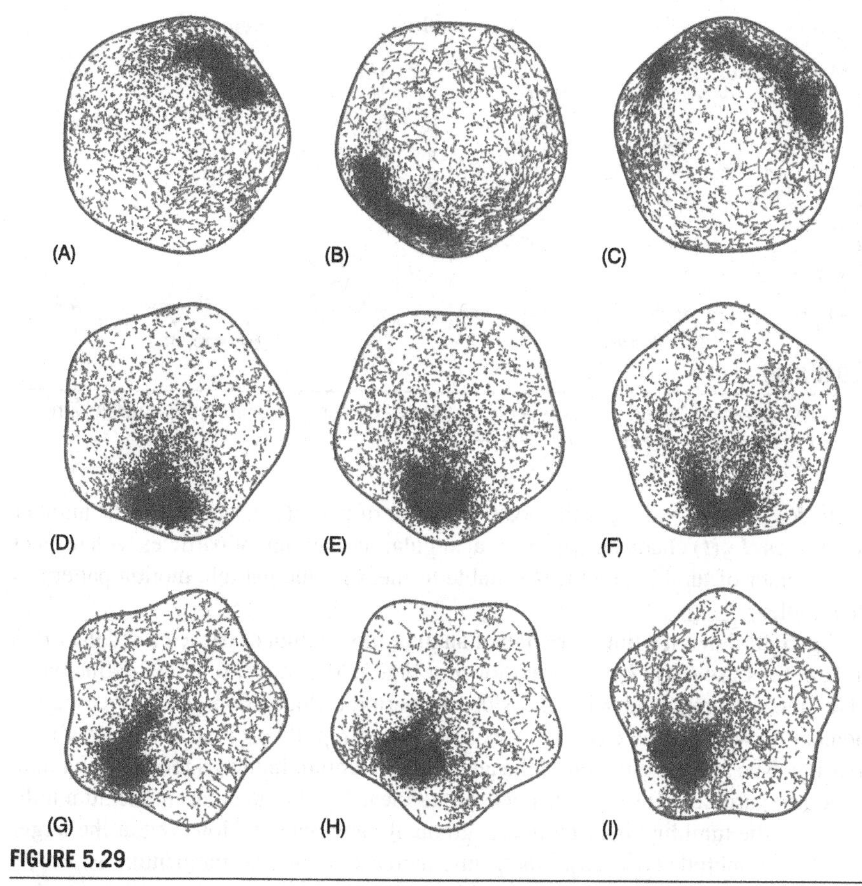

FIGURE 5.29

Snapshots of particle distribution and velocity in the rotating tumbler with the same ω_a (= 5) and different A. (A–C): $A = 3d$; $t = 3.1, 3.6, 4.1$ s. (D–F): $A = 5d$; $t = 3.1, 3.6, 4.1$ s. (G–I): $A = 9d$; $t = 3.1, 3.6, 4.1$ s.

Based on Eq. (5.53), large amplitude can lead to large magnitude of velocity of the wall, which can consequently result in large increment of particle velocity by the particle-wall collision. Then, particles can obtain large kinetic energy from the moving wall by collision. However, the observed results indicate that the large increment of kinetic energy of particles does not lead to the tumbling bulk motion. On the contrary, it leads to bulk motion even in the opposite direction. This contradiction indicates that the motion pattern of particle is not very closely related to the kinetic energy, but related to the patterns of the particle-wall collision. To analyze the particle motion quantitatively, a function $L_p(t)$ is defined here, which is the angular momentum of all particles with respect to the center of tumbler:

$$L_p(t) = |\boldsymbol{L}_p(t)| = \left| \sum_{i=1}^{N_p} \boldsymbol{R}_{p,i} \times m_p \boldsymbol{v}_i \right| \tag{5.57}$$

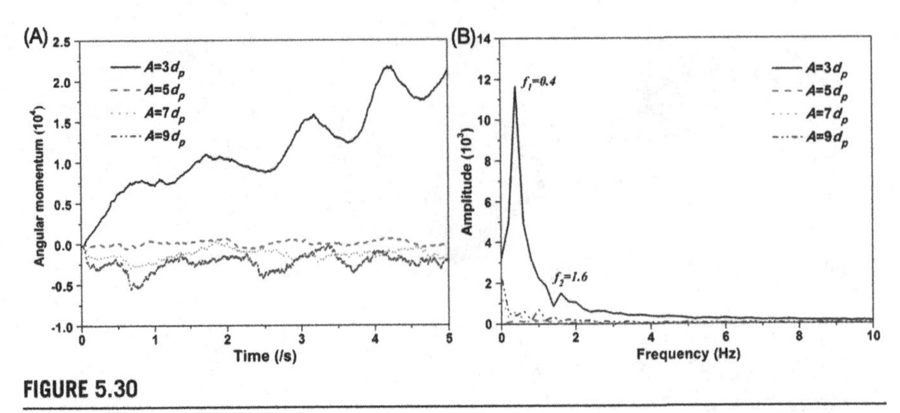

FIGURE 5.30

Temporal variation (A) and Fourier spectrum (B) of angular momentum of particles with respect to the center of tumbler for Case (3).

where $\boldsymbol{R}_{p,i} = x_i\boldsymbol{e}_x + y_i\boldsymbol{e}_y$ is the position vector of particle 'i' and N_p is the number of particles. $L_p(t)$ characterizes the total angular momentum of particles with respect to the center of tumbler, and it is suitable to measure the particle motion pattern in the tumbler.

Fig. 5.30 shows the time variation of angular momentum of particles and the FFTs of them for Case (3) at $k_\lambda = 5$. As shown in Fig. 5.30A, $L_p(t)$ of $A = 3d_p$ increases with time, accompanied with an evidently periodic fluctuation. Thus, the angular momentum with respect to the center of tumbler is rapidly oscillating and slowly increasing. The rapid oscillation of the angular momentum indicates the tumbling bulk motion of particles. The slow increase of the mean level of angular momentum indicates that the tumbling bulk motion is gradually augmented. However, at the larger boundary amplitude ($A = 5d_p$), the gentle increase of angular momentum is greatly suppressed. It fluctuates around zero, which indicates the suppression of the tumbling bulk motion of particle clusters. For even larger boundary amplitudes ($A = 7d_p$ and $9d_p$), a slightly augmented angular momentum takes place in the opposite direction due to the increase of impulse caused by the particle-wall collision. It seems to be a weakly periodic pattern of motion in the direction opposite to the rotation of tumbler. However, the pattern of tumbling bulk motion is still suppressed.

The FFT of $L_p(t)$ validates the above analysis on the bulk motion of periodic tumbling of particle clusters (Fig. 5.30B). For $A = 3d_p$, there exist peak amplitudes at two frequencies, which validate the existence of the tumbling bulk motion. For the largest amplitudes, e.g. $A = 9d_p$, the peak values of amplitudes are much smaller than those of $A = 3d_p$. It indicates the suppression of periodic tumbling motion.

As mentioned before, the primary and secondary collision patterns may coexist. To evaluate the collision patterns quantitatively, some discussions are given on the conditions under which the primary collision pattern ('p.p.' for short) or the secondary collision pattern ('s.p.' for short) may dominate. For this purpose, the angle of the incident velocity ($-\boldsymbol{v}_b$) of colliding particle at the contact point of wall can be divided into two categories, either the p.p. dominated or the s.p. dominated (sketched in

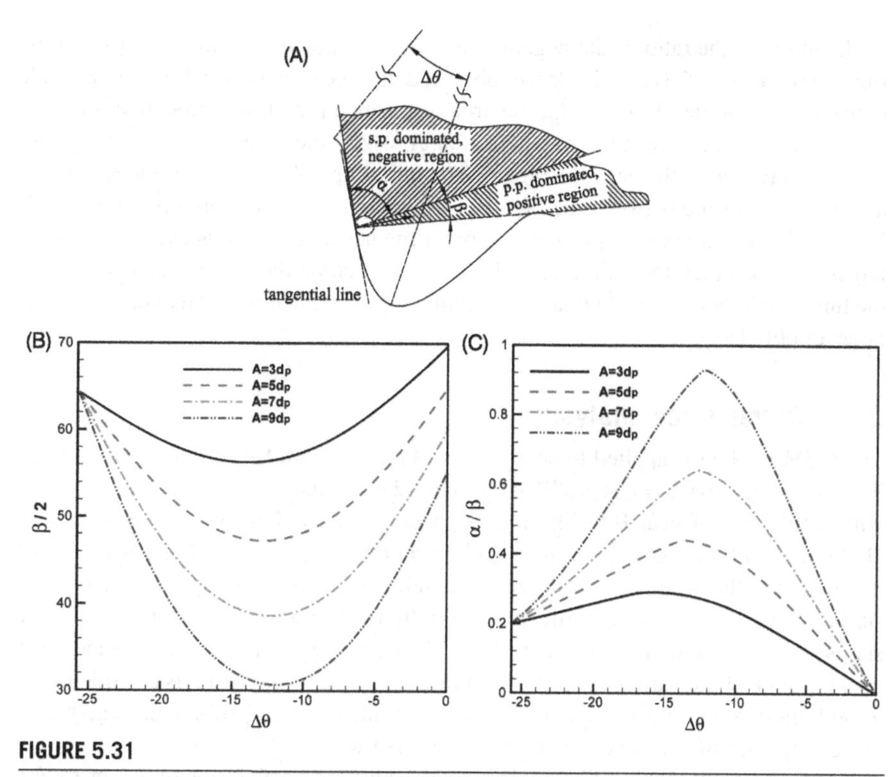

FIGURE 5.31

(A) Sketch of the negative and positive regions. (B) The angle of positive region at a half of wave number. (C) The ratio of the negative region to the positive region at a half wave number.

Fig. 5.31A). The region of p.p. domination is positive to the tumbling motion whereas the region of s.p. domination is negative or less positive to the tumbling motion as the moment of rebound velocity may be negative with respect to the rotation direction of the tumbler. For clarity, the negative and positive regions are evaluated by the angle of α and β, respectively. Then $(\alpha + \beta)$ indicates the region open to the colliding particles. In this way, the area of the positive region and the ratio of negative to positive regions are the most important factors with respect to the tumbling motion.

Due to symmetry, β and (α/β) for a half wave-number are quantified. For example, $k_\lambda = 7$, $\Delta\theta = 0.5 \times (360°/k_\lambda) = 25.7°$, the analysis for other wave-numbers is similar. In Fig. 5.31B, the angle β as well as the area of the positive region is of the minimum value around $\Delta\theta = -12°$, in the middle of the half wave number for all boundary amplitudes. However, a large amplitude corresponds to a small angle β. Larger A corresponds to smaller β. Thus, a small boundary amplitude may increase the area of positive region, and a large boundary amplitude can reduce the positive region. Therefore, the numerical results are well interpreted, i.e. the primary particle-wall collision pattern dominates at low boundary amplitudes, whereas the secondary particle-wall collision pattern dominates at large boundary amplitudes.

In addition, the ratio of the negative region to the positive region is illustrated in Fig. 5.31C. In Fig. 5.31C, it is clearly observed that the ratio of (α/β) increases with boundary amplitude. At $A = 3d_p$, the area of positive region is almost five times as large as that of negative region, whereas at $A = 9d_p$ the area of positive region is almost equivalent to the negative region. The tumbling bulk motion disappears when an equilibrium state between the positive region and negative region is established. When either the area of the positive region or the negative region is overwhelmingly larger than the other, the tumbling bulk motion will certainly take place. This is why the tumbling motion is evident at low amplitudes, whereas it is greatly suppressed at large amplitudes.

5.3.4 Driven force analysis

The DEM method is applied to solve a monolayer of granular particles here. At the beginning, the wavy drum is prefilled by steel spherical particles settled at the bottom, with a fill level of near 0.5. The initial packing state is built by two procedures: (1) Let the particles be located in a regular array with certain small distances above the bottom of the drum. (2) Let the particles fall under gravity freely onto the bottom surface when the drum stay still. After a sufficiently long time, the particles will stay still due to continuous dissipations of kinetic energies and finally a randomly packing state of the particles will be built inside the drum. The drum is considered as a rigid, motion-controlled large particle with a finite radius, of sufficient large mass, presenting a frictional and adiabatic wavy internal wall.

When the drum rotates, it is accelerated at a constant acceleration of $\dot{\omega} = \pi$ rad/s^2 until it reaches the target speed. Three target rotation speeds are simulated, i.e. a very slow rotating speed ($\omega_d = 0.1$ rad/s), an intermediate speed ($\omega_d = 0.5$ rad/s) and a rapid speed ($\omega_d = 1.0$ rad/s). The time step is $dt = 5 \times 10^{-5}$ sec, which is fine enough to maintain numerically stable simulations. In addition, the parameters are listed in Table 5.6.

Moreover, in order to evaluate the particle-particle conduction characteristics inside the tumblers of wavy walls, the particles are divided into two parts. One part is given a high initial temperature $T_\alpha = 100$°C and the other part is given a low initial temperature $T_\beta = 60$°C. As the tumbler rotates, the heat conduction takes place between the particles in contact with others of different temperatures, which could be influenced by the states of particle mixing. The amplitude is fixed, and $k_\lambda = 0, 3, 5, 7, 9, 11$, where $k_\lambda = 0$ denotes the circular drum.

5.3.4.1 Effective driven force (EDF)

In general, the particle motion is driven by the forces from the drum walls. Thus, the mechanisms for mixing improvement should be uncovered by analyzing the driven forces ('DF') from the walls.

As sketched by Fig. 5.22, the driven forces from the wall can be categorized into two classes, i.e. the positive driven force (pDF) and the negative driven force (nDF), which are in the co- and anti-directions of rotation, respectively. Notice that only the

Table 5.6 Simulation parameters.

Radius of gyration, R_0 (m)	1.44
Amplitude of wavy tumbler, A (m)	0.06
Number of particles, (N_α, N_β)	(10115, 10189)
Initial temperate of particles, (T_α, T_β) (°C)	(100, 60)
Particle diameter, d_p (mm)	12
Particle density, ρ_p (kg/m^3)	7800
Particle Specific heat, C_p (J/(kg·K))	460
Particle Thermal conductivity, λ_p (W/(m·K))	46.52
Restitution coefficient, e	0.95
Friction coefficient, γ	0.3
Stiffness factor, k (N/m)	10000
Amplitude of the wavy boundary, A (mm)	60
Wave numbers on drum boundary, k_λ	0, 3, 5, 7, 9, 11
Rotating velocity, ω_d (rad/s)	0.1, 0.5, 1.0
Simulation time step, Δt (s)	5.0 × 10^{-5}
Total simulated time, T_S (s)	150

component in (or projection on) the circumferential direction of rotation is taken into account.

$$pDF = \frac{1}{N_{p,w}} \sum_{j=1}^{N_{p,w}} (\boldsymbol{F}_{j,w} \cdot \boldsymbol{t}^0)\big|_{(\boldsymbol{F}_{j,w} \cdot \boldsymbol{t}^0)>0} \tag{5.58}$$

$$nDF = \frac{1}{N_{p,w}} \sum_{j=1}^{N_{p,w}} (\boldsymbol{F}_{j,w} \cdot \boldsymbol{t}^0)\big|_{(\boldsymbol{F}_{j,w} \cdot \boldsymbol{t}^0)<0} \tag{5.59}$$

where $\boldsymbol{t}^0 = \frac{\boldsymbol{\omega} \times \boldsymbol{\rho}}{|\boldsymbol{\omega} \times \boldsymbol{\rho}|}$ is the unit vector in the circumferential direction, and $\boldsymbol{F}_{j,w}$ is the vector of instantaneous driven force of particle 'j' from the walls. $N_{p,w}$ is the total number of particles contacting the walls. The positive and negative driven forces can be easily computed when the inner product of $(\boldsymbol{F}_{j,w} \cdot \boldsymbol{t}^0)$ is positive (> 0) or negative (< 0).

The positive and negative driven forces for $\omega = 0.1$, 0.5 and 1.0 π/s are shown in Fig. 5.32A–C. For the circular rotating drum, the driven forces cannot be divided into the positive and negative parts. Therefore, the negative parts are always zero. It is clearly seen that the driven forces are always fluctuated with time. It shows an important dynamical feature of the wavy drum, i.e. oscillation. In other words, the most evident feature of particle motion in wavy drums, differing from the mere bulk rotating feature in circular drums, is the bulk oscillating motion. Generally speaking, the mixing improvement by using wavy drums must be closely related to the bulk oscillating feature.

Moreover, it is clearly seen that the absolute magnitudes of either positive or negative driven forces are increased as either wave number k_λ or rotating speed ω

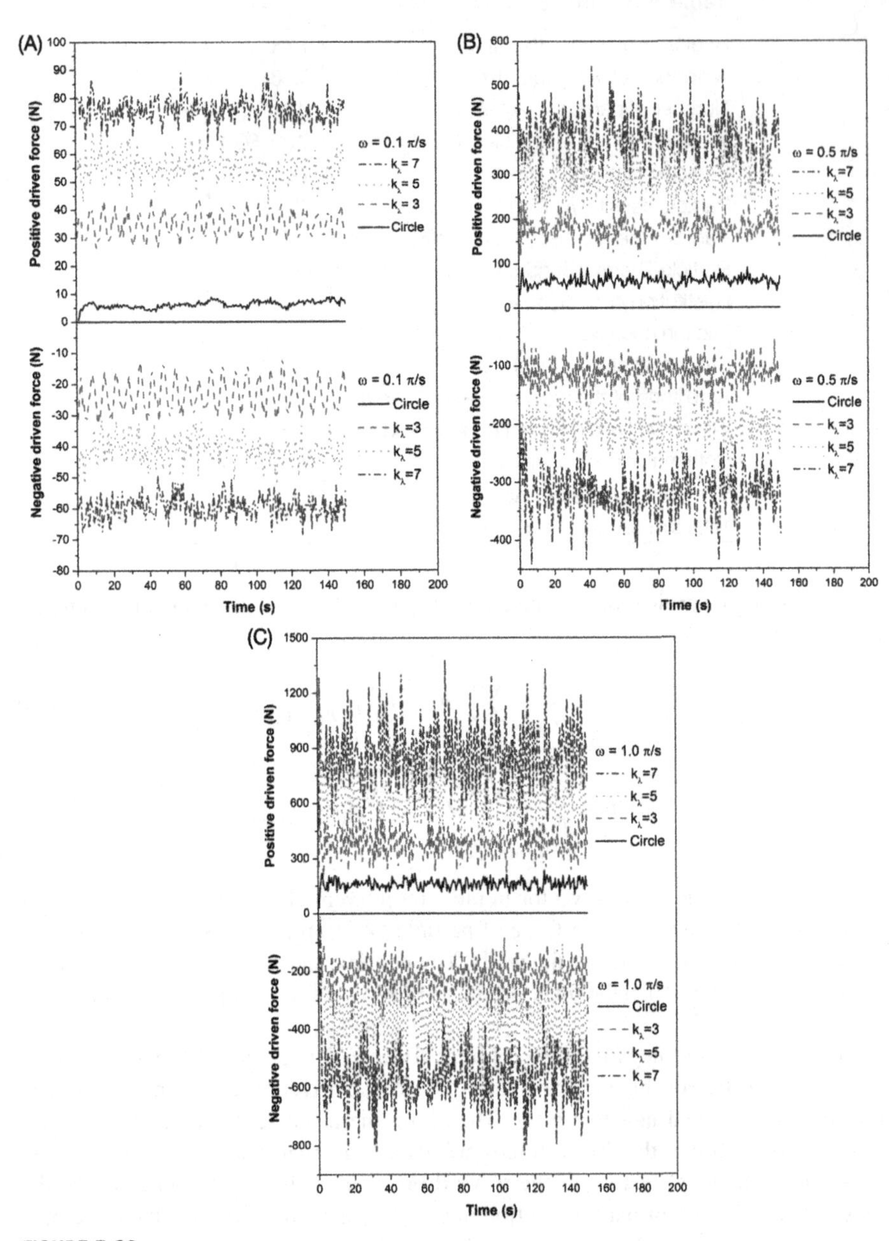

FIGURE 5.32

Positive and negative driven forces for (A) $\omega = 0.1\ \pi/\mathrm{s}$; (B) $\omega = 0.5\ \pi/\mathrm{s}$; (C) $\omega = 1.0\ \pi/\mathrm{s}$, respectively.

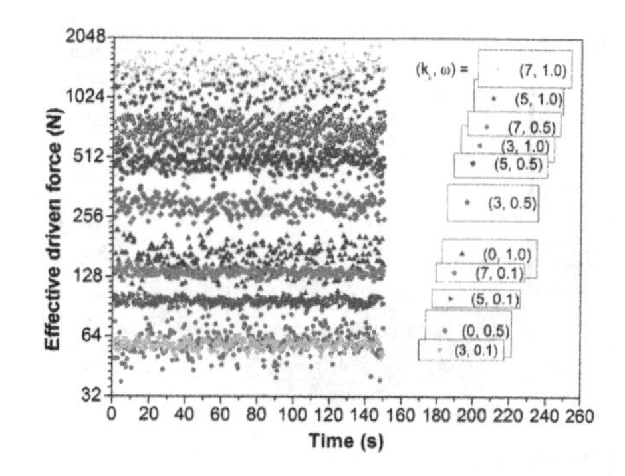

FIGURE 5.33

The distribution of effective driven forces for all wave numbers and rotating speeds, where (k_λ, ω) is the control parameter group.

increases. But, how do the magnitudes of driven forces depend on the wave number k_λ and rotating speed ω together? It seems a bit complicated.

As indicated above, the particle motion in wavy drum is under the influence of both bulk rotating and bulk oscillating effects, which can be characterized by the control parameters of k_λ and ω, respectively. Thus, it is important to show the dependence of driven force on the combined control parameter (k_λ, ω). For this purpose, the concept of effective driven force (EDF) is employed here, which is defined as the difference of the positive and negative driven forces, i.e.

$$\text{EDF} = p\text{DF} - n\text{DF}. \tag{5.60}$$

Fig. 5.33 shows the EDFs for different wave numbers and rotating speeds together. The right side of Fig. 5.33 shows the legends of control parameters (k_λ, ω) corresponding to the EDFs respectively (the case with $(k_\lambda, \omega) = (0, 0.1)$ is not shown). In general, some important characteristics can be found: (1) The EDFs are varied with time, but within clearly restricted banded regions. (2) The bands are overlapping (as indicated by the widths of legends on the right side), but raised in turn as either wave number or rotating speed increases. (3) Similar to the concept of energy levels in quantum mechanisms, the bands of EDFs can, to some extent, be regarded as the energy levels of particles, where the control parameter (k_λ, ω) can be viewed as quantum numbers correspondingly to quantify the levels of particle kinetic energies.

To depict a full diagram of dependence of EDF on wave number and rotating speed, Fig. 5.34A shows a three-dimensional diagram of the mean effective driven force (MEDF) depending on the wave number and rotating speed together, where MEDF is the time averaged EDF. To build Fig. 5.34A, more cases are simulated to get the date for interpolation.

As indicated by Fig. 5.34A, the MEDF is a continuous function of both wave number and rotating speed. But it grows relatively slowly by either increasing the wave number merely or increasing the rotating speed merely (e.g. through route 1 or 2), and cannot reach the maximum point. In contrast, it increases relatively fairly

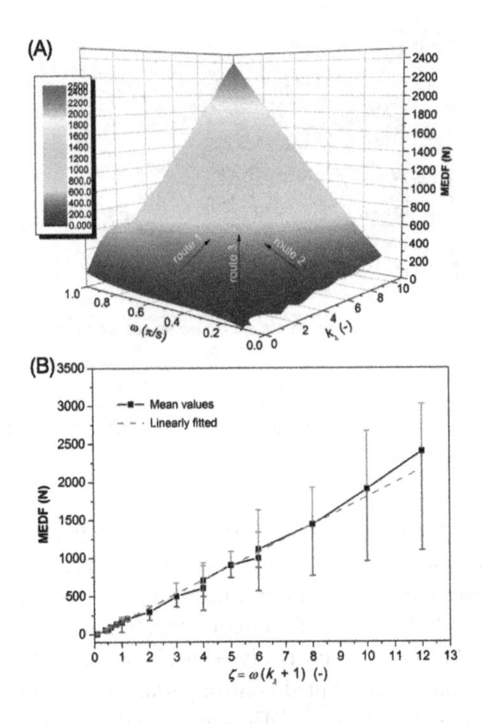

FIGURE 5.34

(A) Dependence of the mean effective driven forces (MEDF) on wave numbers k_λ and rotating speeds ω. (The surface is obtained by data interpolation.) (B) Relation between the mean effective driven force and the combined parameter $\zeta = \omega(k_\lambda + 1)$.

rapid when both of them increases (e.g. through route 3), and can easily achieve the maximum point.

Based on Fig. 5.34A, it is seen that the wave number and rotating speed have the similar effects on increasing the MEDF. It indicates the possibility to combine the wave number and rotating speed together to form a new control parameter. The new control parameter should be used to indicate the level of MEDFs independently, i.e. the function MEDF = f(new control parameter) should be a unary function. In this way, the best form of '*f*' is the linear function. For that purpose, the combination as

$$\zeta = \omega(k_\lambda + 1), \tag{5.61}$$

$$\text{MEDF} = f(\zeta) = a\zeta + b \tag{5.62}$$

is used, where a and b are constant coefficients.

Fig. 5.34B shows the relationship between MEDF and ζ. The upper limit and lower limit of the effective driven force are shown too. It clearly indicates that the relationship between MEDF and ζ is nearly linear. Moreover, as ζ increases, both the upper and lower limits of EDFs are increased almost linearly, which shows the increased widths of the banded region of EDF. Thus the aforementioned energy level function is a continuous linear function depending on the variation of ζ. It can be used for quantifying the levels of particle kinetic energy and estimating the mixing levels of particles.

It is necessary to find some foundation stones to explain ζ. Referring to Eq. (5.51), one has $\zeta = \omega(k_\lambda + 1) = \omega k_\lambda + \omega = \omega_a + \omega$. Notice that k_λ is a dimensionless parameter, whereas ω_a and ω have the same dimension of [$\sec^{-1}$]. ω_a is the angular frequency of the sine wave, and ω is the rotating velocity of the drum. In other words, ω_a characterizes the traveling sine wave on the wall, or the oscillating motion of the wavy walls, whereas ω characterizes the rotating motion of the base circle. Moreover, the effect of traveling sine wave or oscillating wavy wall is restricted within the near wall region, whereas the rotating velocity has an overall effect throughout the drum. In this way, $\zeta = \omega_a + \omega$ can be explained as a superposition of the overall rotating effect and the locally near wall oscillating effect. Thus, the mean effective driven force of particles should be increased almost linearly proportional to the increase of the superposition of the rotating and oscillating effects.

In addition, $\zeta = \omega(k_\lambda + 1)$ can also be explained in another way, i.e. to be viewed as a multiple of ω. For this explanation, by using wave number k_λ (a dimensionless quantity), the rotating speed can be regarded as been increased by $(k_\lambda + 1)$ times. But, in fact, the rotating speed of the whole drum has not been augmented. The augmentation effect is only restricted near the wavy boundary regions. As a result, using a wavy drum can be regarded as using a locally augmented rotating drum, without demanding much power to accelerate the whole drum. Thus, this is an economic way for mixing enhancement.

5.3.5 Heat conduction features

In this section, the thermal conduction of particles in wavy drums are studied by using the soft sphere approach of the DEM model coupled with the particle-particle condition. The coupling process between discrete element model and thermal conduction model is accomplished by solving the particle-particle/particle-wall collision and thermal conduction processes simultaneously. The other simulation conditions are the same as those in the former section on "Driven force analysis".

5.3.5.1 With different wave numbers

To compare the evolution characteristics of thermal conduction of particles inside the drums at different wave numbers, Fig. 5.35 shows the typical snapshots of dimensionless particle temperature under $\omega = 0.1\ \pi$/s for wave numbers $k_\lambda = 0, 3, 5, 7$. At start, the hot and cool particles inside the wavy drums are all separated by a vertical border line (e.g. in Fig. 5.35A at $t = 0.25$ s). After 75 s, the particle temperature field inside the circular drum ($k_\lambda = 0$, Fig. 5.35B) is not changed much. As the drum rotates in the anticlockwise direction, the hot particles immediately near the left-bottom walls are driven by the frictional force of the wall to move from the left side to the right side along the bottom wall of the drum. Thus, the green color in the right side of Fig. 5.35B is caused by the mixing between the hot and cool particles which causes thermal conduction from the driven hot particles to the cool particles in the right

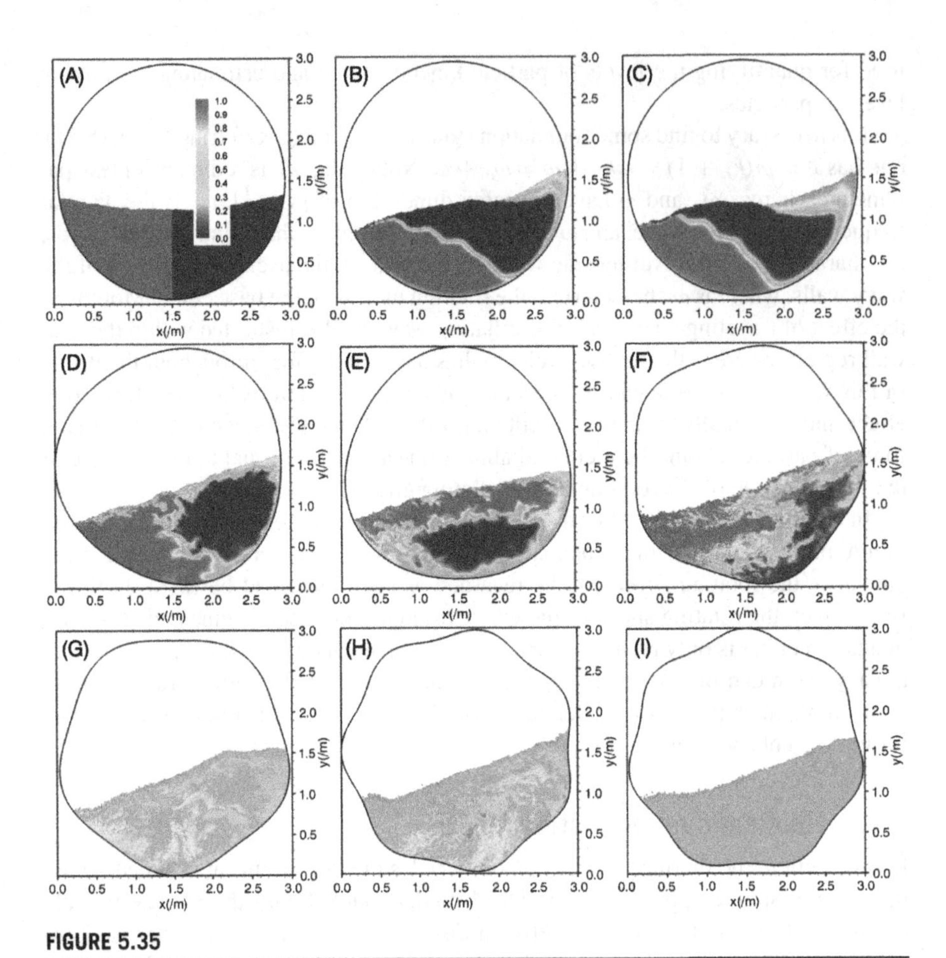

FIGURE 5.35

Snapshots of particle distributions colored by temperature under: $\omega = 0.1\ \pi$/s for $k_\lambda = 0$ (A, B, C); 3 (D, E); 5 (F, G); 7 (H, I) respectively – at $t = 0.25$ s (A), 75 s (B, D, F, H) and 150 s (C, E, G, I), respectively.

side. Otherwise, the temperatures of the rest particles almost keep their initial values. It means that the mixing and thermal conduction under the slow rotating velocity inside the circular drum is fairly weak and low. Furthermore, after about $t = 150$ s, similar results can be observed in Fig. 5.35C, although the particle temperature in the right-up region becomes larger than previous since more hot particles are brought up into this region by the frictional driven force of the wall.

However, for the wavy drum, the particle motions are not only influenced by the frictional driven force but also dominated by the normal force from the wavy wall, since the rotating motion of the wavy wall can impose periodically varied normal force upon the particles. Thus, at $t = 75$ s, the interface between the hot and cool particles becomes jagged (e.g. Fig. 5.35D for $k_\lambda = 3$), which means more green regions with intermediate temperatures are formed, which is caused by the jagged and lengthened interface of heat conduction. Moreover, this trend seems to be much clearer for even larger wave numbers (e.g. Fig. 5.35F for $k_\lambda = 5$ and Fig. 5.35H for $k_\lambda = 7$). It means that the augmented wavy configuration (or increased wave number) may cause more jagged, stratified and lengthened thermal conduction interface, which obviously enhances or speeds up the process of internal thermal conduction of particles. Similar results can be found at $t = 150$ s for $k_\lambda = 3$ (Fig. 5.35E), 5 (Fig. 5.35G) and 7 (Fig. 5.35I), respectively. At about $t = 150$ s with $k_\lambda = 7$ (Fig. 5.35I), the distribution of particle temperature inside the wavy drum is almost uniform, which implies the fairly small temperature differences between the particles, or the near fully-developed states of thermal conduction. In addition, the trends are similar for even larger wave number ($k_\lambda = 9$ and 11), and hence they are omitted in Fig. 5.35.

Thus, it is indicated by Fig. 5.35 that the wavy drum can enhance or speed up the thermal conduction process between particles. The degree of enhancement of thermal conduction depends on the number of waves on the wall. A large wave number implies a more enhanced or a rapider evolution process of thermal conduction. More importantly, the wavy drum can be used for thermal conduction under fairly low speed of rotation, even under the rotating conditions when the circular drum cannot work effectively for mixing and heat transfer.

5.3.5.2 With different rotating speeds

On the other hand, taking the same wave number of $k_\lambda = 9$ for case study, Fig. 5.36 shows the comparison of particle temperature under different rotating speeds. Fig. 5.36A and B are for $\omega = 0.1\ \pi/\text{s}$ at $t = 25$ s and 50 s respectively; Fig. 5.36C and D are for $\omega = 0.5\ \pi/\text{s}$ and Fig. 5.36E and F are for $\omega = 1.0\ \pi/\text{s}$, at the same time points correspondingly.

In general, the thermal conduction is developing in time. Thus the temperature fields in Fig. 5.36A, C, E at $t = 25$ s are less homogeneous or less developed than those in Fig. 5.36B, D, F at $t = 50$ s, respectively. Moreover, at the same time (e.g. Fig. 5.35A, C, E at $t = 25$ s), using a larger rotating speed may result in a more enhanced process of thermal conduction. Thus, the particle temperature fields under low rotating speeds are less developed (or more inhomogeneous) than those under high rotating speeds. It indicates that the rotating speed is another important factor which may speed up or enhance thermal conduction between the particles inside the wavy drum.

In addition, in Fig. 5.36E and F for $\omega = 1.0\ \pi/\text{s}$, the rotating speed is so large that the particles are easily brought up even to the drum top, and the distribution pattern of up-casting is influenced by the wave numbers of the wall. More specifically, the particles near the side of the wavy wall toward the opposite direction of rotation can be propelled by the normal forces from the wavy walls. It easily brings the particles up and throws them at the drum top, causing the 'up-casting' motion pattern. In other words, the wavy drum with very large wave numbers and large rotating speeds can change the motion pattern of particles inside the drum, which subsequently change the thermal conduction process between particles.

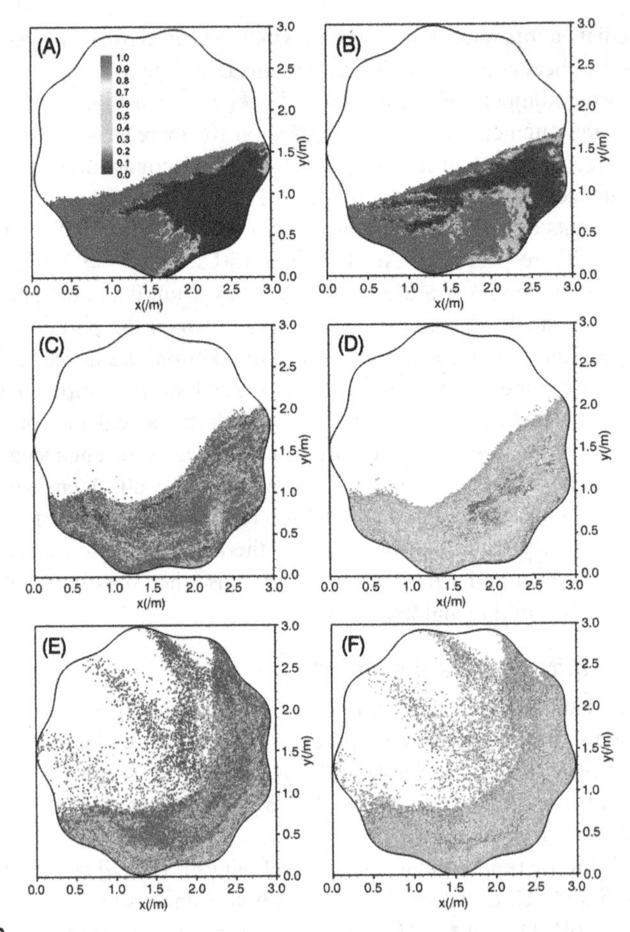

FIGURE 5.36

Snapshots of particle distributions colored by temperature with $k_\lambda = 9$ for rotation velocities $\omega = 0.1\ \pi$/s (A, B); $0.5\ \pi$/s (C, D); $1.0\ \pi$/s (E, F) – at $t = 25$ s (A, C, E) and 50 s (B, D, F), respectively.

5.3.5.3 Probability density function of temperature (T-PDF)

The probability density function of particle temperatures (T-PDF) is used here to evaluate the thermal conduction characteristics inside the wavy drum quantitatively. In general, T-PDF is defined as the number rate of particles with temperature $T^\star \in [T_i^\star, T_i^\star + \delta T^\star]$, $(i = 0, \cdots, N_i,\ \delta T_i^\star = \frac{1}{N_i} = 0.005)$, which is formulated by Eqs. (5.31) and (5.32).

For example, Fig. 5.37 shows the comparison of T-PDFs for $k_\lambda = 0, 3, 5, 7$ at about every 25 s. At start ($t = 0.25$ s), the particles are fully separated by hot ($T_i^\star = 1$) and cool ($T_i^\star = 0$) ones. One can see from Fig. 5.37 that the numbers of particles with intermediate temperatures are increased with time, and meanwhile

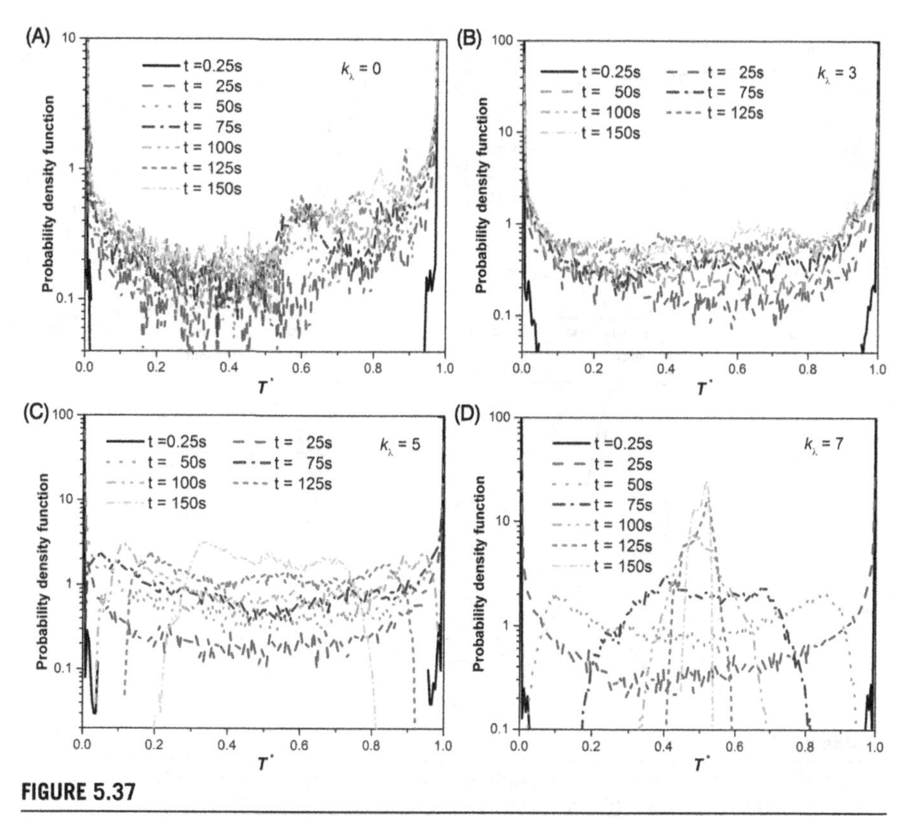

FIGURE 5.37

Probability density functions of dimensionless temperature at different time points for wave numbers $k_\lambda = 0$ (A), 3 (B), 5 (C), 7 (D), respectively, under $\omega = 0.1\ \pi$/s.

the numbers of particles with the smallest and largest temperatures are decreased. Thus, the values of T-PDFs with intermediate temperatures are increased with time, and those with $T_i^\star \approx 0$ and $T_i^\star \approx 1$ are decreased. The values of T-PDFs with intermediate temperatures for $k_\lambda = 3$ are larger (Fig. 5.37B) than those for $k_\lambda = 0$ (Fig. 5.37A) correspondingly. Moreover, the T-PDFs near $T_i^\star \approx 0$ and $T_i^\star \approx 1$ in Fig. 5.37A and B for $k_\lambda = 0$ and 3 are not decreased to zero, whereas they are decreased to near zero after about 100 sec for $k_\lambda = 5$ (Fig. 5.37C) and about 50 sec for $k_\lambda = 7$ (Fig. 5.37D). These results validate the aforementioned conclusions on the relative efficiencies on the enhancement and speedup of thermal conduction by the wavy drum, i.e. E_f $(k_\lambda = 7) > E_f$ $(k_\lambda = 5)$ $> E_f$ $(k_\lambda = 3)$ $> E_f$ $(k_\lambda = 0)$. In addition, for large wave numbers ($k_\lambda = 5$ and 7 in Fig. 5.37C and D respectively), the widths of T-PDFs are greatly reduced with time, especially for $k_\lambda = 7$ where the T-PDFs are almost within the range of $0.45 < T_i^\star < 0.55$ after $t = 150$ s. It means the particle temperature difference is within 0.5 ± 0.05, which corresponds to the fairly homogeneous distributions of particle temperature throughout the drum as indicated in the former section.

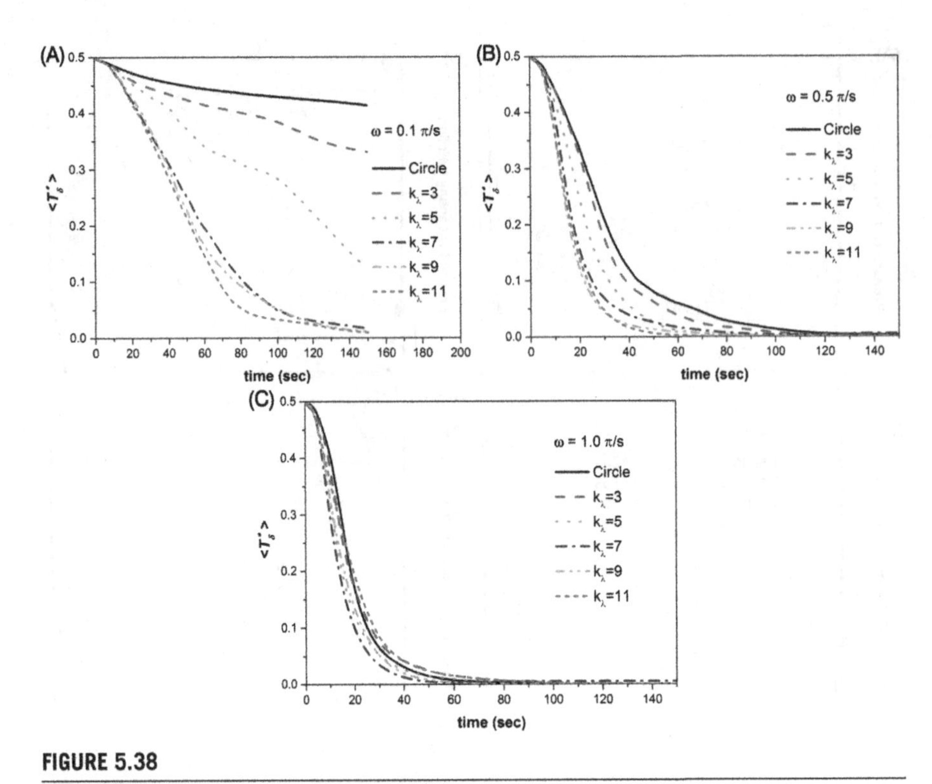

FIGURE 5.38

Temporal evolutions of the mean temperature discrepancy under wave numbers $\omega = 0.1$ (A), 0.5 (B), 1.0 π/s (C), respectively.

5.3.5.4 Mean temperature discrepancy (MTD)

Based on T-PDF, it is possible to define the mean temperature discrepancy (MTD, or $\langle T_\delta^\star \rangle$) as in Eq. (5.33). The MTD evaluates the discrepancy of mean temperature of particles between an instantaneous state and the final fully developed state. For example, Fig. 5.38A–C show the temporal evolutions of MTD distribution for $\omega = 0.1, 0.5$ and $1.0\,\pi$/s, respectively. In Fig. 5.38A, $\langle T_\delta^\star \rangle$ is reduced with time more rapidly under the higher wave numbers at the low rotating velocity ($\omega = 0.1\,\pi$/s). It means that the degree of thermal conduction is greatly enhanced by increasing the wave number. For $\omega = 0.5\,\pi$/s (Fig. 5.38B), the results are similar, i.e. thermal conduction is improved by increasing wave numbers. Moreover, compared to Fig. 5.38A for the low rotating speed at any considered time, the levels of thermal conduction are improved under the intermediate rotating speed since the absolute values of MTD are more reduced than those under the low rotating speed.

With an even larger rotating speed ($\omega = 1.0\,\pi$/s, Fig. 5.38C), the differences between the MTDs for different wave numbers are not as evident as before. It means that the processes of thermal conductions for different wave numbers are speeded up almost to the equivalent levels at any time. Detailed comparison shows that the

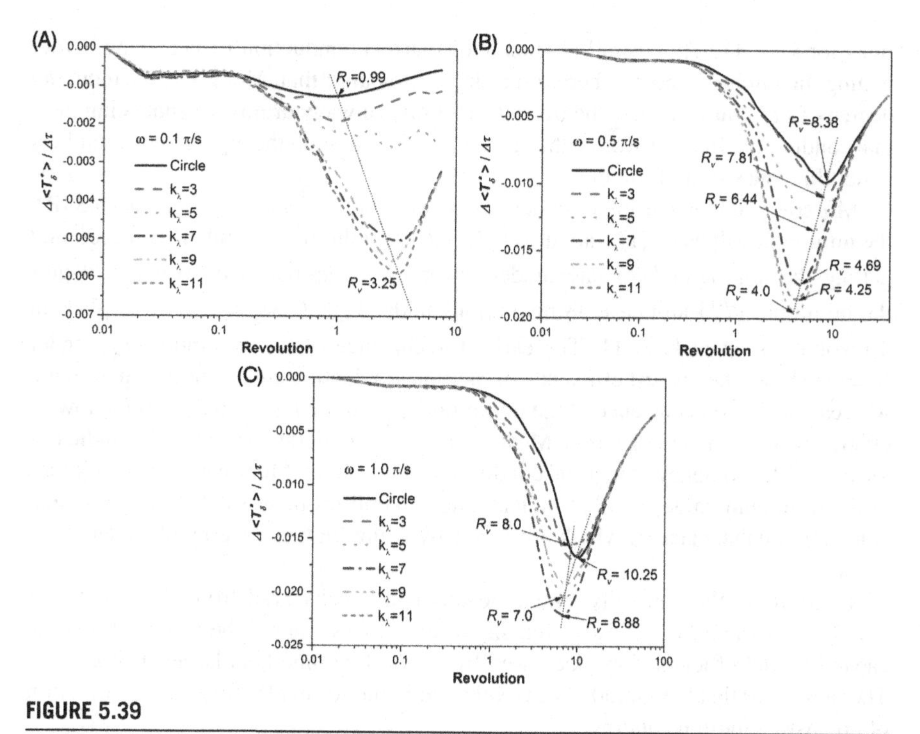

FIGURE 5.39

Time-averaged variation rates of the mean temperature discrepancy under wave numbers $\omega = 0.1$ (A), 0.5 (B), 1.0 π/s (C), respectively.

fastest decrease of $\langle T_\delta^\star \rangle$ occurs at $k_\lambda = 7$. The MTDs are comparatively slower for to decreased at the same time points in the later stage of evolution (e.g. $t > 20$ s) by using even larger wave numbers ($k_\lambda = 9$ and 11) than $k_\lambda = 7$, which means the mean temperature discrepancies are a bit larger than at $k_\lambda = 7$. Referring to the up-casting motion pattern of particles when $k_\lambda = 9$ or 11 under $\omega = 1.0\ \pi$/s, the slower decrease of MTD for $k_\lambda = 9$ and 11 is caused by the up-casting motion. The up-casting motion may weaken the thermal conduction of particles because it makes the up-cast particles to be separated from other particles, and reduces somewhat contact time between the particles. As a result, the thermal conduction process is slightly slowed down and the mean temperature difference is reduced more slowly than in other cases. To say conclusively, using very large rotating velocities under very high wave numbers is detrimental to the enhancement of heat conduction between particles.

The mean reducing rate of MTD can be characterized by $\langle \dot{T}_\delta^\star \rangle$ as formulated in Eq. (5.34). To compare $\langle \dot{T}_\delta^\star \rangle$ under different rotating speeds and wave numbers, Fig. 5.39A, B and C show the $\langle \dot{T}_\delta^\star \rangle$ for $\omega = 0.1$, 0.5 and 1.0 π/s, respectively. As shown by Fig. 5.39A, the $\langle \dot{T}_\delta^\star \rangle$ is reduced at first and then increased slightly. The decreasing rate of $\langle \dot{T}_\delta^\star \rangle$ is enlarged by the increasing wave number k_λ. For $k_\lambda = 0$, the minimum $\langle \dot{T}_\delta^\star \rangle$ occurs around 0.99 revolution, whereas it occurs around 3.25 revolutions for $k_\lambda = 11$. The overall efficiency of thermal conduction is increasing before getting the minimum point whereas it is decreasing after that. Moreover, the reducing degree of $\langle \dot{T}_\delta^\star \rangle$ may indicate the overall capability of wavy drums for enhancing thermal conduction. In other words, the capability is larger with the higher wavy numbers under this low speed of rotation.

Moreover, under the intermediate speed of rotation ($\omega = 0.5\ \pi$/s, Fig. 5.39B), the minimum values of $\langle \dot{T}_\delta^\star \rangle$ are also reduced when the wave number increases, and getting the minimum $\langle \dot{T}_\delta^\star \rangle$ later under lower wave numbers. For example, it gets the minimum $\langle \dot{T}_\delta^\star \rangle$ around 8.38 revolutions under $k_\lambda = 0$ whereas it takes place at 4 revolutions when $k_\lambda = 11$. The earliest occurrence of the minimum $\langle \dot{T}_\delta^\star \rangle$ under $k_\lambda = 11$ shows the highest efficiency of thermal conduction enhancement once again, whereas the latest occurrence of the minimum $\langle \dot{T}_\delta^\star \rangle$ under $k_\lambda = 0$ indicates the lowest efficiency of thermal conduction. Meanwhile, using moderate wave number indicates the moderate efficiency of thermal condition enhancement. Moreover, the magnitudes of the minimum values of $\langle \dot{T}_\delta^\star \rangle$ indicate the capabilities of wavy drums for thermal conduction enhancement, which are larger by using higher wave numbers and vice versa.

In addition, the generally similar results can be observed from Fig. 5.39C for $\omega = 1.0\ \pi$/s when $k_\lambda \leq 7$. However, slight differences can still be observed that the capability and efficiency are decreased by using wave numbers larger than $k_\lambda = 7$. The reason for that has already been explained in the former section on the formation of up-casting motion patterns.

It is noticed that the MTD formulation in Eq. (5.33) is in fact the first-order moment of particle temperature discrepancy $|T^\star - T_\infty^\star|$. In theory, the complete characteristics of thermal conduction can be well explored by the full orders of particle temperatures. In this section, for demonstration, the second-order moments of $\langle T_\delta^\star \rangle$, namely the variance of the mean temperature discrepancy $\sigma(\langle T_\delta^\star \rangle)$, are shown in Fig. 5.40A–C under the three rotating speeds respectively. By using $\sigma(\langle T_\delta^\star \rangle)$, the fluctuating feature as well as the stretching width of T-PDF can be characterized.

For the low rotating velocity ($\omega = 0.1\ \pi$/s, Fig. 5.40A), $\sigma(\langle T_\delta^\star \rangle)$ is not convergent for low wave numbers ($k_\lambda = 0$ and 3). In contrast, for other cases ($k_\lambda = 7 \sim 11$), $\sigma(\langle T_\delta^\star \rangle)$ is decreased to almost zero after about $t = 100$ s. Moreover, $\sigma(\langle T_\delta^\star \rangle)$ for $k_\lambda = 3$ is larger than that for $k_\lambda = 0$. Recalling Fig. 5.37A and B, the nonconvergence of $\sigma(\langle T_\delta^\star \rangle)$ is caused by the temporally increased T-PDFs of intermediate temperatures. On the other side, the convergence of T-PDF is mainly caused by the narrowed width of T-PDF after $t = 100$ s (e.g. referring to Fig. 5.37D for $k_\lambda = 7$), since the narrower PDF can have the smaller value of variance or standard deviation. In addition, for the moderate wave number $k_\lambda = 5$, the T-PDFs becomes much narrower with time and increases almost uniformly (Fig. 5.37C). Thus, $\sigma(\langle T_\delta^\star \rangle)$ is also reduced with time by the narrowed variation of T-PDF, although the rate of variation is a bit slower than those for higher wave numbers ($k_\lambda = 7 \sim 11$).

Based on the before made analysis, it is reasonable to quantify the rate of change of the geometric configuration of T-PDF from the '⊔'-shape to the '_/_'-shape by using $\sigma(\langle T_\delta^\star \rangle)$. The maximum values of $\sigma(\langle T_\delta^\star \rangle)$ may indicate the widest distribution

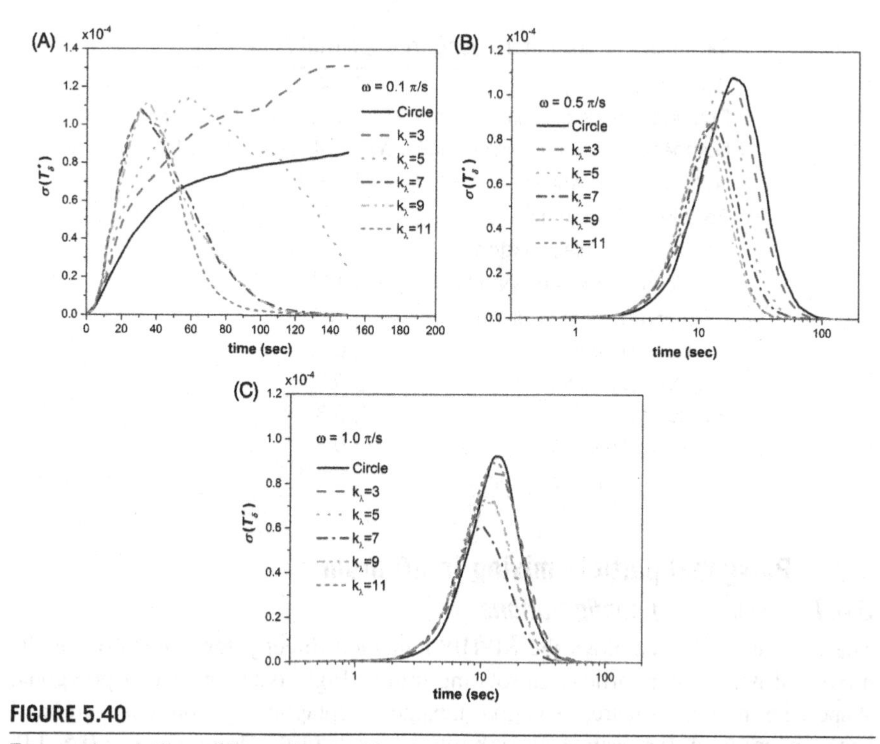

FIGURE 5.40

Temporal evolutions of the variance of temperature discrepancy under wave numbers $\omega = 0.1$ (A), 0.5 (B), 1.0 π/s (C), respectively.

of particle temperature with almost equivalent values of probability throughout all the temperature range, e.g. the relatively widest and most uniform distribution of particle temperature. In other words, the smaller value and earlier occurrence of the maximum $\sigma(\langle T_\delta^\star \rangle)$ is beneficial to the particle thermal conduction, or corresponding to the more effective and higher performance of the thermal conduction enhancement.

Following this criterion, it is clearly demonstrated that using a larger wave number is always more beneficial to thermal conduction of particles when $\omega = 0.5\ \pi$/s (Fig. 5.40B). Moreover, for $\omega = 1.0\ \pi$/s (Fig. 5.40C), the conclusion is similar as for $k_\lambda \leq 7$ where the best wave number for thermal conduction is $k_\lambda = 7$. However, using even larger wave numbers ($k_\lambda = 9$ and 11) may slightly deteriorate the efficiency and capability of the wavy drum for thermal conduction enhancement. It confirms the conclusions and explanations in the former sections.

5.4 Mixing of nonspherical particles

In this section, the SIPHPM methods are used to simulate and explore the mixing of nonspherical particles, e.g. 2D rectangular and polygonal and 3D cubic and tetrahedral particles, in the 2D drum mixer or 3D cylindrical mixers, respectively.

Table 5.7 Parameters used in the simulation.

Bed radius R_d (m)	0.4
Rotating speeds of drum ω (π/s)	0.5, 1, 1.5. 2, 2.5, 3
Number of polygonal particle sides, N_s	3, 4, 5, 6, 7, 8, 9, 10
Number of noncircular particles N_p	3600
Particle area A_p (mm^2)	6.5^2
Density of particle ρ_p (kg/m^3)	2000
Collisional stiffness k_c, (N/m)	10^3
Restoring stiffness k_r, (N/m)	10^5
Restitution coefficient e	0.95
Friction coefficient μ	0.3
Poisson ratio ν	0.3
Simulation time t_s (s)	9
Simulation time step δt (s)	10^{-6}

5.4.1 Polygonal particle mixing in 2D drum

5.4.1.1 Simulation configurations

The present section employs the SIPHPM method directly for simulation of the mixing of noncircular particles in rotating drums. Eight types of regular polygonal shapes, i.e. triangle, square, pentagon, hexagon, heptagon, octagon, enneagon and decagon, are used. For each shape, six rotary speeds of the drum, i.e. $\omega = 0.5$, 1.0, 1.5, 2.0, 2.5, 3.0 π/s are simulated. The drum is circular with a radius of $R_d = 0.4$ m, and the particles have the same area of $A_p = 6.5$ mm^2. In addition, the parameters used in present simulations are listed in Table 5.7.

The initial packing state of particles is obtained by a freely sedimentation process from a certain height. At the beginning, the particles are arranged in a regular array. Then they fall down freely and settle down inside the bed until exhaustion of their kinetic energies because of the damping dissipation in close contact. After settled down, the drum will rotate faster and faster before reaching the predefined target speeds. The period of acceleration of the rotation speed is 0.5^{-1} π/s, and after that, the rotating speeds are kept constants.

5.4.1.2 Mixing processes

For example, the mixing processes of the triangular particles and square particles are shown in Fig. 5.41 ($\omega = 2.5$ π/s) and Fig. 5.42 ($\omega = 3.0$ π/s) respectively. At first, particles are divided into two parts (Fig. 5.41A and 5.42A). As the rotation goes on, the mixing between the two parts develops (Fig. 5.41B–F and 5.41B–F). For $\omega = 2.5$ π/s in Fig. 5.41, the elongated stripes of particles are twisted around each other, which is the basic pattern of particle mixing in the rotating drum. Based on the cascading distribution [303] of particles during the rapid flowing within the top layer from the left-top to the right-bottom of the bed (Fig. 5.41C–F), it indicates mainly a cascading regime at $\omega = 2.5$ π/s.

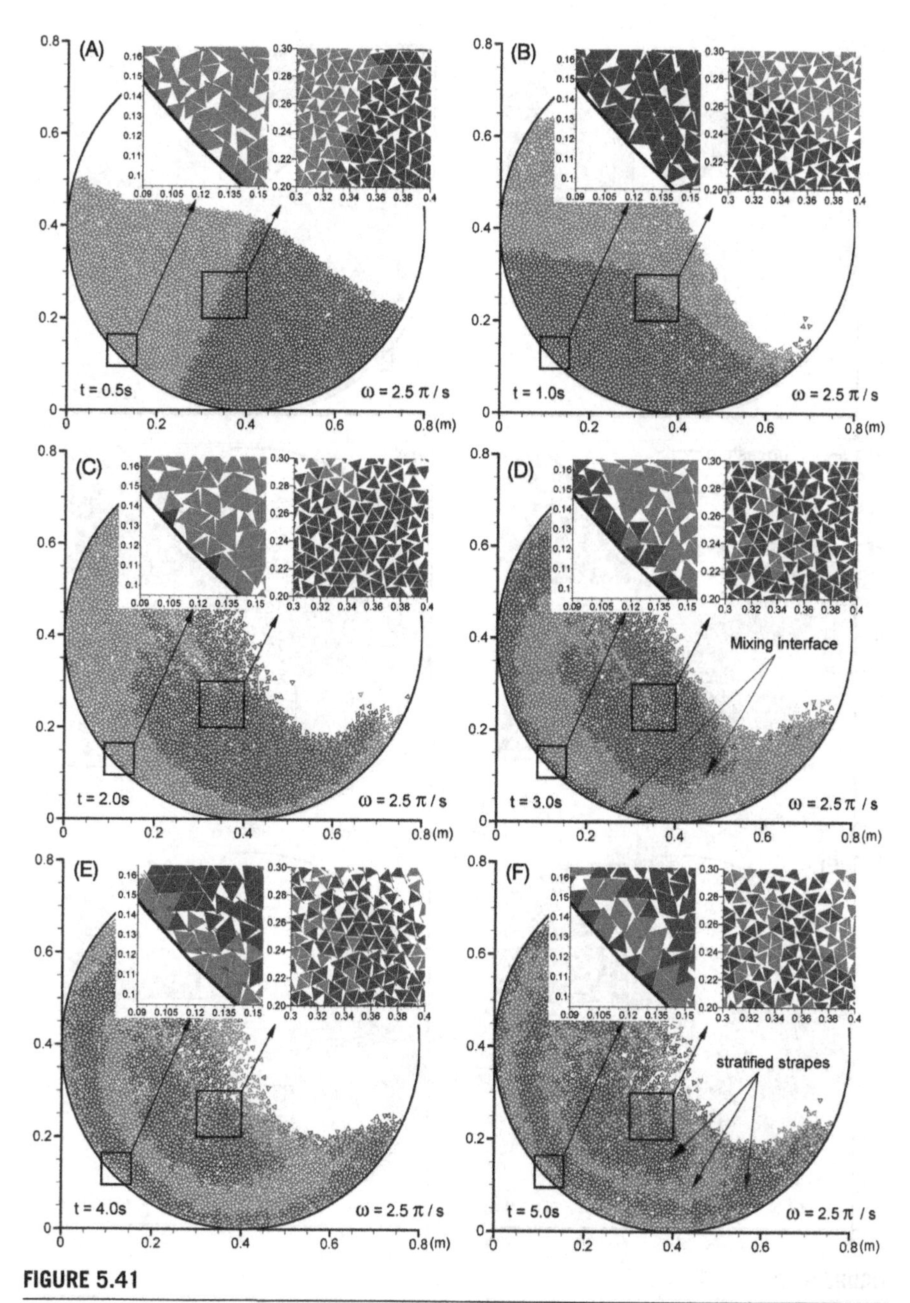

FIGURE 5.41

The process of mixing of the triangular particles at $t = 0.5$, 1.0, 1.5, 2.0, 2.5, 3.0 s (A, B, C, D, E, F, respectively) at the rotating speed $\omega = 2.5\ \pi$/s. The insets show the zoomed in local rectangular regions as designated (0.3 m $< x <$ 0.4 m and 0.2 m $< y <$ 0.3 m).

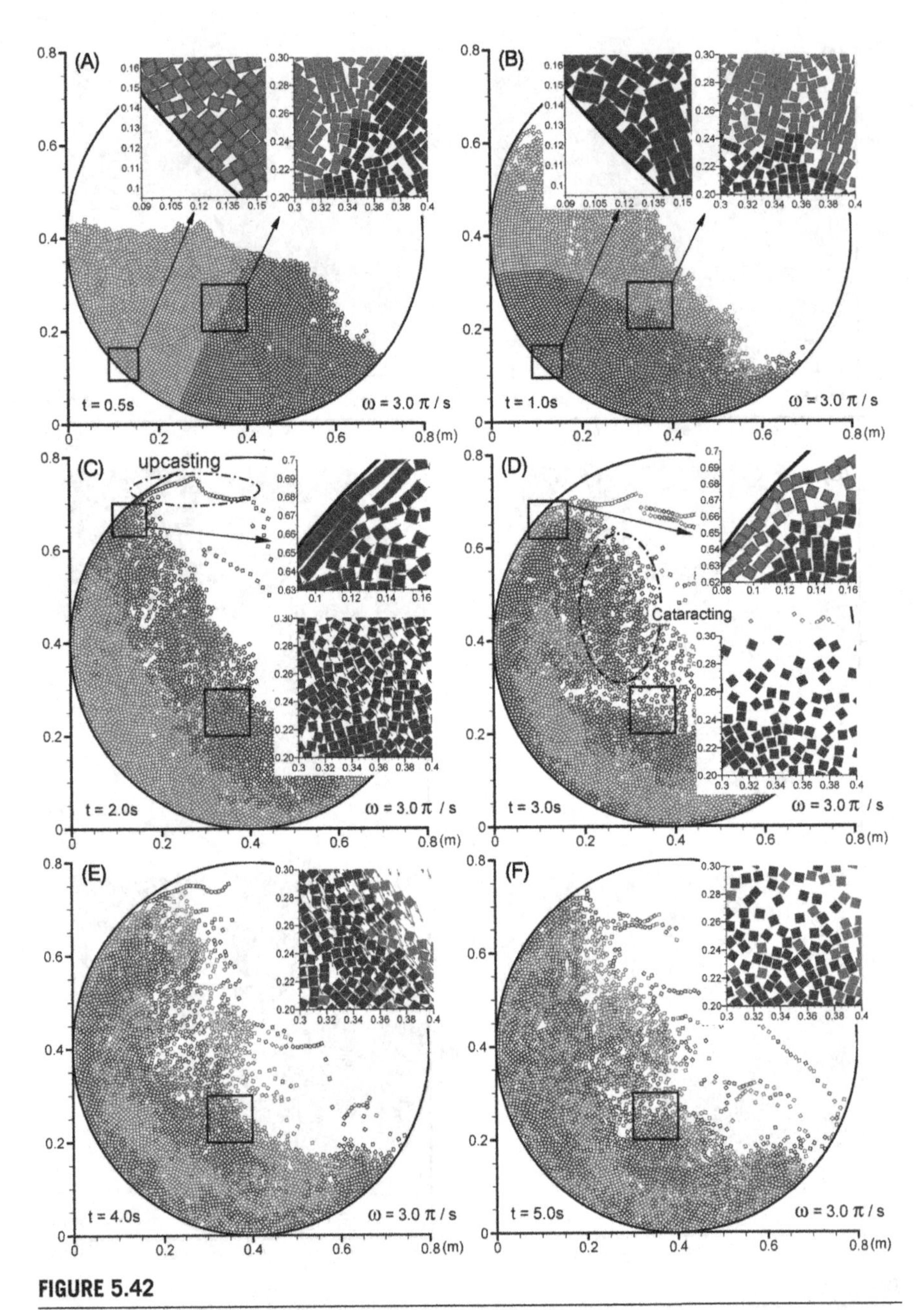

FIGURE 5.42

The process of mixing of the square particles at $t = 0.5$, 1.0, 1.5, 2.0, 2.5, 3.0 s (A, B, C, D, E, F, respectively) at the rotating speed $\omega = 3.0\ \pi$/s. The insets show the zoomed in local rectangular regions as designated (0.3 m < x < 0.4 m and 0.2 m < y < 0.3 m).

When the rotation speeds are very large at $\omega = 3.0\ \pi/s$, the flow seems to be upcasting in the left-top region and starts to go into the cataracting flow regime [296] (Fig. 5.42C–F). The upcasting motion (Fig. 5.42C) of square particles is particularly clearer than other particle shapes. Because, compared to other shapes, it is easier for the square particles at contact with the wall of the drum to form regular arrangements along the wall (e.g. see the right-top inset in Fig. 5.42C). It may result in tight surface-to-surface connection with each other. In this condition, it is also easier to form continuous normal forces between the neighboring particles to cause the cluster-like upcasting motion of square particles. Compared Fig. 5.42 to Fig. 5.41, besides the differences in flow regimes, the mixing processes are clearly different, too. For example, clear stratification structures can be seen in Fig. 5.41 with almost clear mixing interface even in Fig. 5.41F at $t = 5$ s, whereas in Fig. 5.42F still at $t = 5$ s, the mixing interface is no longer as smooth and clear as in Fig. 5.41F. This indicates the necessity for in-depth quantitative explorations of the effect of particle shape on mixing characteristics.

Moreover, Fig. 5.43 shows the mixing states of regular polygonal particles with the number of sides ($N_s = 4$ (A, D), 6 (B, E) and 8 (C, F) at $t = 9$ s for $\omega = 0.5\ \pi/s$ (A, B, C) and $2.5\ \pi/s$ (D, E, F), respectively). At the speed of $\omega = 0.5\ \pi/s$ (Fig. 5.43A–C), the flow regime is rolling, and the particles are not well mixed at $t = 9$ s, although with some dispersions. In contrast, at $\omega = 2.5\ \pi/s$ (Fig. 5.43D–F), the flow regime is cascading, and the particles have been well mixed. By direct observation, although not exactly the same, the mixing degrees of different shapes of particles are not largely different from each other, as the mixing structures of different particle shapes under the same rotating speeds are quite similar. Based on Figs. 5.41–5.43, it is mainly clear that the particle shape will of course influence the mixing process, but the degree of influence seems to be affected by the rotating speeds. Thus, quantitative evaluations on the influence of particle shape should be explored.

5.4.1.3 Mixing index

In this section, the Lacey mixing index [300] (Eq. (5.21)) is used for the purpose of quantitative evaluation of the mixing quality.

Taking the square, hexagonal, octagonal and decagonal shapes for example, the evolutions of the mixing index can be seen in Fig. 5.44. Before $t < 0.5$ s the drum is gently accelerated and the mixing index is almost not changed for different shapes. After that, it increases with time rapidly as the mixing quality improves. In general, the mixing index is higher when the rotating velocity is higher, which indicates a better quality of mixing is achieved under larger rotating speeds. However, when $\omega = 3.0\ \pi/s$, the mixing index is no longer larger than that for $\omega = 2.5\ \pi/s$, and the maximum mixing quality takes place at $\omega = 2.5\ \pi/s$. Recalling the observation in Figs. 5.41–5.42, the flow regime is fast cascading or slight cataracting at $\omega = 3.0\ \pi/s$ and the particles are no longer closely contacted with each other during the upcasting motion. Compared to $\omega = 2.5\ \pi/s$, this phenomenon explains the reason of the slight reduction of mixing degree at $\omega = 3.0\ \pi/s$.

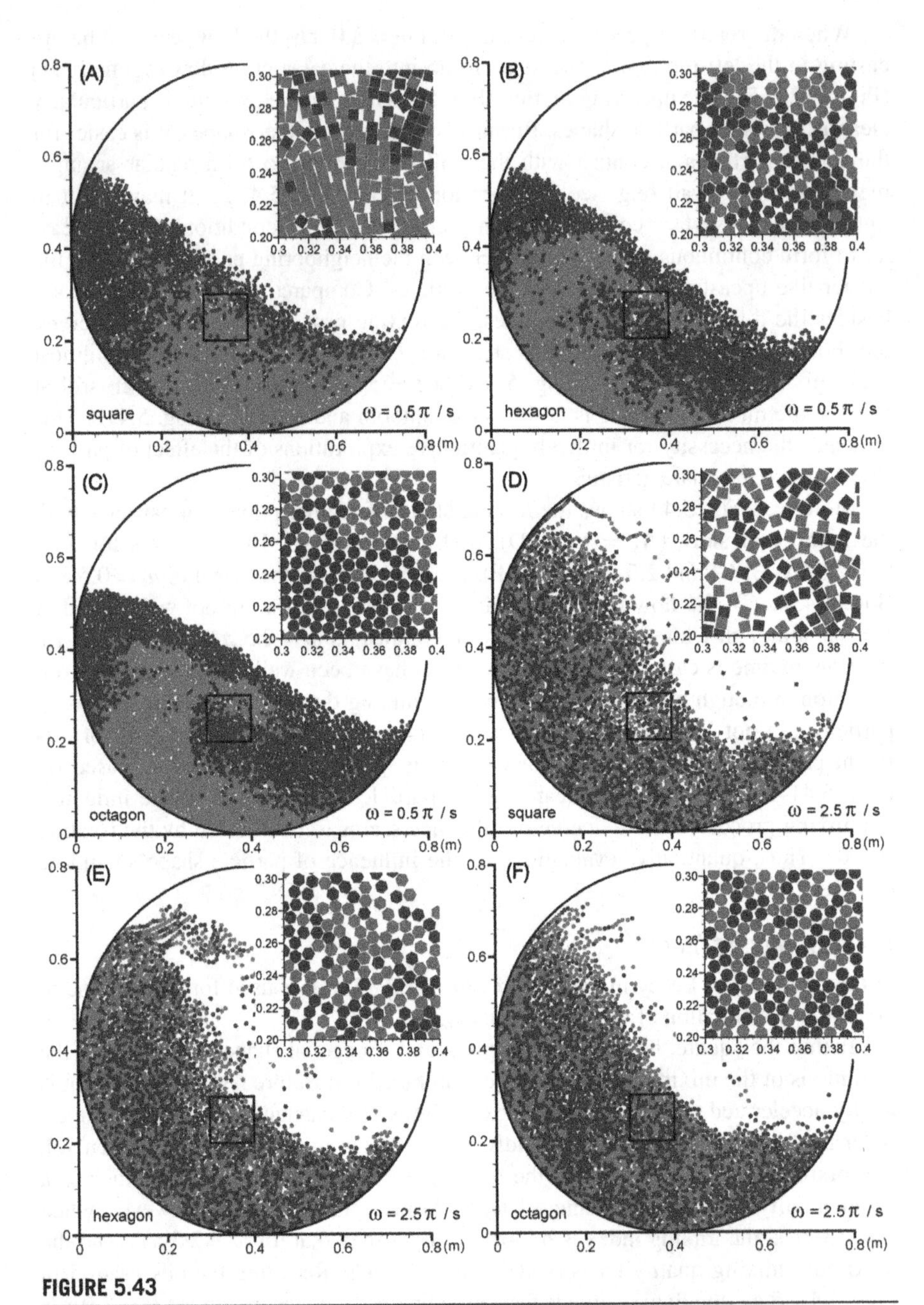

FIGURE 5.43

The mixing of the pentagonal (A), hexagonal (B), heptagonal (C), octagonal (D), enneagonal (E) and decagonal (F) particles at $t = 5$ s at the rotating speed $\omega = 1.5\ \pi$/s.

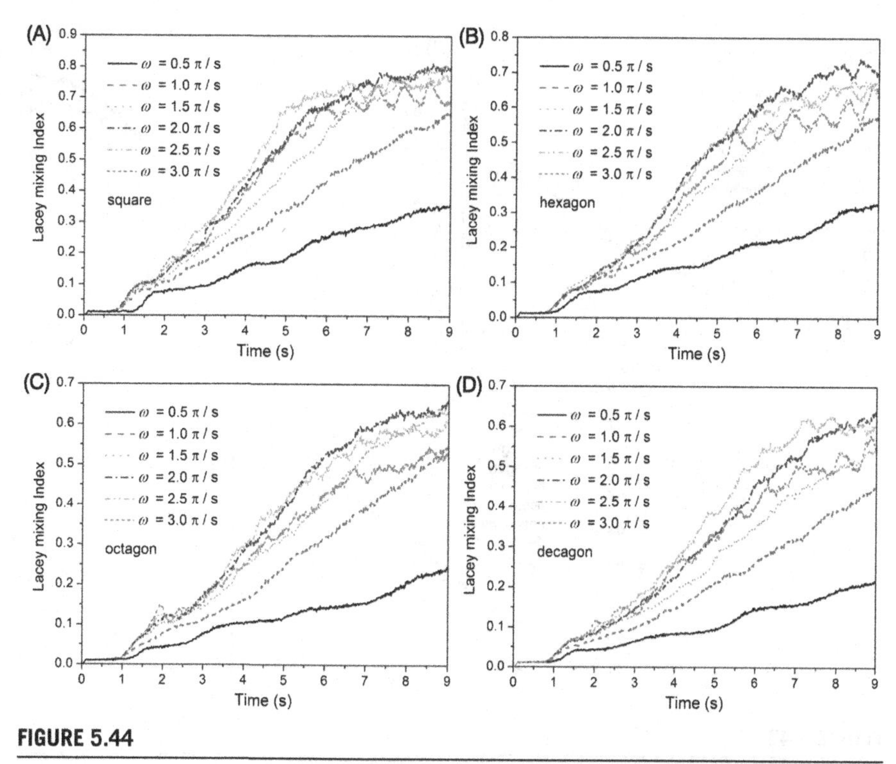

FIGURE 5.44

The variation of the Lacey mixing index with time at different rotation speeds for the square (A), hexagonal (B), octagonal (C), and decagonal (D) particles, respectively.

On the other hand, taking $\omega = 1.0$, 2.0 and 3.0 π/s for example, Fig. 5.45 shows the evolution of mixing index for different particle shapes. In general, it is find the square and hexagonal shaped particles have better mixing quality than other shapes of particles. The square shape is the best, then goes the hexagon, followed with the triangular shape. It may be related to the effect of particle shape on close particle packing. In computational fluid dynamics, uniform triangular shape (Fig. 5.46A), square shape (Fig. 5.46B), and hexagonal shape (Fig. 5.46D) can be used to discretize the flow fields without any disconnection of flaw in meshing, whereas other shapes cannot cover a field perfectly without any disconnection in it (e.g. pentagonal shape in Fig. 5.46C and heptagonal shape in Fig. 5.46E). Thus, the square, hexagon and triangle are good shapes to cover the flow field continuously and form closely packed state with the surface-to-surface connection between neighboring particles. As a result, the degree of static packing and quality of mixing should be better than for other shapes. Moreover, compared to the squares and hexagons, triangles have the lower probability to form perfect continuous covering. Because, for each cell of covering (Fig. 5.46A), it needs at least 6 triangles, which are larger than the square shape (4 squares in Fig. 5.46B) and the hexagonal shape (3 hexagons in Fig. 5.46C).

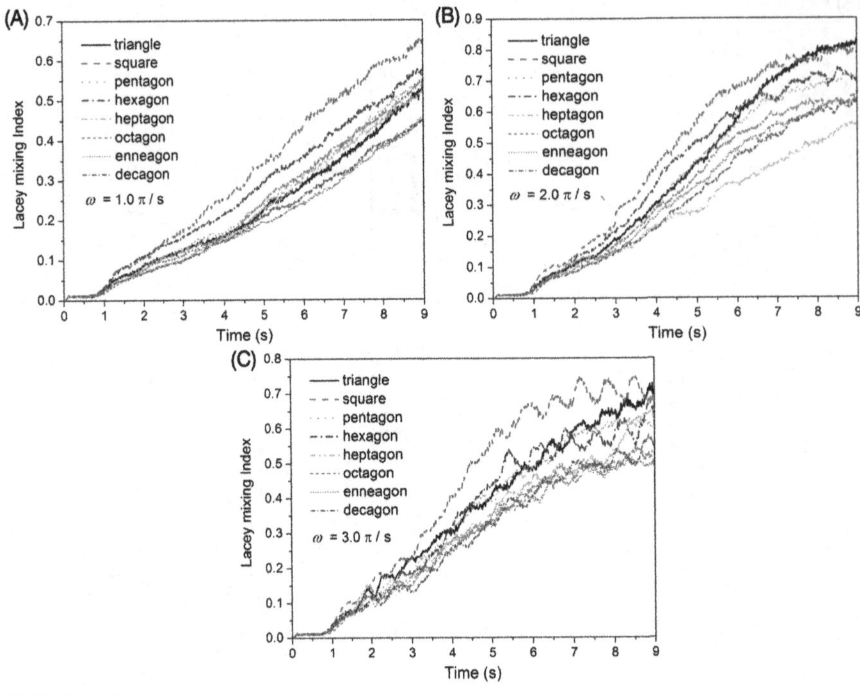

FIGURE 5.45

The variation of the Lacey mixing index with time for different shapes of particles at the rotation speeds of $\omega = 1.0$ (A), 2.0 (B) and $3.0\ \pi$/s (C), respectively.

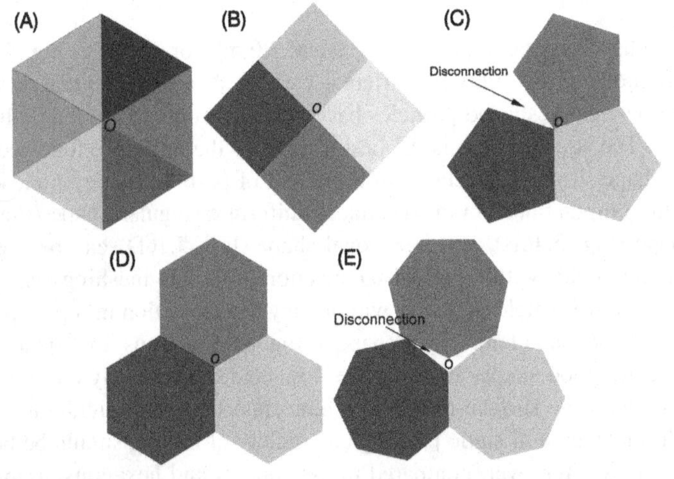

FIGURE 5.46

The sketches for the triangular (A), square (B), pentagonal (C), hexagonal (D) and heptagonal shapes (E) used for meshing a 2D planar field.

Thus, the mixing quality of triangles is lower than that of squares and hexagons. In addition, although it is not rigorous, the general trend of mixing index variation for other shapes is getting lower when N_s becomes large. In other words, it is not expected to get better mixing by using the shape closer to round (namely larger N_s). It is reasonable since, according to the schematic drawing in Fig. 5.46, it may generate more disconnections (voids) between particles when a larger N_s is used.

5.4.1.4 Mixing entropy analysis

On the other hand, the mixing entropy, defined by the mean Shannon information entropy [297] based on the local concentration of particles (Eq. (5.5)), is used here. By integrating over the whole area of the drum, M_E can be used to quantify the degree of particle mixing in the drum.

The mixing entropies for the triangular, square, pentagonal, heptagonal and enneagonal particles are shown in Fig. 5.47. Similar to the variations of mixing index in Fig. 5.44, the mixing entropy is also increasing with time as the rotation goes on.

Generally, it is also larger when the rotation speed is larger, except for the maximum rotation speed $\omega = 3.0\ \pi/\mathrm{s}$. The mechanism is that the particle mixing in the cataracting flow regime is not as good as in the cascading regime, because the particles are upcasting and cataracting, without as close contact with each other as that in the cascading regime. Remember that the mixing entropy in Eq. (5.17) is computed from the local concentration of particles. When the particles become loosely contacted to each other, the local concentration will become lower, which may sometimes decrease slightly the mixing entropy. Because of the similar mechanism, compared to the not fully developed mixing patterns before $t = 5$ s, the mixing entropies in Fig. 5.47 for $\omega \geq 1.5\ \pi/\mathrm{s}$ decrease a little after $t = 5$ s when the normal cascading patterns are fully established, since the cascading flow will also decrease a little the local concentration of particles in the flowing layer. Therefore, the maximum mixing entropies take place before the mixing structures and the cascading/cataracting flow regimes are fully developed (usually around $t = 4$–6 s).

In addition, Fig. 5.48 shows the comparison of mixing entropy between different shapes of particles under the rotating speeds of $\omega = 0.5, 1.5, 2.5\ \pi/\mathrm{s}$. Unlike Fig. 5.45, the mixing entropies of particles of different shapes are not as different from each other as in Fig. 5.45, especially under the high rotating speed at $\omega = 2.5\ \pi/\mathrm{s}$. It is seen that: (1) Before the maximum mixing entropy is reached, the mechanism of particle shape on mixing is similar to that sketched in Fig. 5.46. The square shape and hexagonal shape are always better for particle mixing than the octagonal and decagonal shapes because of the less disconnections and voids especially in the low speed rotation. (2) In contrast, when the maximum mixing entropy is achieved, the normal cascading flow pattern may decrease a little the particle concentration in the flowing layer and cause a little decrease in mixing entropy. Therefore, when the number of particle vertexes or sides N_s are larger, it is closer to the round shape and easier to roll. The more round shaped particles may keep closer to each other during the rolling down process in the flowing layer. On the contrary, it may decrease more concentrations for the nonround shape in the flowing layer and leads to more reduction

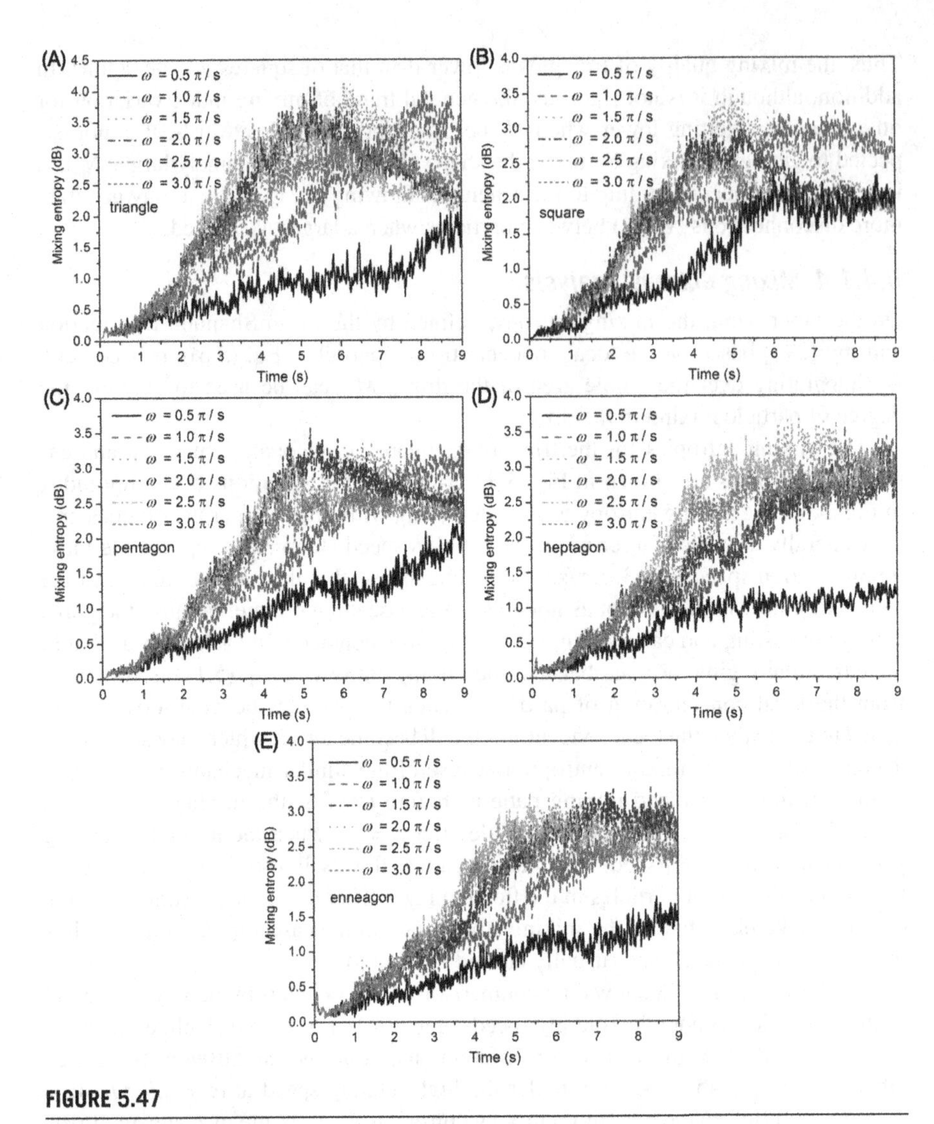

FIGURE 5.47

The variation of the mixing entropy with time at different rotation speeds for the triangular (A), square (B), pentagonal (C), heptagonal (D), and enneagonal (E) particles, respectively.

in mixing entropy. This mechanism is especially clear under the high speed of rotation with evident rolling processes. Therefore, the mixing entropy of square particles decrease the most at high speed rotations $\omega = 2.5\ \pi$/s. (3) The mixing entropies are highly fluctuating whereas the mixing index is not. The reasons are in the following: (i) In Eq. (5.21), the mixing index is calculated from the standard deviation of particle concentration and the overall proportion of particles. Thus, as an overall-level eval-

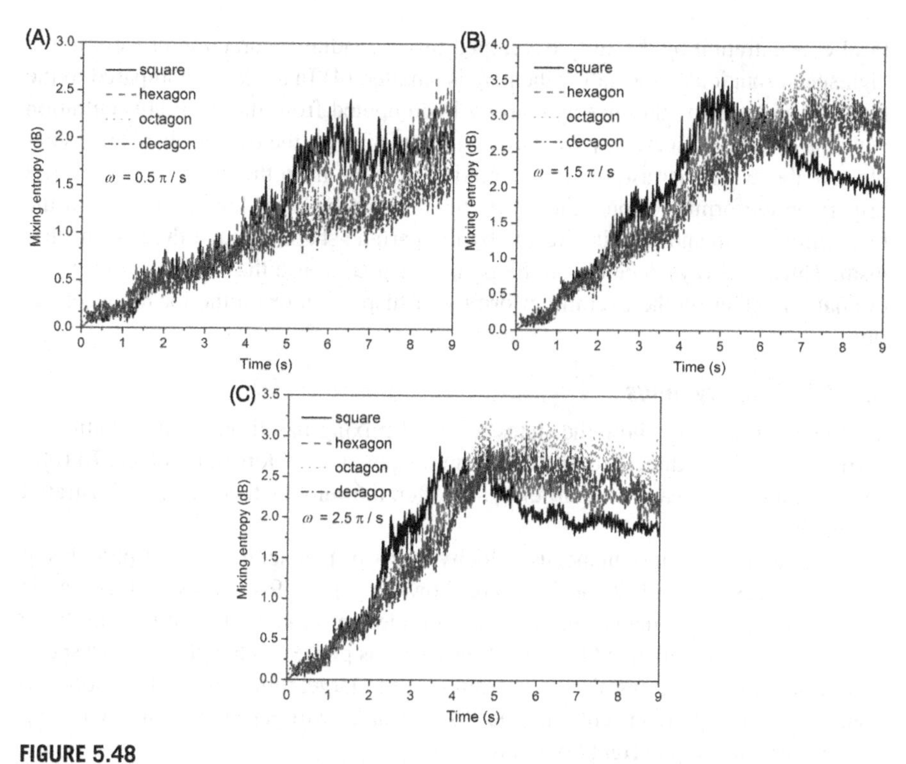

FIGURE 5.48

The variation of the mixing entropy with time for different shapes of particles at the rotation speeds of $\omega = 1.0$ (A), 2.0 (B) and 3.0 π/s (C), respectively.

uation method, the local fluctuations of mixing have been filtered out. (ii) However, in Eq. (5.5), the mixing entropy is directly computed from the local concentrations, where the local fluctuation of particle concentration is rather summed up than filtered out. Therefore, the data points in Fig. 5.48 are scattering more intensively than those in Fig. 5.45. But, it seems that the mixing entropy is still good for evaluation of mixing since it may be closer to the actually local random mixing situation.

To say conclusively, the confirmed results are obtained on the effect of particle shape as follows: (1) The mixing quality or mixing level are mainly influenced by the rotating speed, since it mainly affects the flow regimes. (2) Before reaching the cataracting regime at $\omega = 3\ \pi$/s, the mixing quality and degree are generally improved by increasing the rotating speed. (3) The particle shape may have secondary influences on the mixing level/quality improvement through two kinds of mechanisms: the first is on the formation of disconnections/voids in packing structures, which is especially clear in the low speed conditions. For getting better covering of the flow region, the square, hexagonal and triangular shapes are especially better than others. Consequently, the mixing degree/quality for these shapes are better than other shapes in the rolling or slowly cascading regimes. The second is on the reduction of local concentration by the fully developed fast cascading/cataracting process in the high speed rotations, especially when N_s is smaller. (4) In addition, compared to the Lacey mixing index, the mixing entropy is computed from the local concentration which is certainly affected by the particle shapes through the two mechanisms as explained above. In contrast, the Lacey mixing index is from the viewpoint of overall dispersion uniformity of one kind of particle inside the other kind. It filters out the local information and only is affected by the particle shape through the first mechanism. Thus, it always depends on the particular goal to use the functions of mixing evaluation, either on the overall uniformity of dispersion or on the local concentrations.

5.4.1.5 Velocity fields

The previous sections show the states/levels of mixing based on the distribution of particles, which are directly computed from the position vectors of particles. To fully show the statistical feature of particles, hereafter, the velocity fields will be illustrated and analyzed.

For example, the instantaneous velocity fields of pentagonal shaped particles at rotating speeds of $\omega = 1, 2$ and 3 π/s are shown in Fig. 5.49A–C, respectively. As is well known, the flow fields can be clearly divided into two parts, the flowing layer and the static base, separated by the yellow curve as plotted. When the rotating speed varies, the flowing regions also vary, from small to large, thin to thick, low slopes to high slopes. The particles entering the flowing layer will get kinetic energy during moving from left-top to right bottom regions.

Moreover, the time-averaged mean velocity for decagonal particles of $\omega = 0.5$, 1.5 and 2.5 π/s are shown in Fig. 5.49D–F. Herein the horizontal component of the mean velocity is used for case study, which is defined as $\bar{u}_x = \frac{1}{T}\int_T u_x(t)dt$. As illustrated, the flowing layer and the static base can be well separated by the region with near zero mean velocities ($\bar{u}_x \approx 0$). Thus, in order to evaluate the effect of particle shape on flow fields, the area fractions of the flowing layer and the static base are calculated by the mean horizontal velocities. Herein,

$$\begin{cases} A_f = \dfrac{\text{Area with } \bar{u}_x > \epsilon}{\text{Total area}} & \text{(a)} \\ A_s = \dfrac{\text{Area with } \bar{u}_x < -\epsilon}{\text{Total area}} & \text{(b)} \end{cases} \tag{5.63}$$

are used, where $\epsilon = 0.06$ m/s is a small value, and A_f and A_s are the area fractions of the flowing layer and static base, respectively. The area fractions of the positive and negative $\bar{u}_x$ are shown in Fig. 5.50. In general, the positive velocity area (A_f, flowing layer) is always enlarged as the rotation speed ω increases. In contrary, the negative velocity area (A_s, static base) is generally reduced as ω increases, except for the lowest rotating velocity $\omega = 0.5$ π/s. Note that the total area is a constant. The sum of area fractions of both positive and negative $\bar{u}_x$ at $\omega = 0.5$ π/s is lower than in other cases. Thus, when $\omega = 0.5$ π/s, the mean horizontal velocities are very

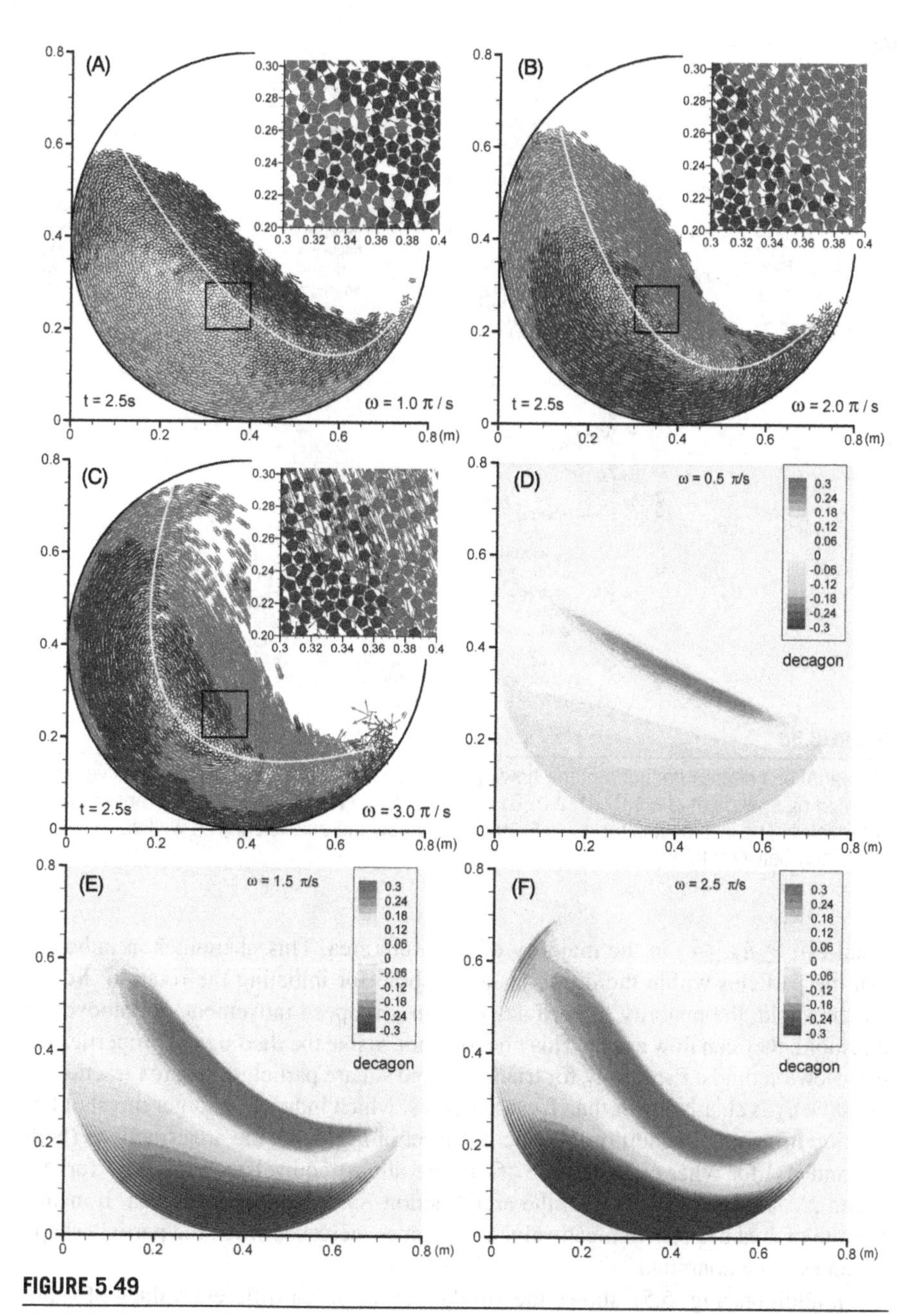

FIGURE 5.49

The instantaneous velocity fields (A–C) for the pentagonal particles at the rotating speeds of $\omega = 1.0$ (A), 2.0 (B) and 3.0 π/s (C), respectively, and the time-averaged horizontal velocity (unit: m/s) for the decagonal particles at the rotating speeds of $\omega = 0.5$ (D), 1.5 (E) and 2.5 π/s (F), respectively.

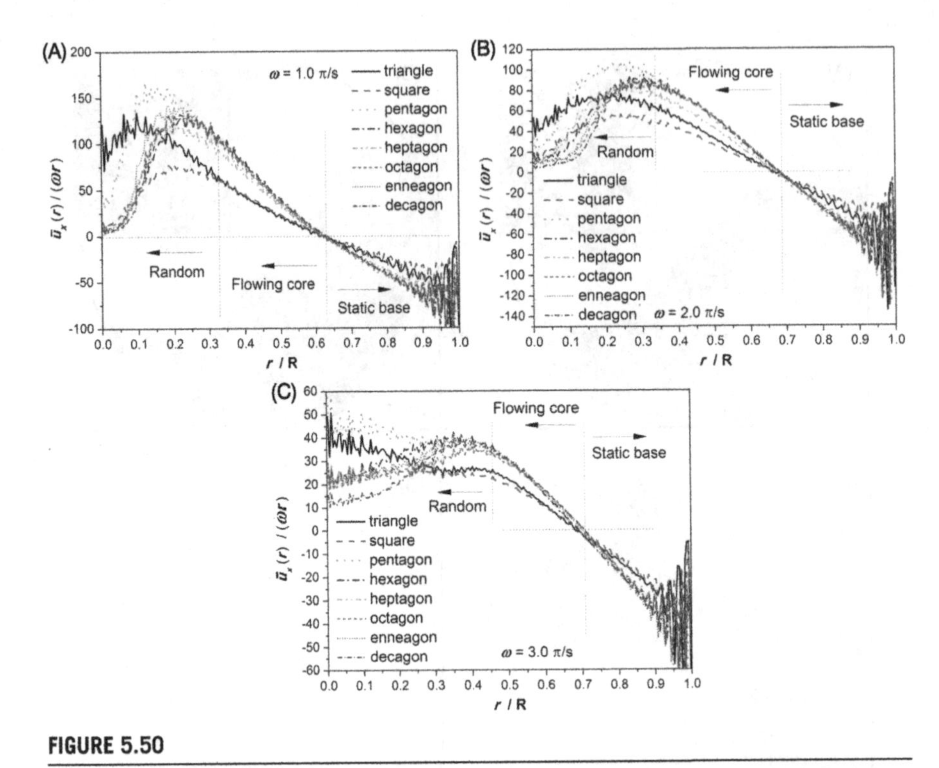

FIGURE 5.50

The variations of area fractions of the flowing layer (positive) and static base (negative) at the rotating speeds of $\omega = 1.0$ (A), 2.0 (B) and 3.0 π/s (C) respectively for different shapes of particles. The last inset is for $N_s = 5$–10 only to show the consistency as well as the difference between them.

small ($-\epsilon \leq \bar{u}_x \leq \epsilon$) in the majority of the drum area. This phenomenon indicates that the particles within the drum has a threshold for initiating the rotation. Below the threshold, the majority of particles keep in low speed movement, and above the threshold, they can flow easier. This phenomenon is like the rheological properties for non-Newton fluids. Especially for triangular and square particles, the area fraction of negative $\bar{u}_x$ is clearly lower than for other cases, which indicates a larger threshold of rotation initiation. In addition, see the last inset of Fig. 5.50, the area fractions (both A_f and A_s) for other cases of $N_s = 5$–10 are almost equivalent, especially for A_s. When N_s reduces from 10 to 5, the area fraction A_f is slightly increased. It means the rate of area of positive mean velocity becomes increased when the particle shape becomes more nonround.

In addition, Fig. 5.51 shows the similar results when different values of ϵ are used to compute the area fractions of A_f and A_s. It is clearly seen that the absolute area fractions of both A_f and A_s may decrease when larger values of ϵ are used. However, the trends still remain similar to that observed in Fig. 5.50 for all shapes. Thus, it consolidates the conclusions obtained from Fig. 5.50.

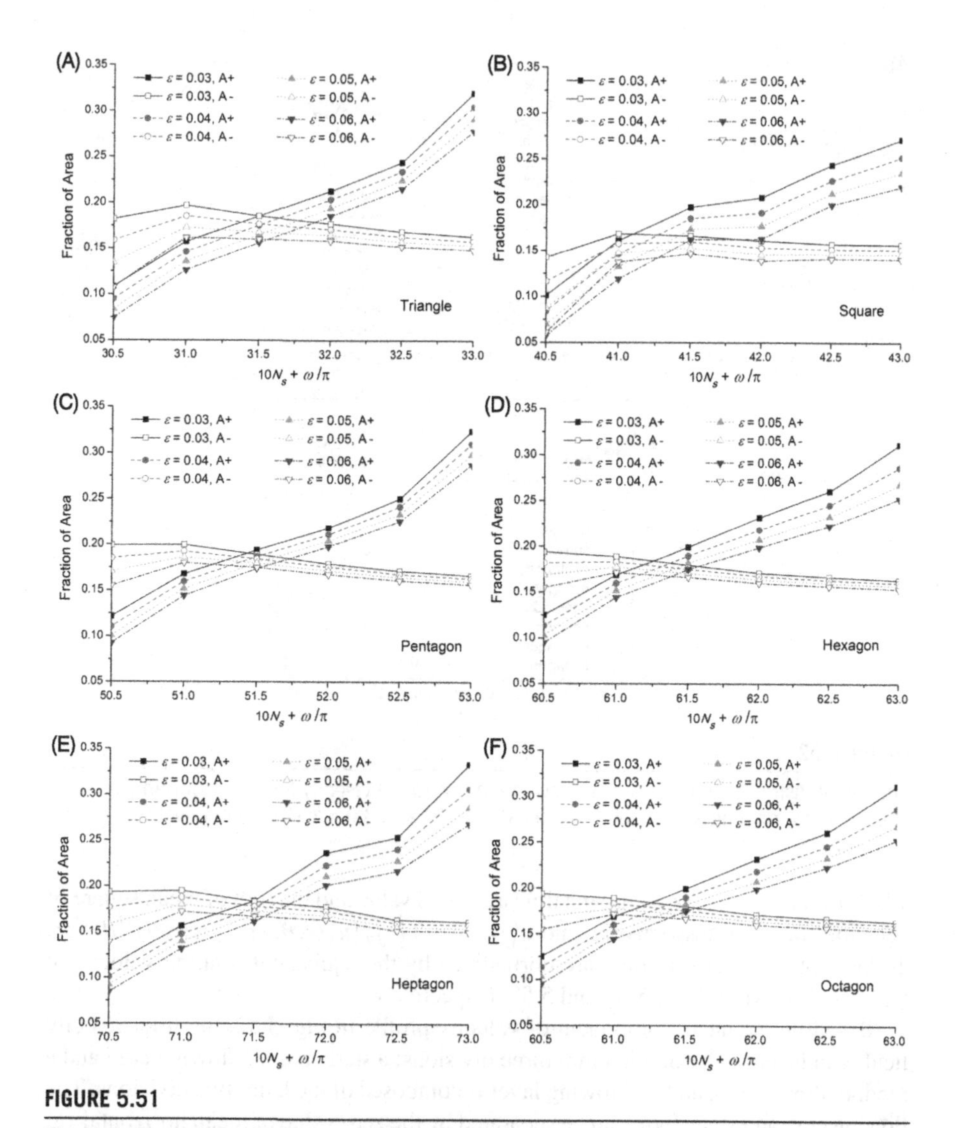

FIGURE 5.51

The variations of area fractions of the flowing layer (A+) and static base (A−) for different ϵ and different shapes: $N_s = 3$ (A), 4 (B), 5 (C), 6 (D), 7 (E) and 8 (F), respectively.

Moreover, because of the circular drum, the radial distribution of velocity profiles can be formulated as follow:

$$\begin{cases} \bar{u}_x(r) = \dfrac{1}{2\pi}\displaystyle\int_0^{2\pi} \bar{u}_x(r,\theta)d\theta & \text{(a)} \\ u'_x(r) = \dfrac{1}{2\pi}\displaystyle\int_0^{2\pi} u'_x(r,\theta)d\theta & \text{(b)} \end{cases} \tag{5.64}$$

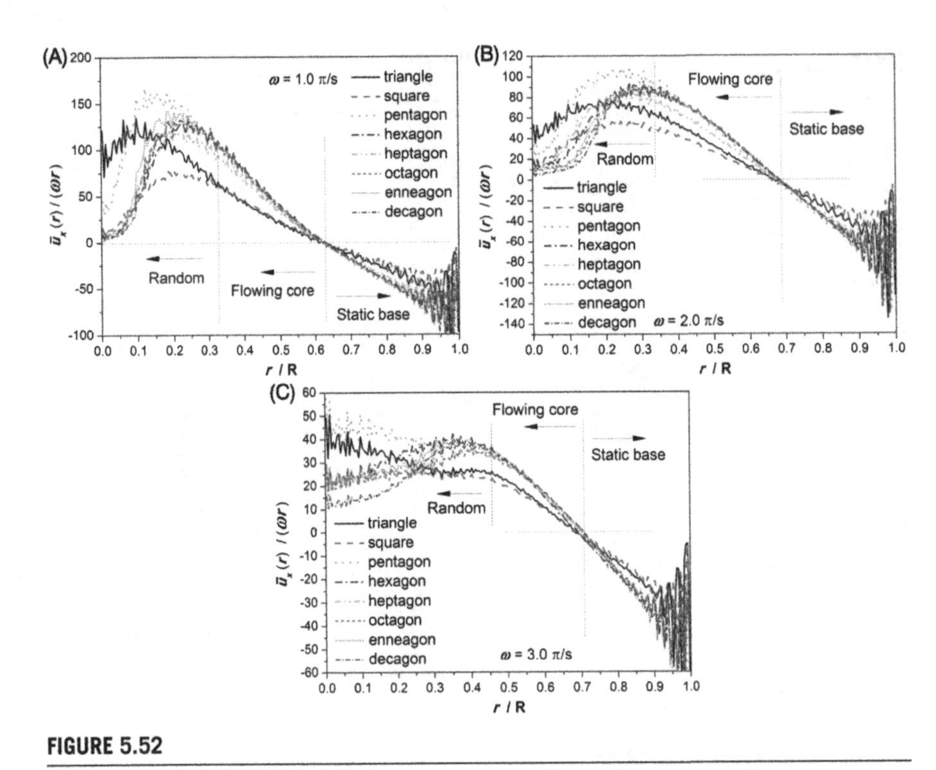

FIGURE 5.52

The radial distribution of the normalized mean horizontal velocity for different particle shapes at the rotating speeds of $\omega = 1.0$ (A), 2.0 (B) and $3.0\ \pi$/s (C), respectively.

where $\bar{u}_x(r,\theta)$ and $u'_x(r,\theta)$ are the time averaged value and the root of mean square of velocities at (r,θ), respectively, and $u'_x(r,\theta) = (\frac{1}{T}\int_T [u_x(r,\theta,t) - \bar{u}_x(r,\theta)]^2 dt)^{1/2}$. To facilitate comparison, they are normalized by the equivalent rotating velocity at (r,θ), and shown in Figs. 5.52 and 5.53, respectively.

Based on the mean component of velocity profile in Fig. 5.52, the flow velocity fields can be basically divided into three divisions: a static base, a flowing core and a random flow region, and the flowing layer is composed of the latter two divisions. The flowing core and static base can be separated by the zero value of mean horizontal velocity. The flowing core and random flow region can be separated by the convergence point of velocity profiles at some radial distance, e.g. $r = 0.325R$ in Fig. 5.52A. It is concluded from Fig. 5.52A that: (1) In the flowing core and static base, the variations of radial distribution of mean velocity follow the consistent orders, i.e.

$$\begin{cases} \bar{u}_x|\text{square} < \bar{u}_x|\text{triangle} < \bar{u}_x|\text{other shapes, in the flowing core} & \text{(a)} \\ \bar{u}_x|\text{square} > \bar{u}_x|\text{triangle} > \bar{u}_x|\text{other shapes, in the static base} & \text{(b)} \end{cases} \quad (5.65)$$

(2) The above orders lead to the same conclusion: the square particles have the smallest magnitudes of moving velocity. It confirms the conclusion observed in

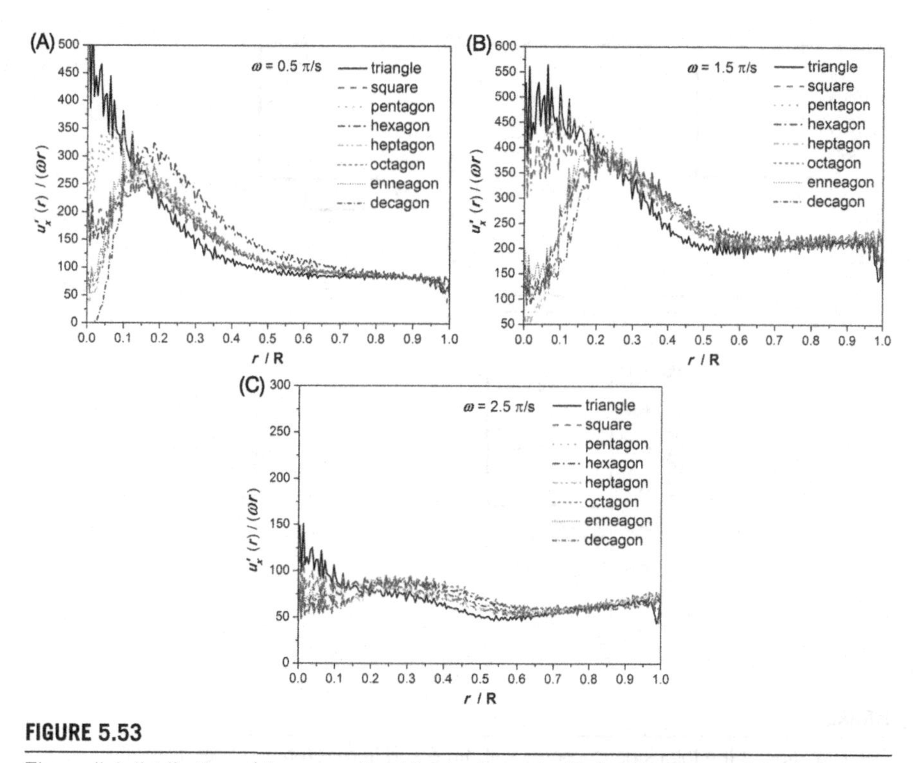

FIGURE 5.53

The radial distribution of the normalized fluctuation velocity for different particle shapes at the rotating speeds of $\omega = 0.5$ (A), 1.5 (B) and 2.5 π/s (C), respectively.

Fig. 5.50 that the square particles have the largest threshold for initiation of rotation. Then, the triangular particles are less likely than the square particles to be driven. For other shapes, they are easier to be driven to flow than the triangular and square shapes. (3) In addition, based on the separation points between the divisions, the thickness of the flowing layer is enlarged as the rotating speed increases. Meanwhile, the thickness of the static base is reduced as ω increases.

On the other hand, from Fig. 5.52B, the fluctuation velocity $u'_x(r)$ is decreasing in radial direction except for the random flow region immediately near the center of the drum. Moreover, $u'_x(r)$ is also reduced as the rotation speed increases. In fact, based on the definition of $u'_x(r)$ in Eq. (5.64b), the reduction of $u'_x(r)$ is mainly caused by the improved consistency in the circumferential variation of particle velocity at large rotating velocity (e.g. see $\omega = 2.5\ \pi/s$ in the flowing layer configuration in Fig. 5.49F).

5.4.1.6 Kinetic energies

At last, as kinetic energy is also an important issue of particles, it is better to analyze the variation of the kinetic energy for different cases. As shown in Fig. 5.54, in general, the total kinetic energies of particles are increasing with time until they

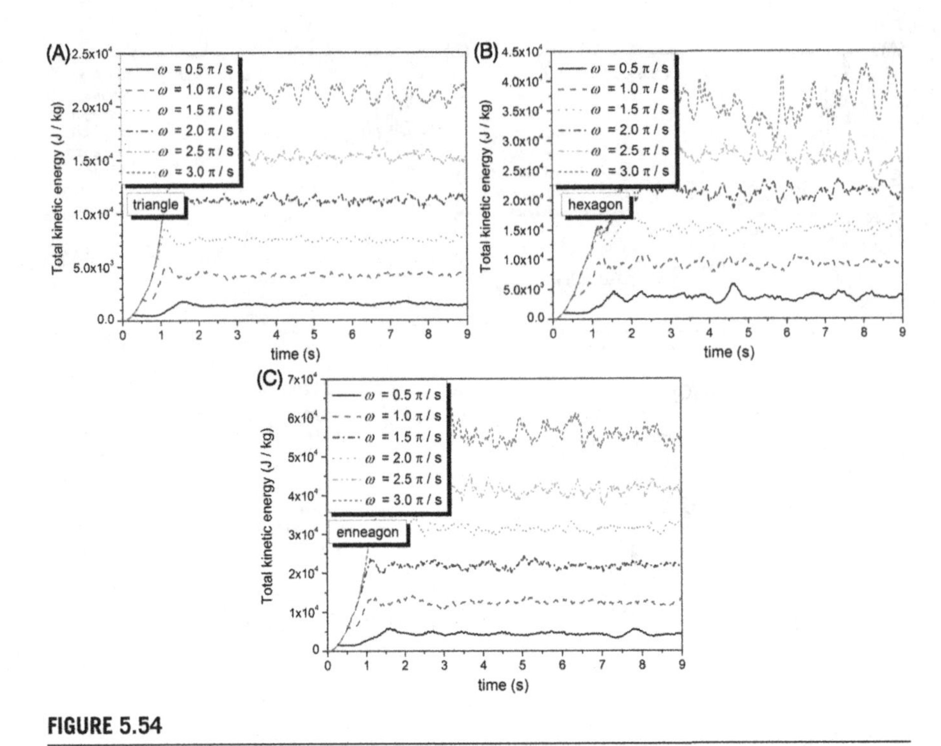

FIGURE 5.54

The evolution of the total kinetic energy with time at different rotating speeds for the triangular (A), hexagonal (B) and enneagonal (C) shapes, respectively.

reach an ultimate level, when the energy input from the rotating drum is in equilibrium with the energy dissipation by collision and friction. Thus, the ultimate level is higher when the rotating speed is higher. This is true for all kinds of particle shapes.

However, under the same rotating speed, the total kinetic energies of particles are also increasing as the number of sides increases (see Fig. 5.55, from triangle to decagon, corresponding to the number of sides $N_S = 3–10$). In this situation, the energy input from the drum could be considered as equivalent. Thus, the difference in the ultimate level of kinetic energy is mainly caused by the different dissipation of kinetic energy during friction and collision. In general, it indicates that the kinetic energies of particles of better roundness (namely with higher N_S) are less likely to be dissipated than those of less round particles (namely with lower N_S). In other words, the less round particles are more prone to cause severe energy dissipation by collision and friction between each other. It is reasonable, since it is certainly more difficult for the less round particles (with smaller angles on the vertex) to slide and rub with each other, and easier to pack and collide tightly. Thus, the kinetic energy is more prone to be dissipated.

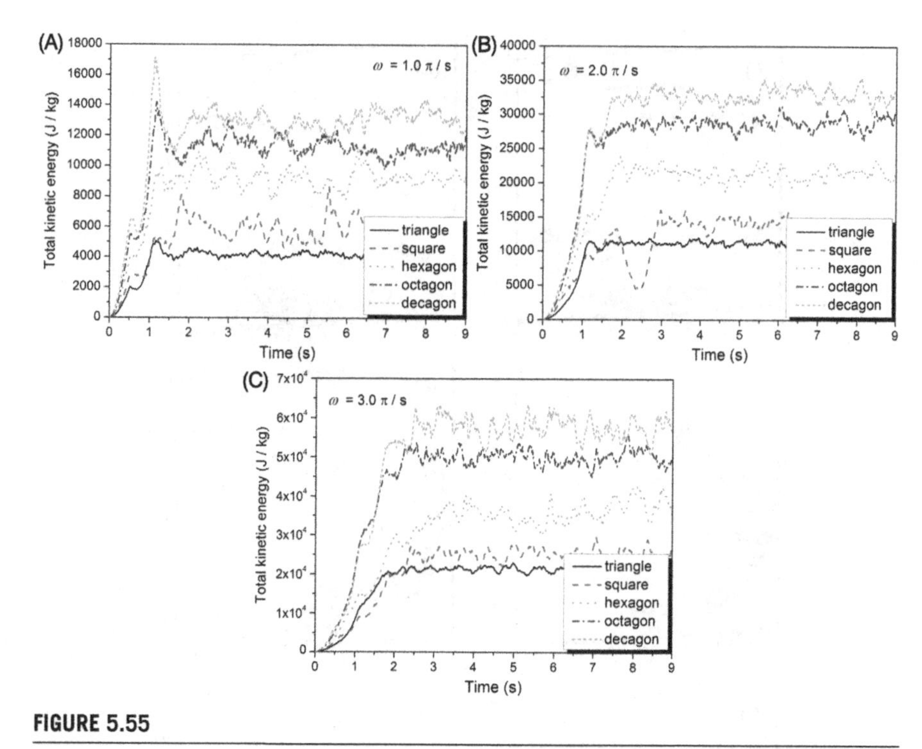

FIGURE 5.55

The evolution of the total kinetic energy with time for different particle shapes at the rotating speeds of $\omega = 1.0$ (A), 2.0 (B) and $3.0\ \pi$/s (C), respectively.

5.4.2 Cubic particle mixing in one layer

5.4.2.1 Numerical setup

As shown in Fig. 5.56, a container is used. It is a three-dimensional thin cylindrical drum with 32 mm in thickness (x-direction) and 800 mm in diameter. The cubic particles have a uniform size of 16.1 mm, covered by 5 subspheres of 3.22 mm in diameter on each edge.

The process of filling is through a freely-falling-by-gravity from a certain height above the bottom, too. After 1 s of sedimentation, let the velocity of particles be zero to keep them stationary inside the drum. After that, the drum is rotated at a constant rate of acceleration (2π/s) before getting the target rotating speeds of $\omega = 0.25$–$5\ \pi$/s, respectively. For clarity, the parameters used in simulation are listed in Table 5.8.

In the rotating cylinder, the friction of rotating walls is a factor for driving particles to rotate, whereas the friction of stationary end-wall may play a negative role in resisting the rotation of particles. Thus, in this work, two comparative categories of cases, with frictional or smooth end-walls respectively, are considered to illustrate the effect of end-walls of the thin cylindrical drum. In addition, the beds of different filling levels are analyzed to show the different features of particle motion. The

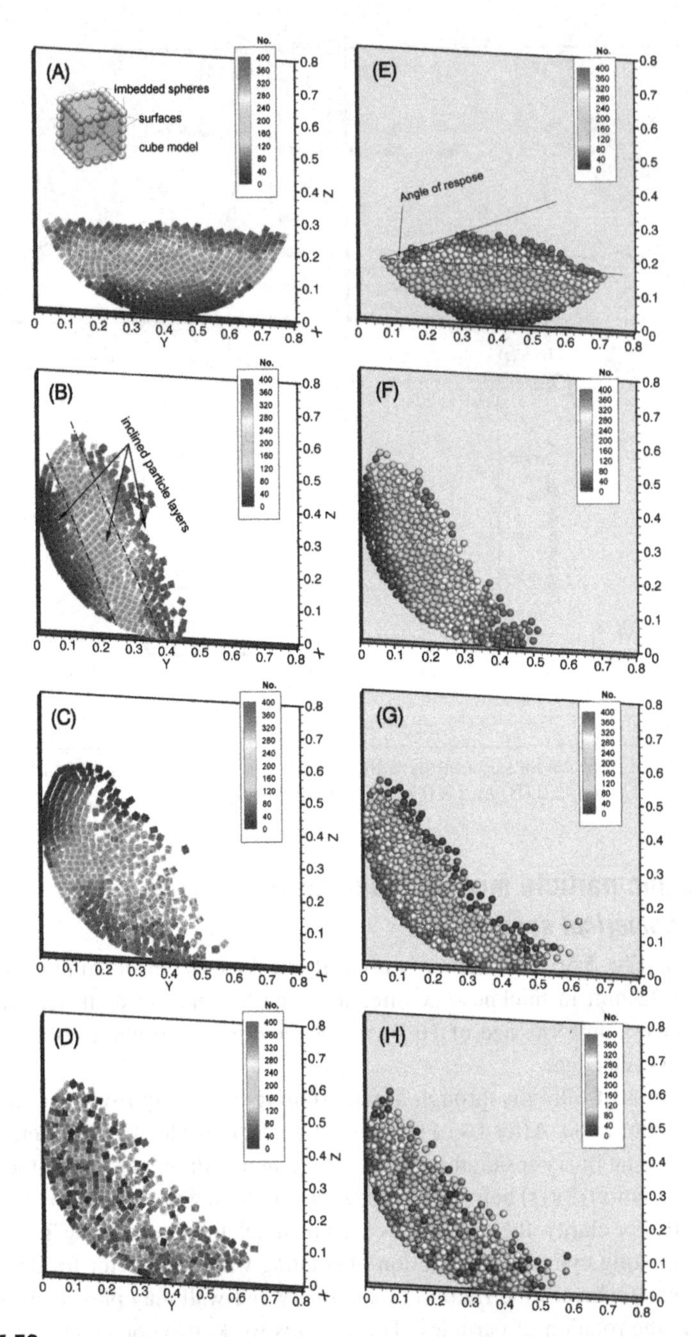

FIGURE 5.56

The process of particle motion in the drum at $t = 0$ s (A, E), 1 s (B, F), 2 s (C, G), 10 s (D, H), respectively. $N_p = 400$ and $\omega = 0.5\ \pi$/s for cubic particles (A–D) and the same for spherical particles (E–H). The inset of the part (A) is the model of cubic particle imbedded with subspheres.

Table 5.8 Parameters used in the simulation.

Radius of bed, R (mm)	400
Bed thickness, H (mm)	32
Bed rotating speed, ω rad/s^{-1}	0.25π–5π
Initially rotational acceleration, a_0 rad/s^{-2}	2π
Cube side length, l (mm)	16.1
Subsphere diameter, d (mm)	3.22
Subsphere density, ρ_s (kg/m^3)	2000
Cubic particle number N_c	400, 600, 989
Subsphere number in each cube N_s	56
Total number of subspheres N_p	24000, 36000, 59340
Stiffness factor k_c, k_r (N/m)	4×10^3, 10^5
Poisson ratio ν	0.3
Restitution coefficient e	0.95
Friction coefficient μ	0.3
Time step δt (s)	10^{-6}
Number of simulation step N_t	10^7

Table 5.9 Flow patterns.

Index	Particle No.	Shape	End-wall	ω (π/s) & flow patterns[a]
C1	400	Cubic	frictional	0.25⊙, 0.5⊙, 0.75⊙, 1⊙, 1.5⊙, 2⊘, 2.5⊘, 3⊘, 4⊘, 5⊘
C2	400	Cubic	smooth	0.5⊗, 1⊗, 2⊗, 3⊗, 4⊕, 5⊕
C3	400	Spherical	frictional	0.5⊙, 1⊙, 2⊙, 3⊘, 4⊘, 5⊘
C4	400	Spherical	smooth	0.5⊙, 1⊙, 2⊙, 3⊙, 4⊙, 5⊙
C5	600	Cubic	frictional	0.5⊙, 1⊙, 2⊘, 3⊘, 4⊘, 5⊘
C6	600	Spherical	frictional	0.5⊙, 1⊙, 2⊘, 3⊘, 4⊘, 5⊘
C7	600	Spherical	smooth	0.5⊙, 1⊙, 2⊙, 3⊙, 4⊙, 5⊘
C8	989	Cubic	frictional	0.5⊙, 1⊙, 2⊘, 3⊘, 4⊘, 5⊘

[a] *⊙ rolling, cascading, or cataracting patterns; ⊗ slipping pattern, ⊕ slumping pattern; ⊘ tumbling pattern.*

operation conditions for these categories from C1 to C8 are listed in Table 5.9 for clarity.

As the spherical particles were used in the majority of existing research work on the drum mixer, it is necessary to compare the motion behavior of cubic particles with that of the spherical particles of the same mass and volume for better understanding. After that, the effect of end-wall friction and filling level are also analyzed in the following sections.

5.4.2.2 Motion process and flow pattern comparisons

Comparing to spherical particles (for C3 at $\omega = 0.5\ \pi$/s in Table 5.9), the motion processes of cubic particles of C1 at $\omega = 0.5\ \pi$/s are shown in Fig. 5.56. At start

($t = 0$ s), the particles stay stationary at the bottom of the drum (Fig. 5.56A and E for cubes and spheres, respectively). The angle of repose of cubic particles is almost zero in the drum (Fig. 5.56A) whereas it is about 20° for spherical particles (Fig. 5.56E).

When the drum rotation begins, the particles are driven by frictional force from the front and rear-walls as well as the lateral wall. At $t = 1$ s (Fig. 5.56B and F), the spherical particles in the top flowing layer quickly flows downside because of the nonzero (20°) angle of repose. Meanwhile, the particles at the bottom are continuously driven upside, causing the particle packing layers to be inclined (Fig. 5.56B). In comparison, the cubic particles seem to respond to the rotation of the drum later than the spheres, which can be shown by the even larger slope of the particle layers for cubic particles in Fig. 5.56B than that in Fig. 5.56F for spherical particles. It indicates the lower flowability or larger flow resistance of cubic particles than the spherical particles. After $t = 10$ s (Fig. 5.56D), the cyclic motion process results in at least two phenomenological features: (1) Mixing develops between the particles, and consequently a randomly mixed state is achieved. (2) The voids inside the particle system increases (see the comparison of particle distributions at the same time points, e.g., Fig. 5.56C and G, or Fig. 5.56D and H). This effect is caused by the nonspherical shape of particles which reduces the particle bulk density and expands the particle distribution.

Moreover, Fig. 5.57 shows the results of particle motion states of C1 at $t = 5$ s with different rotation speeds ($\omega = 0.25$, 0.75, 1.5 and 3.0 π/s, respectively). With low ω (Fig. 5.57A), only a small portion of particles in the top flowing layer moves downside slowly. As the rotation speed increases (Fig. 5.57B and C), the particle velocity increases largely, and all the particles are taking rigorously cyclic motion inside the drum. In comparison, with ω larger enough, a tumbling flow pattern is formed (Fig. 5.57D), and the centrifugal motion is large enough to drive all the particles at a continuous contact with the inner wall of the drum.

Then, to show the effect of frictional end-walls, Fig. 5.58 compares the results of particle states for the cases of frictional front and rear walls (Fig. 5.58A and B for $\omega = 0.5$ and 2 π/s, respectively), or smooth front and rear walls but frictional lateral walls (Fig. 5.58C and D for $\omega = 0.5$ and 2 π/s, respectively too, see C1 and C2). It is striking to see that the particle velocity is significantly larger in the case of frictional end-walls than that of smooth end-walls, and the flow patterns are totally different. With frictional end-walls, the flow patterns are rolling at $\omega = 0.5$ π/s and cataracting or tumbling at $\omega = 2$ π/s, whereas they are slipping/slumping all with smooth end-walls. It indicates the significant effect of friction from the front and rear walls of the thin drums on particle motion pattern, since the wall friction may play a positive role in accelerating or driving particle motion and consequentially result in pure kinetic energy input for the assembly of particles.

5.4.2.3 Effect of cubic shape on velocity distribution

In the above section, the basic features of motion process have been visualized. Thereafter, it is necessary to analyze the various effects influencing on it. In this section, the effect of cubic shape on particle velocity distribution is analyzed. The data

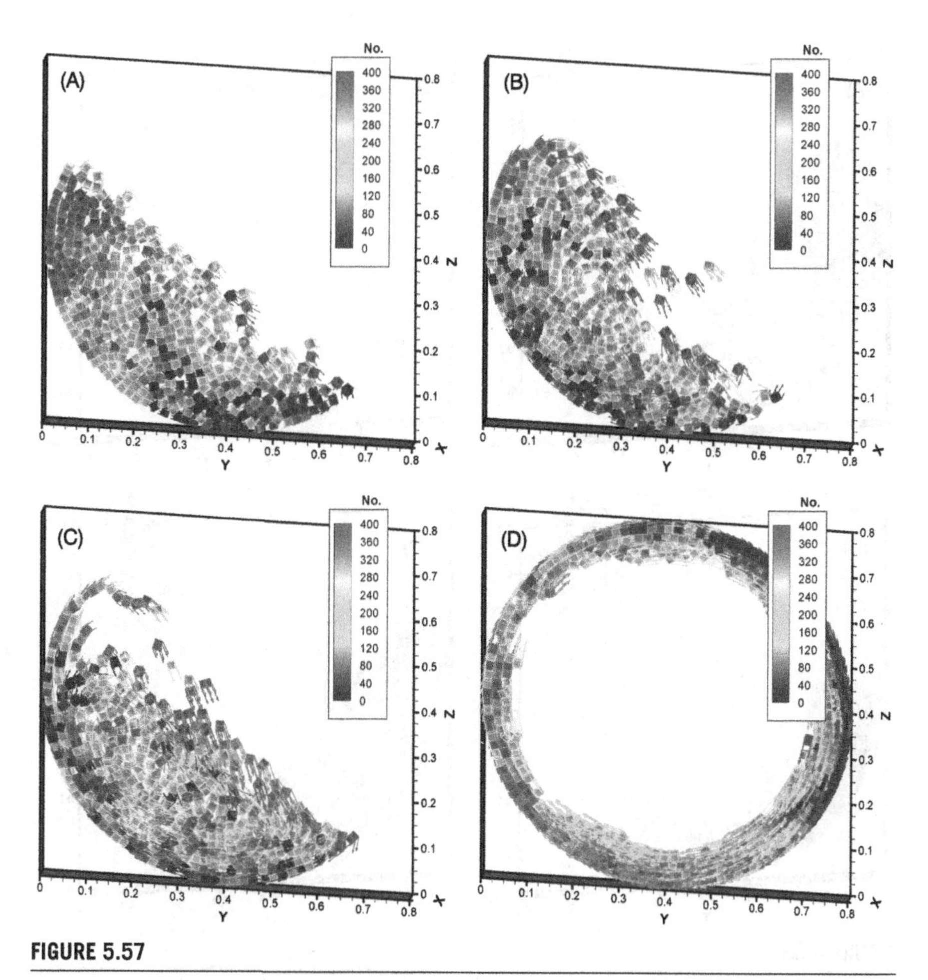

FIGURE 5.57

The process of cubic particle motion of C1 at $t = 5$ s with different rotation speeds ($\omega = 0.25$, 0.75, 1.5 and 3.0 π/s, respectively).

of all particle velocities within a time interval of $\Delta t = 0.1$ s are used, and divided into several subsets by velocity interval Δv. Then, the probability density function (PDF) is defined and calculated numerically as follows:

$$f(v) = \frac{\Delta N(v)}{N_{tot} \cdot \Delta v} \tag{5.66}$$

where $\Delta N(v)$ is the number of data points within the interval of $[v - \frac{\Delta v}{2}, v + \frac{\Delta v}{2}]$, and N_{tot} is the total number of data points in that time period.

For example, Fig. 5.59A and B show the probability density functions of particle velocity components $V_{p,x}$ and $V_{p,y}$, respectively. The velocity range is broadened when the rotating velocity of bed increases, in despite of cubic or spherical particle

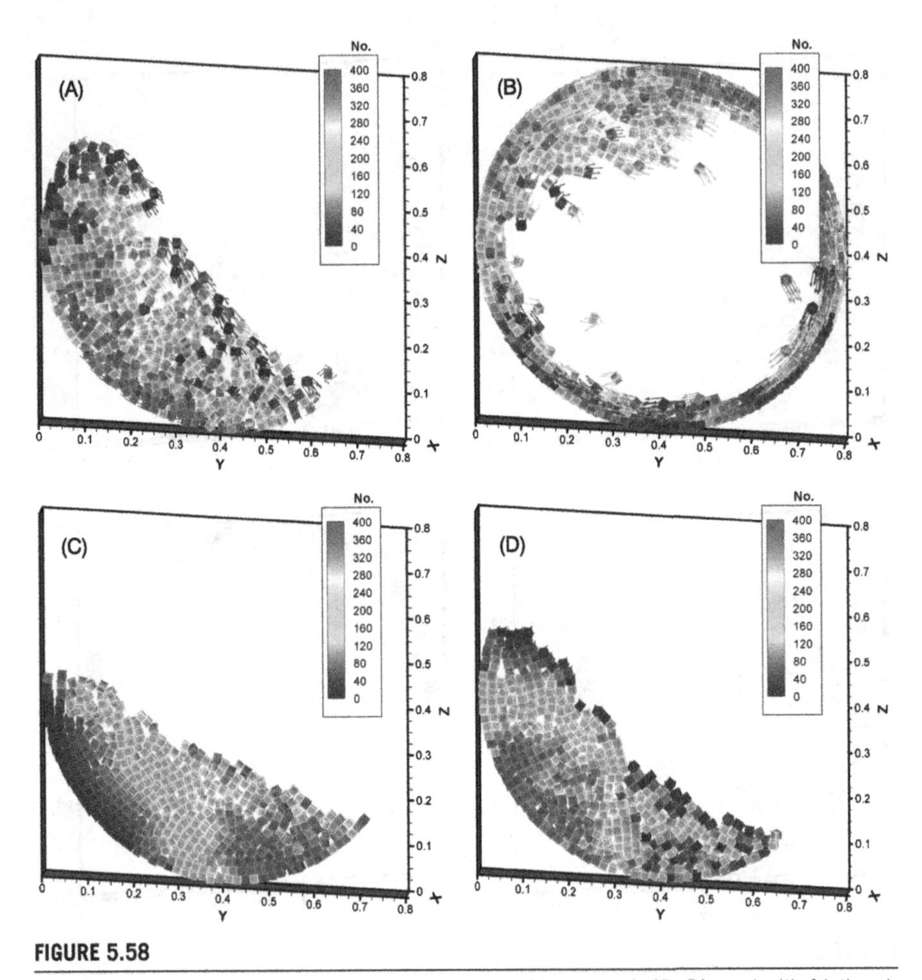

FIGURE 5.58

The states of particle distribution for $\omega = 0.5\ \pi$/s (A, C) and $2\ \pi$/s (B, D), and with frictional (A, B) and smooth (C, D) front and rear walls (see C1 and C2), respectively.

shapes. When ω is small (< 0.5, $1\ \pi$/s), the PDF profile is narrow and concaved down, whereas it is wide and concaved up when ω is large ($> 2\ \pi$/s). According to the flow pattern categories in Table 5.9, $\omega > 2\ \pi$/s corresponds to the tumbling flow patterns for all the cubic and spherical shapes. Thus, it means the velocity profiles in the tumbling flow pattern are significantly different from those not in the tumbling flow pattern. To better understand this feature and based on the PDFs, the evolutions of flatness factors of $v_{p,x}$ and $v_{p,y}$ for cases C1 and C3 are shown in Fig. 5.60. The flatness factor is defined as

$$\beta = \frac{\int (v - \bar{v})^4 f(v)dv}{[\int (v - \bar{v})^2 f(v)dv]^2} \tag{5.67}$$

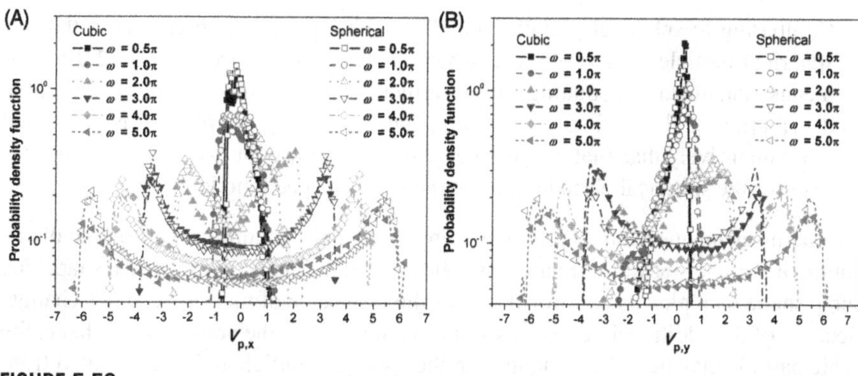

FIGURE 5.59

The probability density functions of the particle velocity components $V_{p,x}$ (A) and $V_{p,y}$ (B), respectively, with $n_p = 400$ cubic and spherical particles in C1 and C3.

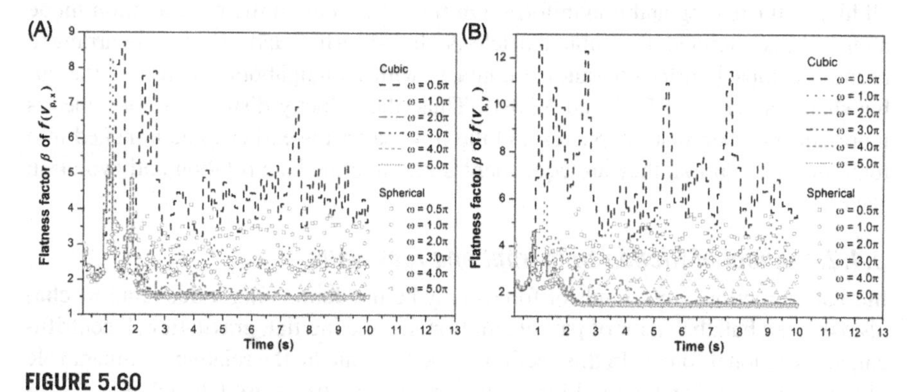

FIGURE 5.60

The evolutions of flatness factor for $v_{p,x}$ (A) and $v_{p,y}$ (B) in case C1 and C3.

The flatness factor can be regarded as a quantitative evaluation on the degree of uniformity of a random distribution. In particular, the flatness factor of the normal distribution is 3. When β is larger, the distribution deviates farther from the uniform distribution. When it is lower, the inverse becomes true.

It is noticed from Fig. 5.60 that:

- For cubic particle (Case C1 in Table 5.9), the flow pattern is evolved into tumbling motion when $\omega = 2\ \pi/s$ (see Fig. 5.58B). For spherical particle, the flow pattern is still cataracting at $\omega = 2\ \pi/s$ and has not reached the tumbling flow pattern (Case C3 in Table 5.9).
- The flatness factors β of either $v_{p,x}$ or $v_{p,y}$ are fluctuating with time when $\omega < 2\ \pi/s$ for cubic particle in C1 and $\omega <= 2\ \pi/s$ for spherical particle in C3. On the contrary, β becomes constant with time for $\omega >= 2\ \pi/s$ for cubic particle and $\omega > 2\ \pi/s$ for spherical particle, which corresponds to the tumbling flow pattern.

- Comparing to spherical particle, the β for cubic particle is higher than that for spherical particle at $\omega = 0.5\ \pi/s$, whereas the reverse is true at $\omega = 2.0\ \pi/s$. It means that, under low rotating speeds, either the time distribution or the ensemble distribution of flow velocity (both $v_{p,x}$ and $v_{p,y}$) of cubic particle is fluctuated more intensively than that of spherical particle. On the contrary, it is more uniform than for the spherical particle in the large rotation speed mode.

The mechanisms for the high fluctuating and highly nonuniform ensemble distribution of particles may be explained as follows: Comparing to the spherical shape, the cubic shape may play an additional role as the resistance of motion behavior change. Because of the additional motion resistance resulted from the nonspherical shape, the cubic particle may be more difficult than the spherical particle to be accelerated from low speed to high speed. It may be easier or more difficult for some cubic particles located in some places at some time points to be accelerated than the spherical ones. Consequently, the ensemble distribution and time distribution of particle velocity are all highly fluctuating and nonuniform. On the other side, in the high rotation mode after fully accelerated, the cubic particle is almost in-line and one-by-one arranged at more compact surface-to-surface contacts with its neighbors because of the cubic shape (see Figs. 5.57D and 5.58B). Thus, the velocity distribution of cubes is more uniform than that of spheres and the flow pattern is earlier to be transited into tumbling. In addition, they are equivalent under intermediate rotation and evolution stages.

5.4.2.4 Effect of friction from front and rear walls

In rotating motions, four kinds of forces may be the dominating influencing mechanisms on the bulk behavior of particle motion, i.e. the inertial, gravitational, centrifugal, and frictional forces. In this section, to better evaluate the relative dominant role of the gravitational and inertial forces, the Froude number is used, which is

$$Fr = \frac{\bar{v}_p^2}{g \cdot R} \tag{5.68}$$

where $\bar{v}_p$ is the ensemble-averaged particle velocity, and g is the acceleration of gravity.

The evolutions of Fr are shown in Fig. 5.61A and B for (C1, C2) and (C5, C7), respectively. For cubic particles in C1 and C2 (Fig. 5.61A), the Fr with friction from front and rear walls is much larger than that without friction. Regular periodic oscillations of Fr exist in the cases under smooth end-walls, which is the basic feature of slipping or slumping flow patterns. The stronger fluctuations of Fr take place in the cases of frictional end-walls before they reach the tumbling flow pattern with near constant Fr. The threshold value between the regular periodic oscillation and strong fluctuation of Fr for cubic particles may be about $Fr \approx 0.02$–0.03.

On the other hand, for spherical particles under smooth end-walls in C7 (Fig. 5.61B), Fr is about less than 0.01 for smooth spherical particles under the rolling/cascading/cataracting flow patterns when $\omega \leq 4\ \pi/s$. Then, the flow evolves

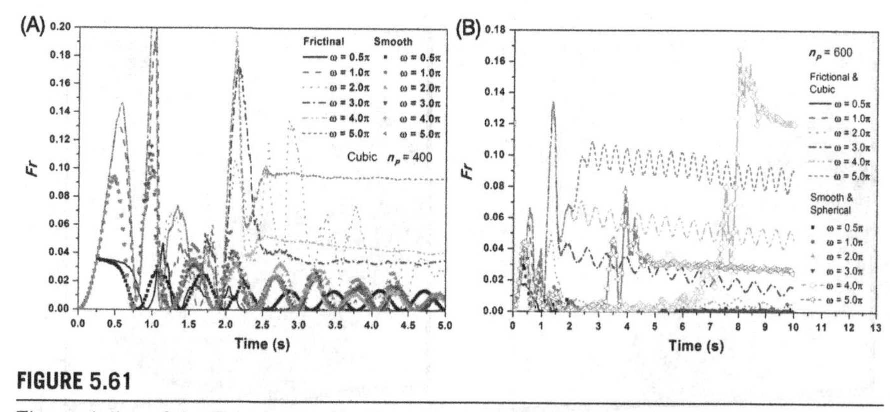

FIGURE 5.61

The evolution of the Froude number for cubic particles with and without friction from front and rear walls in C1 and C2.

into the tumbling pattern later at $\omega = 4\ \pi/\text{s}$ and quickly at $\omega = 5\ \pi/\text{s}$. In contrast, *Fr* for cubic particles under frictional end-walls in C5 are above $0.02 \sim 0.04$ (Fig. 5.61B) when $\omega \geq 3\ \pi/\text{s}$, which generally agree with the flow pattern categories in Table 5.9.

To say conclusively, as is well known, the flow pattern can be affected by a set of influencing parameters, including the particle properties and operation and configuration parameters of the drum mixer. But, subject to the particle and drum used in current study, the transition of flow patterns from slipping to cascading can be indicated by $Fr \leq 0.02$, whereas the tumbling flow pattern can be indicated by $Fr \geq 0.03$. In addition, the transition flow pattern around the cataracting pattern can be indicated by $Fr \approx 0.02 \sim 0.03$. Under the same flow conditions except the frictions from the front and rear walls, *Fr* of cubic particle is highly fluctuating and approximately larger than 0.02 when $\omega \geq 3\ \pi/\text{s}$ with frictions from the end-walls, whereas it is generally less than 0.02 even when $\omega \geq 5\ \pi/\text{s}$ without friction from the front and rear walls.

5.4.2.5 Effect of filling levels and driving modes

In this section, the cubic particle flow patterns with different bed filling levels are shown in Fig. 5.62 at $t = 5$ s and under the same rotation speed. It is seen that the typical flow pattern with $n_p = 400$ and 600 (Fig. 5.62A and B, respectively) are rolling or cascading normally. In contrast, the overloading of particles (Fig. 5.62C) will reduce the area of freely flowing region for particles, making the majority of particles contacted tightly with each other without efficient internal mixing, and consequently result in the worst mixing levels. It indicates that the overloading of particles is unnecessary, since it is power-wasting and inefficient for mixing.

Moreover, based on the evolutions of the ensemble mean velocity (Fig. 5.63A), the flow patterns are developed fully and kept stable after $t = 3$ s. Thus, it is useful to compute the time averaged value of particle velocity after $t = 3$ s as follows:

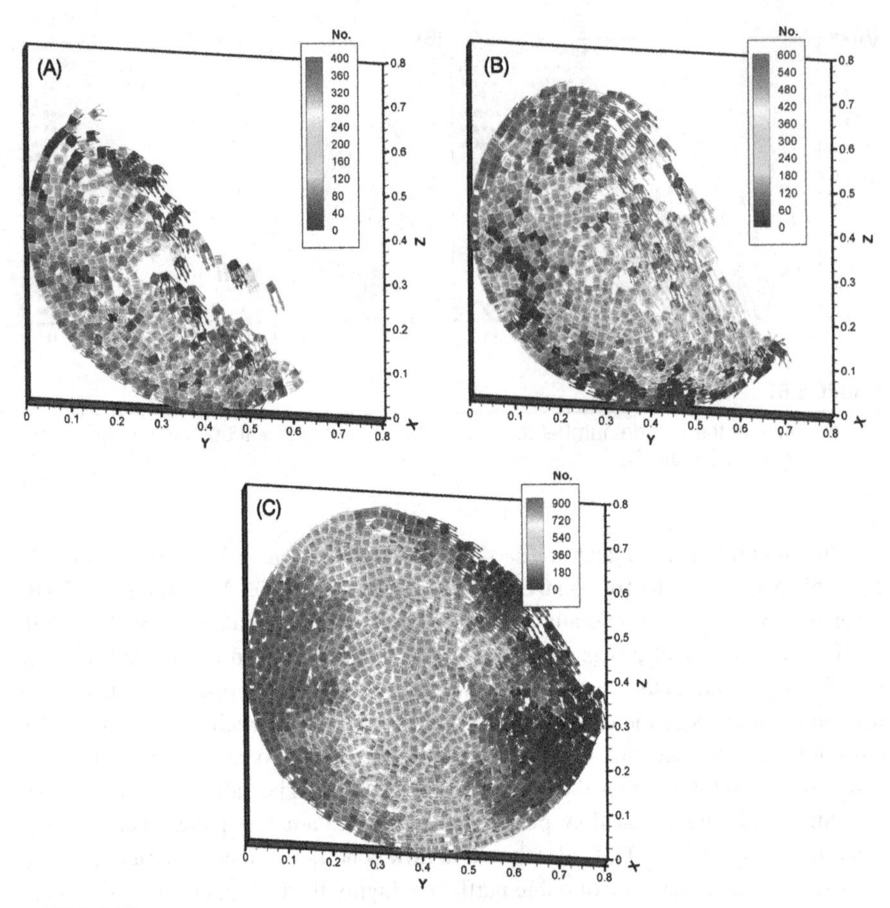

FIGURE 5.62

The processes of cubic particle motion for cases C1, C5 and C8 at $t = 5$ s corresponding to different filling levels ($n_p = 400$ (A), 600 (B) and 989 (C), respectively).

$$\langle |v_p| \rangle = \frac{1}{T_s - 3} \int_3^{T_s} \bar{v}_p dt \tag{5.69}$$

$$\bar{v}_p = \frac{1}{N_p} \sum_{i=1}^{N_p} |\boldsymbol{v}_{p,i}| \tag{5.70}$$

to study the motion features under different operation conditions.

Fig. 5.63 shows the $\langle |v_p| \rangle$ for all the cases from C1 to C8. It is clearly seen that:

1. For all the cases with friction from the end-walls, $\langle |v_p| \rangle$ is almost linearly dependent on the bed rotation velocity ω.
2. For all the cases without friction from the end-walls, when ω is small ($\leq 4\ \pi$/s), $\langle |v_p| \rangle$ is almost weakly dependent on ω, or independent of ω at $n_p = 400$ for

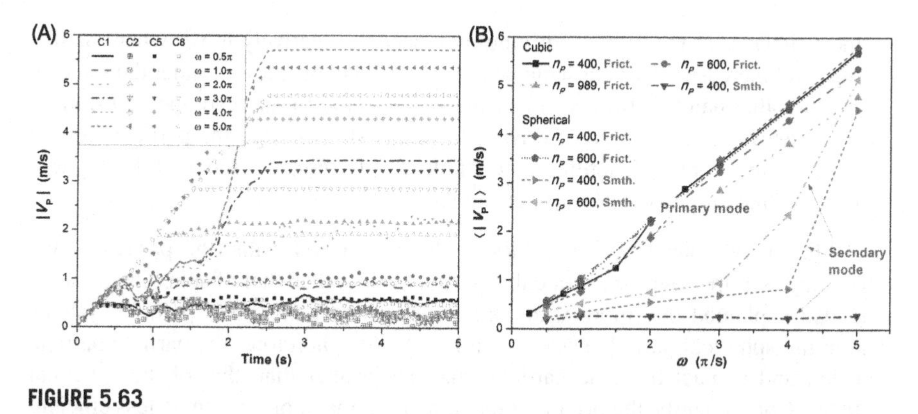

FIGURE 5.63

(A) Time evolution of ensemble mean particle velocity $|v_p|$ in C1, C2, C5, C8. (B) The time averaged mean velocity $\langle |v_p| \rangle$ after 3 s for all cases.

cubic particles. When $\omega = 5\ \pi/\text{s}$, $\langle |v_p| \rangle$ increases significantly, or jumps onto a high level suddenly. The relevant mechanisms can be explained by the primary and secondary modes of driving as follows:

- Primary mode of driving: It is a mode with frictional front and rear walls. In this mode, almost every particle, both cubic and spherical, are straightforwardly contacted or continuously driven by either the front or the rear wall. Under this condition, each particle's motion is rigorously driven and influenced by the wall friction. The particle velocity is easy to be signified when the bed rotation velocity increases. Thus, the particle motion is almost linearly and directly dependent on the bed rotation speed.
- Secondary mode of driving: It is a mode with smooth front and rear walls, and only the lateral wall is frictional for driving particles. In this mode, only a small portion of particles contacting the lateral wall are directly driven by the frictional wall. Other portions of particles are indirectly driven by the lateral wall through particle-particle interactions. Because of the indirect transfer of driving force, there is a delay for particle motion to respond to the variation of bed rotation speed. Thus, when the rotation speed ω is low, particle velocity is only weakly dependent on or independent of ω. On the other hand, when ω is large enough, the particles suffer a sudden flow regime transition to jump directly into the tumbling pattern.

3. With the end-wall friction, $\langle |v_p| \rangle$ becomes smaller when the filling level increases. On the contrary, without the end-wall friction, the reverse becomes true. The mechanisms are explained as follows: based on the above considerations, in the primary mode, the majority of particles are not always driven by the rigorous driving forces when the filling level increases. Because the driving force of particles is the friction from the front, rear and lateral walls. Near the core of the drum, it is always weak, whereas it is rigorous far from the origin of the bed. In other words,

the efficiency of the primary driving may become relatively lower when the filling level increases. As a consequence, $\langle |v_p| \rangle$ will become lower under the same ω. On the other hand, in the secondary mode, when the filling level increases, more particles will play the role of driving force transfer through the particle-particle collision and contact. Therefore, the efficiency of secondary mode of driving will become higher, and then $\langle |v_p| \rangle$ gets larger.

Additionally, considering $n_p = 400$ with smooth end-walls, the particle shape effect may be important. $\langle |v_p| \rangle$ of cubic particle is less than that of spherical particle. It is also explained that more voids are formed within the cubic particle systems than in the spherical particle system (see Fig. 5.56). Therefore, the particle-particle collision and contact of cubic particles may be weaker than that of the spherical particles. Consequently, the driving force transfer in the cubic system is less efficient than in the spherical system. Then, $\langle |v_p| \rangle$ of cubic particle is less than that of spherical particle.

5.4.3 Tetrahedral particle mixing in one layer tumbler

In this section, a drum of diameter $D = 800$ mm is filled with regular tetrahedrons of side $a = 16$ mm. The initially filling state (e.g. Fig. 5.64A) is formed by a freely falling process of all the particles from a certain height above the bottom of the drum. Then, a common initially packing state of particles is generated. After that, the drums rotate at the same acceleration ($2.0\ \pi/s^2$) from the same initial state to different target speeds, ranging from $\omega = 0.25$–$5\ \pi/s$ (Table 5.10). Thus, the drums in all the cases have the common speed of acceleration, and different target speeds in steady rotation.

Two categories (category A with $N_p = 1449$ and category B with $N_p = 924$) of different filling levels of tetrahedral particles are studied, with the drum rotated at $\omega = 0.25, 0.5, 1, 2, 3$ for category A and $\omega = 0.5, 1, 2, 3, 4, 5$ for category B respectively. To compare the behaviors of tetrahedrons with spheres, the rotating drums filled with the same number of spheres of same mass and equivalent diameter are also simulated as referential cases. The parameters are all listed in Table 5.10.

5.4.3.1 *The observation of mixing of tetrahedrons*

At first, Fig. 5.64A shows the initial static packing state which is the start point of rotation. The particles are stratified layer by layer. After about 10 s of rotation, the mixing states of $N_p = 1449$ tetrahedrons in the drums for category A are shown in Fig. 5.64B–F, respectively. In contrast, Fig. 5.64G–H shows the mixing of spheres at the speeds of $\omega = 2$ and $3\ \pi/s$, respectively. At $\omega = 0.25\ \pi/s$ (Fig. 5.64B), the particles are almost unmixed between the stratified structures. For $\omega = 0.5$–$2\ \pi/s$ (Fig. 5.64C–E), the tetrahedrons are better mixed in the rolling and cascading flow states. At $\omega = 3\ \pi/s$ (Fig. 5.64F), a tumbling/centrifugal flow state of the tetrahedrons is established. In contrast, the flow states of spheres (Fig. 5.64G and H) are also the same with those of the tetrahedrons respectively.

In a similar manner, the mixing of $N_p = 924$ tetrahedrons for category B are shown in Fig. 5.65A–F for $\omega = 0.5$–$5\ \pi/s$, respectively. It is noticed that the flow

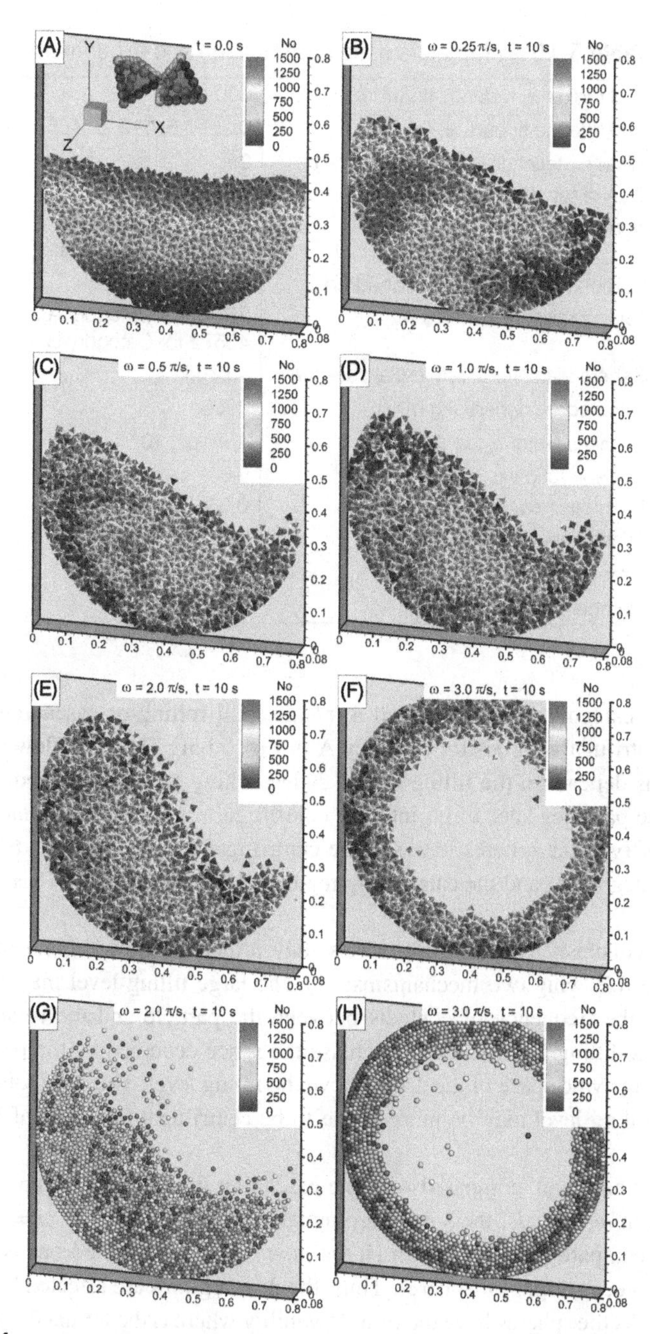

FIGURE 5.64

The initial packing state at $t = 0$ s (A) and the mixing states of tetrahedrons at $t = 10$ s in category A for $\omega = 0.25$ (B), 0.5 (C), 1 (D), 2 (E), 3 π/s (F), respectively. The inset in (A) is a schematic model of the tetrahedron. The mixing states of spheres for $\omega = 2$ (G), 3 π/s (H).

Table 5.10 Parameters used in simulation and experiment.

Parameter	Value
Bed radius $R \times$ depth W, (mm)	800 × 32
Bed rotating speed, ω, (π/s^{-1})	0.25, 0.5, 1, 2, 3, 4, 5
Initially rotational acceleration, β rad/s^{-2}	2π
Side of tetrahedron a, (mm)	16
Number of tetrahedrons N_p	924 for Category A, 1449 for Category B
Number of sphere in a tetrahedron	28
Total number of spheres N_p	25872 for Category A, 40572 for Category B
Sphere diameter in referential cases d_e	13.764
Tetrahedron density (kg·m^{-3})	2000
Stiffness factor k_c, k_r (N/m)	4×10^3, 10^5
Poisson ratio ν	0.3
Restitution coefficient e	0.95
Friction coefficient μ	0.3
Time step δt (s)	10^{-6}
Number of simulation step N_t	10^7

regimes for category B at $\omega = 3$ and 4 π/s are still rolling or cascading, whereas they are centrifugal regimes for category A in Fig. 5.65F. Thus, the flow regimes of tetrahedrons depend on the filling levels. A less filling level of tetrahedrons seems to cause the particles later to go into the centrifugal regime. The similar trend can also be observed for spheres, e.g. see the centrifugal regime in Fig. 5.64H at $\omega =$ 3 π/s for category A, and the cataracting regime in Fig. 5.65G at $\omega = 3$ π/s for category B.

To say conclusively, the filling levels may affect the flow regime of particles through the following two mechanisms: (1) The large filling level may lead to the increase of the frequency of contacts between the particle and the rotating walls. (2) The large filling level may also increase the space occupied by the particles and reduce the free void space of the drum above the filling level. Thus, the rotation drum of a large filling level may be more prone to be centrifugal than that of less filling level.

On the other hand, compared with the cascading flow regime of tetrahedrons in Fig. 5.65D at $\omega = 3$ π/s, the sphere are upcasting in Fig. 5.65G at $\omega = 3$ π/s, too. Moreover, compare Fig. 5.65E with H, the flow regime of tetrahedrons is cascading whereas it is centrifugal for spheres. Thus, the flow regimes are affected by the particle shapes. As the spheres have the best flowability whereas the tetrahedrons have the worst, it is harder for tetrahedrons to go into the centrifugal regimes than for spheres. In other words, it may be favorable for the particles of less sphericity to keep steady regimes, especially to maintain the working regime of cascading.

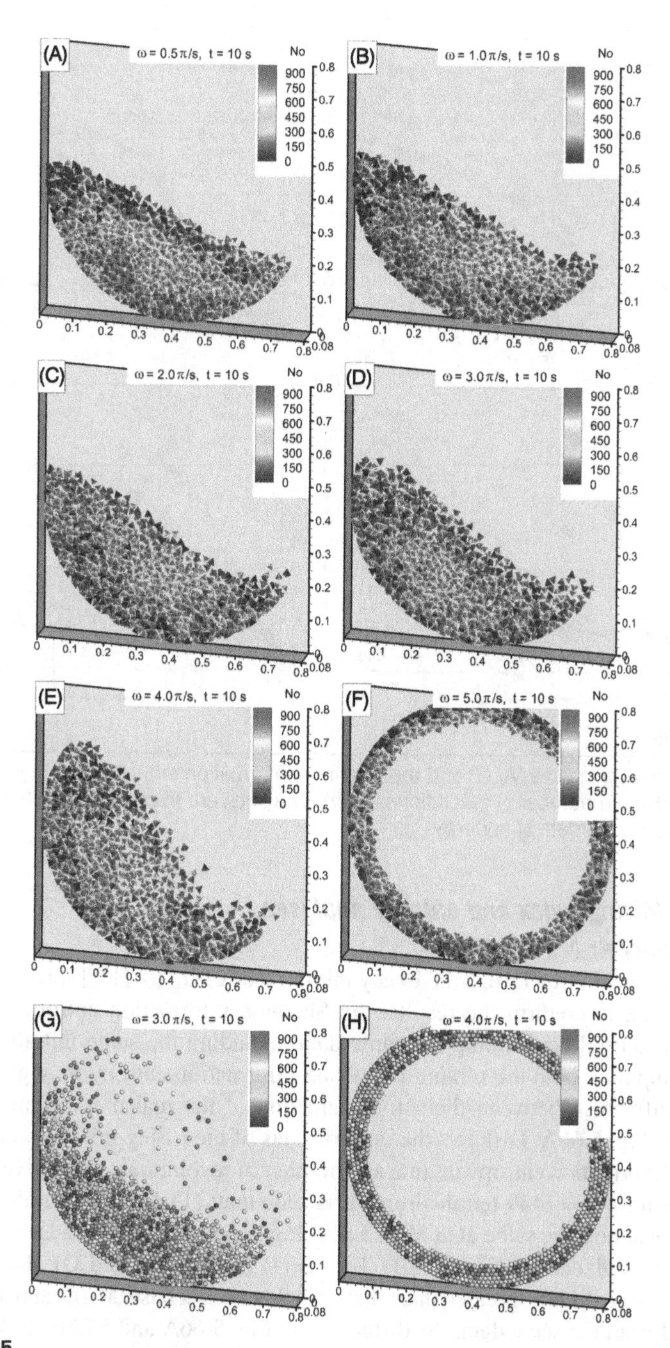

FIGURE 5.65

The mixing states of tetrahedrons at $t = 10$ s in category B for $\omega = 0.5$ (A), 1 (B), 2 (C), 3 (D), 4 (E), 5 (F) π/s, respectively. The mixing states of spheres for $\omega = 3$ (G), 4 π/s (H).

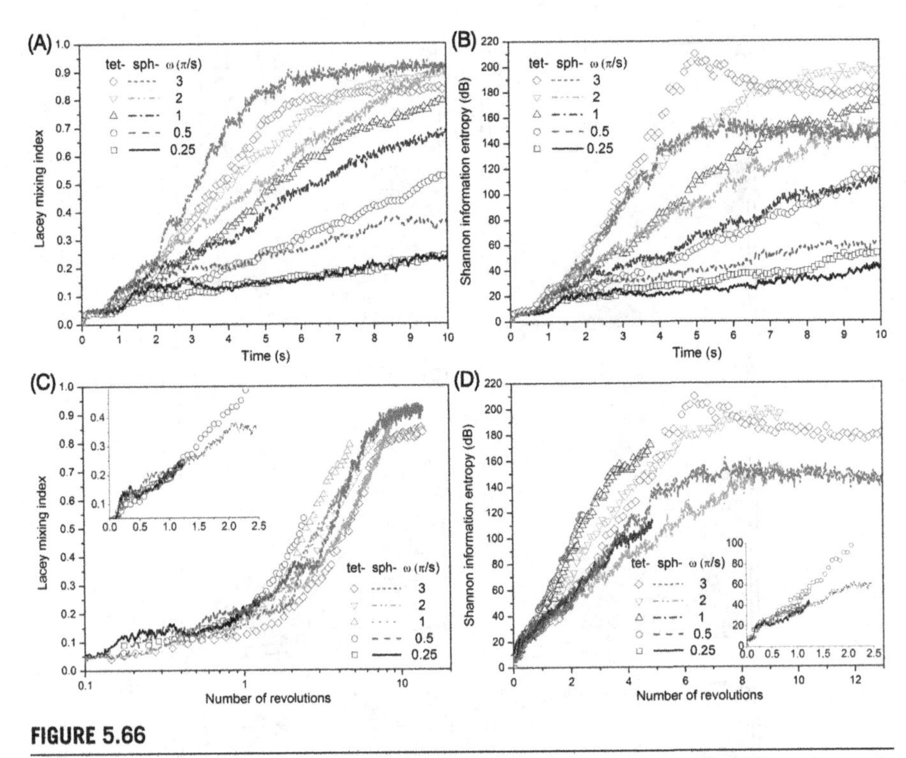

FIGURE 5.66

The Lacey mixing indices (A, C) and the Shannon information entropy (B, D) for category A in time (A, B) and number of revolutions (C, D). The lines are the data of tetrahedrons, and the scatters are the data of spheres.

5.4.3.2 Mixing index and entropy analyses

For category A: $N_p = 1449$

To quantify the mixing levels, the Lacey mixing index (Eq. (5.21)) [300] is used here.

Moreover, to confirm the results, the Shannon information entropy (Eq. (5.5)) [297] is also used here. Using the same data, including the same initially unmixed states, computing both the mixing index and information entropy may give insights into the difference between them for evaluation of the real particle mixing level. Therefore, Fig. 5.66A–D shows the comparisons of the Lacey mixing index and the Shannon information entropy in time and number of revolutions, respectively, for category A with $N_p = 1449$ tetrahedrons. It is seen that: (1) $M_l(t)$ of tetrahedrons and spheres are almost the same at $\omega = 0.25\ \pi/\mathrm{s}$ (Fig. 5.66A and C), because the particles are almost unmixed (see Fig. 5.64B). By $M_E(t)$ (Fig. 5.66B and D), the mixing of tetrahedrons is slightly greater than that of spheres. Because $M_l(t)$ and $M_E(t)$ are computed from the same data, the difference in Fig. 5.66A and 5.66B (or Fig. 5.66C and D) for $\omega = 0.25\ \pi/\mathrm{s}$ may indicate the difference between the mixing index and entropy for mixing evaluation. $M_l(t)$ is based on the viewpoint of fluctuation of particle concentration and indicates the overall uniformity of dispersion. $M_E(t)$ is based

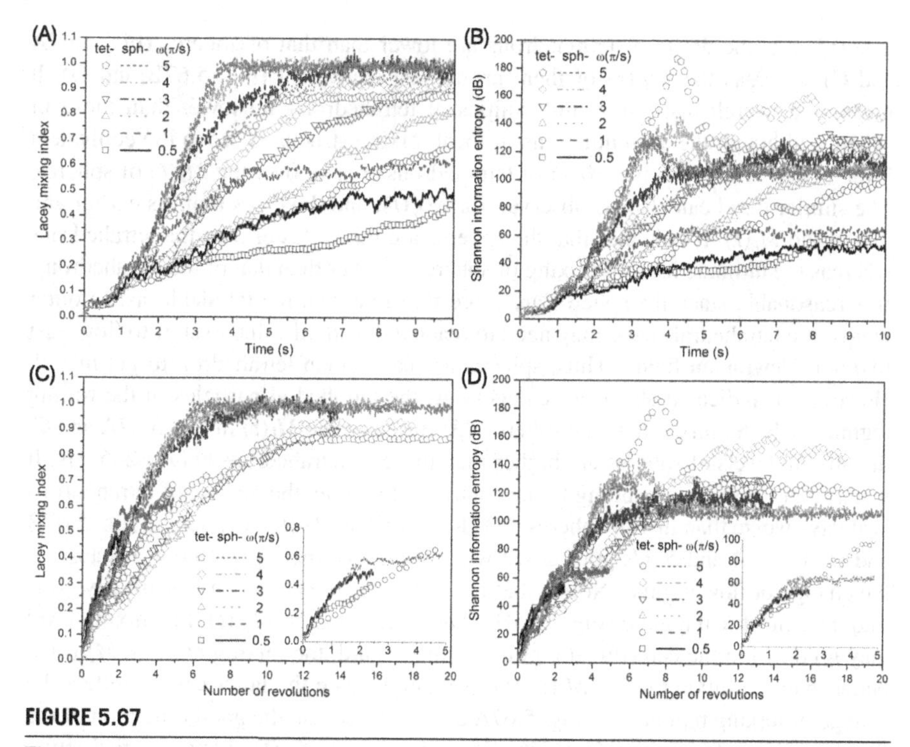

FIGURE 5.67

The Lacey mixing indices (A, C) and the Shannon information entropy (B, D) for category B in time (A, B) and number of revolutions (C, D). The lines are the data of tetrahedrons, and the scatters are the data of spheres.

on the accumulation of local mixing structure. Thus, when $M_l(t)$ is the same between the data of tetrahedrons and spheres, $M_E(t)$ is still not necessary the same. It means that, under the same level of overall dispersion, the local mixing structure may still differ from case to case. (2) For $0.5 \leq \omega \leq 2\ \pi/s$, the $M_l(t)$ of tetrahedrons in both time, Fig. 5.66A, B, and number of revolutions, Fig. 5.66C, D, are greater than the $M_l(t)$ of spheres. So are the $M_E(t)$. Thus, both mixing index and entropy indicate the same conclusion that the tetrahedral shape is better for mixing than the spherical shape in the rolling and cascading regimes. It would be caused by that the interactions between the tetrahedrons are more complicated than those between the spheres. (3) For $\omega = 3\ \pi/s$, the inverse becomes true for the $M_l(t)$ (Fig. 5.66A and C). The tetrahedrons have lower $M_l(t)$ than the spheres. It means that, in the centrifugal regime, the spherical spheres are better uniformly distributed near the wall of the rotating drum than the tetrahedrons. However, for the mixing entropy in Fig. 5.66B and D, the local mixing of tetrahedral particles is still better than that of spheres.

For category B: $N_p = 924$

Moreover, for category B, the mixing index and entropy are shown in Fig. 5.67A and B in time and Fig. 5.67C and D in revolutions, respectively. It is found that: (1) At

$\omega = 0.5\ \pi/\text{s}$, the $M_I(t)$ of tetrahedrons are lower than that of spheres (Fig. 5.67A and C), whereas the $M_E(t)$ of them are almost equivalent (Fig. 5.67B and D). It means the overall dispersion uniformities of tetrahedrons are lower than those of spheres, although the accumulations of local mixing structures are in the equivalent level. (2) At $\omega = 1\ \pi/\text{s}$, the $M_I(t)$ of tetrahedrons crosses over the $M_I(t)$ of spheres. The similar trend can also be observed for $M_E(t)$, and the cross point is earlier and that of the $M_I(t)$. It indicates that the spheres are mixed faster than the tetrahedrons whereas the ultimate level of mixing of spheres is lower than that of the tetrahedrons. It is reasonable since the spheres are more flowable than the tetrahedrons. In other words, the tetrahedral shape may need to reach a threshold value to start to flow, just like non-Newtonian fluids. Thus, spheres are easier than tetrahedron to get mixed. However, as indicated above for category A, the tetrahedral particles in the rolling regime are better mixed than the spheres. (3) Based on the $M_I(t)$ in Fig. 5.67A and C, the mixing levels of spheres are higher than those of tetrahedrons at $\omega = 2\text{–}5\ \pi/\text{s}$. It means that, from the cascading to the centrifugal regime, the mixing of tetrahedrons is always worse than that of spheres. (4) However, the $M_E(t)$ of sphere in Fig. 5.67B and D at $\omega = 4$ and $5\ \pi/\text{s}$ goes down when it reaches the highest point because of the change of flow regime. Notice that for spheres at $\omega = 4\ \pi/\text{s}$, it has already gone into the centrifugal regime (Fig. 5.65H). Thus, indicated by $M_E(t)$, the mixing level may go down in the centrifugal regime. In this regard, the mixing entropy $M_E(t)$ is better than the mixing index $M_I(t)$ for evaluation since the latter cannot reflect the change of mixing regime (see Fig. 5.67A and C). Based on the good reflection of the flow regime evolution by $M_E(t)$, it is considered that $M_E(t)$ shows the real situation of mixing states. (5) Compared with Fig. 5.66 for the most useful flow regimes from rolling to cataracting, the mixing of nonspherical particles is better than that of spherical particles at large filling levels. In contrast, still for the same flow regimes, the mixing of nonspherical particles is lower than that of the spherical particles at small filling levels. Therefore, the influences of filling level and particle shape as well as rotating speed on mixing levels are correlated together.

5.4.3.3 Probability density function of velocity

In this section, the probability density function (PDF) is calculated as follows to explore the characteristics of velocities ($u_{p,x}$, $u_{p,y}$, $\omega_{p,z}$):

$$f(\phi,t)|_{\Delta T} = \left.\frac{\delta N_\phi(\tau)}{N_\phi \Delta T}\right|_{\tau\in[t-\frac{1}{2}\Delta T,\, t+\frac{1}{2}\Delta T]} \tag{5.71}$$

where N_ϕ is the total number of data points of ϕ, and $\delta N_\phi(t)$ is the number of data points located within time interval $[t-\frac{1}{2}\Delta T, t+\frac{1}{2}\Delta T]$. Here $\phi = u_{p,x}, u_{p,y}, \omega_{p,z}$, and $\Delta T = 1$ s for category A and 0.1 s for category B, respectively.

Fig. 5.68 shows the $f(u_{p,x})$ (A, D), $f(u_{p,y})$ (B, E), $f(\omega_{p,z})$ (C, F), with $\Delta T = 1$ s for category A at $t = 1$ s (A, B, C), and $t = 10$ s (D, E, F), respectively. At $t = 1$ s, the $f(u_{p,x})$ (or $f(u_{p,y})$) of tetrahedrons and spheres do not have evident differences (Fig. 5.68A and B) except for $f(u_{p,x})$. It confirms that the flowability

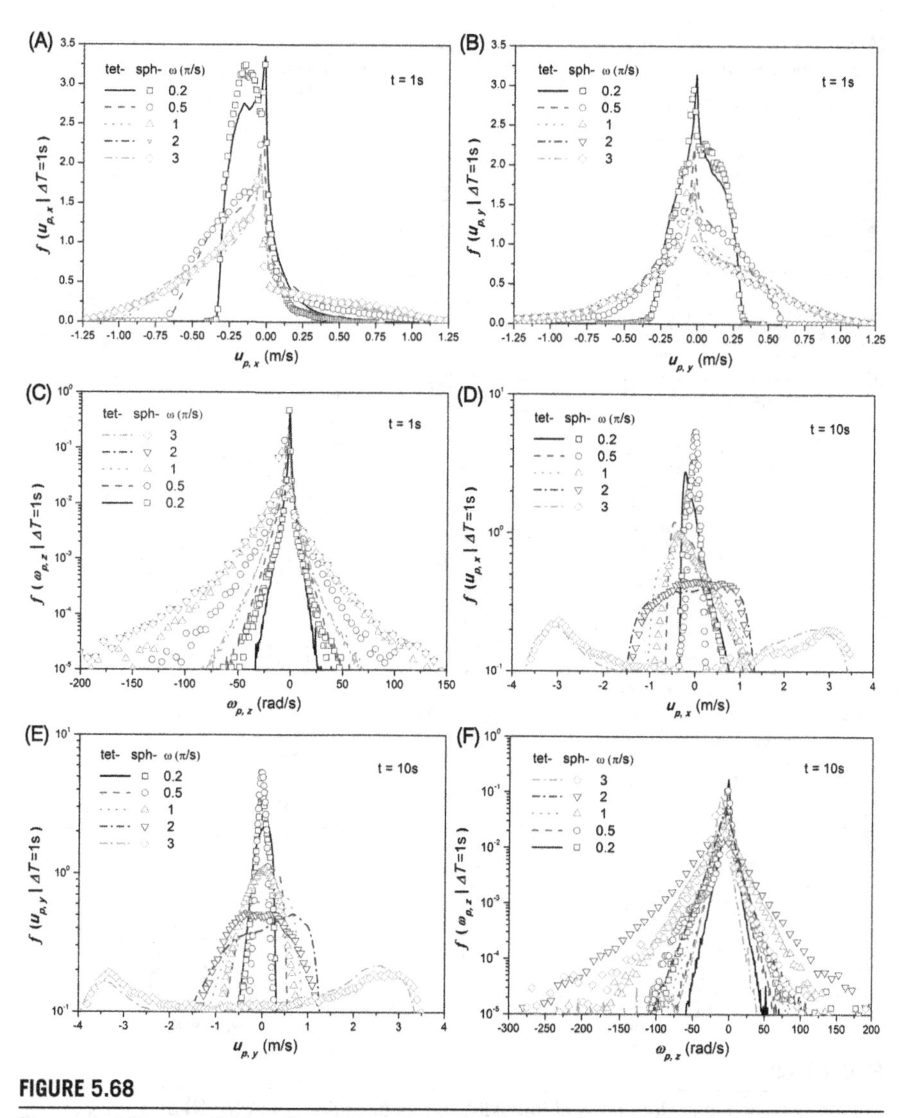

FIGURE 5.68

The probability density functions of translational velocity ($f(u_{p,x})$ in (A, D) and $f(u_{p,y})$ in (B, E)) and rotational velocities ($f(\omega_{p,z})$ in (C, F)) for category A at $t = 1$ s (A, B, C) and 10 s (D, E, F), respectively. Here, $\Delta T = 1$ s.

of tetrahedral particles is lower than that of spherical particles, since the $f(u_{p,x})$ of tetrahedron always has the maximum at $u_{p,x} = 0$, whereas the $f(u_{p,x})$ of sphere has the maximum about $u_{p,x} \approx -0.15$ m/s (Fig. 5.68A). Moreover, the $f(\omega_{p,z})$ of tetrahedrons and spheres are quite different (Fig. 5.68C). The spheres have larger $f(\omega_{p,z})$ at larger $|\omega_{p,z}|$, whereas the tetrahedrons always have lower $f(\omega_{p,z})$ at nonzero $|\omega_{p,z}|$. It means that the rotational motions of tetrahedrons and spheres at

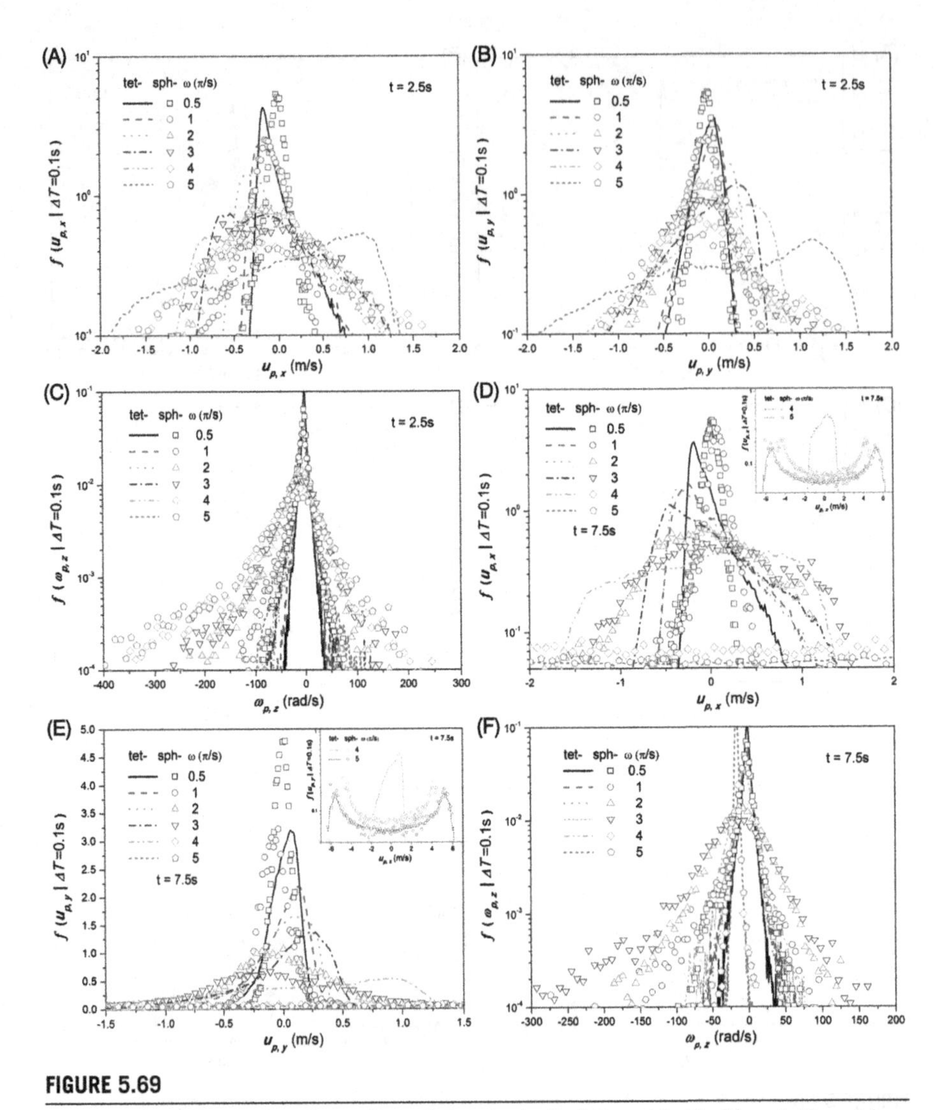

FIGURE 5.69

The probability density functions of translational velocity ($f(u_{p,x})$ in (A, D) and $f(u_{p,y})$ in (B, E)) and rotational velocities ($f(\omega_{p,z})$ in (C, F)) for category B at $t = 2.5$ s (A, B, C) and 7.5 s (D, E, F), respectively. Here, $\Delta T = 0.1$ s.

the early stage are quite different (Fig. 5.68F). In other words, it confirms again the previous assumption that the tetrahedrons are the most nonspherical particles which have the lowest flowability. This is also true at $t = 10$ s.

However, at $t = 10$ s, the $f(u_{p,x})$, $f(u_{p,y})$ or the translational motions of the tetrahedrons are quite different from those of spheres at almost all rotation speeds. At low rotating speeds ($\omega \leq 1$), the peak values of $f(u_{p,x})$ and $f(u_{p,y})$ of spheres

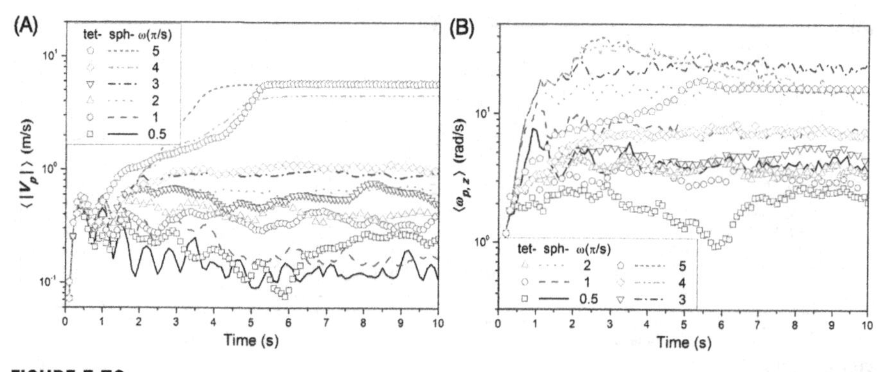

FIGURE 5.70

The mean values of particle velocity ($\langle|\boldsymbol{V}_p|\rangle$) (A) and rotational velocity ($\langle\omega_{p,z}\rangle$) (B) for category A, respectively.

are larger and more symmetrical than those of $f(u_{p,x})$ and $f(u_{p,y})$ of tetrahedrons, respectively.

In addition, in Fig. 5.69, the time interval ΔT is reduced from 1 s to 0.1 s to see the PDF characteristics of instantaneous velocities. Similar results of $f(u_{p,x})$, $f(u_{p,y})$, $f(\omega_{p,z})$ can also be obtained for category B. For translational motions, the velocity PDFs of tetrahedrons are more widely distributed than that for the spheres, with larger skewness, which means the velocities of tetrahedrons are more asymmetrically distributed throughout the bed than for spheres. For rotational velocity, the PDFs of tetrahedrons are narrower than that of spheres. In other words, most of tetrahedrons have smaller rotational velocities than the spheres. In addition, at $\omega = 5\ \pi$/s, the $f(\omega_{p,z})$ of tetrahedrons and spheres both become narrower, which are caused by the centrifugal regime with almost near zero rotational rotation of both tetrahedral or spherical particles.

5.4.3.4 Mean and variance of velocity

With the aid of PDFs, more characteristics of particle velocity can be quantified. For example, the mean and variance of velocity can be directly computed, respectively, as follows:

$$\langle\phi\rangle(t) = \int_{-\infty}^{+\infty} f(\phi, t)\phi(t)d\phi \tag{5.72}$$

$$\sigma(\phi) = \int_{-\infty}^{+\infty} f(\phi, t)(\phi(t) - \langle\phi\rangle)^2 d\phi \tag{5.73}$$

where ϕ is $|\boldsymbol{V}_p|$ ($|\boldsymbol{V}_p| = (V_{p,x}^2 + V_{p,y}^2)^{1/2}$) or $\omega_{p,z}$.

Taking category B for example, Figs. 5.70A and B show $\langle|\boldsymbol{V}_p|\rangle$ and $\langle\omega_{p,z}\rangle$, respectively. It is found that: (1) At low rotating speeds ($\omega = 0.5$, $1\ \pi$/s), $\langle|\boldsymbol{V}_p|\rangle$ of tetrahedrons is equivalent or greater than that of spheres. (2) In contrast, at larger

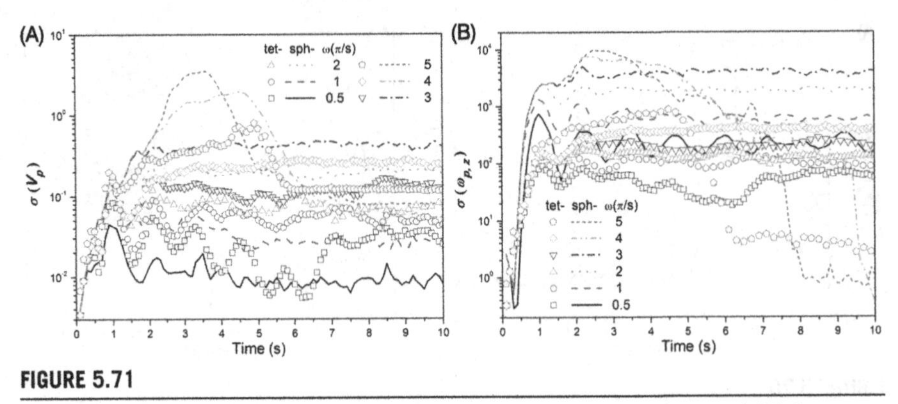

FIGURE 5.71

The variances of particle velocity ($\sigma(\boldsymbol{V}_p)$) (A) and rotational velocity ($\sigma(\omega_{p,z})$) (B) for category A, respectively.

rotating speeds ($\omega \geq 3\ \pi$/s), $\langle|\boldsymbol{V}_p|\rangle$ of tetrahedrons is lower than that of spheres. (3) Additionally, $\langle\omega_{p,z}\rangle$ of tetrahedrons is always lower than that of spheres. The reasons would be that: (i) The rolling of spherical particles near the wall, which is caused by the friction from the drum, can evidently increase the rotational velocity. Thus, the spherical particles of the greater flowability always have larger rotational velocity than the tetrahedral particles. (ii) On the other hand, the greater rolling of particles near the wall may sometimes attenuate the efficiency of transfer of kinetic energy from the rotating wall to the spherical particle assembly. In other words, the tetrahedral particle assembly would have larger inner interaction than the spherical particle assembly. Thus, when the drum rotates, the friction from the wall of the drum would be more efficiently transferred throughout the tetrahedral particle assembly than throughout the spherical particle assembly, especially at small rotating speeds when there are sufficient time for energy transfer through the particle-particle interaction. As a result, the mean level of translational speed of tetrahedral particles is greater than that of the spherical particles at lower rotating speeds.

Finally, Fig. 5.71 shows the corresponding variances of $\boldsymbol{V}_p$ and $\omega_{p,z}$. The trends seen from Fig. 5.71 are similar to those observed from Fig. 5.70 on the mean values of $\boldsymbol{V}_p$ and $\omega_{p,z}$. These mean values and the variances of $\boldsymbol{V}_p$ for tetrahedrons are greater than those for spheres at lower rotating speeds ($\omega < 2\ \pi$/s), whereas the opposite is true at larger rotating speeds ($\omega \geq 3\ \pi$/s). The variance of $\omega_{p,z}$ of tetrahedrons is always lower than that of spheres. The mechanisms are also similar as explained above. In addition, when the flow regime gets into the centrifugal regime, the mean rotational velocity and variances of translational and rotational velocity may go down because of the uniform low level of rotational velocity for all particles in the centrifugal regime.

5.4.4 Cubic particle mixing in 3D cylinder

As shown in Fig. 5.72, the container is a three-dimensional cylinder of 320 mm in diameter (y and z directions) and 640 mm in depth (x direction or axial direction,

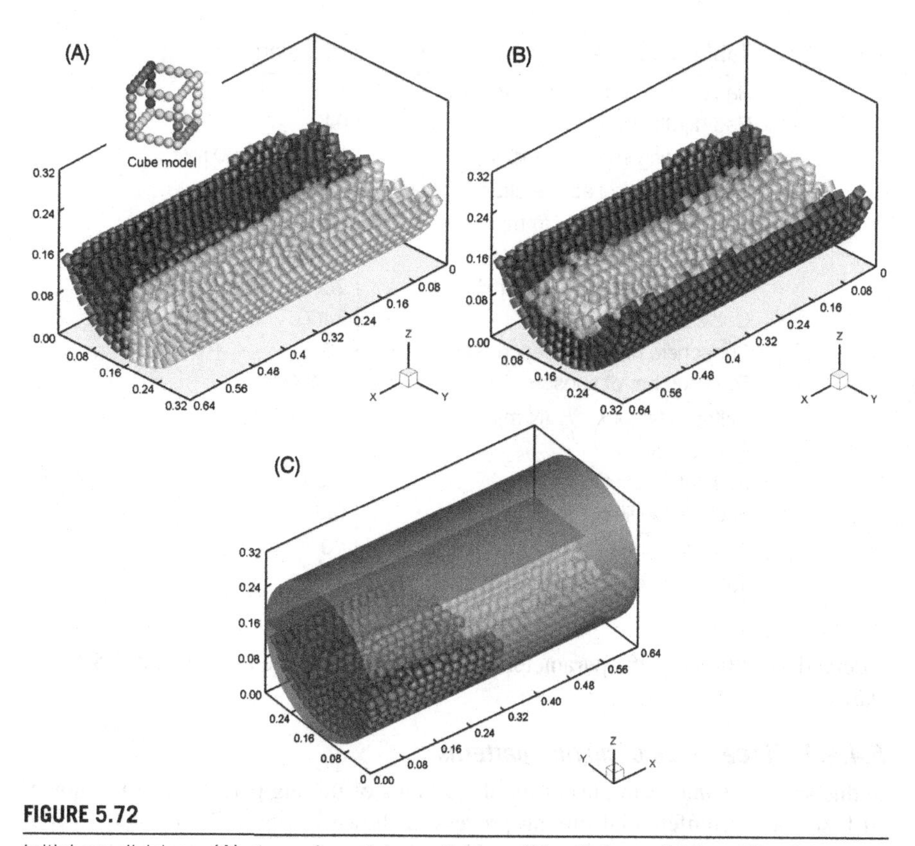

FIGURE 5.72

Initial pre-divisions: (A) circumferential pre-division; (B) radial pre-division; (C) axial pre-division.

see in Fig. 5.72C). The cubic particles have a uniform size of 16.1 mm, covered with 5 subspheres of 3.22 mm in diameter on each edge (see the inset in Fig. 5.72A). For comparative analysis, auxiliary cases which use spherical particles of the same volume, mass and filling level are filled into the same cylinder under the same rotation speeds respectively to show the effect of cubic particle shape on 3D mixing features.

The filling process is through a freely sedimentation process of an array of 2400 cubes driven by gravity to fall down at a certain height above the cylindrical wall. After 1 s, the velocities of particles are set to zero to keep stationary states of them in the cylinder. Then, the cylinder is rotated at a constant rate of acceleration ($5\ \pi/s\ ^{2}$) before achieving the target rotating speeds, e.g. $\omega = 15$–60 rpm, respectively. Moreover, Fig. 5.72 shows the pre-divisions of the bulk particle assembly into two parts in the circumferential (A), radial (B) and axial (C) directions respectively to facilitate the comparison of mixing patterns. When the cylinder rotates, the features of circumferential, radial and axial mixing between the two pre-divided parts can be clearly

Table 5.11 Parameters used in the simulation.

Radius of cylindrical bed, R (mm)	160
Bed depth, D_B (mm)	640
Bed rotating speed, ω rad/s^{-1}	15, 30, 45, 60 rpm
Initially rotational acceleration, a_0 rad/s^{-2}	5π
Cube edges length, l (mm)	16.1
Subsphere diameter, d (mm)	3.22
Subsphere density, ρ_s (kg/m^3)	2000
Cubic particle number N_c	2400
Subsphere number in each cube N_s	56
Total number of subspheres N_p	134400
Stiffness factor k_c, k_r (N/m)	2×10^3, 10^5
Poisson ratio ν	0.3
Restitution coefficient e	0.95
Friction coefficient μ	0.3
Time step δt (s)	10^{-6}
Total number of simulation steps N_t	10^7

observed. Additionally, the parameters used in simulation are listed in Table 5.11 for clarity.

5.4.4.1 Three kinds of mixing patterns

In this section, some examples of the three kinds of mixing patterns will be shown. At first, the circumferential mixing process is shown in Fig. 5.73 for $\omega = 15$ rpm at $t = 1, 3, 5, 10$ s, respectively. When the cylinder rotates clockwise, the blue part of cubic particles on the top of red part begins to roll down at $t = 1$ s (Fig. 5.73A), and the rolling-down process of the blue part is ongoing for a while (up to $t = 5$ s (Fig. 5.73C)). After that, the alternative rolling-down process of the red part takes place (Fig. 5.73D), although this process is fast or slow on different axial locations. Therefore, the repeated rolling-down processes of the red and blue parts of cubic particles construct the basis of the circumferential mixing pattern.

In a similar manner, the radial mixing pattern and axial mixing pattern are shown in Fig. 5.74 ($\omega = 30$ rpm) and Fig. 5.75 ($\omega = 60$ rpm), respectively. Differently from the circumferential mixing pattern, the lower particles of the red part (i.e. the inner part of the radial direction) flows downward firstly at the early stage of $t = 1$ s (Fig. 5.74A). As the filling level is not above a half, a steady inner core located around the cylinder axis and wholly below the flow layer does not exist. As a result, after the initial flow-down (a bulk displacement from the inner to the outer), the radial mixing is further evolved and developed into a manner similar to the circumferential mixing pattern (Fig. 5.74B and C). Because of the initial flow-down process, the separation interface between the red and yellow part is severely changed as well as the structure of radial mixing. As a consequence, radial mixing efficiency seems to be better than the circumferential mixing, which will be quantified latter in following sections.

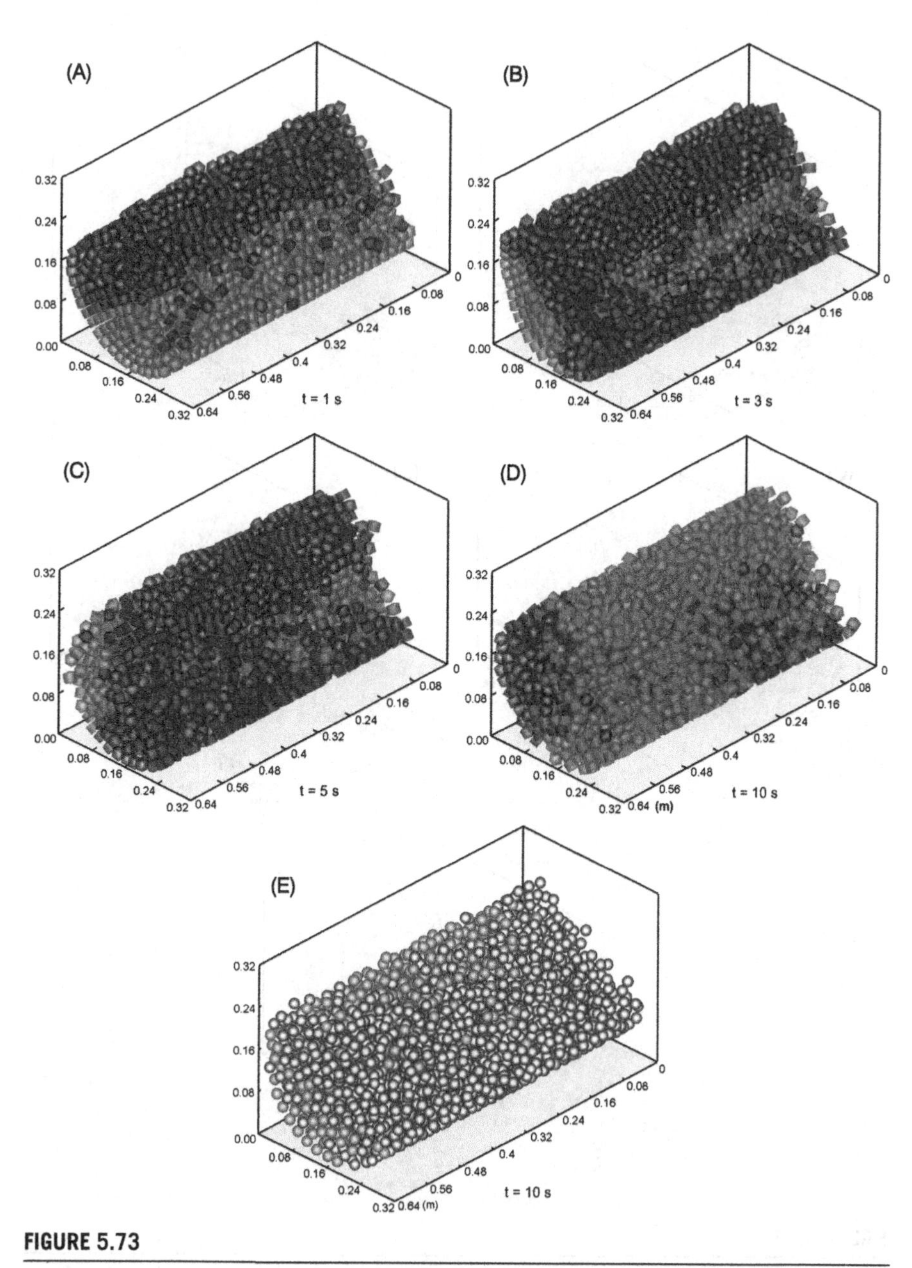

FIGURE 5.73

The mixing states of cubic particles with circumferential pre-divisions for $\omega = 15$ rpm at $t = 1$ s (A), 3 s (B), 5 s (C) and 10 s (D), respectively.

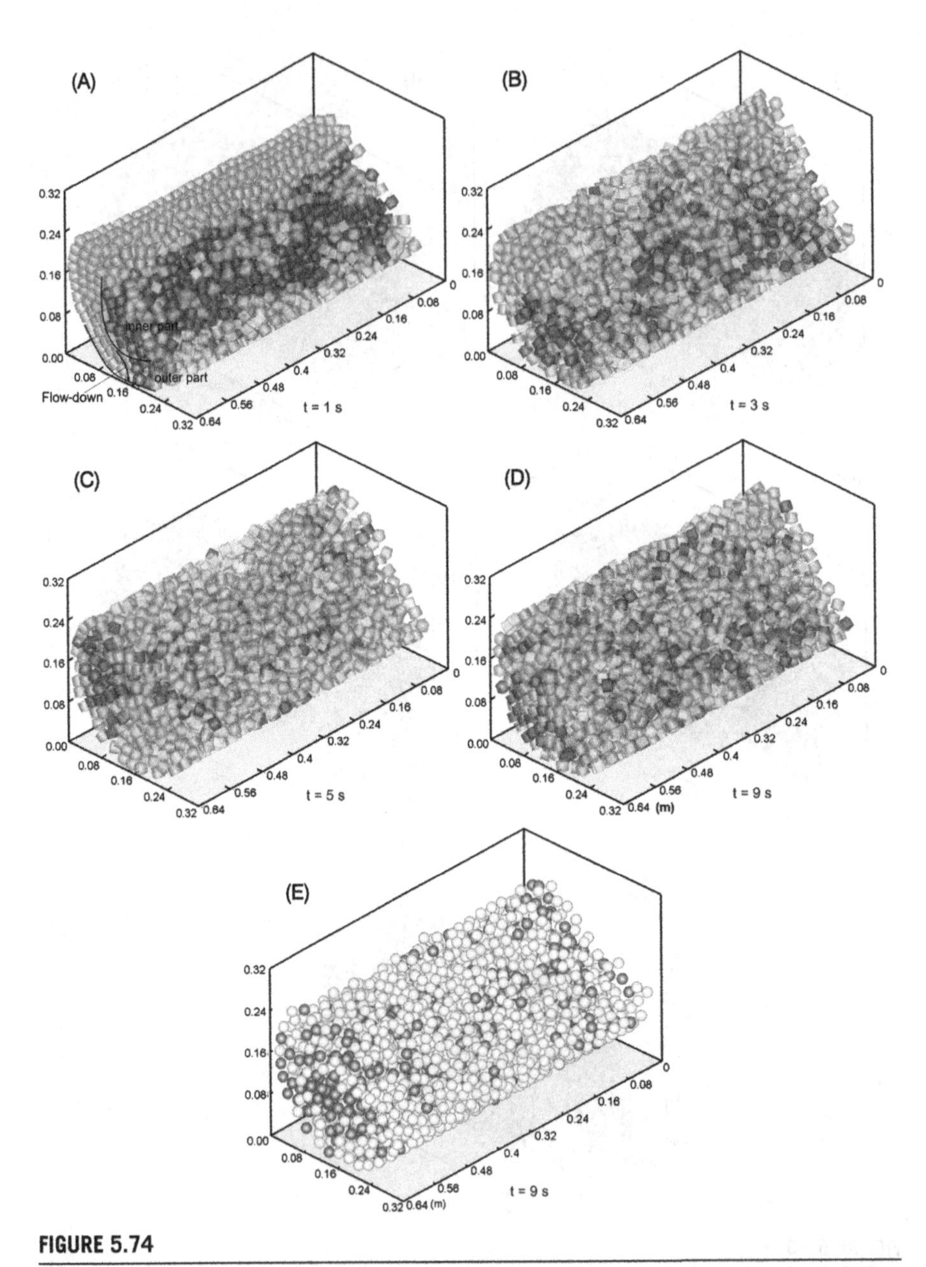

FIGURE 5.74

The mixing states of cubic particles with radial pre-divisions for $\omega = 30$ rpm at $t = 1$ s (A), 3 s (B), 5 s (C) and 9 s (D), respectively.

In comparison, the axial mixing (Fig. 5.75) seems to be much slower than the radial and circumferential mixing. Although with the fastest rotating speed ($\omega =$ 60 rpm), the axial mixing still has not taken place as efficiently as the radial and circumferential mixing after $t = 10$ s. As the axial mixing is mostly caused by the axially asymmetrical rolling-down process, the basic mixing interface is always kept in the planer shape located on the same axial location in Fig. 5.75. Thus, it is preferred to regard the axial mixing as axial dispersion.

In addition, Fig. 5.73D and 5.73E show that the cubes and spheres are mixed at clearly different paces of circumferential mixing, although the initial pre-divisions for them are the same. However, differently from that in Fig. 5.73D and 5.73E, the radial mixing features of cubic shaped particles in Fig. 5.74D or the axial mixing features in Fig. 5.75D look like similar with that of spheres of radial mixing in Fig. 5.74E or axial mixing in Fig. 5.75E, respectively. Thus, it seems that the cubic shape may have different influences on the three mixing patterns of particles, which implies that detailed and quantitative analyses for the effect of cubic shape on mixing patterns and levels are very needed, which will be shown in following sections.

5.4.4.2 Mixing indices

The well-known Lacey mixing index [300] is applied to characterize the degree of mixing.

The mixing indices of the circumferential ($M_{l,C}$), radial ($M_{l,R}$) and axial ($M_{l,A}$) mixing processes for cubes and spheres under all rotating speeds are shown in Fig. 5.76A–C, respectively. It is seen that the circumferential mixing pattern is clearly dependent on the rotating speed (Fig. 5.76A). When the rotating seed is increased, the circumferential efficiency is increased and the mixing degree is intensified. However, for the radial mixing, the mixing efficiency and degree will not be evidently dependent on the rotating speeds (see close values of $M_{l,R}$ for $\omega = 30$, 45 and 60 rpm in Fig. 5.76B), provided the rotating speed is sufficiently large. In contrast, when the rotating speed $\omega = 15$ rpm, it is too low that the rolling-down process has not been prevalently built up. At this stage, $M_{l,R}$ may become lower than the former cases of $\omega = 30$, 45 and 60 rpm. For the axial dispersion, it is indicated in Fig. 5.76C that the axial dispersion degree ($M_{l,A}$) is also increased as the rotating speed ω increases. In addition, the absolute degrees of axial dispersion $M_{l,A}$ are much lower than the radial mixing $M_{l,R}$ and circumferential mixing $M_{l,C}$.

On the other hand, comparing the mixing indices of sphere mixing and cube mixing, it is shown that: (1) $M_{l,C}$ of spheres are lower than $M_{l,C}$ of cubes, especially under large rotating speeds of $\omega = 30$, 45 and 60 rpm (Fig. 5.76A). (2) On the contrary, $M_{l,R}$ of spheres are mostly equivalent to $M_{l,R}$ of cubes, and $M_{l,A}$ of spheres are larger than $M_{l,A}$ of cubes (Fig. 5.76C). This confirms the aforementioned qualitative observation in Figs. 5.73 to 5.75 that the cubic shape on circumferential mixing levels are different from that on radial and axial mixing. In other words, the cubic shape may improve the circumferential mixing whereas reduce the axial mixing of particles in cylinder.

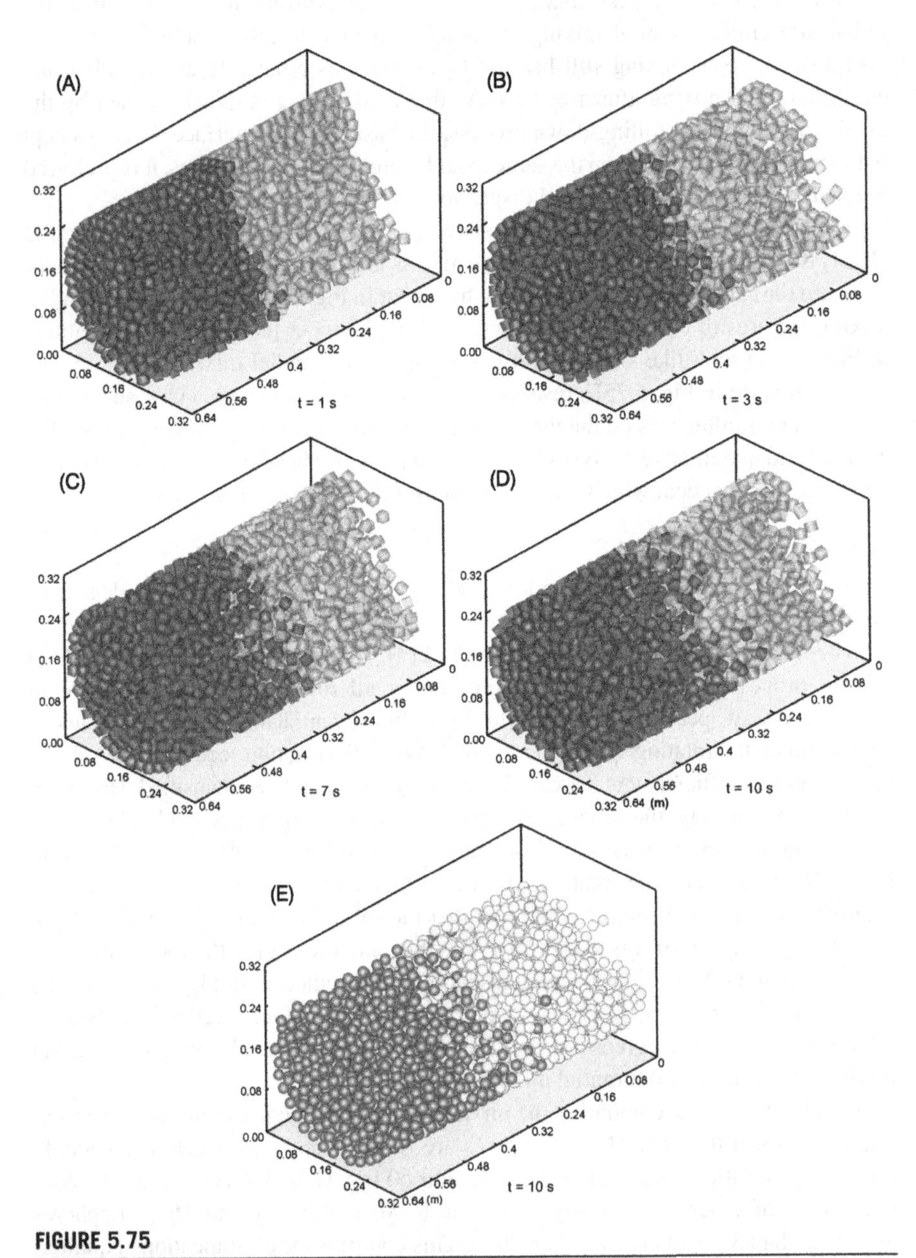

FIGURE 5.75

The mixing states of cubic particles with axial pre-divisions for $\omega = 30$ rpm at $t = 1$ s (A), 3 s (B), 7 s (C) and 10 s (D), respectively.

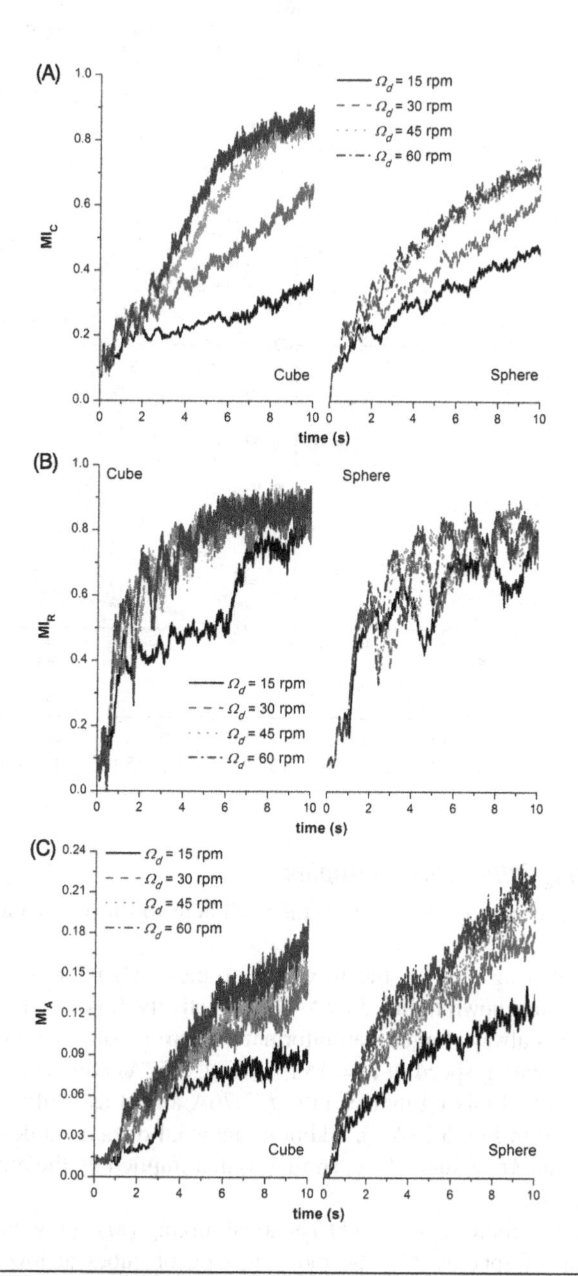

FIGURE 5.76

The mixing indices $M_{I,C}$ (A), $M_{I,R}$ (B), $M_{I,A}$ (C) at different rotating speeds for circumferential, radial and axial pre-divisions, respectively.

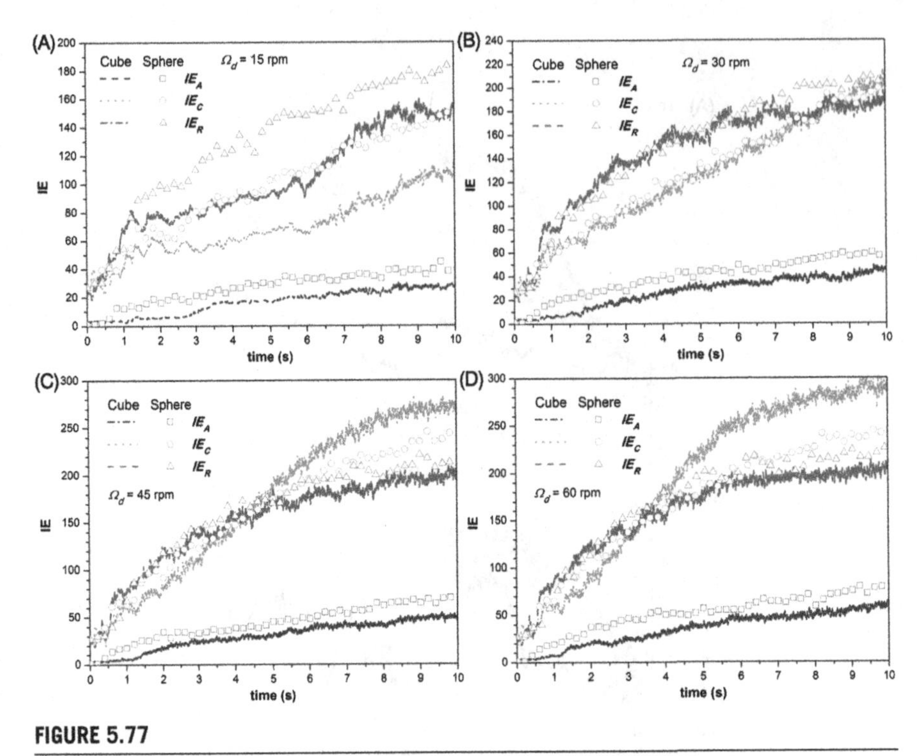

FIGURE 5.77

The mixing entropy M_E (named 'IE' in the figure) for circumferential (with subscript '$_C$'), radial ('$_R$') and axial ('$_A$') pre-divisions at $\omega = 15$ (A), 30 (B), 45 (C) and 60 (D) rpm, respectively.

5.4.4.3 Mixing information entropies

The Shannon information entropy [297] is used here to further quantify the mixing degree.

Similar to mixing indices, the mixing entropies (M_E) for rotating speeds of $\omega = 15$–60 rpm are shown in Fig. 5.77A–D, respectively. It is seen that M_E of radial mixing ($M_{E,R}$) is always larger than information entropy of circumferential mixing $M_{E,C}$ at a low rotating speed of $\omega = 15$ rpm (Fig. 5.77A) which confirms the comparison of quantified mixing indices in Fig. 5.76A and B as well as the qualitative analysis sketched in Fig. 5.74A. In addition, the axial dispersion degrees $M_{E,A}$ are always lower than $M_{E,C}$ and $M_{E,R}$, similar to that implied by the mixing indices in the former section.

Moreover, it is confirmed that: (1) The axial mixing ($M_{E,A}$) of cubes are always lower than that of spheres. (2) The radial mixing of cubes at low rotating speed ($\omega = 15$ rpm in Fig. 5.77A) is still lower than that of spheres, whereas they are almost equivalent at other rotating speeds ($\omega = 30$–60 rpm in Fig. 5.77B–D). (3) On the contrary, although the circumferential mixing of cubes ($M_{E,C}$ of cubes) is still lower than that of spheres ($M_{E,C}$ of spheres) at low rotating speed of $\omega = 15$ rpm in Fig. 5.77A, and equivalent to $M_{E,C}$ of spheres at $\omega = 30$ rpm in Fig. 5.77B, it becomes larger than that for spheres at larger rotating speeds of $\omega = 45$–60 rpm in Fig. 5.77C–D. Therefore, it once again details the different effects of cubic shape on improving the circumferential mixing and reducing the axial mixing.

5.4.4.4 Normalized values of entropy

It is worth noticing that, when the rotating speed accelerates $\omega = 45$–60 rpm, the circumferential mixing degree indicated by $M_{E,C}$ becomes larger than the radial mixing degree $M_{E,R}$. Remember that the definition of information entropy is based on local microscopic mixing states, it must indicate some particular microscopic mixing features. Notice that $(N_{\text{cir, pA}}, N_{\text{cir, pB}}) = (1143, 1257)$ in the initial circumferential pre-division and $(N_{\text{rad, pA}}, N_{\text{rad, pB}}) = (1848, 552)$ in the initial radial pre-division, the better microscopic mixing state of circumferential pre-division over the radial pre-division may be caused by difference between the nearly equivalent initial circumferential pre-divisions and the nonequivalent initial radial pre-divisions. To say specifically, for the fully-developed state of circumferential pre-division, the microscopic mixing entropy in a local cell is about

$$-P_a\log_2 P_a - P_b\log_2 P_b = -0.52\log_2 0.52 - 0.48\log_2 0.48 \approx 0.98, \tag{5.74}$$

where $P_a = \frac{N_{pA}}{N_{pA}+N_{pB}}$ is the number fraction of part A particles to the total number of particles, and $P_b = 1 - P_a$ is the number fraction of part B particles. However, because of the nonequivalent initial radial pre-division, the microscopic mixing entropy in one fully developed local cell is about

$$-P_a\log_2 P_a - P_b\log_2 P_b = -0.77\log_2 0.77 - 0.23\log_2 0.23 \approx 0.778. \tag{5.75}$$

As $0.778 < 0.98$, the quantified final fully developed microscopic mixing state of radial pre-division is lower than that of fully developed microscopic mixing state of circumferential pre-division. Therefore, the absolute value of M_R is not suitable for indicating the mixing degree. As a result, a normalized M_R is proposed as follows:

$$M_R^n = \sum_{i,j,k} \frac{-c_a\log_2 c_a - c_b\log_2 c_b}{-P_a\log_2 P_a - P_b\log_2 P_b} \Delta x_i \Delta y_j \Delta z_k, \tag{5.76}$$

where the factor $(-P_a\log_2 P_a - P_b\log_2 P_b)$ indicates the information entropy of constructing a cell of P_a fraction from part A and P_b fraction from part B when the mixing state is fully developed. If $P_a = P_b = \frac{1}{2}$, $(-P_a\log_2 P_a - P_b\log_2 P_b) = 1$, which has the largest value and does not change the absolute value of M_R if the two parts have equivalent numbers of particles. Otherwise, if $P_a > P_b$ or vice versa, $(-P_a\log_2 P_a - P_b\log_2 P_b) < 1$ and $M_{E,R}^n$ should be increased since the overall probability to construct a cell of P_a fraction of part A particles and P_b fraction of part B particles is larger than that with $P_a = P_b$.

For example, the normalized information entropies M_E^n corresponding to Fig. 5.77C–D are shown in Fig. 5.78A–C, respectively. It is clearly seen that the

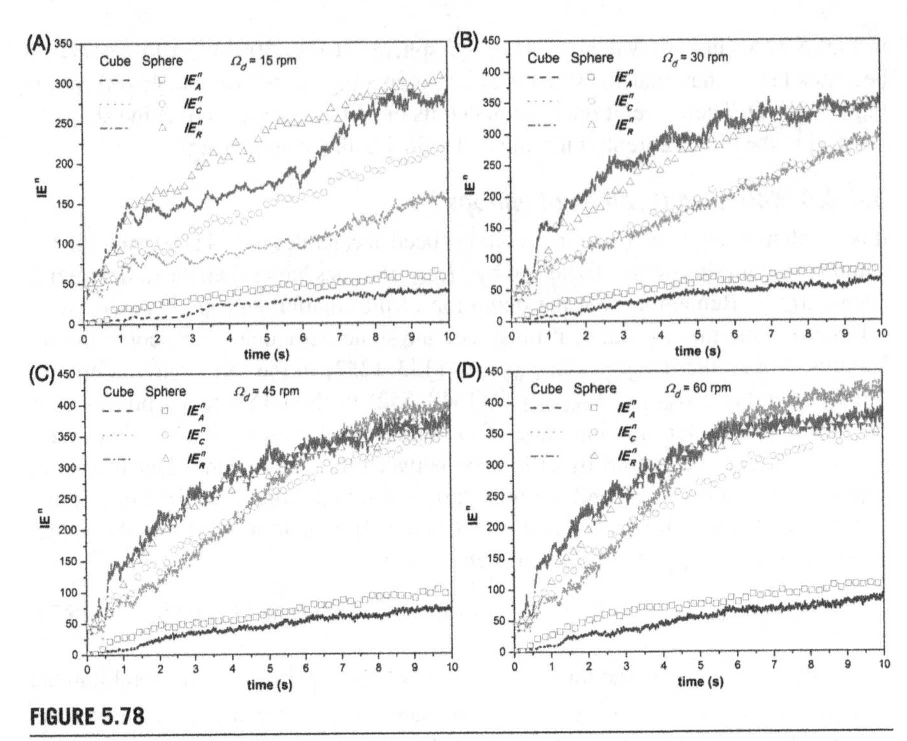

FIGURE 5.78

The normalized mixing entropy M_E^n (named 'IE^n' in the figure) for circumferential (with subscript '$_C$'), radial ('$_R$') and axial ('$_A$') pre-divisions at $\omega = 15$ (A), 30 (B), 45 (C) and 60 (D) rpm, respectively.

normalized $M_{E,R}^n$ becomes much larger than the original $M_{E,R}$. By M_E^n in Fig. 5.78, it indicates that $M_{E,R}^n$ increases with time much faster than $M_{E,C}^n$, and the final level of $M_{E,R}^n$ after fully developed state is still somewhat larger than $M_{E,C}^n$. Thus, the general agreements between the normalized information entropy M_E^n and the mixing index M_I (see Fig. 5.76) confirm the conclusion that the efficiency and degree of radial mixing are higher than those of circumferential, and their final degrees are comparable and always larger than for the axial mixing.

To say conclusively, the normalized mixing entropy M_E^n may be more indicative than the absolute information entropy since it may be used to compare the degree of mixing between universal cases (not limited to present section), particularly with nonequivalent numbers of particles.

5.4.4.5 Particle trajectories analysis

In this section, particular attention is paid on the axial dispersion characteristics of cubic particles in the cylinder. Firstly, take No. 1 particle for example, the trajectories of it under different rotation speeds are shown in Fig. 5.79. Comparing Fig. 5.79A with Fig. 5.79B, the particle motion may be in opposite axial directions (backward dispersion in $\omega = 15$ rpm vs. forward dispersion in $\omega = 30$ rpm in the axial direction)

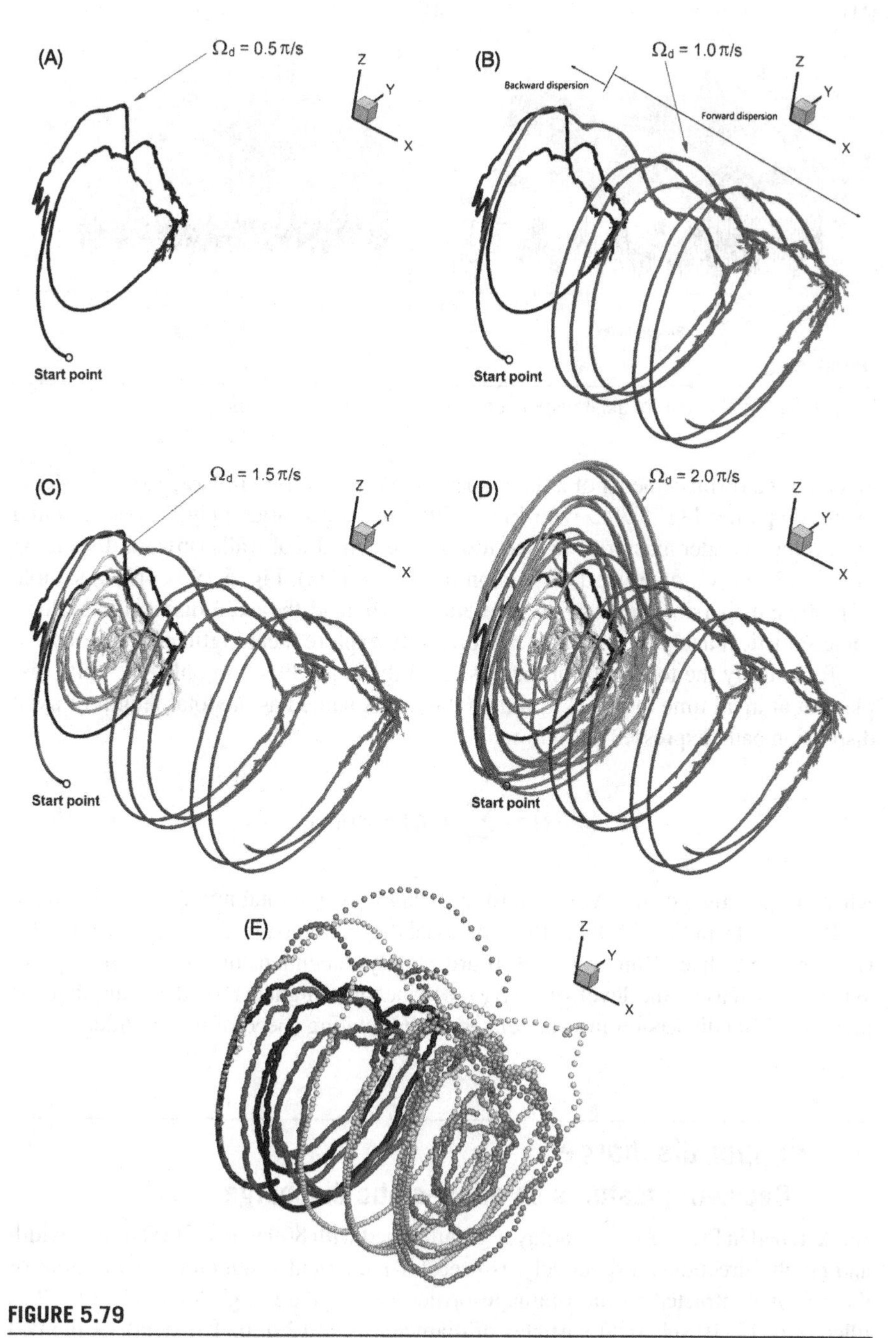

FIGURE 5.79

The sampled trajectories of No. 1 particle at different rotating speeds ω.

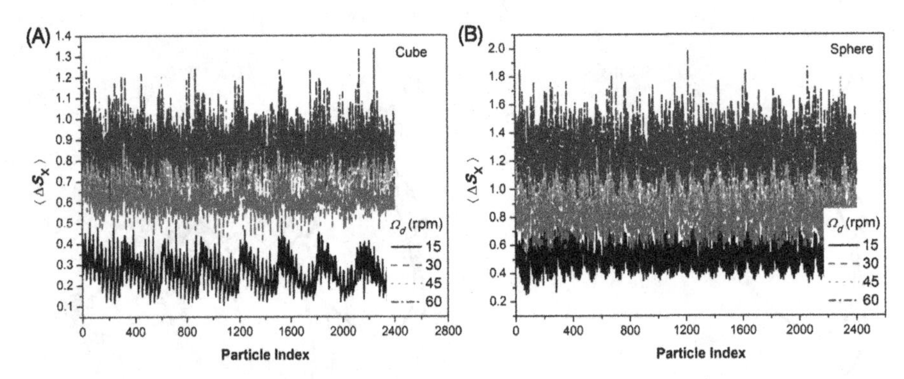

FIGURE 5.80

The total integral length of axial dispersion paths $\langle S_X \rangle$ for all particles.

and the total displacements of axial dispersion may be varied in large extents. Moreover, comparing Fig. 5.79C with Fig. 5.79D, the trajectories may be concentrated toward the cylinder axis or dispersed toward the cylindrical walls (inward dispersion in $\omega = 45$ rpm vs. outerward dispersion in $\omega = 60$ rpm). Fig. 5.79 is only a sample of particle dispersion trajectories. To better understand the axial dispersion feature, some statistical analysis should be performed to explore meaningful conclusions.

To quantify the total path of dispersion in the axial direction only, the axial displacement in all time intervals is summed up. It is named as the total length of axial dispersion path, expressed as follows:

$$\langle \Delta S_X \rangle = \sum_{i=1}^{N_t} |\boldsymbol{x}(t_i) - \boldsymbol{x}(t_{i-1})| \tag{5.77}$$

where $\boldsymbol{x}(t_i)$ is the position vector at time t_i, and N_t is the total number of time steps.

It is seen from Fig. 5.80 that the total axial dispersion path $\langle \Delta S_X \rangle$ for all particles (for particle indices from 1 to 2400) are mainly dependent on the rotating speed. When ω increases, the level of $\langle \Delta S_X \rangle$ will become higher, too. It means that the particle's axial dispersion may be related to the rotating speed of the cylinder.

5.5 Hopper discharge

5.5.1 Geometric features on quasi-static discharge

As sketched in Fig. 5.81, a monolayer granular bed with 800 mm × 2000 mm in width and height directions, respectively, is used for numerical simulation. The bottom of the bed is contracted to the drainage orifice with a flare angle of $\theta = 120°$. It is filled with 12210 spherical particles of diameter $d_p = 12$ mm. The depth of the bed is equal to the particle diameter, which is consequently termed as a monolayer bed. However, as it is a monolayer bed wherein the particles are set on the same plane in

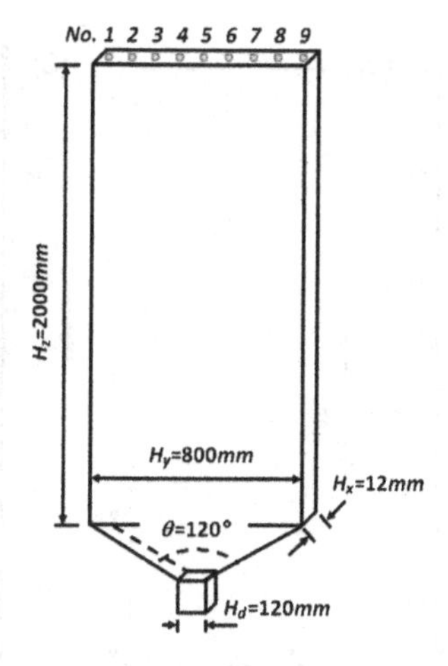

FIGURE 5.81

The sketch of numerical setup for quasi-static discharge.

Table 5.12 Parameters used in the simulation.

Dimensions of bed $H_y \times H_z$ (mm)	800 × 2000
Diameter of removal orifice D_o (mm)	120
Base flare angle θ (°)	120
Particle diameter d_p (mm)	12
Particle number N_p	12210
Friction coefficient γ	0.3
Restitution coefficient e	0.9
Stiffness factor K_n (N·m^{-1})	1×10^4
Poisson rate	0.3
Time step (μs)	50
Total simulated time (s)	500
Number rate of circulation (s^{-1})	20

the depth direction, the friction and normal forces from the front and back walls are not considered. The parameters used in simulation are included in Table 5.12.

The facility runs in a recirculation mode. The particles are removed one by one at a slow rate from the drainage orifice with 120 mm wide in the bottom center, and reloaded simultaneously onto the bed through nine fixed reloading locations at the top of the bed (Nos. 1–9 in Fig. 5.81).

At the beginning, the particles are randomly packed inside the bed, where the randomly packing state can be obtained by a 'sedimentation' and 'deposition' procedure. For recirculation, the particles contacting the bottom and falling outside the orifice are removed and reloaded. The processing of recirculation of particles establishes a dynamically steady packing state inside the bed soon. Thus, a reliable statistical anal-

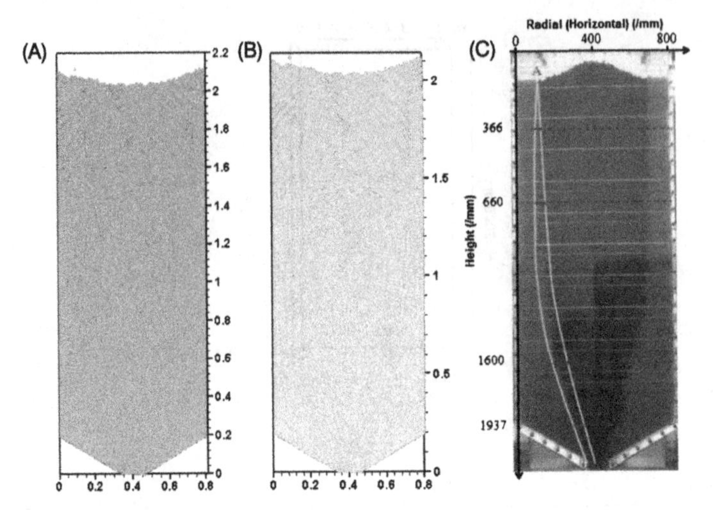

FIGURE 5.82

(A) Particle distribution state at the final time of simulation. (B) Particle trajectories originated from the nine fixed points at the bed top. (C) Experimental snapshots of particles in the granular bed.

ysis can be conducted based on such a dynamically-equilibrium or time-stationary process.

5.5.1.1 Particle trajectories

For demonstration, Fig. 5.82A shows a snapshot of particle distribution state inside the bed at the final time of simulation. Fig. 5.82B shows all the particle trajectories originated from the nine fixed points, respectively (see Fig. 5.81). The trajectories form a band area from the origination point to the end. It is named as trajectory band hereafter. By observation of the banded distribution of particle trajectories, it is appropriate to consider that: (1) Somewhat stochastic movement of particles in the horizontal direction takes place when the bulk motion of gravity-driven drainage in the vertical direction is going on. (2) This stochastic movement leads to ambiguity in determining particle trajectories and cause lateral self-mixing between the particles. (3) This stochastic motion can be defined as a specific type of 'dispersion' behavior which is important for granular flows, since it may cause fluxes of mass, heat and mixing within granular flow.

In addition, experiments (Fig. 5.82C) were carried out by the authors to validate the numerical results. The details of the experimental method and procedures can be found in Ref. [89]. The experimental bed is in the same scales of the numerical bed, including width, height, base angle, particle diameter and etc. Moreover, the movements of particles in experiment are traced similarly as in simulation. However, a slight difference between experiments and simulations exists in the reloading manner of particles. In experiment, it is reloaded onto the bed through the central area for convenience, whereas it is uniformly reloaded onto the bed through the nine fixed points in simulation. Because of this difference, the top edge of the piling state of particles in experiment are somewhat raised in the central region, which is different from that in simulation where it is a bit sunken in the central region. However, this difference near the starting points of the trajectories doesn't affect the horizontal dispersion characteristics inside the bed, which will be confirmed by the agreement of dispersion distributions between them in the following sections.

In simulation, the particle trajectories are confined regularly in narrow bands from top to bottom. The narrow band region indicates a fairly weak dispersion in the horizontal direction. However, as accurate determination of horizontal dispersion is very important, it is necessary to quantify in detail the particle dispersion characteristics in the horizontal directions.

5.5.1.2 Probability density function

The horizontal dispersion characteristics of particles can be shown by probability density distribution (PDF) of particle displacements. The PDF can be calculated as follows: (1) Fix an interrogation area of the trajectory band at a height h. (2) Superimpose a ruler grid of Δr onto the area, computing the number of trajectories $n(r)$ through them. (3) Normalize the number of trajectories through each grid by dividing the total number of trajectories $N(r)$ through the whole interrogation area and Δr to obtain the PDF, i.e.

$$f(r) = \frac{n(r)}{(N(r) \cdot \Delta r)} \tag{5.78}$$

For example, Fig. 5.83 shows the PDFs of particle horizontal distribution for bands No. 2 (A), 5 (B) and 8 (C), respectively, in the middle height of the bed. It is clearly seen that distribution of particles across the bands follows a Gaussian-form distribution within the narrow area. For the bands in the central of the bed (No. 5), the PDFs are symmetrical. On the two sides (No. 2 on the left-hand side and No. 8 on the right-hand side of the bed), the tails of PDFs toward the walls are longer than that toward the center. It means that the particle dispersion in the horizontal direction is not very uniform. To say specifically, the symmetry of particle dispersion is relevant to locations inside the bed. For the locations on the side regions, particles flow faster on the side toward the bed center whereas they flow more slowly on the side toward the bed wall. This asymmetry of particle motion is caused mainly by the contracted configuration of drainage.

5.5.1.3 Comparison with experiments

To validate the above numerical results of particle horizontal dispersion, the normalized probability density functions (nPDF) for bands No. 2 and No. 8 are compared to their experimental counterparts, respectively. Because the numerical bed is a monolayer bed while the experimental bed is a multilayer bed, the depths of the numerical and experimental beds are different. Therefore, it is better use the normalized PDFs for comparison since the difference in bed depth may affect the absolute level of particle dispersion.

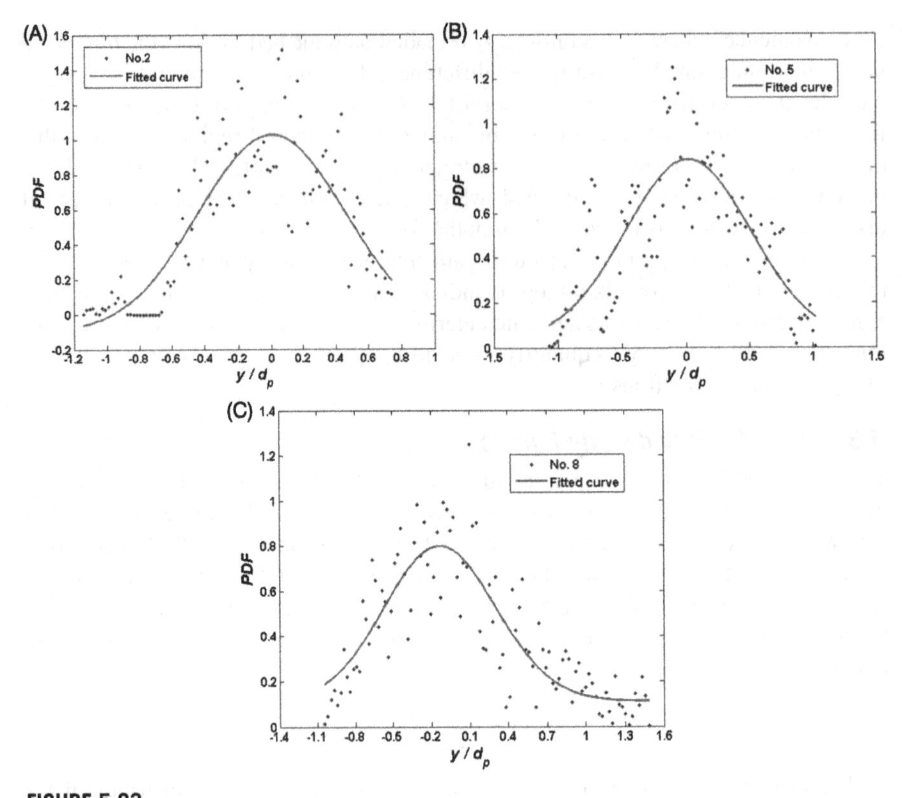

FIGURE 5.83

The probability density function of particle distribution at height $H_z = 1$ m for the bands of No. 2 (A), No. 5 (B) and No. 8 (C), respectively. The '0' position in the abscissa axis represents the center of the bands.

The normalization of PDF follows the procedures as follows: (1) Fit the numerical data by a Gaussian function to obtain $h = f(y)$. (2) Divide y by the full width at half maximum σ_δ (FWHM) of the Gaussian function. (3) Divide f by the maximum value of it to obtain nPDF.

After comparison, as shown in Fig. 5.84, it is found that the numerically fitted nPDFs agree very well with the experimental counterparts, for both in the bed central region (Fig. 5.84A) and the side region (Fig. 5.84B). Thus, it is convinced that, in general, the particle dispersion in the horizontal direction follows a Gaussian form of dispersion. Thus, the aforementioned results in Fig. 5.83 on particle dispersion are validated.

5.5.1.4 Geometrical models for explanation

It is necessary to explore the nature of particle dispersion within such a gravity driven slow particle flow. As is mentioned before, it is natural to analyze the dispersion

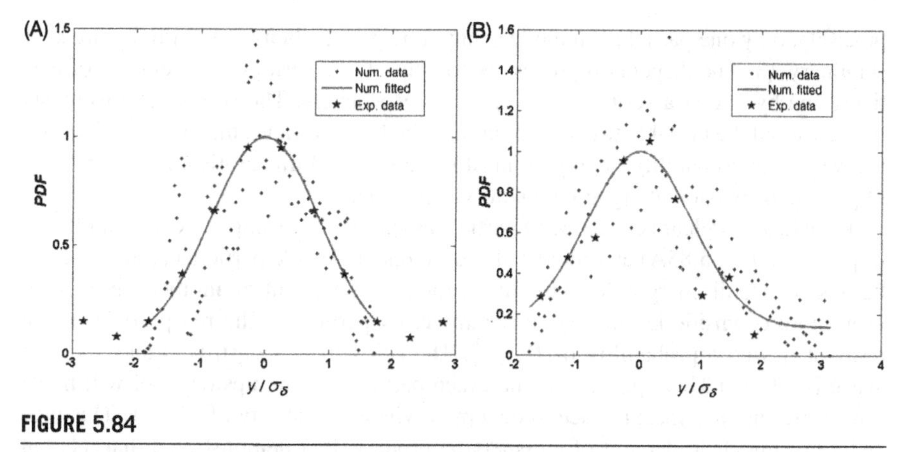

FIGURE 5.84

Comparison of normalized PDFs between numerical and experimental results for Nos. 5 (A) and 8 (B), respectively.

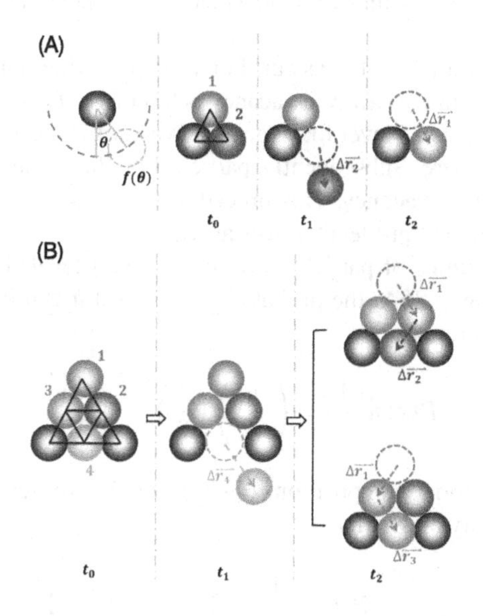

FIGURE 5.85

Illustration of typical diffusion process under two levels (A) and three levels (B) of particles.

behavior in slow flows by geometrical techniques. This section demonstrates the geometrical analysis for explanation of particle dispersion characteristics.

In general, particles can move downward via any direction angle of motion θ with a probability $f(\theta)$ (Fig. 5.85A). Let's take an ideal case for procedure analysis. It is assumed that: (1) The particles are regularly packed, forming a regular triangle-shaped feature of particle lattice. (2) The discharge or movement processes of particles are independent, i.e. they are not correlated to each other. To say more precisely, only one particle can move at any time point, when other particles are kept stationary. (3) The dispersion process is composed of a series of movements of particles one by one at a series of time points, respectively. The series of movements may cover all the possible trajectories along which the particles may follow. Thus, by analyzing the probability on any point of these series of trajectories, the probability of particle dispersion along any trajectory can be obtained.

Consider a nuclear cell lattice of particle assembly for example, i.e. a triangle cell of particles (Fig. 5.85A) as follows: (1) At t_0, one downside particle can move away because of the discharging of particles or the pre-movement of another particles on even more downside levels. (2) For example, let particle 2 (the red particle) move away at t_1 with a probability of $P_2 = \frac{1}{2}$. The cell lattice is destroyed and a vacant space is left. (3) Then, particle 1 (the green particle) on the upside level will move downward into the vacant space to occupy it with a probability of $P_1 = 1$. This procedure is a nuclear element of a dispersion process. It is demonstrated that: (1) On the initial top level, the probability of movement is $P_1 = 1/1$. (2) On the next level, the probabilities are $P_2 = \frac{1}{2}$ and $P_3 = \frac{1}{2}$. Then, what is next? How is this procedure going on?

Fig. 5.85B shows a possible answer. For example, when particle 4 (the orange one) on the third level moves away, a vacancy is left there. Then, both particle 2 (red) and 3 (blue) on the second level may move downward to occupy the vacancy, and leave a new vacancy there. Subsequently, particle 1 on the top level may move downward to occupy the new vacancy. This procedure analysis can be repeated similarly if another particle on the third level moves away.

Thus, the probabilities of particle movements on all levels can be illustrated by Fig. 5.86A. From Fig. 5.86A, the probabilities on level n can be fully predicted by the binomial distribution, i.e.

$$P(k; n, \frac{1}{2}) = \binom{n}{k} \left(\frac{1}{2}\right)^{n-k} \left(\frac{1}{2}\right)^{k} \tag{5.79}$$

where k denotes the horizontal positions $(k \cdot d_p)$ from left to right, and $\binom{n}{k} = \frac{n!}{k!(n-k)!}$. The overall probability on level n is

$$1 = \left(\frac{1}{2} + \frac{1}{2}\right)^{n} = \sum_{k=0}^{n} P(k; n, \frac{1}{2}) = \sum_{k=0}^{n} \binom{n}{k} \left(\frac{1}{2}\right)^{n-k} \left(\frac{1}{2}\right)^{k} \tag{5.80}$$

It is noticed that the above procedure is just a special case of the general binomial distribution with $p = \frac{1}{2}$. The general expression of the binomial distribution $B(n, p)$ is

$$B(k; n, p) = \binom{n}{k} p^{n-k} q^{k}, \text{ where } q = 1 - p. \tag{5.81}$$

As is well known, the binomial distribution is a good approximate distribution of the Gaussian distribution when n is large (e.g. $n > 20$) and p is not near 0 or 1 (e.g. $p \approx \frac{1}{2}$).

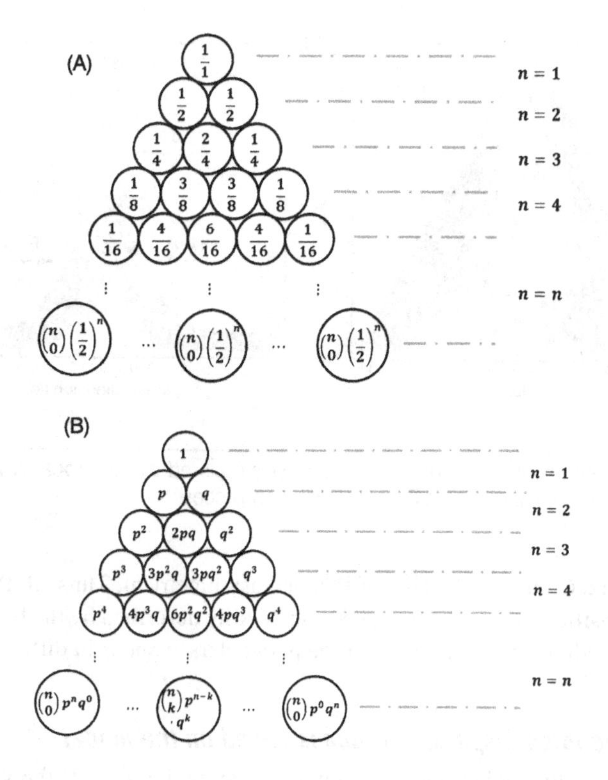

FIGURE 5.86

The probability distribution of the diffusion process on all levels and positions.
(A) Distribution of a symmetrical diffusion probability. (B) Distribution of a general diffusion probability.

Based on the above analysis, it is straightforward to deduce the explanation for the mechanism of a Gaussian-form distribution of dispersion characteristics inside the bed. In fact, the Gaussian-form distribution is just an approximate form of the binomial distribution. Although, the ideal dispersion process may not take place exactly since the aforementioned assumptions on the ideal dispersion are too strict. But the general trends are similar. In other words, the ideal dispersion process as indicated above plays the dominant role, although the practical dispersion process is much more complex than this.

In addition, Fig. 5.86B shows the dispersion probability for more general but still ideal dispersion processes. Ifn this case, the dispersion is not symmetrical when $p \neq q \neq \frac{1}{2}$. For example, when $p > q$, the particles may move more likely toward the left side than the right side. It represents the asymmetrical condition when the velocity of particles on the left side are larger than the right. Therefore, the probability density function on the left side should have larger values than that on the right side. Meanwhile, it could possibly have a longer tail on the left side than on the right one. Moreover, the opposite may be true when $p < q$. It explains the features of PDFs in

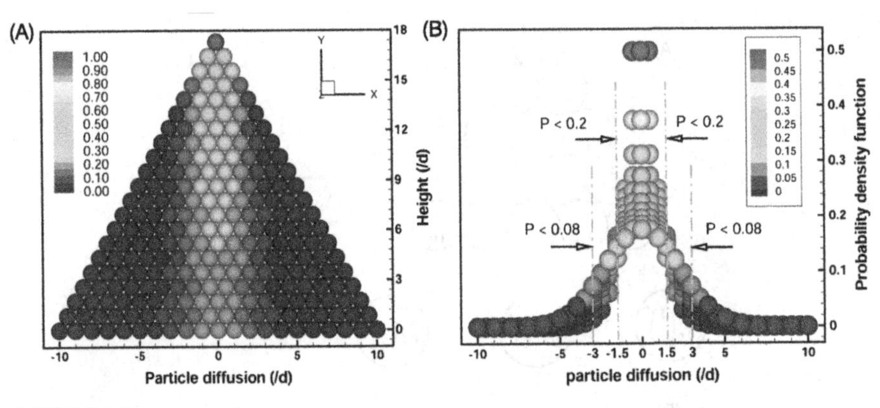

FIGURE 5.87

(A) The ideal diffusion process on the first 20 levels (colored by the probability on each level). (B) The probability view of the ideal diffusion process.

Fig. 5.83A and C on why the tails of PDF are asymmetrical. Thus, all the dispersion behavior of particles throughout the bed can be fully described by the binomial distributions, depending on the variation of the parameters p and q in different horizontal regions of the bed.

5.5.1.5 The ideal dispersion process based on the model

In the above section, when the level number is large, i.e. $n \gg 1$, the width of ideal horizontal dispersion must be fairly large, too. But in practice, the width of dispersion band is always small, because the probability toward the two ends of the dispersion band decreases rapidly.

For example, Fig. 5.87A shows an ideal dispersion map which illustrates all possible trajectories for one particle to move from level 1 at the top to level 20 at the bottom. The map is colored by the local probabilities for the particle to move from level 1 to this position. Based on the basic procedure analysis in Fig. 5.85, the probabilities for one particle to move from the initial point on level 1 to those regions on level n with the same color are the same. Statistically, for a sufficient large number of particle trajectories originated from the same point on level 1, the numbers of trajectories passing through those regions of same probabilities or colored in the same color are equal to each other. In general, it is clearly seen that the regions with large probabilities are restricted in a very narrow band region, e.g. within about $\pm 3d_p$ when $(P > 0.1)$, and about $\pm 1 - 2(d_p)$ when $P > 0.2$.

More clearly, Fig. 5.87B shows the 3-dimensional visualization of this ideal dispersion map, where the regions with dispersion $< 3d_p$ have $P > 0.08$, and the regions with dispersion $< 1.5d_p$ have $P > 0.2$. These ideal and restricted regions with approximated probabilities confirm the values of PDFs in Fig. 5.82 and 5.83, where the fitted curves of simulation results are generally for $P > 0.2$ within about $\pm 1dp$. In other words, on the confidence level of $P > 0.92$, the confidence interval of particle

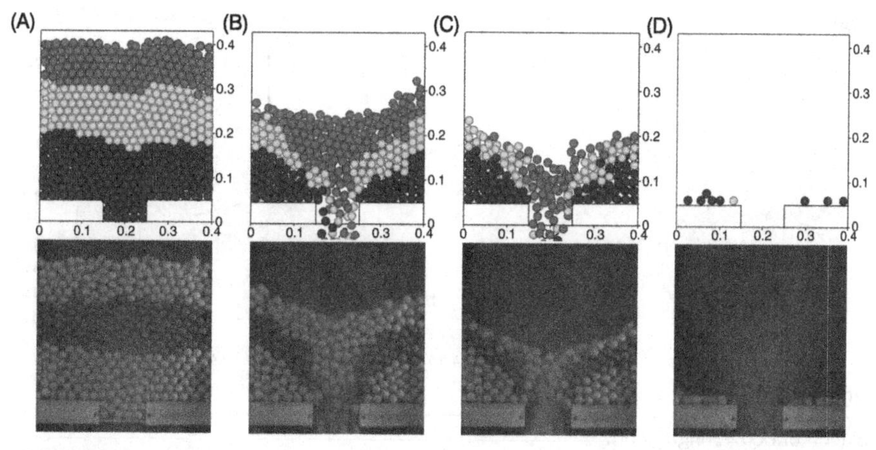

FIGURE 5.88

Comparison of particle discharge process in a rectangular hopper between experimental observations and numerical simulations at $t = 0$ (A), 1 (B), 1.5 (C) and 8 s (D), respectively.

dispersion is within $\pm 3d_p$. On the level of $P > 0.8$, the confidence interval of particle dispersion is within about $\pm 1.5d_p$. Thus, in practice, the actual dispersion area is a fairly narrow band. It explains the formation of narrow band configuration of particle dispersion in Fig. 5.82.

5.5.2 Shaken discharge

5.5.2.1 Model validation and simulation conditions

In this section, the particle discharge process in a shaken hopper is studied by using the DEM model. For qualitative validation of the DEM collision model, experimental observations of the discharge of spherical particles from a rectangular hopper (0.4×1.4 m in width and height, respectively) through an orifice (0.1 m in width) at the bottom are used. According to experimental data [63], the restitution coefficient $e = 0.55$ and the friction coefficient $\mu = 0.39$ are used in simulation. The numerical results of particle distribution at four time points ($t = 0$, 1, 1.5 and 8 s) are compared to the experimental counterparts respectively (Fig. 5.88). After comparison, it is clearly seen that the discharging process can be rebuilt by the DEM simulation to a great degree of resemblance. Therefore, this study will use the DEM model to predict the discharging process of particles in the hoppers.

The hopper bed is 0.8 m $\times$ 1.8 m $\times$ 0.04 m in width (W), height (H), and depth (D_e) direction, respectively. The flare angle of the bed base is 120° (with 30° sidewalls). The orifice width is $D_o = 120$ mm. The bed is filled with 9297 particles of about $H_0 = 1.2$ m pile height (Fig. 5.89A). The particle diameter is $d_p = 12$ mm. With such kind of bed scales, the constraints of (1) $D_o > 6d_p$; (2) $H_0 > D_o$; (3) $W > 2.5D_o$ are met. As a result, the discharging rate should be independent of the domain size [304]. The bed is rocking sinusoidally following the governing functions

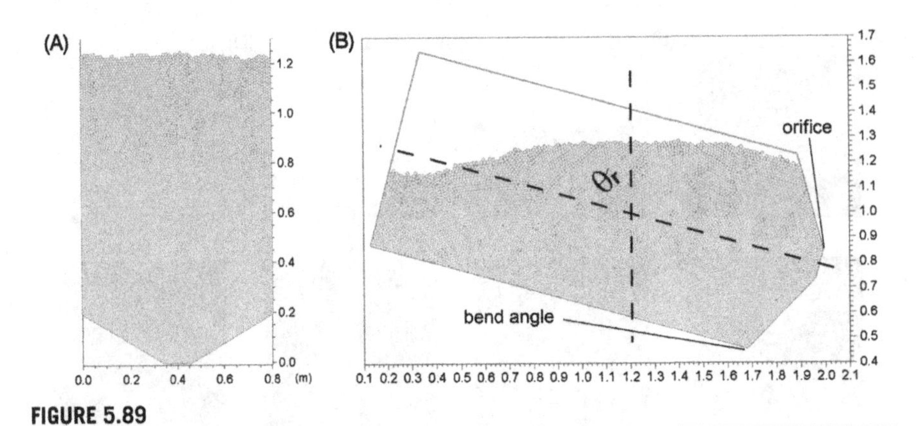

FIGURE 5.89

Sketch of bed configuration.

Table 5.13 Parameters used in the simulation.

The bed's width, height and depth (mm)	800 × 1800 × 40
The flare angle of the bed's base (°)	120
Orifice width D_o (mm)	120
Particle diameter d_p, (m)	12×10^{-3}
Number of particles, N_p	9297
Particle density ρ_p, (kg/m^3)	2281
Restitution coefficient e	0.95
Friction coefficient μ	0.3
Collision stiffness k, (N/m)	1000
Rocking frequency f_r, (Hz)	0, 0.1, 0.2, 0.3, 0.4, 0.5
Rocking amplitude θ, (°)	0, 5, 15, 30, 45, 60, 75
Rocking initial phase angle θ_0, (°)	$\pi/2$
Time step Δt, (s)	1.0×10^{-4}
Simulation time T_s, (s)	20

of

$$\theta(t) = \theta_r \sin(2\pi f_r t + \theta_0) \tag{5.82}$$

where $\theta(t)$ is the rocking included angle between the bed symmetrical axis and the right vertical direction. θ_r is a constant coefficient, namely the rocking amplitude or the maximum included angle of rocking. $\theta_0 = \frac{\pi}{2}$ is the initial phase angle. f_r is the frequency of rocking.

To show the effect of rocking amplitude and frequency on the particle discharging flow, six rocking amplitudes are used ($\theta_r = 5°, 15°, 30°, 45°, 60°, 75°$, respectively); and for each amplitude, five frequencies of $f_r = 0.1$, 0.2, 0.3, 0.4 and 0.5 Hz are simulated, respectively. The parameters used in the current simulation are listed in Table 5.13.

The initial states of particle in the vertical bed are obtained by a sedimentation process of all the particles, and let them sediment for some time to get stationary (Fig. 5.89A). Then, the beds are tilted gradually to reach the angle θ_r of rocking amplitudes, respectively. For example, Fig. 5.89B shows the initial distribution states of particles in the beds of $\theta_r = 75°$. After that, the bottom orifice of the bed is opened and the particles can be discharged from the orifice.

In addition, it is necessary to mention that, based on the simulation results of the stationary bed without rocking, the particles move downward almost uniformly. The particles near walls do not stay stagnant. Thus, under current flow condition, the flow regime in stationary bed without rocking is a mass flow.

5.5.2.2 Discharge process under rocking and flow pattern

At first, for phenomenological comparison, the velocity fields of particles are visualized to show the representative process of particle discharge under rocking. For example, Fig. 5.90A–F shows the particle velocities at $t = 0.1$ s (A), 0.4 s (B), 0.8 s (C), 1.2 s (D), 1.6 s (E), and 2.0 s (F), respectively, in the rocking beds with $\theta_r = 60°$ and $f_r = 0.3$ Hz.

At the beginning, only the particles near the orifice begin to discharge (Fig. 5.90A). After that, the particles near the left-top wall and right-bottom side of the bed begin to move rapidly, forming a large scale 'vortex' from the left-top to the right-bottom (Fig. 5.90B). As a consequence, the particles near the right-bottom side will move to the orifice faster than the particles in the center and near the opposite side wall. When the bed is almost upright (Fig. 5.90C and D), the particles near the left-bottom wall of the bed get large velocities which is mainly caused by the rebouncing from the wall. As a result, the bulk particle flows from the left-bottom and right-bottom walls converge in the bottom-center, together with the particles from the upside. Therefore, the particles near the bottom-center get large velocities. Similar trends will repeat when the bed move toward the opposite side (Fig. 5.90E and F).

Thus, in such a typical process of rocking, the discharged particles may sometimes come from the bottom center, and sometimes come from the near side-walls regions. Furthermore, large scale bulk motions may exist within the bed, which is caused by the rocking motion of the bed. The particles will sometimes stay near the wall with low velocities (Fig. 5.90B, the region near the left-center wall), and sometimes may depart from the wall with large velocities (Fig. 5.90F, the same region near the left-center wall). As a result, the particle flow in the bed is complex which is apparently different from the flow pattern within the stationary hopper. Neither the steady mass flow, where the particles move downward with steady and uniform velocity, nor the steady funnel flow pattern, where the particles near the wall may be stagnated, will take place in such kind of rocking beds. Instead, it is mainly characterized by the bulk flow osçillation within the bed superposed with the oscillated discharge flows out of the bed.

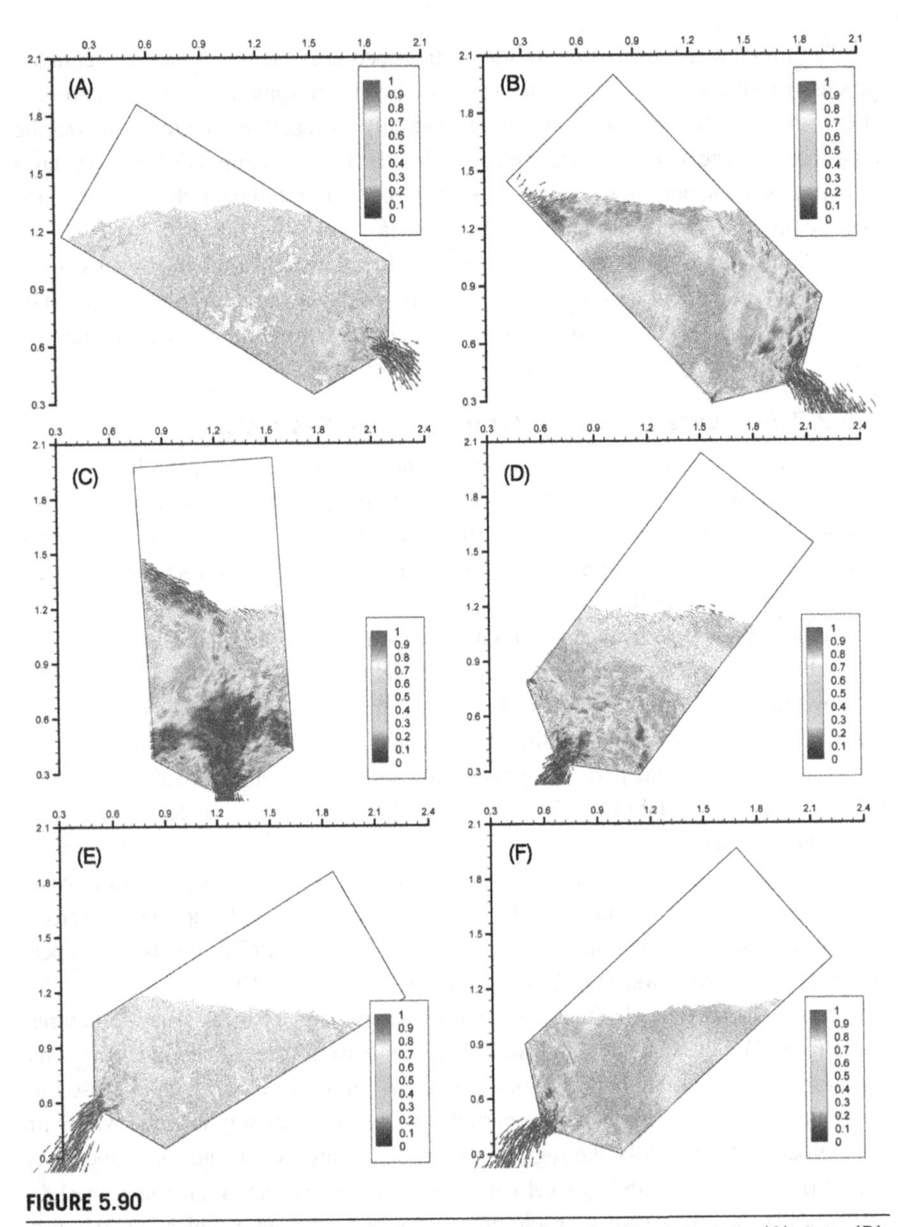

FIGURE 5.90

The discharge process of particles under $\theta_r = 60°$ and $f_r = 0.3$ Hz at $t = 0.1$ s (A), 0.4 s (B), 0.8 s (C), 1.2 s (D), 1.6 s (E), and 2.0 s (F), respectively, after the discharge process begins.

5.5.2.3 Effect of rocking amplitude

In this section, the representative particle distributions are compared for the beds under different rocking amplitudes but the same rocking frequency. For example,

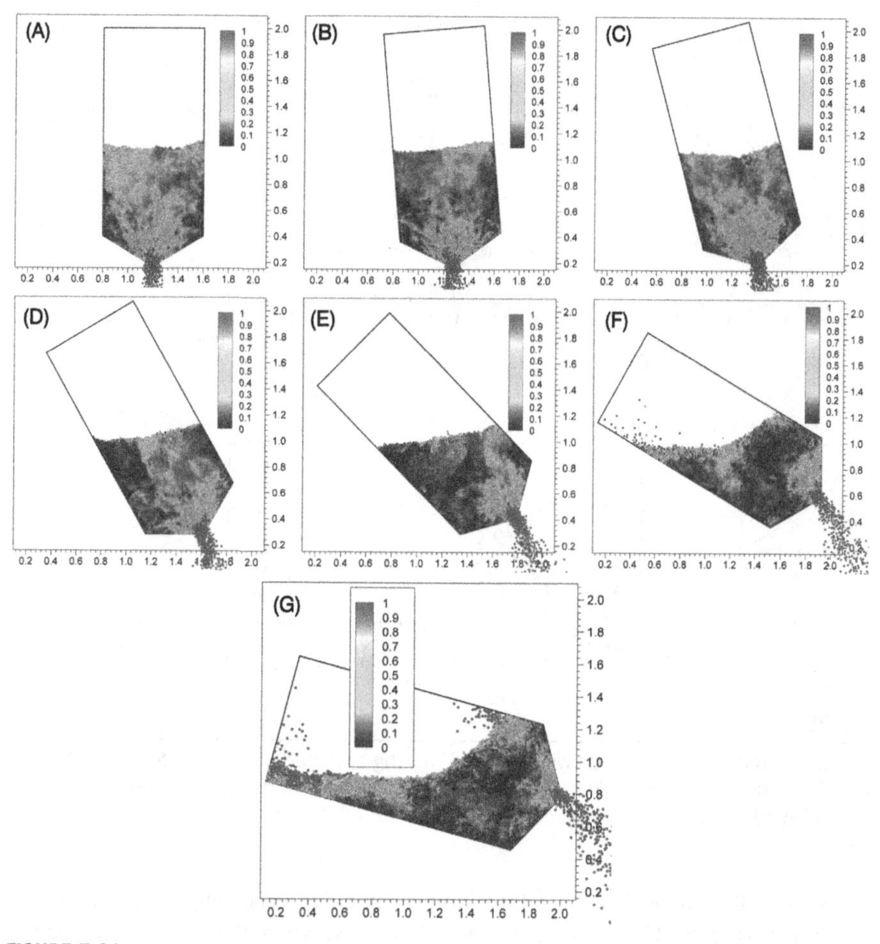

FIGURE 5.91

Particle distributions for the beds with different rocking amplitudes $\theta_r = 60° = 0°$ (A), 5° (B), 15° (C), 30° (D), 45° (E), 60° (F), 75° (G) at $t = 2$ s and $f_r = 0.5$ Hz.

Fig. 5.91A–G shows the particle distribution states for the beds with $\theta_r = 0°, 5°, 15°, 30°, 45°, 60°, 75°$, respectively, when $t = 2$ s and $f_r = 0.5$ Hz.

When the rocking amplitude is increased, the particles on the top surface will move toward the west just like fluids. Meanwhile, the particles near the bottom orifice are still discharging. When $\theta_r < 30°$ (Fig. 5.91A–B), the particles near the orifice are always the lowest particles in the vertical direction, as the flare angle of the base is 120° (with 30° sidewalls). The particles near the bottom center always reach the largest velocities (Fig. 5.91A and B, where particles are colored by their velocity magnitudes). In other words, most discharged particles may move from the bottom center, just like in funnel flows. Otherwise, when $\theta_r > 30°$ (Fig. 5.91E–G), the par-

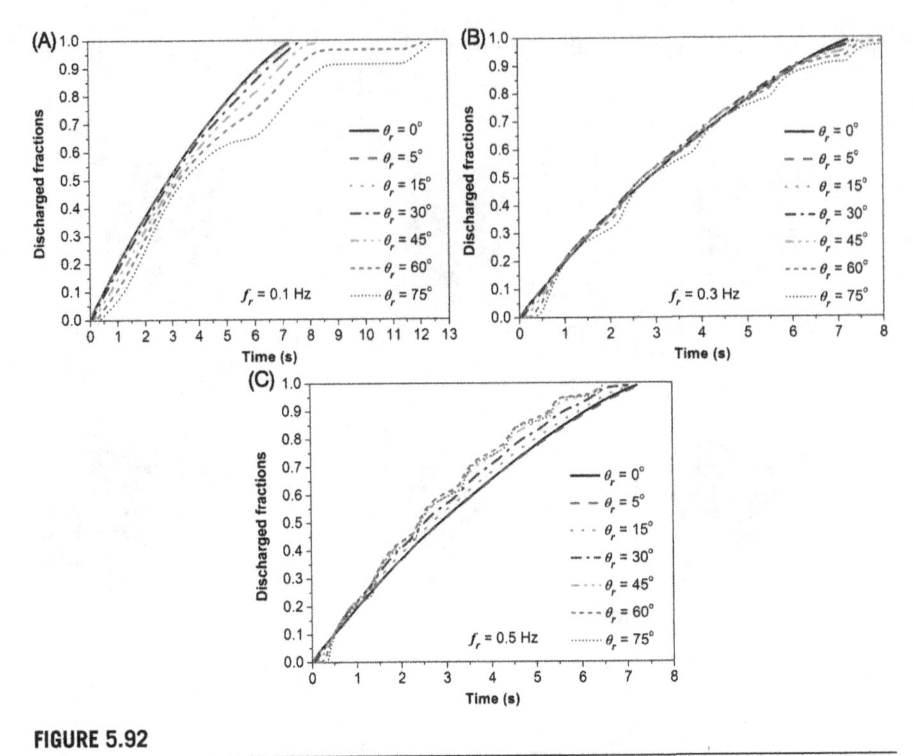

FIGURE 5.92

The discharge rate for the beds with different rocking amplitudes ($\theta_r = 0°$, $5°$, $15°$, $30°$, $45°$, $60°$, $75°$) under the fixed rocking frequencies ($f_r = 0.1$ Hz (A), 0.3 Hz (B), and 0.5 Hz (C), respectively).

ticles along the side walls may get the largest velocities, especially for large rocking amplitudes (Fig. 5.91F and G, see the particle colors) where the side wall can be upright. Then, the discharged particles are mostly from the particles near the side wall, which is just like anti-funnel flow as mentioned in the introduction. Furthermore, the particles near the orifice may occupy higher vertical positions than the particles near the bend angle (Fig. 5.89B and Fig. 5.91E–G). As a result, the discharging flows of the particles near the orifice may be attenuated when $\theta_r > 30°$, which should be quantified in detail.

For that purpose, the discharged fraction defined as

$$R_d(t) = \frac{\text{number of discharged particles}}{\text{total number of particles}} \tag{5.83}$$

is shown in Fig. 5.92 for the beds with different rocking amplitudes ($\theta_r = 0°, 5°, 15°, 30°, 45°, 60°, 75°$) under fixed rocking frequencies of $f_r = 0.1$ Hz (A), 0.3 Hz (B), and 0.5 Hz (C), respectively. It is indicated from Fig. 5.92 that: (1) When the rocking frequency is low (Fig. 5.92A, for $f_r = 0.1$ Hz), the particle discharged fraction is decreased as the rocking amplitude increases. Especially for $\theta_r = 60°$ and $75°$, the particles are completely discharged after $t = 13$ s, about more than 5 seconds later than the normal bed without rocking. The delay of discharge is caused by the plateaus in the discharging plot (e.g. from $t = 8.5$ s to 11.5 s) where the particles cannot be discharged continuously. (2) For the intermediate rocking frequency (Fig. 5.92B, for $f_r = 0.3$ Hz), the delay of particle discharge is not evident as in the former case. In other words, the discharged fractions in rocking beds are quite near that of normal bed without rocking. (3) For the high rocking frequency (Fig. 5.92C, for $f_r = 0.5$ Hz), the opposite trend becomes true. The discharged fractions at large rocking frequencies, although still having the plateaus, are increased faster than those at small rocking frequencies, as well as the normal bed without rocking. It means that the larger rocking amplitude is beneficial for particle discharge under the large rocking frequency, whereas the larger rocking amplitude is detrimental to particle discharge under the small rocking frequency. In addition, with even larger rocking frequency (e.g. $f_r = 1$ Hz or more), the particles inside the bed will be agitated so heavily that they cannot flow continuously and behave in partially gas-like dispersion.

Furthermore, the flow rate defined as the derivative of the discharged fraction is used to indicate the mass flow through the bed orifice as follows

$$n_d(t) = \frac{dR_d(t)}{dt} \tag{5.84}$$

which is shown in Fig. 5.93 for $f_r = 0.1$ Hz (A), 0.3 Hz (B), and 0.5 Hz (C), respectively. The following can be indicated from Fig. 5.93: (1) The flow rate is gradually decreasing with time in the normal beds without rocking. In contrast, according to [305], when the thickness of the layer is greater than 1.2 times the silo diameter, the pressure at the silo bottom saturates due to the Janssen effect, and the flow rate is independent of the height of granular layer (a mass flow pattern). In current case, the layer thickness is one particle diameter ($d_p = 12$ mm, Table 5.13), only $\frac{1}{10}$ of the silo orifice ($D_o = 120$ mm). The layer thickness is far less than $1.2D_o$. It is a two-dimensional silo and hence, the flow rate variation should be different from the three-dimensional experimental data where the flow rate is independent of the height of granular layer. Therefore, although the flow pattern is near the mass flow, the flow rate may still be decreasing gradually. The gradual decrease of flow rate is caused by the gradual decrease of bed pressure when the particles are continuously discharged. (2) Under the small rocking frequency (Fig. 5.93A), the flow rate becomes gradually oscillated when the rocking amplitude is larger than 30°. (3) Under the intermediate rocking frequency (Fig. 5.93B), the trends of oscillation become clearer than for the former case, even at the critical rocking amplitude of $\theta_r = 30°$. Moreover, the oscillation periods of various rocking amplitudes ($\theta_r = 45°, 60°, 75°$) are almost the same, and the peak values occur at almost the same time, which corresponds to the state similar to Fig. 5.90C with largest velocities moving downward during rocking, that state is near the upright position of the hopper. This phenomenon shows the forced oscillation with the same pace of particle discharge in such kinds of flows. (4) Under an even larger rocking frequency (Fig. 5.93C), the trends of forced oscillation for

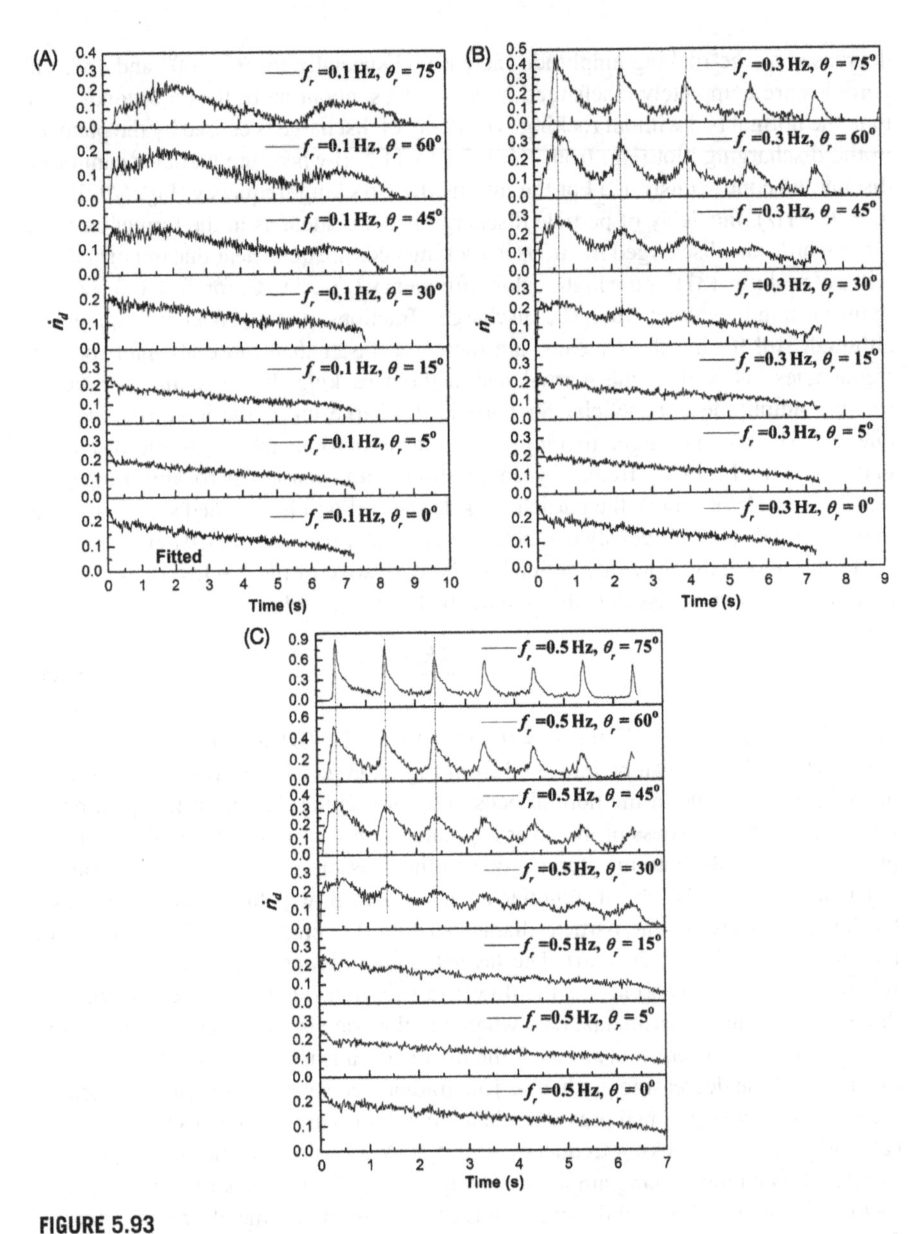

FIGURE 5.93

The temporal variations of flow rates for different rocking amplitudes when $f_r = 0.1$ Hz (A), 0.3 Hz (B), and 0.5 Hz (C), respectively.

$\theta_r \geq 30°$ are increasingly clearer than before, following still the same pace of oscillation. In addition, although with large rocking frequency, the flow rates under small rocking amplitudes ($\theta_r < 30°$) are not evidently changed.

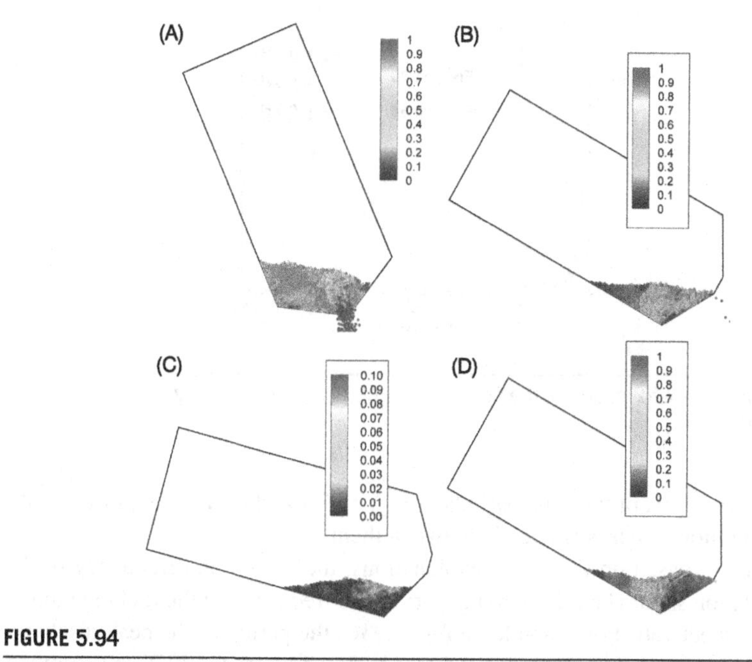

FIGURE 5.94

Particle distributions for the beds of $f_r = 0.1$ Hz and $\theta_r = 75°$ at $t = 8$ s (A), 9 s (B), 10 s (C), 11 s (D), respectively.

Based on Fig. 5.93, three important questions may be raised as follows: (1) Why the critical threshold value is 30°? (2) When $\theta_r \geq 30°$, why is the flow rate oscillated with the same pace, i.e. why forced oscillation occurs? How does it depend on the mode of rocking? (3) When $\theta_r < 30°$, is the flow rate exactly not changed by the rocking, or how will it be affected by the rocking? We will discuss and answer the three questions one by one in the following. The first and second questions are explained as follows. Recalling Fig. 5.92A, there is a plateau from about $t = 9$ s to $t = 10$ s for $\theta_r = 75°$ and $f_r = 0.1$ Hz. For this case, the particle distributions at $t = 8$ s, 9 s, 10 s, 11 s are shown in Fig. 5.94A, B, C, D, respectively. At $t = 8$ s (Fig. 5.94A), the bed is being tilted toward the west and the particles are still discharging from the orifice. At $t = 9$ s (Fig. 5.94B), as the orifice is higher than the particles near it when the bed goes on tilting. The particles are ceased to be discharged. Similar trends can be observed in Fig. 5.94C and 5.94D where the temporary cease of particle discharge is ongoing. As the flare angle of the bed's base is 120° (with 30° sidewalls), when the bed is tilted 30° toward either west or east, the lowest points of the bed are below the bed orifice (recalling Fig. 5.91E–G), where the particles may stay lower than the bed orifice and the discharging flow may be ceased temporarily. That is why the critical rocking amplitude (maximum angle of rocking) is 30°. When the rocking amplitude is larger than the critical value, after most of the particles are discharged, they will be temporally and periodically ceased to be discharged since they may stay near the corner and too low to reach the orifice. That is why the near zero flow rates (the plateaus)

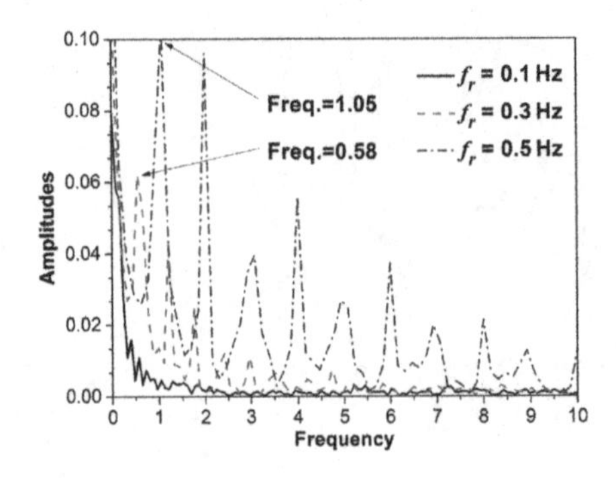

FIGURE 5.95

The Fast Fourier Transformation of the flow rate for the bed of $f_r = 0.1$ Hz, 0.3 Hz, 0.5 Hz under $\theta_r = 30°$.

occur in the most part of the time periods in Fig. 5.93B and C, and the pulsed peaks of discharging flow occur periodically between them.

Furthermore, based on the above explanations, the period and frequency of the forced oscillation should be related to the period and frequency of the rocking motion of the bed, respectively. For example, in Fig. 5.93C, the period of the peak discharging flow is about 1 s ($= \frac{1}{2} \times \frac{1}{0.5\ \text{Hz}} = 0.5 \times 2$ s). In addition, the FFT spectrums of the flow rate oscillation in the beds of $\theta_r = 30°$ at $f_r = 0.1$ Hz, 0.3 Hz and 0.5 Hz are shown in Fig. 5.95. For $f_r = 0.1$ Hz, there is no evident peak of frequency. For $f_r = 0.3$ Hz, the fundamental frequency is about $f_p = 0.58 \approx 0.6$ Hz ($= 2 \times 0.3$ Hz). For $f_r = 0.5$ Hz, it is about $f_p = 1.05 \approx 1$ Hz ($= 2 \times 0.5$ Hz). Note that, in one period of rocking, the flow rates are peaked two times. Thus, the period between two consecutive peaks of the discharge flow rate is only a half of the period of rocking. Meanwhile, the frequency (f_r) of the oscillated discharging flow is two times of the frequency (f_p) of rocking. That is why the flow discharge is forced to oscillate with the same pace.

For the third question, as the relation between the flow rate and time is almost linear, the linear fitting technique (linear regression) is used to analyze the influence of rocking on particle discharge. As shown in Fig. 5.96, the following points are indicated: (1) At the rocking frequency of $f_r = 0.5$ Hz, the slope of the flow rate is increased when the rocking amplitudes increase. (2) Similar trends are shown in other rocking frequencies, although the degrees of influences of the rocking amplitudes on the flow rates are different, i.e. it becomes clearer when the rocking frequency increases. (3) In the beds with a large rocking frequency and large rocking amplitude, the flow rates are initially larger than the flow rates with a small rocking frequency and small rocking amplitude, whereas they are finally smaller than that with small rocking frequency and amplitude, since the flow rates are decreasing faster under the large rocking frequency and amplitude. This confirms the observations that the beds with large rocking frequency and amplitude have larger flow rates than those with small rocking frequency and amplitude.

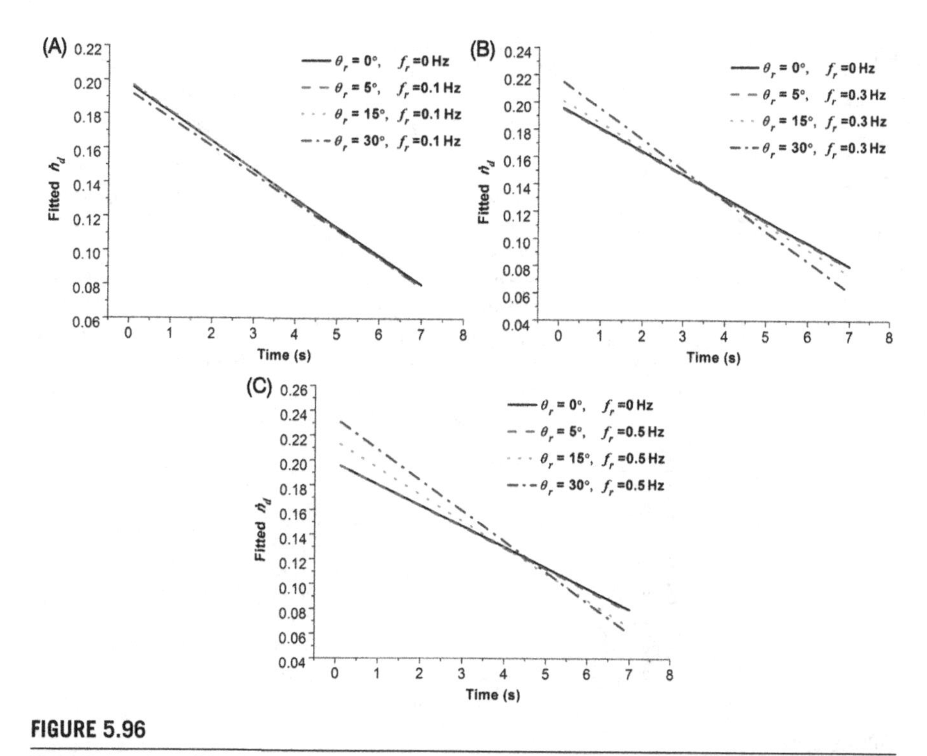

FIGURE 5.96

The linear regressions of the flow rates for the beds of $f_r = 0.1$ Hz (A), 0.3 Hz (B), 0.5 Hz (C) for $\theta_r < 30°$.

5.5.2.4 Effect of rocking frequency

In the above section, the effects of rocking frequency on particle discharge have already been partially explained. In this section, we will explain it complementarily.

Firstly, for example, the discharged fractions of particles in the beds with different rocking frequencies at the rocking amplitudes of $\theta_r = 15°$ (A), 45° (B) and 75° (C) are shown in Fig. 5.97A–C, respectively. Under the low rocking amplitude ($\theta_r = 15°$, Fig. 5.97A), the effect of rocking frequencies on the discharged fraction is negligible. Under the intermediate rocking amplitude ($\theta_r = 45°$, Fig. 5.97B), the discharged fractions are decreased mildly under small rocking frequencies ($f_r = 0.1$ and 0.2 Hz), whereas they are increased mildly under large rocking frequencies ($f_r = 0.3$–0.5 Hz). Under the large rocking amplitude ($\theta_r = 75°$, Fig. 5.97C), the trends are similar but clearer than the former case.

Recalling and comparing this to Fig. 5.92, it is clear that the effect of rocking frequency on the discharging rate is bifurcated, i.e. the low frequency is detrimental to particle discharge whereas the large frequency is beneficial, no matter what the rocking amplitude is. However, the rocking amplitude can influence or determine the degree of the rocking frequency's effect on particle discharge. In other words, the rocking frequency may be regarded as the main influencing factor on particle dis-

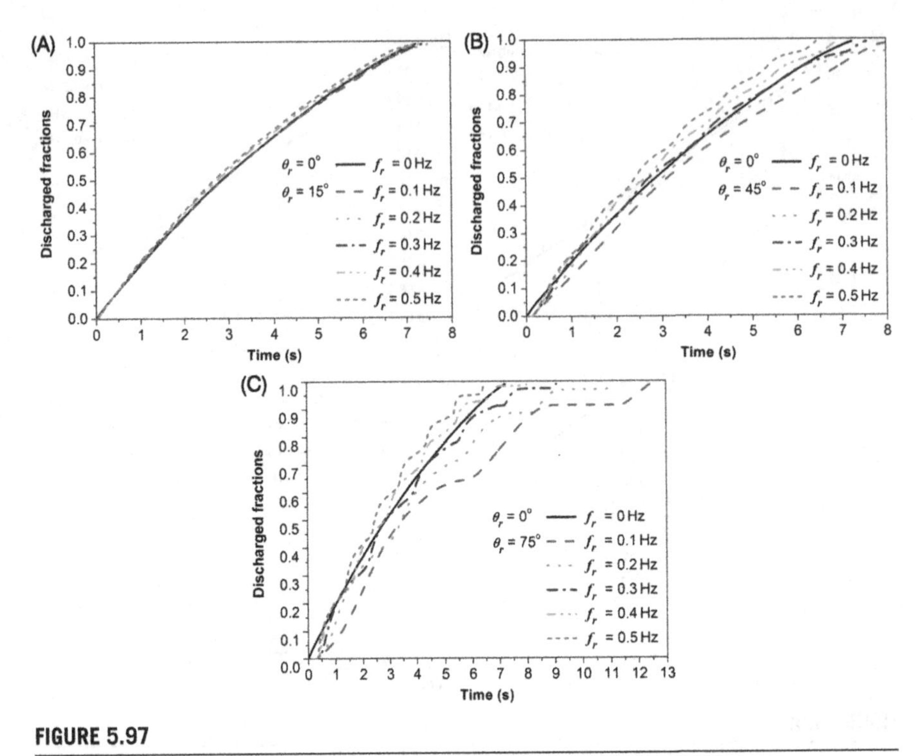

FIGURE 5.97

The discharged fractions for the beds with different rocking frequencies at $\theta_r = 15°$ (A), 45° (B) and 75° (C), respectively.

charge whereas the rocking amplitude is the secondary influencing factor. The main influencing factor determines the influence direction on particle discharge characteristics by the rocking, either enhancing or attenuating the particle discharging flow rates; the secondary influencing factor determines the degree of the main influencing factor's influence on particle discharge. Besides, the secondary influencing factor (rocking amplitude θ_r) also determines the mode of influence, i.e. either with a linear mode of enhancement when $\theta_r < 30°$ (critical value), or a oscillation mode of enhancement when $\theta_r \geq 30°$.

Secondly, with regard to the flow rate, Fig. 5.98A and B shows the comparison of flow rates between $\theta_r = 15°$ and $\theta_r = 75°$ under different rocking frequencies. It is seen that the influence of rocking frequency on particle flow rate $\dot{n}_d$ is indistinguishable when the rocking amplitude is small (Fig. 5.98A, $\theta_r = 15°$). In contrast, when the rocking amplitude is large (Fig. 5.98B, $\theta_r = 75°$), the influences of rocking frequency on flow rate are evident, i.e. the flow rates become pulsated, and the pulsation frequency and amplitude are determined by the rocking frequency. Additionally, the discharging rate is determined by the integral of the pulsated flow rate. As a result, a large rocking frequency may result in large and fast pulsation in flow rate and the integral of the pulsated flow rate may be larger than the normal bed without rocking. On

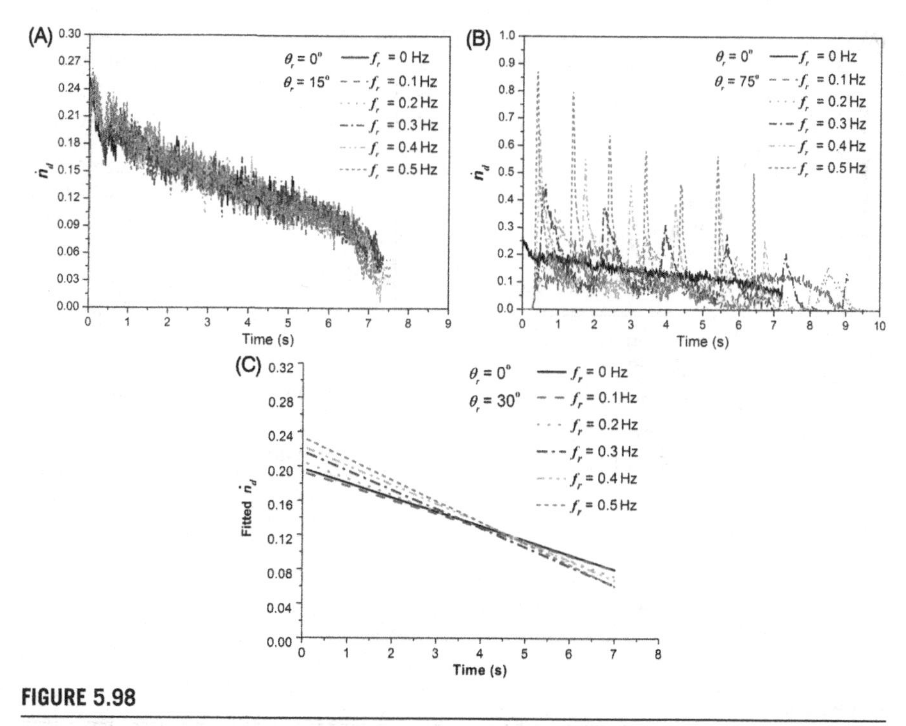

FIGURE 5.98

The temporal variations of flow rates for different rocking frequencies when $\theta_r = 15°$ (A) and 75° (B), and the linearly fitted flow rates for different rocking frequencies at $\theta_r = 30°$ (C), respectively.

the other hand, a small rocking frequency may result in small and slow pulsation in flow rates, causing the integral of the pulsated flow rate to be smaller than the normal bed without rocking. This explains the mechanism of the role of rocking frequency on influencing the discharging rate of particles, either beneficial or detrimental. In addition, at the intermediate and critical rocking amplitude (Fig. 5.98C, $\theta_r = 30°$), although the pulsating variation of the flow rates are not very clear, the fitted lines obtained by linear regressions indicate that the flow rates are changed by increasing the slope of variation as the rocking frequency increases. However, although the slopes are varied, the flow rates are initially increased and finally decreased and the areas covered by the lines are almost equivalent, i.e. the integral of the flow rates are very similar. That explains the reason of negligible variations in discharged fractions at $\theta_r = 30°$ in Fig. 5.92A–C.

5.5.2.5 Kinetic energy of the resident particles

As is mentioned before, the flow rate of particles is pulsating at large rocking frequency and amplitude. It is necessary to discuss the reason of pulsation of flow rates. In this section, the kinetic energy of the resident particles – the particles inside the bed before discharged, is to be discussed. The total kinetic energy and mean kinetic

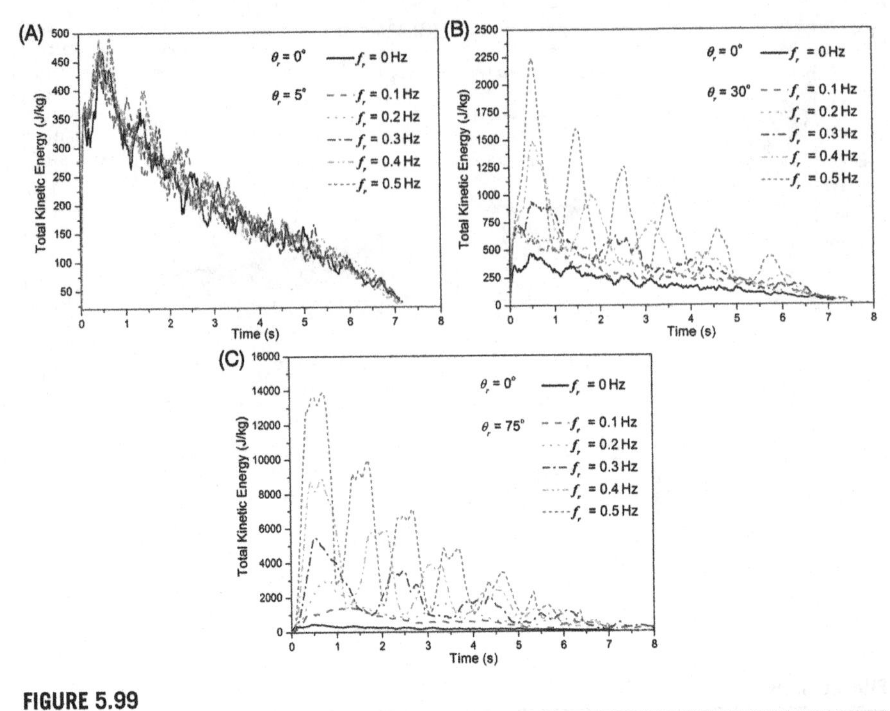

FIGURE 5.99

The temporal variation of the total kinetic energy of the particles inside the beds with different rocking frequencies at $\theta_r = 5°$ (A), 30° (B) and 75° (C), respectively.

energy is defined in the following, respectively:

$$E_{TOT}(t) = \sum_{j=1}^{n_r(t)} (E_{Tr,j}(t) + E_{Ro,j}(t)) \tag{5.85}$$

$$E_{ME}(t) = \frac{1}{n_r(t)} E_{TOT}(t) = \frac{1}{n_r(t)} \sum_{j=1}^{n_r(t)} (E_{Tr,j}(t) + E_{Ro,j}(t)) \tag{5.86}$$

where $E_{Tr,j}(t)$ is the translational kinetic energy and $E_{Ro,j}(t)$ is the rotational kinetic energy, and $n_r(t)$ is the number of particles still not discharged at time t. As all the particles have the same mass (i.e. $m_p = Cons.$), for analyzing the relative magnitudes and trends of the kinetic energies in Figs. 5.99 and 5.100, the mass can be omitted during calculating $E_{Tr,j}(t)$ and $E_{Ro,j}(t)$.

As shown in Fig. 5.99, the general trends of the total kinetic energy are decreasing with time, whereas the decreasing processes are pulsating with time, too. Under the low rocking amplitude (Fig. 5.99A, $\theta_r = 5°$), the decreasing process of the total kinetic energies under different rocking frequencies are almost indistinguishable from each other. However, for the intermediate and large rocking amplitudes, the pul-

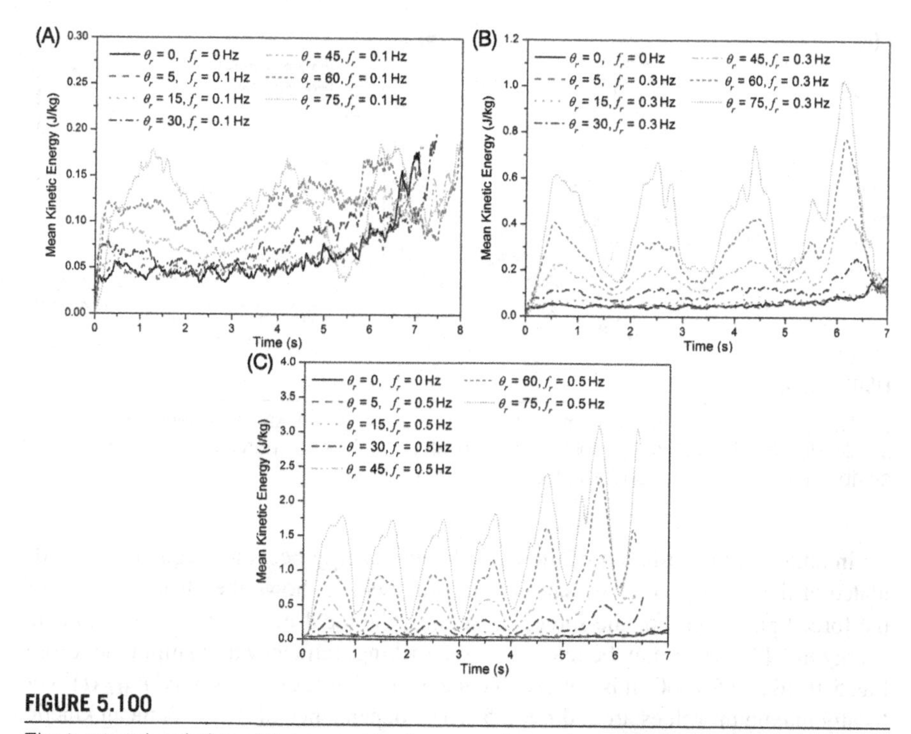

FIGURE 5.100

The temporal variation of the mean kinetic energy for each particle inside the beds with different rocking amplitudes at $f_r = 0.1$ Hz (A), 0.3 Hz (B) and 0.5 Hz (C), respectively.

sating variations of the total kinetic energy are augmented as the rocking frequency increases. Moreover, the frequency of pulsation of the total kinetic energy is determined by the frequency of rocking as well. On the other hand, comparing Fig. 5.99B to 5.99C and 5.99A, the pulsation amplitudes are augmented by the increase of the rocking amplitude, too. Thus, the influences of the rocking amplitude and frequency on the flow rates are mainly caused by the augmentation of pulsation amplitude and frequency of the total kinetic energy of particles.

More clear trends can be indicated by Fig. 5.100 with the mean kinetic energy of particles. In Fig. 5.100A, in general, the mean kinetic energy increases mildly as the rocking amplitude increases under the low rocking frequency of $f_r = 0.1$ Hz. But the variation seems irregular. In contrast, in Fig. 5.100B and C, the amplitudes of pulsations of the mean kinetic energies are increased with time and also the increase of rocking amplitudes of the beds. This explains that: (1) From the mean point of view, the particles may be agitated more heavily by large rocking frequencies than by small rocking frequencies. (2) Under same rocking frequency, the total amount of agitation energy which is possible to be transported from the rocking hopper to the particles can be regarded to be of the same level. Hence, when the discharging proceeds, the particle number is continuously decreasing, and the mean energy obtained by one particle from the rocking hopper will increase gradually.

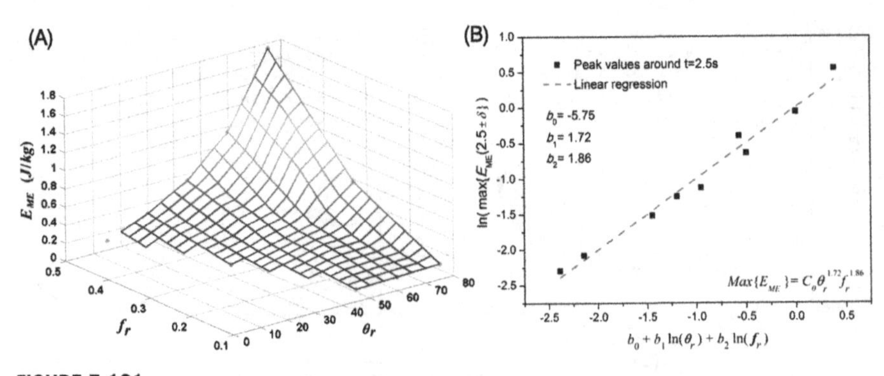

FIGURE 5.101

(A) The dependence of the peak mean kinetic energy on the rocking frequency and amplitude. (B) The relationship between the peak mean kinetic energy and a linear combination of frequency and amplitude in the log-log plot.

In addition, the variation of the mean kinetic energy becomes regular and modulated at the same pace of pulsation with time, which shows the clear evidence of the forced pulsation. For the forced pulsations, the local peaks of the mean kinetic energy $\max\{E_{ME}(t)\}$ may be relevant to the rocking frequency and amplitude. From Fig. 5.100B and 5.100C, it is noticed that the mean kinetic energies $\max\{E_{ME}(t)\}$ get locally maximum values around $t = 2.5$ s. The dependence of the peak mean kinetic energy on the rocking frequency and amplitude is shown in Fig. 5.101A. It is seen that the local peak values of $E_{ME}(t)$, i.e. $\max\{E_{ME}(t)\}$, increase faster than either the increase of rocking frequency or the increase of rocking amplitude. For quantitative analysis, the relationship between them is shown in Fig. 5.101B through the log-log plot. The evident linearity in Fig. 5.101B indicates the following relationship:

$$\max\{E_{ME}(t)\} = C_0\theta_r^{\alpha} f_r^{\beta} \tag{5.87}$$

where the exponents α and β are about 1.7 and 1.9, respectively. Referring to the motion equation of the bed (Eq. (5.82)), the rocking angular acceleration is $|\ddot{\theta}(t)| = |-4\pi^2\theta_r f_r^2 \sin(2\pi f_r t + \theta_0)| \leq |4\pi^2\theta_r f_r^2| \propto \theta_r f_r^2$, or $\max\{\ddot{\theta}(t)\} \propto \theta_r f_r^2$. As $I_p|\ddot{\theta}(t)|$ can be regarded as the 'driven force' added upon the particles and can be viewed as the 'displacement' driven by the force, the product of the force and displacement, i.e. $I_p|\ddot{\theta}(t)|\theta_r \propto \theta_r^2 f_r^2$, can be regarded as the external power input from the rocking beds. With

$$\begin{cases} \max\{E_{ME}(t)\} \propto \theta_r^{1.7} f_r^{1.9} \\ \max\{\text{rocking energy input from the bed}\} \propto \theta_r^2 f_r^2 \end{cases} \tag{5.88}$$

and $\alpha = 1.7 \approx 2$, $\beta = 1.9 \approx 2$, it is appropriate to conclude that: (1) The forced pulsating mean kinetic energy of particle is mainly determined by the 'external power

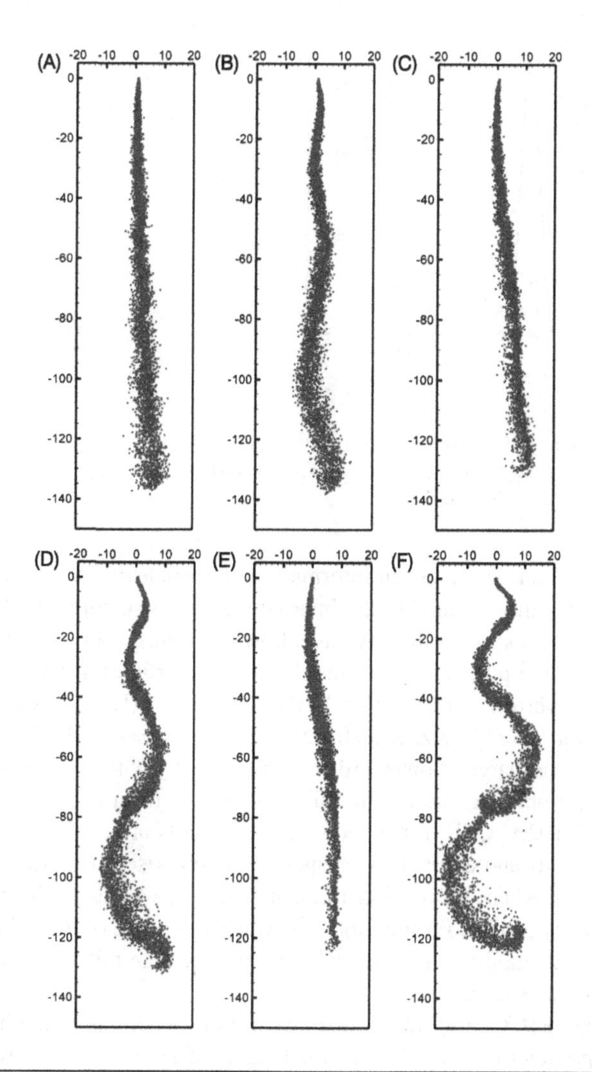

FIGURE 5.102

The distribution of discharged particles for: (A) $\theta_r = 15°$, $f_r = 0.1$ Hz; (B) $\theta_r = 15°$, $f_r = 0.5$ Hz; (C) $\theta_r = 45°$, $f_r = 0.1$ Hz; (D) $\theta_r = 45°$, $f_r = 0.5$ Hz; (E) $\theta_r = 75°$, $f_r = 0.1$ Hz; (F) $\theta_r = 75°$, $f_r = 0.5$ Hz.

input' from the rocking bed. (2) $\alpha < 2$ and $\beta < 2$ are partially because of the incomplete elastic collision nature ($e < 1$) which causes the dissipation of kinetic energy transfer from the rocking bed to particle, and partially caused by the damping and friction effects in the particle system.

5.5.2.6 Trajectory of the discharged particles

It is necessary to show the motion trajectories of discharged particles. Although the particles are driven only by gravity after being discharged, the distributions of all the

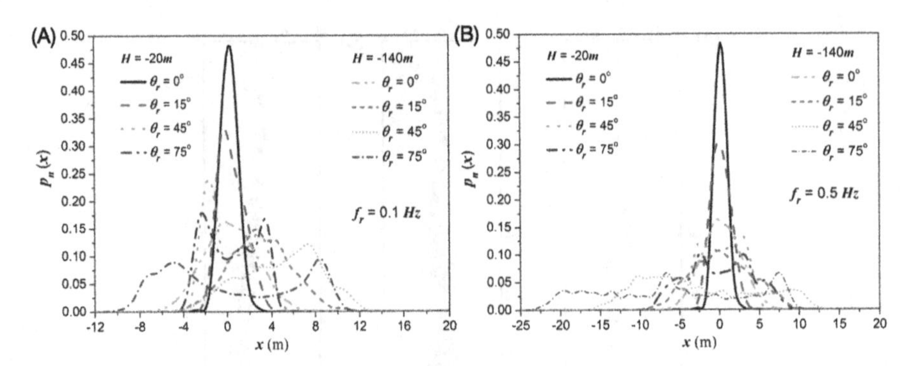

FIGURE 5.103

The probability density functions of particle density in the horizontal direction for all the particles passing through the heights of -20 and -140 m with $\theta_r = 0°, 15°, 45°, 75°$ at (A) $f_r = 0.1$ Hz and (B) $f_r = 0.5$ Hz.

particles may indicate some useful information on particle discharge, such as the history of flow continuity (related to the internal bridging structure and flow blockage) and the effects of rocking on the coverage, flow uniformity and continuity, etc.

For example, the positions of all the discharged particles at $t = 5$ s for $\theta_r = 5°$, 45° and 75° are shown in Fig. 5.102(A, B), (C, D), and (E, F), respectively, where (A, C, E) are for $f_r = 0.1$ Hz, and (B, D, F) are for $f_r = 0.5$ Hz. Without rocking, the particles are discharged downward straightforwardly with only mild spread in the horizontal direction, just like a long tail dragged out from the bed orifice. With a rocking (Fig. 5.102A, C, E, for $f_r = 0.1$ Hz), the tails are curved and rocked, too, featured by various and evident '*S*' shapes. By comparison, the wave length of the '*S*' shape is long when the rocking frequency is low, whereas it is short when the rocking frequency is high. Furthermore, the wave amplitude of the '*S*' shape is large when the rocking amplitude is large, whereas the wave amplitude is small when the rocking amplitude is small.

Finally, Fig. 5.103 shows the probability density functions of the horizontal distributions of particles passing through the heights of $H = -20$ m and -140 m. The probability density function $p_n(x)$ is defined as the number density $n(x)$ of all the particle trajectories passing though the horizontal location x on a fixed height. For each case, the distribution of particle trajectories on $H = -140$ m spreads wider and becomes smaller than that on $H = -20$ m, respectively. Moreover, similar trends can be seen when the rocking amplitude and frequency increases. Thus, the rocking motion makes the particles to discharge widely and become more averaged throughout the covering range. The point is that the rocking bed provides a way to possibly design an optimal discharging rocking bed to have wider covering area of discharge and a nearly even number density distribution throughout the covering area, which can be used in some powder industries. Indicated by Fig. 5.103, especially for the case of $\theta_r = 75°$ and $f_r = 0.5$, using a larger rocking amplitude and frequency may be a good choice to achieve that purpose.

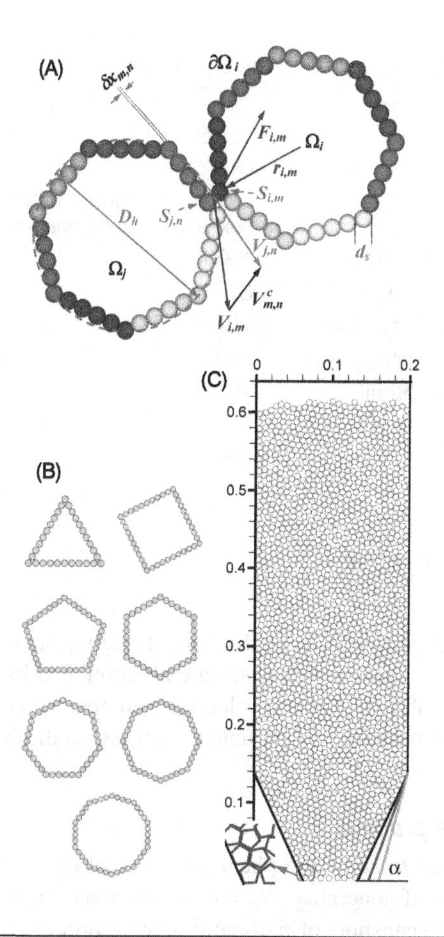

FIGURE 5.104

(A) Sketch of the particle-particle collision. (B) Nonspherical particles composed of small spherical particles. (C) The initially packing state of pentagonal particles inside the beds of $\alpha = 65°$, and bed configurations of different base angles $\alpha = 65°, 70°, 75°$.

5.5.3 Discharge of 2D polygonal particles

The SIPHPM method is used here to capture the dynamic discharge and static arching structure in the silo.

The different configurations of nonspherical shapes are shown in Fig. 5.104A and B. The particles of different shapes have the same area of $|\Omega| = A_p = 6.5\ \text{mm}^2$. The hopper bed is two-dimensional with 0.2 m wide and 0.65 m high (Fig. 5.104C). Three base angles of $\alpha = 65°, 70°, 75°$ are used, which are filled with 1800 particles. For clarity, the parameters used in simulation are listed in Table 5.14.

The process of filling is through a freely sedimentation process falling down from a certain height under gravity. After 2 s of sedimentation, let the velocity of particles be zero to keep them stationary inside the bed. After that, the bottom orifice of the

Table 5.14 Parameters used in the simulation.

Bed width and height (mm)	200 × 650
Height of the base (mm)	138.56
Bed base angle α (°)	65, 70, 75
Polygonal particle shape	circular, triangular, rectangular, pentagonal, hexagonal, heptagonal, octagonal, decagonal
Polygonal particle number N_p	1800
Polygonal particle area A_p (mm²)	6.5^2
Polygonal particle side a_p, (mm)	–, 9.88, 6.50, 4.96, 4.03, 3.42, 2.96, 2.34
Polygonal particle density ρ_p (kg/m³)	2000
Stiffness factor k_c, k_r (N/m)	10^4, 10^6
Poisson ratio ν	0.3
Restitution coefficient e	0.95
Friction coefficient μ	0.3
Time step δt (s)	10^{-6}
Number of simulation step N_t	3×10^6

hopper will open. The particles can be discharged freely from the bottom orifice. The collision, motion and discharge processes are all simulated by the discrete element method as mentioned above. With the velocities and positions of all the particles, the discharge feature and the effect of particle roundness on discharge can be analyzed as well.

5.5.3.1 Discharge process

At first, it is necessary to show the phenomenological observation of the discharge process. Without loss of generality, $N_s = 4$, 6, and 8 are taken for case study firstly. Fig. 5.105 shows the snapshots of particle discharge process for rectangular (A–C), hexagonal (D–F) and octagonal particles (G–I) at $t = 0.4$ s (A, D, G), 0.6 s (B, E, H) and 0.8 s (C, F, I) respectively, within the bed of $\alpha = 70°$. In general, at any time, the heights of packed rectangular particles (Fig. 5.105A–C) are higher than the heights of packed hexagonal particles (Fig. 5.105D–F), respectively. Similarly, the heights of packed hexagonal particles (Fig. 5.105D–F) are higher than the heights of packed octagonal particles (Fig. 5.105G–I), respectively. It means that the discharge speed is increasing from the rectangular particle case to the octagonal particle case. Moreover, for the rectangular particles, it seems more easily than for the other two cases to form internal arching structure (see the inset of Fig. 5.105A). As the boundary condition and parameters are the same, I think this may be related to the degree of roundness of particle shapes. Compared to the hexagonal and octagonal shapes under the same area, the square particle has the worst degree of roundness and the longest side-to-side contact length between neighboring particles. Thus, it may produce the largest motion resistance between contacting particles, including compression, rotation, and friction. Due to the continuous discharge, the arching structure is destroyed soon, which causes a sudden burst of increase in particle velocity and kinetic energies (see

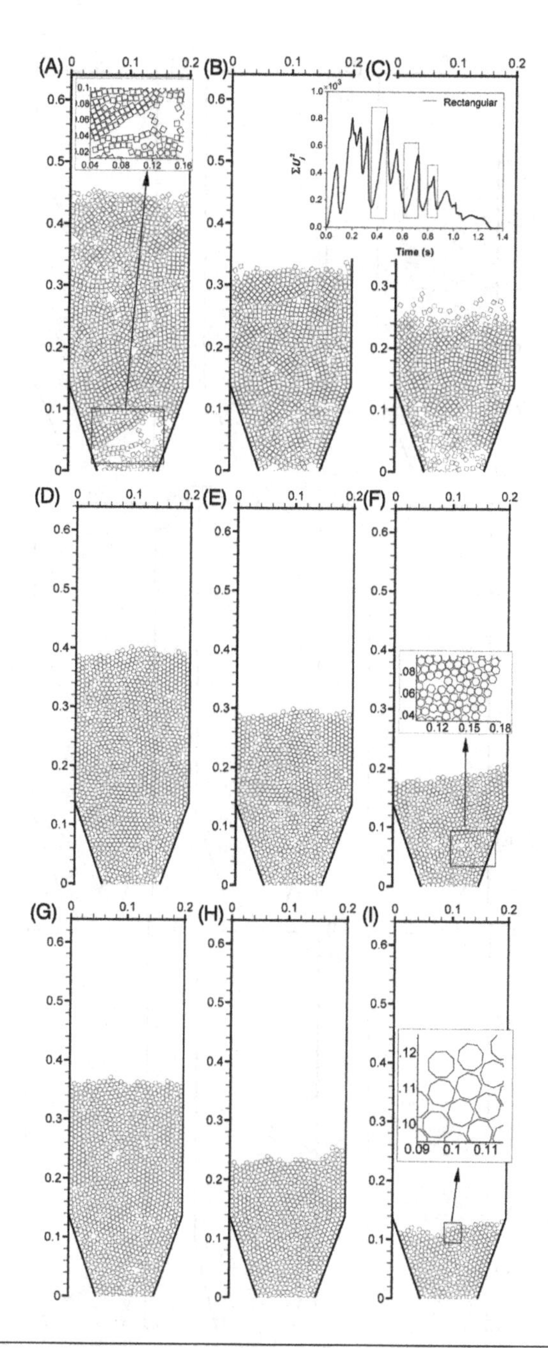

FIGURE 5.105

Snapshots of discharge of rectangular (A–C), hexagonal (D–F), and octagonal (G–I) particles at $t = 0.4$ s (left column), 0.6 s (middle column) and 0.8 s (right column), respectively, within the bed of $\alpha = 70°$. The insets show the amplified packing state of particles in the windowed local region, and the variation of sum of square velocities of all particles inside the bed.

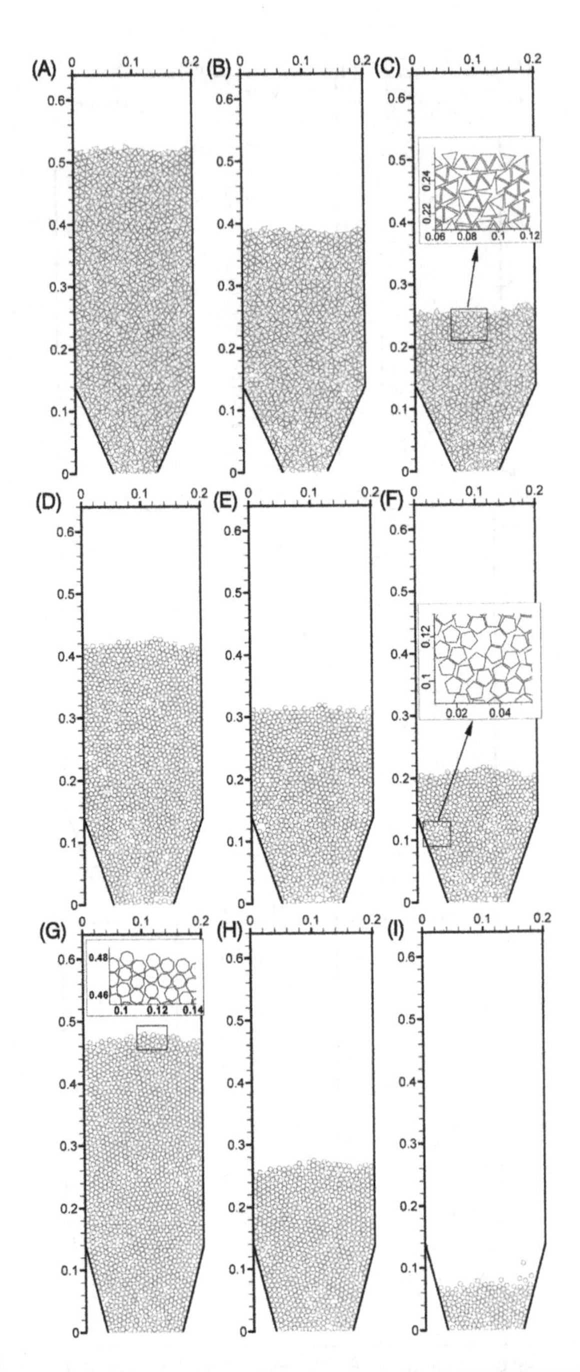

FIGURE 5.106

Snapshots of discharge of triangular particles at $t = 0.1$, 1.0 and 1.5 s (A–C), respectively, within the beds of $\alpha = 65°$. Snapshots of pentagonal particles at $t = 0.4$, 0.6 and 0.8 s (D–F), respectively, within the beds of $\alpha = 70°$. Snapshots of heptagonal particles at $t = 0.2$ s, 0.4 s and 0.6 s (G–I), respectively, within the beds of $\alpha = 75°$.

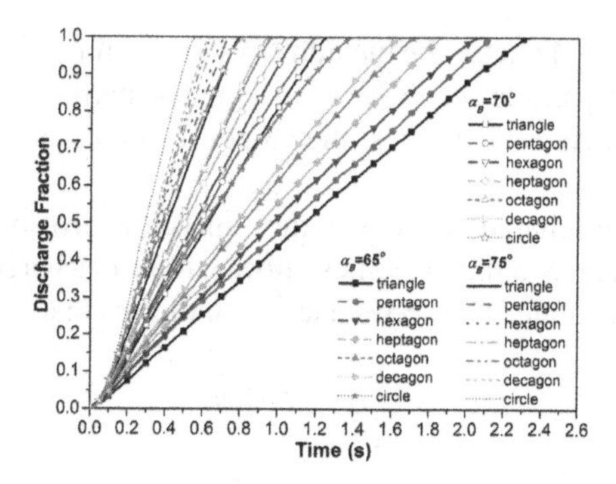

FIGURE 5.107

The time variations of discharge fraction of the particles of different shapes from the beds of different base angles.

the inset in Fig. 5.105B and C). As a consequence, the particle distribution seems to be disturbed by the formation and sudden destroy of internal arches (see the intensive fluctuation of the square of total particle velocities in the inset). However, for the other two cases, the formation of large scale arching structure seems not very clear.

Secondly, taking $N_s = 3, 5$, and 7 for example and considering the effect of different base angles, Fig. 5.106 shows the comparison of discharge process of triangular ((A–C), for $t = 0.1$, 1.0 and 1.5 s, respectively), pentagonal ((D–F), for $t = 0.4$, 0.6 and 0.8 s, respectively) and heptagonal particles ((G–I), for $t = 0.2$, 0.4 and 0.6 s, respectively) in the beds of $\alpha = 65°$ (A–C), 70° (D–F), 75° (G–I). It is seen that the triangular particles in the bed of $\alpha = 65°$ (A–C) are discharged more slowly than the pentagonal particles in the bed of $\alpha = 70°$ (D–F). The similar trend of even accelerated discharge process can be seen in Fig. 5.106G–I for heptagonal particles in the bed of $\alpha = 75°$. It means the base angle will also have great impact on the discharge process, which is certainly reasonable since the base angle determines the size of bed orifice for discharge.

5.5.3.2 Discharge fraction

The above section shows the qualitative comparison of particle discharge processes for differently shaped particles inside the beds of different base angles. For a quantitative evaluation, the number of discharged particles is divided by the total number of particles to give a variable, called discharge fraction:

$$r_D = \frac{\text{number of discharge particles}}{\text{total number of particles}}. \tag{5.89}$$

As shown in Fig. 5.107, it is seen that: (1) The discharge fraction is increasing with time almost linearly. (2) The slope of the linear increase depends on both the base angle and particle shape. The dependence on base angle shows the primary effect whereas the particle shape is of secondary effect. (3) In general, when the base angle increases, or equivalently, as the orifice size increases, the discharge fraction will certainly increase. (4) Let n_s be the number of sides for each shape, for example, $n_s = 0$ for a circular particle, $n_s = 3$ for a triangular particle, ..., and $n_s = 10$ for a decagonal particle. As n_s increases, the slope of the discharge fraction will also increase.

As the slope of discharge fraction is the time derivative of discharge fraction, which can be regarded as a dimensionless flow rate of discharge, i.e. $\dot{r}_D =$ dimensionless discharge flow rate. The above analysis indicates that

$$\dot{r}_D = \tilde{f}(\alpha, n_s) \tag{5.90a}$$

$$\text{or } \dot{n}_D = f(\alpha, n_s) \tag{5.90b}$$

where $\dot{n}_D = N_p \dot{r}_D$ is the discharging number flow rate and N_p is the total number of particles. $f(\alpha, n_s)$ is an increasing function of both α and n_s.

5.5.3.3 Discharge number flow rate

In order to give some useful expressions for $f(\alpha, n_s)$, it is necessary to notice that: For nonspherical particles, the effect of nonspherical shape may play a role of resistance to flowability. For regular polygonal shapes, when n_s increases, the shape will get closer to the spherical/circular shape, which leads to a reduction of flow resistance and increase of flowability. Thus, $f(\alpha, n_s)$ is an increasing function.

But, for more general conditions, i.e. for nonregular polygonal shape or even arbitrary shape, how to take into account the effect of particle shape? For this purpose, a general variable of roundness is used to replace n_s. The roundness is defined as the ratio Φ of the superficial area A_b of a sphere to the superficial area A_{nb} of a nonspherical shape under the condition of same volume $V_b = V_{nb}$, i.e.

$$\Phi = \frac{A_b}{A_{nb}}|_{V_b = V_{nb}}. \tag{5.91}$$

For spherical shape, $\Phi = 1$. For nonspherical shape, $\Phi < 1$.

Moreover, referring to the relation given by [57],

$$\dot{M}_D = C \rho_B g^{1/2} (D_o - kd)^{2.5} \tag{5.92a}$$

$$\text{or } \dot{n}_D = C \frac{\rho_B}{m_p} g^{1/2} (D_o - kd)^{2.5} \tag{5.92b}$$

where $\dot{M}_D$, $\dot{n}_D$, m_p, ρ_B, g, D_o, d are discharging mass flow rate and number flow rate, particle mass, bulk density, gravity, orifice size, particle size, respectively. k is a dimensionless coefficient. It indicates how the discharge flow rate $\dot{n}_D$ is proportional to $(D_o - kd)^{2.5}$.

The roundness is not considered in former studies, whereas the present work indicates that it has a great impact on the flowability and the discharge fraction of particles. In this section, an expression of $f(\alpha, \Phi)$ or $f(D_o, d, \Phi)$ is given through improving Eq. (5.92) by considering the effect of roundness Φ.

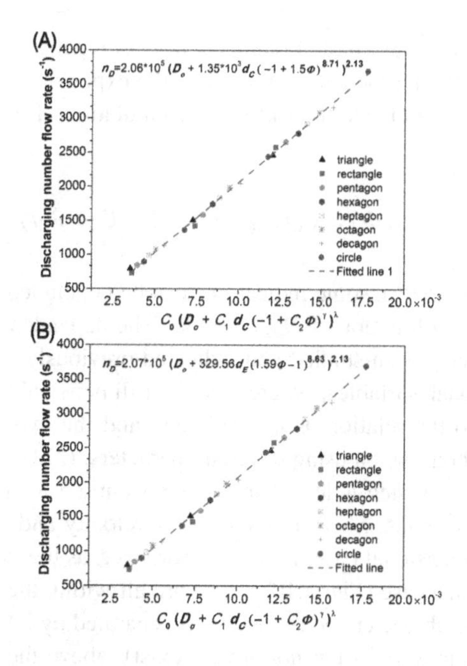

FIGURE 5.108

The data of discharge flow rates and the nonlinear regressions based on a model function of $\dot{n}_D = C_0(D_o + C_1 * d_C(-1 + C_2\Phi)^\gamma)^\lambda$, using the diameter of circumcircle d_C (A) and the equivalent diameter d_E (B), respectively.

In general, a model function of

$$\dot{n}_D = C_0(D_o + C_1 * d(C_2\Phi + C_3)^\gamma)^\lambda \tag{5.93}$$

is used to fit the data to give the expression of $f(\alpha, \Phi)$, where C_0, C_1, C_2, C_3 are fitting coefficients and γ, λ are exponents. d is particle diameter. As the discharging number flow rate $\dot{n}_D$ has the dimension of s^{-1}, C_0 is a dimensional coefficient with dimension $s^{-1} \cdot m^{-2.5}$. For nonspherical shapes, it can either use the diameter of circumcircle d_C of the polygonal shapes or the equivalent diameter d_E (the diameter of the circle of same area) to represent the particle diameter. In fact, both of them are used here.

For example, Fig. 5.108A shows the results of nonlinear regression using the diameter of circumcircle d_C. The obtained expression for $f(D_o, d_C, \Phi)$ is

$$\dot{n}_D = 2.06 * 10^5 * (D_o + 1.35 * 10^3 * d_C(-1 + 1.5\Phi)^{8.71})^{2.13}. \tag{5.94}$$

Instead, using the equivalent diameter d_E of the nonspherical particles, the relation is (Fig. 5.108B):

$$\dot{n}_D = 2.07 * 10^5 * (D_o + 329.56 * d_E(-1 + 1.59\Phi)^{8.63})^{2.13}. \tag{5.95}$$

When using different variables of diameter, the discrepancies only exist in the fitting coefficients, including the exponent γ. But the exponent λ is almost the same. Thus, the above results indicate the following formulation of $f(D_o, d, \Phi)$ for non-spherical particles:

$$\dot{n}_D = f(\alpha, \Phi) = C_0(D_o + C_1(-1 + C_2\Phi)^{\gamma} d)^{2.13}. \tag{5.96}$$

From Eqs. (5.93)–(5.96), some indications can be concluded as follows: (1) The discharge flow rate is a function of D_o, d and Φ. The dependence of $\dot{n}_D$ on the former two variables have been studied and indicated previously, e.g. [57] etc. (2) D_o and d are dimensional variables, whereas Φ is a dimensionless parameter. Thus, it can be added into the relation as a coefficient, and the exponent γ of the term $(-1 + C_2\Phi)$ is different when using different diameters. (3) However, the exponent λ is always near 2.13, which is less than the exponent 2.5 as in Eq. (5.92). Notice that the flow rate $\dot{n}_D \propto \bar{v}A$ where $\bar{v}$ is the mean velocity and A is the area of the silo. In the three-dimensional case, $A \propto D_o^2$. Moreover, as the flow rate is kept constant with time in each case (Fig. 5.107), they are all within the mass flow regimes. Thus, the bed can be characterized by two parts separated by internal arches (the internal arch is possible to exist but not always exist): above the arch, it is regarded as a main body of steady flow; below the arch, it is a discharge part where the particles can be regarded as discharging under gravity from an initially stationary or near stationary state. In the discharge part, there should exist a characteristic length scale, assumed to be $H^\star \propto D_o$. A particle falling in a height of $H^\star$, will get the velocity of $\bar{v} \propto \sqrt{2gH^\star} \propto D_o^{1/2}$. Thus, we get $\dot{n}_D \propto \bar{v}A \propto D_o^{5/2}$. Similarly, in the two-dimensional case, $A \propto D_o$, and $\dot{n}_D \propto \bar{v}D_o \propto D_o^{3/2}$. For the present case, it is a two-dimensional bed with a depth of d_p, where d_p is the diameter of subspheres. Thus, it could be regarded as a pseudo-three-dimensional bed. Thus, the exponent $\lambda = 2.13$ is between 5/2 and 3/2. Thus, the exponent λ depends on the configuration of hopper or macroscopic features. As a result, it is always kept constant for different roundness, i.e. independent of the particle-scale characteristics. (4) In addition, the coefficients, including the exponents γ and λ, are all obtained by a nonlinear curve fit mathematical function (using "nlinfit" function in MATLAB®). The fitting results, in particular the values of fitting coefficients and exponents, may be case-sensitive and case-dependent. That is one of the limitation of the present study as we cannot cover all possible cases to determine a general relation for actual industrial application. But the results and analyses are very indicative for practical application in engineering.

5.5.3.4 Mean voidages in discharge

At last, this section will show the effects of roundness of particles and the base angle of bed on the voidage variations in discharge. For comparison, the voidages are horizontally and temporally averaged to show the vertical variation them in steady

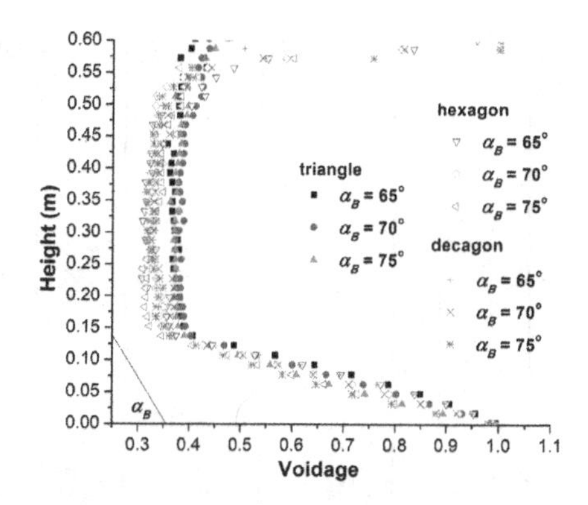

FIGURE 5.109

The temporally and horizontally averaged voidages for the beds discharging particles – triangular, hexagonal and decagonal – under various base angles.

discharge process, where the mean voidage $\bar{\epsilon}_v(y)$ is defined as

$$\bar{\epsilon}_v(y) = \frac{1}{N_t}\sum_{j=1}^{N_t}\frac{1}{W_B(y)}\int^{W_B(y)} A_y(x, t_j)dx, \tag{5.97}$$

where N_t, $W_B(y)$, A_y are the number of time steps, the width of the bed at height y, the area of the bed at (x, y) occupied by particles, respectively, and x, y and t_j are horizontal, vertical and temporal variables respectively.

As shown in Fig. 5.109, taking the triangular, hexagonal and decagonal shapes for example, it is found that: (1) In the main body of the bed ($0.138\text{ m} < y < 0.6\text{ m}$), the voidages are almost not evidently varied in the vertical direction. Compared to triangular particles, the voidage becomes gently smaller when n_s is larger. For different base angles, the voidage distributions are very similar and hard to be distinguished from each other. It means that the base angle difference has little effect on the voidage distribution in the main body, whereas the roundness of particles has a bit larger but still slight influence on the voidage distribution in the main body of the bed. (2) In the wedged base of the hopper ($0 \leq y \leq 0.138$ m), the voidage becomes very gently smaller when α_B or n_s is larger. But the differences are still negligible.

In conclusion, it is found that either the base angle or particle roundness has negligible effects on the vertical distribution of mean voidages of the bed. In addition, as the mean voidages of different cases are close to each other, the fitting coefficient C_0 corresponding to $C\rho_B g^{1/2}$ in Beverloo's relation where ρ_B is the bulk density is equivalent to that using real density of particles.

5.5.3.5 Effect of initial packing condition

What regards the effect of initial condition of packing, it comes through a sedimentation process as shown in Fig. 5.110. At first, the particles are initially distributed in a regular array above the same height of 0.138 m (immediately above the height

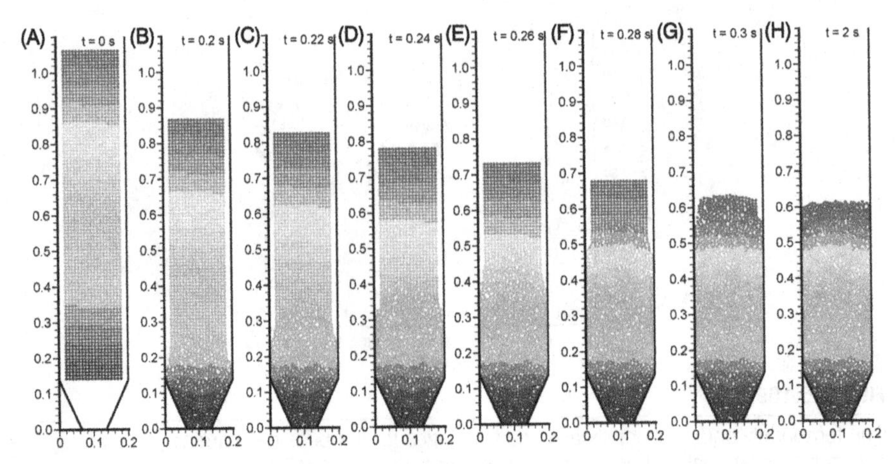

FIGURE 5.110

The process of forming initial packing states (herein the pentagonal particle within the bed of base angle $\alpha_B = 65°$ is taken for example).

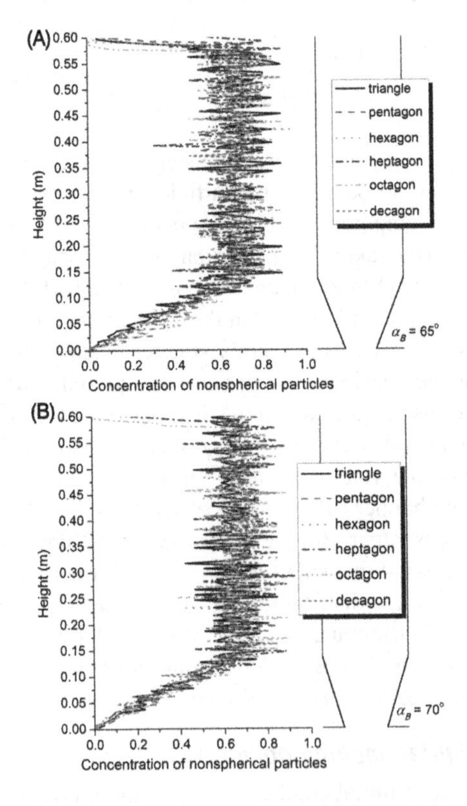

FIGURE 5.111

The initial concentration of large noncircular particles for the beds of base angle $\alpha_B = 65°$ (A) and 70° (B).

of the wedged base, Fig. 5.110A). Then, they start to fall into the bed under gravity. The particles fill in the bed rapidly from $t = 0.2$ s to $t = 0.3$ s (Fig. 5.110B–G). After about $t = 2$ s, the particles have already formed a randomly packing stationary state inside the bed (Fig. 5.110H). As the particle velocities are almost close to zero at $t = 2$ s because of the contact dissipation, they can be directly set to zero before being used as the initial state of random packing in the following discharge process.

As the velocities of particles are all zero before discharge, the effect of initial condition of packing may be mainly determined by the distribution of void fractions, or concentrations. The initial concentration of nonspherical particles can be calculated as $\epsilon_s(y) = 1 - \epsilon_v(y)$, and shown in Fig. 5.111A and B for the beds of $\alpha_B = 65°$ and $70°$, respectively. The ϵ_s of nonspherical particles (different n_s and α_B) are almost within the same order of magnitudes and hard to distinguish from each other. Thus, in general, the distributions of voids, or equivalently the initial randomly packing states, are similar to each other for different n_s and α_B. In conclusion, it is appropriate to consider that the effect of initial packing state for different cases is negligible in the present study.

5.6 Summary

In this chapter, the application of the SIPHPM and DEM models have been demonstrated for exploring the basic features on granular flow behaviors, especially on the drum mixing and hopper discharge characteristics. Based on the application results, it is certain that the models have powerful capability for various scientific studies and engineering applications. In the future, the SIPHPM or EHPM-DEM model may be further developed into a hybrid model, e.g. coupled with the finite element method for considering the deformation of particles.

Bibliography

[1] Phase, https://en.wikipedia.org/wiki/Phase.

[2] U. Frisch, Turbulence, Cambridge University Press, Cambridge, UK, 1995.

[3] S. Tian, Y. Gao, X. Dong, C. Liu, A definition of vortex vector and vortex, https://arxiv.org/abs/1712.03887.

[4] L. Chen, D. Tang, P. Lu, C. Liu, Evolution of the vortex structures and turbulent spots at the late-stage of transitional boundary layers, Science in China. Physics, Mechanics and Astronomy 54 (5) (2011) 986–990.

[5] Y. Yang, Y. Yan, C. Liu, ILES for mechanism of ramp-type MVG reducing shock induced flow separation, Science in China. Physics, Mechanics and Astronomy 59 (12) (2016) 124711.

[6] Y. Yan, C. Chen, H. Fu, C. Liu, DNS study on Λ-vortex and vortex ring formation in flow transition at Mach number 0.5, Journal of Turbulence 15 (1) (2014) 1–21, https://doi.org/10.1080/14685248.2013.871023.

[7] N. Mordant, P. Metz, O. Michel, J.F. Pinton, Measurement of Lagrangian velocity in fully developed turbulence, Physical Review Letters 87 (2001) 214501.

[8] L. Chevillard, S.G. Roux, E. Levêque, N. Mordant, J.F. Pinton, A. Arneodo, Lagrangian velocity statistics in turbulent flows: effects of dissipation, Physical Review Letters 91 (2003) 214502.

[9] G. Voth, K. Satyanarayan, E. Bodenschatz, Lagrangian acceleration measurements at large Reynolds numbers, Physics of Fluids 10 (1998) 2268–2280.

[10] S.B. Pope, Lagrangian PDF methods for turbulent flows, Annual Review of Fluid Mechanics 26 (1) (1994) 23–63.

[11] A. La Porta, G.A. Voth, A.M. Crawford, J. Alexander, E. Bodenschatz, Fluid particle accelerations in fully developed turbulence, Nature 409 (6823) (2001) 1017–1019.

[12] G. Falkovich, K. Gawedzki, M. Vergassola, Particles and fields in fluid turbulence, Reviews of Modern Physics 73 (4) (2001) 913–975.

[13] N. Mordant, A.M. Crawford, E. Bodenschatz, Three-dimensional structure of the Lagrangian acceleration in turbulent flows, Physical Review Letters 93 (2004) 214501.

[14] Y.L. Shi, M.B. Ray, A.S. Mujumdar, Effects of Prandtl number of impinging jet heat transfer under a semi-confined laminar slot jet, International Communications in Heat and Mass Transfer 30 (4) (2003) 455–464.

[15] J.K. Harvey, Some observations of the vortex breakdown phenomenon, Journal of Fluid Mechanics 14 (4) (2006) 585–592.

[16] T.B. Benjamin, Theory of the vortex breakdown phenomenon, Journal of Fluid Mechanics 87 (2) (2006) 518.

[17] T. Sarpkaya, On stationary and travelling vortex breakdowns, Journal of Fluid Mechanics 45 (3) (1971) 545–559.

[18] M.P. Escudier, N. Zehnder, Vortex-flow regimes, Journal of Fluid Mechanics 115 (115) (2006) 105–121.

[19] M. Escudier, Vortex breakdown: observations and explanations, Progress in Aerospace Sciences 25 (2) (1988) 189–229.

[20] S. Vladimir, F. Hussain, Collapse, symmetry breaking, and hysteresis in swirling flows, Annual Review of Fluid Mechanics 31 (31) (1999) 537–566.

[21] O. Lucca-Negro, T. O'Doherty, Vortex breakdown: a review, Progress in Energy and Combustion Science 27 (4) (2001) 431–481.

[22] J.E. Ruppert-Felsot, O. Praud, E. Sharon, H.L. Swinney, Extraction of coherent structures in a rotating turbulent flow experiment, Physical Review E 72 (2) (2005) 016311.

[23] M.G. Hall, Vortex breakdown, Annual Review of Fluid Mechanics 4 (4) (1972) 195–218.

[24] S. Leibovich, The structure of vortex breakdown, Annual Review of Fluid Mechanics 10 (10) (1978) 221–246.
[25] S.K. Robinson, Coherent motions in the turbulent boundary layer, Annual Review of Fluid Mechanics 23 (1) (2003) 601–639.
[26] D.J. Peake, M. Tobak, On issues concerning flow separation and vortical flows in three dimensions, in: On Issues Concerning Flow Separation & Vortical Flows in Three Dimensions, 1983.
[27] K.R. Sreenivasan, Fluid turbulence, Reviews of Modern Physics 71 (2) (1999) 644–645.
[28] S.I. Shtork, N.F. Vieira, E.C. Fernandes, On the identification of helical instabilities in a reacting swirling flow, Fuel 87 (10–11) (2008) 2314–2321.
[29] C.F. Edwards, N.R. Fornaciari, C.M. Dunsky, K.D. Marx, W.T. Ashurst, Spatial structure of a confined swirling flow using planar elastic scatter imaging and laser Doppler velocimetry, Fuel 72 (8) (1993) 1151–1159.
[30] Q. Shang, J. Zhang, Simulation of gas-particle turbulent combustion in a pulverized coal-fired swirl combustor, Fuel 88 (1) (2009) 31–39.
[31] K.R. Dinesh, K. Jenkins, M. Kirkpatrick, W. Malalasekera, Modelling of instabilities in turbulent swirling flames, Fuel 89 (1) (2010) 10–18.
[32] C. Crowe, On the relative importance of particle-particle collisions in gas-particle flows, in: Proceedings of the Conference on Gas Borne Particles, 1981, pp. 135–137.
[33] R. Gore, C. Crowe, Effect of particle size on modulating turbulent intensity, International Journal of Multiphase Flow 15 (2) (1989) 279–285.
[34] J.D. Kulick, J.R. Fessler, J.K. Eaton, Particle response and turbulence modification in fully developed channel flow, Journal of Fluid Mechanics 277 (1994) 109–134.
[35] S. Sundaram, L.R. Collins, A numerical study of the modulation of isotropic turbulence by suspended particles, Journal of Fluid Mechanics 379 (2000) 105–143.
[36] S. Elghobashi, G. Truesdell, Direct simulation of particle dispersion in a decaying isotropic turbulence, Journal of Fluid Mechanics 242 (1992) 655–700.
[37] S. Sundaram, L.R. Collins, Collision statistics in an isotropic particle-laden turbulent suspension. Part 1. Direct numerical simulations, Journal of Fluid Mechanics 335 (1) (2000) 75–109.
[38] L.P. Wang, Y. Zhou, A.S. Wexler, Statistical mechanical description and modelling of turbulent collision of inertial particles, Journal of Fluid Mechanics 415 (2000) 117–153.
[39] K. Hadinoto, J.S. Curtis, Reynolds number dependence of gas-phase turbulence in particle-laden flows: effects of particle inertia and particle loading, Powder Technology 195 (2) (2009) 119–127.
[40] V.S. L'Vov, G. Ooms, A. Pomyalov, Effect of particle inertia on turbulence in a suspension, Physical Review E 67 (4 Pt 2) (2003) 046314.
[41] Y. Yu, L. Zhou, B. Wang, Modeling of fluid turbulence modification using two-time-scale dissipation models and accounting for the particle wake effect, Chinese Journal of Chemical Engineering 14 (3) (2006) 314–320.
[42] R.C. Hogan, J.N. Cuzzi, Cascade model for particle concentration and enstrophy in fully developed turbulence with mass-loading feedback, Physical Review E 75 (2007) 056305.
[43] M. Tanaka, Y. Maeda, Y. Hagiwara, Turbulence modification in a homogeneous turbulent shear flow laden with small heavy particles, International Journal of Heat and Fluid Flow 23 (5) (2002) 615–626.
[44] Y. Sato, U. Fukuichi, K. Hishida, Effect of inter-particle spacing on turbulence modulation by Lagrangian PIV, in: Turbulence and Shear Flow Phenomena 1, International Journal of Heat and Fluid Flow 21 (5) (2000) 554–561.
[45] S.L. Rani, C. Winkler, S. Vanka, Numerical simulations of turbulence modulation by dense particles in a fully developed pipe flow, Powder Technology 141 (1) (2004) 80–99.
[46] J. Fan, K. Luo, Y. Zheng, H. Jin, K. Cen, Modulation on coherent vortex structures by dispersed solid particles in a three-dimensional mixing layer, Physical Review E 68 (2003) 036309.
[47] Y. Pan, T. Tanaka, Y. Tsuji, Turbulence modulation by dispersed solid particles in rotating channel flows, International Journal of Multiphase Flow 28 (4) (2002) 527–552.
[48] G. Ooms, C. Poelma, Comparison between theoretical predictions and direct numerical simulation results for a decaying turbulent suspension, Physical Review E 69 (2004) 056311.

[49] G. Hetsroni, Particles-turbulence interaction, International Journal of Multiphase Flow 15 (5) (1989) 735–746.

[50] T. Tanaka, J.K. Eaton, Classification of turbulence modification by dispersed spheres using a novel dimensionless number, Physical Review Letters 101 (2008) 114502.

[51] K.D. Squires, J.K. Eaton, Preferential concentration of particles by turbulence, Physics of Fluids. A 3 (5) (1991) 169–209.

[52] C. Crowe, M. Sommerfeld, Y. Tsuji, Multiphase Flows With Droplets and Particles, CRC Press, 1998.

[53] P. Cundall, O. Strack, A discrete numerical model for granular assemblies, Géotechnique 29 (1979) 47–65.

[54] P.G. Saffman, J.S. Turner, On the collision of drops in turbulent clouds, Journal of Fluid Mechanics 1 (1) (2006) 16–30.

[55] J. Abrahamson, Collision rates of small particles in a vigorously turbulent fluid, Chemical Engineering Science 30 (11) (1975) 1371–1379.

[56] So much more to know, Science 309 (5731) (2005) 78–102, http://science.sciencemag.org/content/309/5731/78.2.full.pdf, http://science.sciencemag.org/content/309/5731/78.2.

[57] W.A. Beverloo, H.A. Leniger, J.V.D. Velde, The flow of granular solids through orifices, Chemical Engineering Science 15 (3) (1961) 260–269.

[58] A. Anand, J.S. Curtis, C.R. Wassgren, B.C. Hancock, W.R. Ketterhagen, Predicting discharge dynamics of wet cohesive particles from a rectangular hopper using the discrete element method (DEM), Chemical Engineering Science 64 (24) (2009) 5268–5275.

[59] S. Albaraki, S.J. Antony, How does internal angle of hoppers affect granular flow? Experimental studies using digital particle image velocimetry, Powder Technology 268 (1) (2014) 253–260.

[60] P.A. Langston, U. Tüzün, D.M. Heyes, Discrete element simulation of granular flow in 2D and 3D hoppers: dependence of discharge rate and wall stress on particle interactions, Chemical Engineering Science 50 (6) (1995) 967–987.

[61] H.P. Zhu, K.J. Dong, A.B. Yu, R.P. Zou, G. Roach, Numerical Simulation of the Flow of Fine Particles in a Hopper, AIP Conference Proceedings, vol. 1145, 2009, pp. 661–664.

[62] D. Höhner, S. Wirtz, V. Scherer, A numerical study on the influence of particle shape on hopper discharge within the polyhedral and multi-sphere discrete element method, Powder Technology 226 (8) (2012) 16–28.

[63] D. Höhner, S. Wirtz, V. Scherer, Experimental and numerical investigation on the influence of particle shape and shape approximation on hopper discharge using the discrete element method, Powder Technology 235 (2) (2013) 614–627.

[64] R.M. Nedderman, Statics and Kinematics of Granular Materials, Cambridge University Press, 1992.

[65] M. Gentzler, G.I. Tardos, Measurement of velocity and density profiles in discharging conical hoppers by NMR imaging, Chemical Engineering Science 64 (22) (2009) 4463–4469.

[66] J. Tang, R.P. Behringer, How granular materials jam in a hopper, Chaos. An Interdisciplinary Journal of Nonlinear Science 21 (4) (2011) 041107.

[67] Y. Yu, H. Saxén, Segregation behavior of particles in a top hopper of a blast furnace, Powder Technology 262 (2014) 233–241.

[68] W.R. Ketterhagen, J.S. Curtis, C.R. Wassgren, B.C. Hancock, Modeling granular segregation in flow from quasi-three-dimensional, wedge-shaped hoppers, Powder Technology 179 (3) (2008) 126–143.

[69] P.A. Langston, U. Tüzün, D.M. Heyes, Discrete element simulation of internal stress and flow fields in funnel flow hoppers, Powder Technology 85 (2) (1995) 153–169.

[70] Q.J. Zheng, A.B. Yu, Finite element investigation of the flow and stress patterns in conical hopper during discharge, Chemical Engineering Science 129 (2015) 49–57.

[71] R. Balevičius, R. Kačianauskas, Z. Mróz, I. Sielamowicz, Analysis and DEM simulation of granular material flow patterns in hopper models of different shapes, in: Special Issue of the 6th World Congress on Particle Technology, Advanced Powder Technology 22 (2) (2011) 226–235.

[72] C. González-Montellano, F. Ayuga, J.Y. Ooi, Discrete element modelling of grain flow in a planar silo: influence of simulation parameters, Granular Matter 13 (2) (2011) 149–158.

[73] C. González-Montellano, A. Ramírez, E. Gallego, F. Ayuga, Validation and experimental calibration of 3D discrete element models for the simulation of the discharge flow in silos, Chemical Engineering Science 66 (21) (2011) 5116–5126.

[74] R.C. Hidalgo, C. Lozano, I. Zuriguel, A. Garcimartin, Force analysis of clogging arches in a silo, Granular Matter 15 (6) (2013) 841–848.

[75] I. Oldal, I. Keppler, B. Csizmadia, L. Fenyvesi, Outflow properties of silos: the effect of arching, Advanced Powder Technology 23 (3) (2012) 290–297.

[76] R.O. Uñac, A.M. Vidales, O.A. Benegas, I. Ippolito, Experimental study of discharge rate fluctuations in a silo with different hopper geometries, Powder Technology 225 (7) (2012) 214–220.

[77] H. Ahn, Computer simulation of rapid granular flow through an orifice, Journal of Applied Mechanics 74 (1) (2007) 111–118.

[78] A.A. Boateng, Rotary Kilns: Transport Phenomena and Transport Processes, Butterworth Heinemann USA, 2008.

[79] G.H. Lohnert, H. Reutler, The modular HTR-a new design of high temperature pebble bed reactor, Nuclear Energy 22 (3) (1982) 197–200.

[80] H. Frewer, W. Keller, R. Pruschek, The modular high-temperature reactor, Nuclear Science and Engineering 90 (4) (1985) 4.

[81] A. Koster, H. Matzner, D. Nicholsi, PBMR design for the future, in: HTR-2002 1st International Topical Meeting on High Temperature Reactor Technology, Nuclear Engineering and Design 222 (2) (2003) 231–245.

[82] W. Terry, H. Gougar, A. Ougouag, Direct deterministic method for neutronics analysis and computation of asymptotic burnup distribution in a recirculating pebble-bed reactor, Annals of Nuclear Energy 29 (11) (2002) 1345–1364.

[83] A.C. Kadak, M.V. Berte, Advanced modularity design for the MIT pebble bed reactor, Nuclear Engineering and Design 236 (5–6) (2006) 502–509.

[84] R. Schulten, Pebble bed HTRs, in: Thorium and Gas Cooled Reactors, Annals of Nuclear Energy 5 (8) (1978) 357–374.

[85] D. Wang, Y. Lu, Roles and prospect of nuclear power in China's energy supply strategy, Nuclear Engineering and Design 218 (1) (2002) 3–12.

[86] Z. Zhang, Z. Wu, Y. Sun, F. Li, Design aspects of the Chinese modular high-temperature gas-cooled reactor HTR-PM, Nuclear Engineering and Design 236 (5–6) (2006) 485–490.

[87] C. Li, X. An, R. Yang, R. Zou, A. Yu, Experimental study on the packing of uniform spheres under three-dimensional vibration, Powder Technology 208 (3) (2011) 617–622.

[88] N. Gui, X. Yang, J. Tu, S. Jiang, Effect of bed configuration on pebble flow uniformity and stagnation in the pebble bed reactor, Nuclear Engineering and Design 270 (5) (2014) 295–301.

[89] X. Yang, W. Hu, S. Jiang, K. Wong, J. Tu, Mechanism analysis of quasi-static dense pebble flow in pebble bed reactor using phenomenological approach, Nuclear Engineering and Design 250 (Supplement C) (2012) 247–259.

[90] S. Jiang, X. Yang, Z. Tang, W. Wang, J. Tu, Z. Liu, J. Li, Experimental and numerical validation of a two-region-designed pebble bed reactor with dynamic core, Nuclear Engineering and Design 246 (Supplement C) (2012) 277–285.

[91] A.C. Kadak, M.Z. Bazant, Pebble flow experiments for pebble bed reactors, in: Proceedings of International Topical Meeting on High Temperature Reactor Technology, 2002.

[92] J. Choi, A. Kudrolli, R.R. Rosales, M.Z. Bazant, Diffusion and mixing in gravity-driven dense granular flows, Physical Review Letters 92 (17) (2004) 174301.

[93] J. Choi, A. Kudrolli, M.Z. Bazant, Velocity profile of granular flows inside silos and hoppers, Chemical Engineering Science 17 (24) (2005) S2533–S2548.

[94] Y. Li, Y. Xu, S. Jiang, DEM simulations and experiments of pebble flow with monosized spheres, in: Special Issue: Discrete Element Methods: the 4th International Conference on Discrete Element Methods, Powder Technology 193 (3) (2009) 312–318.

[95] A. Shams, F. Roelofs, E. Komen, E. Baglietto, Optimization of a pebble bed configuration for quasi-direct numerical simulation, Nuclear Engineering and Design 242 (Supplement C) (2012) 331–340.

[96] A. Shams, F. Roelofs, E. Komen, E. Baglietto, Quasi-direct numerical simulation of a pebble bed configuration. Part I: flow (velocity) field analysis, Nuclear Engineering and Design 263 (Supplement C) (2013) 473–489.
[97] A. Shams, F. Roelofs, E. Komen, E. Baglietto, Large eddy simulation of a nuclear pebble bed configuration, Nuclear Engineering and Design 261 (Supplement C) (2013) 10–19.
[98] Y.M. Ferng, K.Y. Lin, Investigating effects of BCC and FCC arrangements on flow and heat transfer characteristics in pebbles through CFD methodology, Nuclear Engineering and Design 258 (Supplement C) (2013) 66–75.
[99] X. Yang, W. Hu, S. Jiang, Experimental investigation on feasibility of two-region-designed pebble-bed high-temperature gas-cooled reactor, Journal of Nuclear Science and Technology 46 (4) (2009) 374–381.
[100] N. Gui, X.T. Yang, J.Y. Tu, S.Y. Jiang, A simple geometrical model for analyzing particle dispersion in a gravity-driven monolayer granular bed, Powder Technology 254 (2) (2014) 432–438.
[101] Y. Li, N. Gui, X. Yang, J. Tu, S. Jiang, Experimental research and DEM simulations on stagnant region in pebble bed reactor, in: International Conference on Nuclear Engineering, 2013.
[102] Y. Li, N. Gui, X. Yang, J. Tu, S. Jiang, Effect of wall structure on pebble stagnation behavior in pebble bed reactor, Annals of Nuclear Energy 80 (2015) 195–202.
[103] C. Crowe, Review—numerical models for dilute gas-particle flows, Journal of Fluids Engineering 104 (3) (1982) 297–303.
[104] Y. Tsuji, Activities in discrete particle simulation in Japan, in: Neptis Symposium on Fluidization—Present and Future, Powder Technology 113 (3) (2000) 278–286.
[105] C.S. Campbell, Granular shear flows at the elastic limit, Journal of Fluid Mechanics 465 (465) (2002) 261–291.
[106] C.S. Campbell, Stress-controlled elastic granular shear flows, Journal of Fluid Mechanics 539 (539) (2005) 273–297.
[107] C.S. Campbell, Granular material flows – an overview, Powder Technology 162 (3) (2006) 208–229.
[108] C.G. Speziale, Analytical methods for the development of Reynolds-stress closures in turbulence, Annual Review of Fluid Mechanics 23 (1) (2003) 107–157.
[109] W. Rodi, A new algebraic relation for calculating the Reynolds stresses, in: Gesellschaft Angewandte Mathematik und Mechanik, Workshop Paris France, 1976, pp. T 219–T 221.
[110] S. Wallin, A.V. Johansson, An explicit algebraic Reynolds stress model for incompressible and compressible turbulent flows, Journal of Fluid Mechanics 403 (2000) 89–132.
[111] P. Spalart, S. Allmaras, A one-equation turbulence model for aerodynamic flows, La Recherche Aérospatiale 439 (1) (1992) 5–21.
[112] F.R. Menter, Two-equation eddy-viscosity turbulence models for engineering applications, AIAA Journal 32 (2) (1994) 1598–1605.
[113] S.B. Pope, A more general effective-viscosity hypothesis, Journal of Fluid Mechanics 72 (2) (1975) 331–340.
[114] L. Prandtl, Über die ausgebildete Turbulenz, Zeitschrift für Angewandte Mathematik und Mechanik 5 (21) (1925) 36–39.
[115] L. Prandtl, K. Wieghardt, Über ein neues Formelsystem für die ausgebildete Turbulenz, Nachrichten der Akademie der Wissenschaften in Göttingen. 2, Mathematisch-Physikalische Klasse (1945) 6–19.
[116] A.N. Kolmogorov, Equations of turbulent motion of an incompressible fluid, Izvestiâ Akademii Nauk SSSR. Seriâ Fizičeskaâ 6 (6) (1942) 56–58.
[117] J.L. Lumley, Computational modeling of turbulent flows, Advances in Applied Mechanics 18 (1979) 123–176.
[118] J. Hinze, Turbulence, 2nd ed., McGraw-Hill, New York, N.Y., 1975.
[119] J. Smagorinsky, General circulation experiments with the primitive equations, Monthly Weather Review 91 (3) (1963) 99–164.
[120] S.B. Pope, Turbulent Flows, Cambridge University Press, Cambridge, UK, 2000.
[121] M. Germano, U. Piomelli, P. Moin, W.H. Cabot, A dynamic subgrid-scale eddy viscosity model, Physics of Fluids. A, Fluid Dynamics 3 (3) (1991) 1760–1765.
[122] S. Orszag, Analytical theories of turbulence, Journal of Fluid Mechanics 41 (1970) 363–386.

[123] X. Shan, G. Doolen, Multicomponent lattice-Boltzmann model with interparticle interaction, Journal of Statistical Physics 81 (1) (1995) 379–393.
[124] Y.H. Qian, D. D'Humières, P. Lallemand, Lattice BGK models for Navier-Stokes equation, Europhysics Letters 17 (6) (1992) 479.
[125] L. Giraud, D. D'Humières, P. Lallemand, A lattice Boltzmann model for Jeffreys viscoelastic fluid, Europhysics Letters 42 (6) (1998) 625.
[126] D. D'Humières, Generalized lattice-Boltzmann equations, in: AIAA Rarefied Gas Dynamics: Theory and Applications, in: Progress in Astronautics and Aeronautics, 1992.
[127] S. Chen, G.D. Doolen, Lattice Boltzmann method for fluid flows, Annual Review of Fluid Mechanics 30 (1) (2012) 329–364.
[128] X. Shan, H. Chen, Lattice Boltzmann model for simulating flows with multiple phases and components, Physical Review E 47 (1993) 1815–1819.
[129] M.R. Swift, W.R. Osborn, J.M. Yeomans, Lattice Boltzmann simulation of nonideal fluids, Physical Review Letters 75 (1995) 830–833.
[130] X. He, X. Shan, G.D. Doolen, Discrete Boltzmann equation model for nonideal gases, Physical Review E 57 (1998) R13–R16.
[131] J. Wang, X. Zhang, A.G. Bengough, J.W. Crawford, Domain-decomposition method for parallel lattice Boltzmann simulation of incompressible flow in porous media, Physical Review E 72 (2005) 016706.
[132] C.U. Hatiboglu, T. Babadagli, Lattice-Boltzmann simulation of solvent diffusion into oil-saturated porous media, Physical Review E 76 (2007) 066309.
[133] M.C. Sukop, D.T. Thorne, Lattice Boltzmann Modeling an Introduction for Geoscientists and Engineers, Springer-Verlag Berlin Heidelberg, 2006.
[134] G. Molaeimanesh, H.S. Googarchin, A.Q. Moqaddam, Lattice Boltzmann simulation of proton exchange membrane fuel cells – a review on opportunities and challenges, International Journal of Hydrogen Energy 41 (47) (2016) 22221–22245.
[135] M. van der Hoef, M. van Sint Annaland, J. Kuipers, Computational fluid dynamics for dense gas–solid fluidized beds: a multi-scale modeling strategy, Chemical Engineering Science 59 (22) (2004) 5157–5165.
[136] N. Gui, J.R. Fan, S. Chen, Numerical study of particle-particle collision in swirling jets: a DEM-DNS coupling simulation, Chemical Engineering Science 65 (10) (2010) 3268–3278.
[137] S. Kriebitzsch, M. van der Hoef, J. Kuipers, Fully resolved simulation of a gas-fluidized bed: a critical test of DEM models, Chemical Engineering Science 91 (Supplement C) (2013) 1–4.
[138] S. Das, N.G. Deen, J. Kuipers, A DNS study of flow and heat transfer through slender fixed-bed reactors randomly packed with spherical particles, Chemical Engineering Science 160 (Supplement C) (2017) 1–19.
[139] H. Zhou, G. Flamant, D. Gauthier, DEM-LES of coal combustion in a bubbling fluidized bed. Part I: gas-particle turbulent flow structure, Chemical Engineering Science 59 (20) (2004) 4193–4203.
[140] N. Gui, J.R. Fan, K. Luo, DEM-LES study of 3-D bubbling fluidized bed with immersed tubes, Chemical Engineering Science 63 (14) (2008) 3654–3663.
[141] N. Gui, J. Fan, Numerical simulation of pulsed fluidized bed with immersed tubes using DEM-LES coupling method, Chemical Engineering Science 64 (11) (2009) 2590–2598.
[142] H. Zhou, G. Mo, J. Zhao, K. Cen, DEM-CFD simulation of the particle dispersion in a gas–solid two-phase flow for a fuel-rich/lean burner, Fuel 90 (4) (2011) 1584–1590.
[143] S. Yang, K. Luo, M. Fang, J. Fan, LES-DEM investigation of the solid transportation mechanism in a 3-D bubbling fluidized bed. Part II: solid dispersion and circulation properties, Powder Technology 256 (Supplement C) (2014) 395–403.
[144] X. Chen, W. Zhong, X. Zhou, B. Jin, B. Sun, CFD-DEM simulation of particle transport and deposition in pulmonary airway, Powder Technology 228 (Supplement C) (2012) 309–318.
[145] T. Tang, Y. He, A. Ren, Y. Zhao, Investigation on wet particle flow behavior in a riser using LES-DEM coupling approach, in: SI:Particle2015, Powder Technology 304 (Supplement C) (2016) 164–176.

[146] S. Wang, K. Luo, S. Yang, C. Hu, J. Fan, Parallel LES-DEM simulation of dense flows in fluidized beds, Applied Thermal Engineering 111 (Supplement C) (2017) 1523–1535.
[147] K. Chu, B. Wang, A. Yu, A. Vince, CFD-DEM modelling of multiphase flow in dense medium cyclones, in: Special Issue: Discrete Element Methods: the 4th International Conference on Discrete Element Methods, Powder Technology 193 (3) (2009) 235–247.
[148] K. Chu, B. Wang, D. Xu, Y. Chen, A. Yu, CFD-DEM simulation of the gas–solid flow in a cyclone separator, Chemical Engineering Science 66 (5) (2011) 834–847.
[149] S. Mondal, C.H. Wu, M.M. Sharma, Coupled CFD-DEM simulation of hydrodynamic bridging at constrictions, International Journal of Multiphase Flow 84 (Supplement C) (2016) 245–263.
[150] B. Krause, B. Liedmann, J. Wiese, P. Bucher, S. Wirtz, H. Piringer, V. Scherer, 3D-DEM-CFD simulation of heat and mass transfer, gas combustion and calcination in an intermittent operating lime shaft kiln, International Journal of Thermal Sciences 117 (Supplement C) (2017) 121–135.
[151] H. Wu, N. Gui, X. Yang, J. Tu, S. Jiang, A smoothed void fraction method for CFD-DEM simulation of packed pebble beds with particle thermal radiation, International Journal of Heat and Mass Transfer 118 (Supplement C) (2018) 275–288.
[152] Z. Miao, Z. Zhou, A. Yu, Y. Shen, CFD-DEM simulation of raceway formation in an ironmaking blast furnace, in: Special Issue on Simulation and Modelling of Particulate Systems, Powder Technology 314 (Supplement C) (2017) 542–549.
[153] S. Wang, K. Luo, C. Hu, J. Fan, CFD-DEM study of the effect of cyclone arrangements on the gas-solid flow dynamics in the full-loop circulating fluidized bed, Chemical Engineering Science 172 (Supplement C) (2017) 199–215.
[154] A.C. Varas, E. Peters, J. Kuipers, CFD-DEM simulations and experimental validation of clustering phenomena and riser hydrodynamics, Chemical Engineering Science 169 (Supplement C) (2017) 246–258.
[155] K. Zhou, J. Hou, Q. Sun, L. Guo, S. Bing, Q. Du, C. Yao, An efficient LBM-DEM simulation method for suspensions of deformable preformed particle gels, Chemical Engineering Science 167 (Supplement C) (2017) 288–296.
[156] X. Cui, J. Li, A. Chan, D. Chapman, Coupled DEM-LBM simulation of internal fluidisation induced by a leaking pipe, Powder Technology 254 (Supplement C) (2014) 299–306.
[157] X. Cui, J. Li, A. Chan, D. Chapman, A 2D DEM-LBM study on soil behaviour due to locally injected fluid, in: Advances in Characterization and Modeling of Particulate Processes, Particuology 10 (2) (2012) 242–252.
[158] X. Sun, M. Sakai, Y. Yamada, Three-dimensional simulation of a solid–liquid flow by the DEM-SPH method, Journal of Computational Physics 248 (Supplement C) (2013) 147–176.
[159] S. Natsui, A. Sawada, K. Terui, Y. Kashihara, T. Kikuchi, R.O. Suzuki, DEM-SPH study of molten slag trickle flow in coke bed, Chemical Engineering Science 175 (Supplement C) (2018) 25–39.
[160] M. Sinnott, P. Cleary, R. Morrison, Combined DEM and SPH simulation of overflow ball mill discharge and trommel flow, Minerals Engineering 108 (Supplement C) (2017) 93–108.
[161] C.H. Ke, S. Shu, H. Zhang, H.Z. Yuan, LBM-IBM-DEM modelling of magnetic particles in a fluid, in: Special Issue on Simulation and Modelling of Particulate Systems, Powder Technology 314 (Supplement C) (2017) 264–280.
[162] H. Zhang, Y. Tan, S. Shu, X. Niu, F.X. Trias, D. Yang, H. Li, Y. Sheng, Numerical investigation on the role of discrete element method in combined LBM-IBM-DEM modeling, Computers & Fluids 94 (Supplement C) (2014) 37–48.
[163] H. Zhang, A. Yu, W. Zhong, Y. Tan, A combined TLBM-IBM-DEM scheme for simulating isothermal particulate flow in fluid, International Journal of Heat and Mass Transfer 91 (Supplement C) (2015) 178–189.
[164] Z. Wang, J. Fan, K. Luo, K. Cen, Immersed boundary method for the simulation of flows with heat transfer, International Journal of Heat and Mass Transfer 52 (19) (2009) 4510–4518.
[165] N.G. Deen, M. van Sint Annaland, J. Kuipers, Direct numerical simulation of complex multi-fluid flows using a combined front tracking and immersed boundary method, Chemical Engineering Science 64 (9) (2009) 2186–2201.

[166] Y. Kim, C.S. Peskin, 3-D Parachute simulation by the immersed boundary method, Computers & Fluids 38 (6) (2009) 1080–1090.
[167] Z. Wang, J. Fan, K. Cen, Immersed boundary method for the simulation of 2D viscous flow based on vorticity–velocity formulations, Journal of Computational Physics 228 (5) (2009) 1504–1520.
[168] J. Wu, C. Shu, Implicit velocity correction-based immersed boundary-lattice Boltzmann method and its applications, Journal of Computational Physics 228 (6) (2009) 1963–1979.
[169] D. Le, J. White, J. Peraire, K. Lim, B. Khoo, An implicit immersed boundary method for three-dimensional fluid–membrane interactions, Journal of Computational Physics 228 (22) (2009) 8427–8445.
[170] S. Luding, E. Clement, J. Rajchenbach, J. Duran, Simulations of pattern formation in vibrated granular media, Europhysics Letters 36 (4) (2012) 247–252.
[171] B.P.B. Hoomans, J.A.M. Kuipers, W.J. Briels, W.P.M.V. Swaaij, Discrete particle simulation of bubble and slug formation in a two-dimensional gas-fluidised bed: a hard-sphere approach, Chemical Engineering Science 51 (1) (1996) 99–118.
[172] J.J. Moreau, Some numerical methods in multibody dynamics: application to granular materials, European Journal of Mechanics. A, Solids 13 (1) (1994) 93–114.
[173] M. Jean, The non-smooth contact dynamics method, Computer Methods in Applied Mechanics and Engineering 177 (3–4) (2016) 235–257.
[174] N. Gui, J. Fan, K. Cen, Effect of particle-particle collision in decaying homogeneous and isotropic turbulence, Physical Review. E, Statistical, Nonlinear, and Soft Matter Physics 78 (2) (2008) 046307.
[175] N. Gui, J. Fan, Numerical simulation of motion of rigid spherical particles in a rotating tumbler with an inner wavelike surface, Powder Technology 192 (2) (2009) 234–241.
[176] F. Jin, H. Xin, C. Zhang, Q. Sun, Probability-based contact algorithm for non-spherical particles in DEM, Powder Technology 212 (1) (2011) 134–144.
[177] W. Cai, G. McDowell, G. Airey, Discrete element visco-elastic modelling of a realistic graded asphalt mixture, in: Special Issue on 2nd International Conference on Transportation Geotechnics IS-Hokkaido2012, Soil and Foundation 54 (1) (2014) 12–22.
[178] G. Yan, H.-sui Yu, G. McDowell, Simulation of granular material behaviour using DEM, Procedia Earth and Planetary Science 1 (1) (2009) 598–605.
[179] J. de Bono, G. McDowell, An insight into the yielding and normal compression of sand with irregularly-shaped particles using DEM, Powder Technology 271 (2015) 270–277.
[180] C. Wensrich, A. Katterfeld, Rolling friction as a technique for modelling particle shape in DEM, Powder Technology 217 (2012) 409–417.
[181] N. Cho, C. Martin, D. Sego, A clumped particle model for rock, International Journal of Rock Mechanics and Mining Sciences 44 (7) (2007) 997–1010.
[182] N. Gui, J. Gao, Z. Ji, Numerical study of mixing and thermal conduction of granular particles in rotating tumblers, AIChE Journal 59 (6) (2013) 1906–1918.
[183] C. Boon, G. Houlsby, S. Utili, A new contact detection algorithm for three-dimensional non-spherical particles, in: Discrete Element Modelling, Powder Technology 248 (2013) 94–102.
[184] A. Wachs, L. Girolami, G. Vinay, G. Ferrer, Grains3D, a flexible DEM approach for particles of arbitrary convex shape — Part I: numerical model and validations, Powder Technology 224 (2012) 374–389.
[185] N. Gui, X. Yang, J. Tu, S. Jiang, An extension of hard-particle model for three-dimensional non-spherical particles: mathematical formulation and validation, Applied Mathematical Modelling 40 (4) (2016) 2485–2499.
[186] F.Y. Fraige, P.A. Langston, G.Z. Chen, Distinct element modelling of cubic particle packing and flow, Powder Technology 186 (3) (2008) 224–240.
[187] K.L. Johnson, Contact Mechanics, Cambridge University Press, New York, 1985.
[188] Y. Tsuji, T. Kawaguchi, T. Tanaka, Discrete particle simulation of two-dimensional fluidized bed, Powder Technology 77 (1993) 79–87.
[189] P.A. Cundall, Formulation of a three-dimensional distinct element model—Part I. A scheme to detect and represent contacts in a system composed of many polyhedral blocks, International Journal of Rock Mechanics and Mining Sciences & Geomechanics Abstracts 25 (1988) 107–116.

[190] E.G. Nezami, Y. Hashash, D. Zhao, J. Ghaboussi, A fast contact detection algorithm for 3-D discrete element method, Computers and Geotechnics 31 (2004) 575–587.
[191] B.S. Jin, H. Tao, W.Q. Zhong, Flow behaviors of non-spherical granules in rectangular hopper, Chinese Journal of Chemical Engineering 18 (2010) 931–939.
[192] G.K. Batchelor, R.W. O'Brien, Thermal or electrical conduction through a granular material, Proceedings of the Royal Society A. Mathematical, Physical and Engineering Sciences 355 (355) (1977) 313–333.
[193] W.L. Vargas, J.J. McCarthy, Heat conduction in granular materials, AIChE Journal 47 (5) (2001) 1052–1059.
[194] D. Shi, W.L. Vargas, J. McCarthy, Heat transfer in rotary kilns with interstitial gases, Chemical Engineering Science 63 (18) (2008) 4506–4516.
[195] I. Figueroa, W.L. Vargas, J.J. Mccarthy, Mixing and heat conduction in rotating tumblers, Chemical Engineering Science 65 (2) (2010) 1045–1054.
[196] B. Chaudhuri, F.J. Muzzio, M.S. Tomassone, Modeling of heat transfer in granular flow in rotating vessels, Chemical Engineering Science 61 (19) (2006) 6348–6360.
[197] N. Gui, J. Yan, W. Xu, G. Liang, D. Wu, Z. Ji, J. Gao, S. Jiang, X. Yang, DEM simulation and analysis of particle mixing and heat conduction in a rotating drum, Chemical Engineering Science 97 (7) (2013) 225–234.
[198] M. Lesieur, O. Metais, New trends in large-eddy simulations of turbulence, Annual Review of Fluid Mechanics 28 (1) (1996) 45–82.
[199] J. Ding, D. Gidaspow, A bubbling fluidization model using kinetic theory of granular flow, AIChE Journal 36 (4) (1990) 523–538.
[200] D.L. Koch, R.J. Hill, Inertial effects in suspension and porous-media flows, Annual Review of Fluid Mechanics 33 (1) (2001) 619–647.
[201] J. Link, L. Cuypers, N. Deen, J. Kuipers, Flow regimes in a spout–fluid bed: a combined experimental and simulation study, Chemical Engineering Science 60 (13) (2005) 3425–3442.
[202] Y. Tsuji, T. Tanaka, T. Ishida, Lagrangian numerical simulation of plug flow of cohesionless particles in a horizontal pipe, Powder Technology 71 (3) (1992) 239–250.
[203] Z.Y. Zhou, A.B. Yu, P. Zulli, Particle scale study of heat transfer in packed and bubbling fluidized beds, AIChE Journal 55 (4) (2009) 868–884.
[204] S.B. Kuang, A.B. Yu, Z.S. Zou, Computational study of flow regimes in vertical pneumatic conveying, Industrial & Engineering Chemistry Research 48 (14) (2009) 6846–6858.
[205] Z. Zhou, S. Kuang, K. Chu, A. Yu, Discrete particle simulation of particle-fluid flow: model formulations and their applicability, Journal of Fluid Mechanics 661 (661) (2010) 482–510.
[206] K. Chu, S. Kuang, A. Yu, A. Vince, Particle scale modelling of the multiphase flow in a dense medium cyclone: effect of fluctuation of solids flowrate, in: Computational Modelling, Minerals Engineering 33 (Supplement C) (2012) 34–45.
[207] S.B. Kuang, R.P. Zou, R.H. Pan, A.B. Yu, Gas–solid flow and energy dissipation in inclined pneumatic conveying, Industrial & Engineering Chemistry Research 51 (43) (2012) 14289–14302.
[208] K. Chu, S. Kuang, A. Yu, A. Vince, G. Barnett, P. Barnett, Prediction of wear and its effect on the multiphase flow and separation performance of dense medium cyclone, Minerals Engineering 56 (Supplement C) (2014) 91–101.
[209] T. Eppinger, K. Seidler, M. Kraume, DEM-CFD simulations of fixed bed reactors with small tube to particle diameter ratios, Chemical Engineering Journal 166 (1) (2011) 324–331.
[210] O. Ayeni, C. Wu, K. Nandakumar, J. Joshi, Development and validation of a new drag law using mechanical energy balance approach for DEM–CFD simulation of gas–solid fluidized bed, Chemical Engineering Journal 302 (Supplement C) (2016) 395–405.
[211] Q. Hou, S. Kuang, A. Yu, A DEM-based approach for analyzing energy transitions in granular and particle-fluid flows, Chemical Engineering Science 161 (Supplement C) (2017) 67–79.
[212] Q. Hou, D. E, S. Kuang, Z. Li, A. Yu, DEM-based virtual experimental blast furnace: a quasi-steady state model, in: Special Issue on Simulation and Modelling of Particulate Systems, Powder Technology 314 (Supplement C) (2017) 557–566.
[213] D. Gidaspow, Multiphase Flow and Fluidization, Academic Press, San Diego, 1994.

[214] H. Zhu, Z. Zhou, R. Yang, A. Yu, Discrete particle simulation of particulate systems: theoretical developments, in: Frontier of Chemical Engineering – Multi-Scale Bridge Between Reductionism and Holism, Chemical Engineering Science 62 (13) (2007) 3378–3396.

[215] Y. Zhao, M. Jiang, Y. Liu, J. Zheng, Particle-scale simulation of the flow and heat transfer behaviors in fluidized bed with immersed tube, AIChE Journal 55 (12) (2009) 3109–3124.

[216] Q.F. Hou, Z.Y. Zhou, A.B. Yu, Computational study of heat transfer in a bubbling fluidized bed with a horizontal tube, AIChE Journal 58 (5) (2012) 1422–1434.

[217] Y. Guo, C.Y. Wu, C. Thornton, Modeling gas-particle two-phase flows with complex and moving boundaries using DEM-CFD with an immersed boundary method, AIChE Journal 59 (4) (2013) 1075–1087.

[218] D. Wang, L.S. Fan, Bulk coarse particle arching phenomena in a moving bed with fine particle presence, AIChE Journal 60 (3) (2014) 881–892.

[219] Z. Peng, E. Doroodchi, C. Luo, B. Moghtaderi, Influence of void fraction calculation on fidelity of CFD-DEM simulation of gas-solid bubbling fluidized beds, AIChE Journal 60 (6) (2014) 2000–2018.

[220] K. Luo, Z. Wang, J. Fan, A modified immersed boundary method for simulations of fluid–particle interactions, Computer Methods in Applied Mechanics and Engineering 197 (1–4) (2007) 36–46.

[221] Q. Zhou, L.S. Fan, A second-order accurate immersed boundary-lattice Boltzmann method for particle-laden flows, Journal of Computational Physics 268 (2014) 269–301.

[222] Y. Tang, E.A.J.F. Peters, J.A.M. Kuipers, S.H.L. Kriebitzsch, M.A. van der Hoef, A new drag correlation from fully resolved simulations of flow past monodisperse static arrays of spheres, AIChE Journal 61 (2) (2015) 688–698.

[223] Y. Tang, E.A.J.F. Peters, J.A.M. Kuipers, Direct numerical simulations of dynamic gas-solid suspensions, AIChE Journal 62 (6) (2016) 1958–1969.

[224] S.B. Kuang, K.W. Chu, A.B. Yu, Z.S. Zou, Y.Q. Feng, Computational investigation of horizontal slug flow in pneumatic conveying, Industrial & Engineering Chemistry Research 47 (2) (2008) 470–480.

[225] S. Kuang, C. LaMarche, J. Curtis, A. Yu, Discrete particle simulation of jet-induced cratering of a granular bed, Powder Technology 239 (Supplement C) (2013) 319–336.

[226] S. Kuang, K. Li, R. Zou, R. Pan, A. Yu, Application of periodic boundary conditions to CFD-DEM simulation of gas–solid flow in pneumatic conveying, Chemical Engineering Science 93 (Supplement C) (2013) 214–228.

[227] K. Li, S. Kuang, R. Pan, A. Yu, Numerical study of horizontal pneumatic conveying: effect of material properties, Powder Technology 251 (Supplement C) (2014) 15–24.

[228] T. Tsuji, K. Higashida, Y. Okuyama, T. Tanaka, Fictitious particle method: a numerical model for flows including dense solids with large size difference, AIChE Journal 60 (5) (2014) 1606–1620.

[229] B.E. Griffith, C.S. Peskin, On the order of accuracy of the immersed boundary method: higher order convergence rates for sufficiently smooth problems, Journal of Computational Physics 208 (1) (2005) 75–105.

[230] D. Fu, Y.A. Ma, High order accurate difference scheme for complex flow fields, Journal of Computational Physics 134 (1) (1997) 1–15.

[231] S.K. Lele, Compact finite difference schemes with spectral-like resolution, Journal of Computational Physics 103 (1) (1992) 16–42.

[232] A. Jameson, W. Schmidt, Some recent development in numerical methods for transonic flow, Computer Methods in Applied Mechanics and Engineering 51 (s1–3) (1985) 467–493.

[233] U. Ananthakrishnaiah, R. Manohar, J.W. Stephenson, Fourth-order finite difference methods for three-dimensional general linear elliptic problems with variable coefficients, Numerical Methods for Partial Differential Equations 3 (3) (2010) 229–240.

[234] P. Moin, K. Mahesh, Direct numerical simulation: a tool in turbulence research, Annual Review of Fluid Mechanics 30 (1) (1998) 539–578.

[235] M.R. Maxey, J.J. Riley, Equation of motion for a small rigid sphere in a nonuniform flow, Physics of Fluids 26 (1983) 883–889.

[236] G.G. Stokes, On the Effect of the Internal Friction of Fluids on the Motion of Pendulums, Transactions of the Cambridge Philosophical Society, vol. 9, 1851, p. 8.
[237] S. Poisson, Mémoire sur les mouvements simultanes d'un pendule et de l'air environnemant, Mémoires Del' Academie des Sciences 9 (1831) 521–523.
[238] A.B. Basset, Treatise on hydrodynamics, Nature 40 (1889) 412–413.
[239] P.G. Saffman, The lift on a small sphere in a slow shear flow, Journal of Fluid Mechanics 22 (1965) 385–400.
[240] P.G. Saffman, The lift on a small sphere in a slow shear flow-corrigendum, Studies in Applied Mathematics 50 (1968) 93–101.
[241] R. Mei, An approximate expression for the shear lift force on a spherical particle at finite Reynolds number, International Journal of Multiphase Flow 18 (1) (1992) 145–147.
[242] J.B. Mclaughlin, Inertial migration of a small sphere in linear shear flow, Journal of Fluid Mechanics 224 (2006) 261–274.
[243] G. Magnus, A note on the rotary motion of the liquid jet, Annalen der Physik und Chemie 63 (1861) 363–365.
[244] C.T. Crowe, M. Sommerfeld, Y. Tsuji, Fundamentals of Gas-Particle and Gas-Droplet Flows, CRC Press, Boca Raton, 1998, pp. 95–99.
[245] B. Oesterlé, T.B. Dinh, Experiments on the lift of a spinning sphere in a range of intermediate Reynolds numbers, Experiments in Fluids 25 (1) (1998) 16–22.
[246] S.I. Rubinow, J.B. Keller, The transverse force on a spinning sphere moving in a viscous fluid, Journal of Fluid Mechanics 11 (3) (2006) 447–459.
[247] F.G. Mclaren, A.F.C. Sherratt, A.S. Morton, Effect of free stream turbulence on the drag coefficient of bluff sharp-edged cylinders, Nature 223 (5208) (1969) 828–829.
[248] P. Bearman, T. Morel, Effect of free stream turbulence on the flow around bluff bodies, Progress in Aerospace Sciences 20 (2) (1983) 97–123.
[249] A. Clamen, W.H. Gauvin, Effects of turbulence on the drag coefficients of spheres in a supercritical flow regime, AIChE Journal 15 (2) (2010) 184–189.
[250] R. Clift, W.H. Gauvin, The motion of particles in turbulent gas streams, Revista Argentina De Cardiología 82 (3) (1970) 14–28.
[251] W.D. Warnica, M. Renksizbulut, A.B. Strong, Drag coefficients of spherical liquid droplets. Part 2: turbulent gaseous fields, Experiments in Fluids 18 (4) (1995) 265–276.
[252] E.E. Michaelides, Particles, Bubbles and Drops: Their Motion, Heat and Mass Transfer, World Scientific, 2006, i pp.
[253] S.K. Friedlander, Smoke, Dust, and Haze: Fundamentals of Aerosol Dynamics, Wiley, New York, 1977.
[254] L. Talbot, R.K. Cheng, R.W. Schefer, D.R. Willis, Thermophoresis of particles in a heated boundary layer, Journal of Fluid Mechanics 101 (1980) 737–758.
[255] J.R. Brock, On the theory of thermal forces acting on aerosol particles, Journal of Colloid Science 17 (8) (1962) 768–780.
[256] G. Sarton, On the theories of the internal friction of fluids in motion, Transactions of the Cambridge Philosophical Society 8 (2) (2004) 153–169.
[257] C.W. Oseen, Ueber die Stokes'sche formel und uber eine verwandte aufgabe in der hydrodynamik, Arkiv för Matematik, Astronomi Och Fysik 6 (2) (1910) 154–160.
[258] R. Clift, J.R. Grace, M.E. Weber, Bubbles, Drops and Particles, Academic Press, New York, 1978, 111 pp.
[259] S.C.R. Dennis, S.N. Singh, D.B. Ingham, The steady flow due to a rotating sphere at low and moderate Reynolds numbers, Journal of Fluid Mechanics 101 (2) (1980) 257–279.
[260] W.C. Hinds, Aerosol Technology, Wiley & Sons, 2000.
[261] S. Sundaram, L.R. Collins, Numerical considerations in simulating a turbulent suspension of finite-volume particles, Journal of Computational Physics 124 (1996) 337–350.
[262] K. Yamamoto, N. Takada, M. Misawa, Combustion simulation with lattice Boltzmann method in a three-dimensional porous structure, Proceedings of the Combustion Institute 30 (1) (2005) 1509–1515.

[263] L.C. Burmeister, Convective Heat Transfer, Wiley, 1983, pp. 132–138.

[264] N. Gui, X.U. Wenkai, G.E. Liang, J. Yan, LBE-DEM coupled simulation of gas-solid two-phase cross jets, Science China. Technological Sciences 56 (6) (2013) 1377–1386.

[265] Z. Guo, B. Shi, C. Zheng, A coupled lattice BGK model for the Boussinesq equations, International Journal for Numerical Methods in Fluids 39 (4) (2002) 325–342.

[266] R. Nourgaliev, T. Dinh, T. Theofanous, D. Joseph, The lattice Boltzmann equation method: theoretical interpretation, numerics and implications, International Journal of Multiphase Flow 29 (1) (2003) 117–169.

[267] T. Lee, C.L. Lin, Pressure evolution lattice-Boltzmann-equation method for two-phase flow with phase change, Physical Review E 67 (2003) 056703, https://doi.org/10.1103/PhysRevE.67.056703.

[268] P.A. Cundall, O.D.L. Strack, Discussion: a discrete numerical model for granular assemblies, Géotechnique 30 (3) (1980) 331–336.

[269] C. Shu, N. Liu, Y. Chew, A novel immersed boundary velocity correction–lattice Boltzmann method and its application to simulate flow past a circular cylinder, Journal of Computational Physics 226 (2) (2007) 1607–1622.

[270] L. Zhu, G. He, S. Wang, L. Miller, X. Zhang, Q. You, S. Fang, An immersed boundary method based on the lattice Boltzmann approach in three dimensions, with application, in: Mesoscopic Methods for Engineering and Science — Proceedings of ICMMES-09, Computers & Mathematics with Applications 61 (12) (2011) 3506–3518.

[271] A. ten Cate, C.H. Nieuwstad, J.J. Derksen, H.E.A.V. den Akker, Particle imaging velocimetry experiments and lattice-Boltzmann simulations on a single sphere settling under gravity, Physics of Fluids 14 (11) (2002) 4012–4025, https://doi.org/10.1063/1.1512918.

[272] V. Eswaran, S.B. Pope, An examination of forcing in direct numerical simulations of turbulence, Computers & Fluids 16 (3) (1988) 257–278.

[273] R.S. Rogallo, Numerical Experiments in Homogeneous Turbulence, NASA TM81315, 1981.

[274] N.N. Mansour, A.A. Wray, Decay of isotropic turbulence at low Reynolds number, Physics of Fluids 6 (2) (1994) 808–814.

[275] S. Elghobashi, On predicting particle-laden turbulent flows, Applied Scientific Research 52 (4) (1994) 309–329.

[276] L. Biferale, G. Boffetta, A. Celani, B.J. Devenish, A. Lanotte, F. Toschi, Multifractal statistics of Lagrangian velocity and acceleration in turbulence, Physical Review Letters 93 (6) (2004) 064502.

[277] S.A. Stanley, S. Sarkar, J.P. Mellado, A study of the flow-field evolution and mixing in a planar turbulent jet using direct numerical simulation, Journal of Fluid Mechanics 450 (450) (2002) 377–407.

[278] Y.M. Chung, K.H. Luo, N.D. Sandham, Numerical study of momentum and heat transfer in unsteady impinging jets, International Journal of Heat and Fluid Flow 23 (5) (2002) 592–600.

[279] I. Orlanski, A simple boundary condition for unbounded hyperbolic flows, Journal of Computational Physics 21 (3) (1976) 251–269.

[280] K. Luo, Z. Wang, J. Fan, A modified immersed boundary method for simulations of fluid-particle interactions, Computer Methods in Applied Mechanics and Engineering 197 (1) (2007) 36–46.

[281] M. Klein, A. Sadiki, J. Janicka, Investigation of the influence of the Reynolds number on a plane jet using direct numerical simulation, International Journal of Heat and Fluid Flow 24 (6) (2003) 785–794.

[282] A.E. Davies, J.F. Keffer, W.D. Baines, Spread of a heated plane turbulent jet, Physics of Fluids 18 (7) (1975) 770–775.

[283] B.R. Ramaprian, M.S. Chandrasekhara, LDA measurements in plane turbulent jets, Journal of Fluids Engineering 107 (2) (1985) 264–271.

[284] A.J. Chorin, Numerical solution of the Navier–Stokes equations, Mathematics of Computation 22 (1968) 745–762.

[285] J.H. Williamson, Low-storage Runge-Kutta schemes, Journal of Computational Physics 35 (1) (1980) 48–56.

[286] H. Nasr, G. Ahmadi, The effect of two-way coupling and inter-particle collisions on turbulence modulation in a vertical channel flow, International Journal of Heat and Fluid Flow 28 (6) (2007) 1507–1517.

[287] D. Modarress, S. Elghobashi, H. Tan, Two-component LDA measurement in a two-phase turbulent jet, AIAA Journal 22 (5) (2013) 624–630.
[288] P. Billant, J. Chomaz, P. Huerre, Experimental study of vortex breakdown in swirling jets, Journal of Fluid Mechanics 376 (376) (1998) 183–219.
[289] J. Jeong, F. Hussain, On the identification of a vortex, Journal of Fluid Mechanics 285 (285) (1995) 69–94.
[290] H. Zhang, M. Liu, T. Li, Z. Huang, X. Sun, H. Bo, Y. Dong, Experimental investigation on gas-solid hydrodynamics of coarse particles in a two-dimensional spouted bed, Powder Technology 307 (2017) 175–183.
[291] G.Q. Liu, S.Q. Li, X.L. Zhao, Q. Yao, Experimental studies of particle flow dynamics in a two-dimensional spouted bed, Chemical Engineering Science 63 (4) (2008) 1131–1141.
[292] L.A.P. Freitas, O.M. Dogan, C.J. Lim, J.R. Grace, D. Bai, Identification of flow regimes in slot-rectangular spouted beds using pressure fluctuations, Canadian Journal of Chemical Engineering 82 (1) (2004) 60–73.
[293] G.E. Mueller, Radial void fraction distributions in randomly packed fixed beds of uniformly sized spheres in cylindrical containers, Powder Technology 72 (3) (1992) 269–275.
[294] J. Theuerkauf, P. Witt, D. Schwesig, Analysis of particle porosity distribution in fixed beds using the discrete element method, Powder Technology 165 (2) (2006) 92–99.
[295] R. Hillborn, Mechanics and Thermodynamics, Oxford University Press, New York, 1994.
[296] G. Finnie, N. Kruyt, M. Ye, C. Zeilstra, J. Kuipers, Longitudinal and transverse mixing in rotary kilns: a discrete element method approach, Chemical Engineering Science 60 (15) (2005) 4083–4091.
[297] C.E. Shannon, A mathematical theory of communication, The Bell System Technical Journal 27 (3) (1948) 379–423.
[298] O.O. Ayeni, C.L. Wu, J.B. Joshi, K. Nandakumar, A discrete element method study of granular segregation in non-circular rotating drums, Powder Technology 283 (2015) 549–560.
[299] C. Shannon, W. Weaver, The Mathematical Theory of Communication, University of Illinois Press, Urbana, 1949.
[300] P.M.C. Lacey, Developments in the theory of particle mixing, Journal of Chemical Technology and Biotechnology 4 (5) (2010) 257–268.
[301] G.R. Chandratilleke, A.B. Yu, J. Bridgwater, K. Shinohara, A particle-scale index in the quantification of mixing of particles, AIChE Journal 58 (4) (2012) 1099–1118.
[302] T. Shinbrot, A. Alexander, F.J. Muzzio, Spontaneous chaotic granular mixing, Nature 397 (6721) (1999) 675–678.
[303] J. Mellmann, The transverse motion of solids in rotating cylinders—forms of motion and transition behavior, Powder Technology 118 (3) (2001) 251–270.
[304] C.R. Wassgren, M.L. Hunt, P.J. Freese, J. Palamara, C.E. Brennen, Effects of vertical vibration on hopper flows of granular material, Physics of Fluids 14 (10) (2002) 3439–3448.
[305] C. Mankoc, A. Janda, R. Arévalo, J.M. Pastor, I. Zuriguel, A. Garcimartín, D. Maza, The flow rate of granular materials through an orifice, Granular Matter 9 (6) (2007) 407–414.

Index

A

Algebraic stress model (ASM), 15
Ambient gas phase, 113
Anisotropic turbulences, 16, 167
Augmented cataracting motion, 241

B

Bed filling levels, 295
Bed orifice, 335, 337, 346, 351
Binary collisions, 15
Bonded particle method (BPM), 53
Boundary amplitudes, 252, 254, 255
Boundary discretization (BD), 53
Brick host particle, 58
Brick particles, 52
Brownian motion, 98
Brownian motion particles, 10
Bulk motion, 252–254, 322, 331
Bulk motion tumbling, 253, 254, 256
Bulk oscillation, 250

C

Cataracting flow patterns, 294
Cataracting flow regime, 273, 277
Central spout, 199
Centrifugal motion, 290
Centrifugal regime, 300, 303, 304, 307, 308
Circular
  drum, 256, 257, 261–263, 283
  drum mixers, 217
  particle, 352
  tumbler, 207, 246
Circumferential distribution, 183, 184
Circumferential distribution collision, 183, 184
Circumferential mixing, 310, 313, 316, 317
  degree, 317
  pattern, 310, 313
Clumped particle, 53
Colliding particles, 22, 29, 65, 80, 101, 115, 181, 249, 254, 255
Collision
  circumferential distribution, 183, 184
  detection, 30, 59, 60, 70, 71, 125, 244
  dominated flow, 69
  duration, 39
  dynamics, 9, 21
  force, 44, 125
  frequency, 9
  governing equations, 43
  impulse, 73
  interparticle, 123–126, 133, 145
  model, 83, 132, 329
  parameters, 101
  particles, 183
  patterns, 247
  PDFs, 172
  point, 98
  probability, 172
  process, 22, 27, 46, 50, 59, 60, 71, 112, 127
  rate, 9, 14, 127–129, 133
  schemes, 76
  solutions, 77
  subspheres, 53, 54
  time, 8
Collisional, 58, 59
  force, 23, 45, 71, 180, 181, 183
  impulse, 25, 28, 29, 34, 70, 98
  parameters, 62
  point, 72
  stiffness, 62
  torque, 34, 114, 116
Colored particles, 74
Condensed particle, 132
Contacting particles, 348
Convective heat transfer, 113, 114
Cubic
  particle mixing, 287, 308
  particles, 32, 33, 50, 62, 287, 289, 290, 294, 295, 297, 298, 309, 310, 318
  shaped particles, 313
Cyclic motion, 227, 229, 230, 251
Cyclic motion particles, 228, 230
Cyclic motion process, 290

D

Decagonal particles, 280, 352
Decaying fluid turbulence, 124
Decaying gas turbulence, 127
Dilute flow regime, 14
Dimensionless
  fundamental frequency, 155
  governing equations, 92
  particle temperature, 261
Direct numerical simulation (DNS), 17, 87
Discharge fraction, 351, 352
Discharged particles, 331, 333, 334, 345, 346, 351

Discharged particles motion trajectories, 345
Discharging hopper, 74
Discrete element method (DEM), 9, 19, 86, 114
  model, 21, 54, 69, 70, 77, 80, 91, 329
Discrete particle, 6
Dispersed particles, 126
Drag force, 12, 85–87, 89, 91, 92, 96, 97, 100, 124, 125, 158, 168, 174, 176, 177, 180–183, 193, 194
Drag force particles, 18, 92, 123
Drainage hopper flow, 75
Drainage orifice, 320, 321
Drum, 208, 232, 256, 261, 263, 265, 270, 273, 277, 282, 285–287, 290, 295, 297, 298, 300, 308
  area, 282
  circular, 256, 257, 261–263, 283
  mixer, 269, 289, 295
  rotation, 290, 300
  top, 263
  walls, 256
  wavy, 256, 257, 261–265, 268

E

East particle, 33, 34
Effective driven force (EDF), 256, 259
Element method, 9, 19, 21, 58, 67, 86, 95, 114
Enneagonal particles, 277

F

Fictitious particles, 17, 87
Filling level, 289, 295, 297, 298, 300, 304, 309, 310
Filling level tetrahedrons, 300
Finest particle, 91
Flow
  pattern, 11, 20, 74, 75, 190–192, 200, 217, 277, 289, 290, 292–295, 331, 335
  regime, 11, 14, 15, 69, 82, 83, 230, 236, 238–241, 243, 273, 279, 300, 304, 308, 331
  regime transition, 297
  turbulence, 9
Flowing layer, 217, 218, 229, 232, 277, 280, 284, 285
Fluctuated velocities, 135
Fractal dimension, 4, 139, 140, 142, 143, 219–221, 235
Free motion, 8, 30
Free motion particles, 7, 125
Freely flowing layer, 229
Friction, 22, 28, 41, 43, 57, 68, 73–75, 286, 287, 290, 294–297, 308, 321, 348
  coefficient, 33, 36, 43, 44, 58, 67, 76, 77, 329
  contact, 246
  effects, 345
Frictional, 225, 256, 262, 287, 290, 297
  force, 261, 290, 294
  motion, 73
  torque, 72
  wall, 297
Fundamental frequency, 155, 157, 179, 183, 196, 204, 231, 249, 338

G

Gas
  phase, 85–87, 89, 92, 98, 114, 128, 173, 186, 187, 190, 195, 197–201, 204
    behavior, 195
    motion, 195, 197, 203
    velocity, 200, 204
  turbulence, 130, 131
  velocities, 201, 203
  voidages, 202
Generalized hard particle model (GHPM), 21, 60
Governing equations, 21, 59, 66, 70, 85, 89, 96, 105, 116, 123, 168
  collision, 43
  dimensionless, 92
  motion, 71
Granular particles, 256
  heat transfer, 11
  motion, 218
  spherical, 225
Gyration motion, 226

H

Hard sphere model (HSM), 9
Heat transfer granular particles, 11
Heptagonal particles, 351
Hexagonal shaped particles, 275
Homogeneous turbulence, 123, 124
Hopper, 36, 74, 76, 320, 329, 335, 348, 354, 355, 357
Hopper flows, 74, 75
Hopper rocking, 343

I

Imbedded subspheres, 54, 55
Immersed boundary method (IBM), 87, 114
Immersed particle boundary, 116
Immersed tubes, 177, 183–185
Inelastic collision, 100, 168, 246, 250
Inertia particles, 3
Inertial motion, 156
Inflow velocities, 177
Information entropy, 208–211, 213, 216, 217, 221–223, 225, 238, 239, 302, 317, 318

- definition, 209
- function, 211
- Instantaneous particle velocity vector, 192
- Instantaneous velocities, 307
- Interparticle
  - collision, 123–127, 133, 145
    - frequency, 124
    - process, 125
    - rate, 127
  - displacements, 114
- Isotropic turbulence, 9, 167

K

- Kinematic collision models, 21
- Kinetic energies in collision, 34
- Kinetic energy
  - particles, 124, 129, 253, 259, 261, 285, 286, 343, 344
  - rotational, 34, 73, 342
  - transfer, 345
  - turbulence, 123, 170

L

- Lacey mixing index (LMI), 212, 273, 280, 302
- Laden particles, 145–147
- Lagrangian points, 20, 87, 93, 94
- Large eddy simulation (LES), 16
- Lattice Boltzmann method (LBM), 17
- Layer tumbler, 298
- Lifting hopper, 36
- Local concentration, 7, 14, 207–214, 219, 277, 279, 280
- Local turbulence, 7, 14

M

- Mass flow pattern, 335
- Mass flow regimes, 354
- Mass loading, 8, 14, 126, 128, 131, 132, 145–147, 158, 161, 164–169
- Mean effective driven force (MEDF), 259
- Mean temperature discrepancy (MTD), 216, 266
- Microscopic particles, 6
- Modulated gas turbulence, 203
- Modulated subparticle turbulent fluctuation, 203
- Molecular dynamics (MD), 21
- Motion
  - characteristics, 226
  - correction, 54
  - equation, 44, 97, 117, 125, 344
  - features, 296
  - frictional, 73
  - gas phase, 195, 197, 203
  - governing equations, 71
  - granular particles, 218
  - oscillation, 156
  - particles phase, 192
  - pattern, 244, 263
  - periodic, 249
  - prediction, 54
  - process, 289
  - rocking, 331, 338, 346
  - rotation, 48
  - rotational, 4, 33, 34, 50
  - subspheres, 54
  - trajectories, 35
  - tumbler, 244
  - tumbling, 217, 252, 255, 256, 293

N

- Neighboring particles, 14, 91, 273, 275, 348
- Noncircular particles, 270
- Nonideal collision, 74
- Nonpulsed fluidization, 179, 182, 183, 185
- Nonsliding collision, 28, 67
- Nonspherical
  - particle, 50, 53, 62, 67, 91, 269, 304, 306, 352–354, 357
  - particle collision, 36
  - particle flows, 62
  - particle shape, 72
  - shape, 21, 57, 58, 294, 347, 352, 353

O

- Octagonal particles, 348
- Orifice, 36, 62, 321, 329, 331, 333, 334, 337
- Orifice bed, 335, 337, 346, 351
- Orifice size, 62, 351, 352
- Orifice width, 329
- Oscillating motion, 257, 261
- Oscillation, 153–157, 167, 177, 179, 182, 200, 202, 203, 223–225, 252, 257, 331, 335, 336, 338
  - amplitudes, 153, 223
  - frequency, 177
  - mode, 340
  - motion, 156
  - period, 335
  - periodic, 155–157, 294
  - periodicities, 154
- Overlapping spheres, 57

P

- Packed rectangular particles, 348
- Particles
  - agglomeration, 133
  - behavior, 7, 158
  - border, 69
  - Brownian motion, 10
  - centroid, 117

characteristics, 129
cluster, 133, 181, 245, 250, 252, 254
clustering, 131
clusters kinetic energy, 249
clusters tumbling bulk motion, 254
coagulation, 101, 124
collision, 183, 245
collision rates, 9, 127
concentration, 7, 14, 18, 127, 128, 185, 200, 207, 245, 277–279, 302
condensation, 131
cool, 217, 231, 233, 238–241, 261, 262
cubic, 32, 33, 50, 62, 287, 289, 290, 294, 295, 297, 298, 309, 310, 318
cyclic motion, 228, 230
cyclic motion characteristics, 225
deformation, 69, 178
density, 8, 123, 126, 181
deposition, 76
deposition process, 77
diameter, 7, 9, 11, 15, 80, 86, 88, 91, 114, 127, 129, 147, 188, 190, 320, 322, 329, 335, 353
discharge, 335, 339, 340, 346
dispersion, 124, 190, 323, 324, 326, 329
dispersion characteristics, 161, 323, 325
dispersion trajectories, 320
displacements, 323
distribution, 7, 36, 208, 252, 290, 322, 329, 332, 333, 337, 351
distribution function, 133
downhill motion, 227
drag, 144
drag force, 18, 92, 123
dynamics, 130
dynamics governing equations, 21
dynamics in turbulence, 8
evolution, 174
falling, 354
flow, 3, 64, 70, 75, 76, 80, 82, 83, 324, 331
  dynamics, 11
  fast, 13
  patterns, 295
  regime, 300
flowability, 11
fluctuating motion, 185, 186
fluctuation, 177, 185
free motion, 7, 125
heat transferring, 11
horizontal distribution, 323
hot, 217, 231, 233, 238–241, 261, 262
in spouted beds, 194
index, 89, 320
inertia, 8
interaction, 15
kinetic energy, 124, 129, 253, 259, 261, 285, 286, 343, 344
kinetics, 9
loadings, 159
local concentration, 14, 207, 208, 210, 277
mass, 65, 88, 98, 166, 352
mass loading, 14, 128, 164
mean velocity, 133, 195, 203
methods, 19
mixing, 215, 229, 230, 237, 256, 270, 277
mixing information entropy, 212
mixing level, 302
motion
  equation, 97, 100, 101, 144, 178
  pattern, 246, 247, 253, 263, 290
  trajectories, 72, 87, 178
  velocity, 204
movements, 116, 326
nonspherical, 50, 53, 62, 67, 91, 269, 304, 306, 352–354, 357
nonspherical shape, 290
number, 126, 130, 133, 245, 249, 343
number density, 127, 130, 131, 133, 245
packing, 290
pairs, 9
phase, 114, 158, 168, 189, 195, 197, 204
  flow pattern, 204
  fluctuation, 185
  motion, 192
position, 117
properties, 9, 167, 194, 295
radius, 68
relative motion, 171
response properties, 166
Reynolds number, 8, 86, 99, 100
rotation, 125
rotational motions, 115, 125
roundness, 348, 355
scale, 18, 87
sedimentation processes, 117
shape, 20, 29, 59, 75, 83, 273, 275, 277, 279, 280, 282, 286, 298, 300, 304, 309, 348, 351, 352
spouted flow pattern, 191
states, 290
statistics, 131, 185
surface, 18, 20, 41, 65, 68, 90, 202
systems, 14, 21, 213, 214, 298, 345
temperature, 80, 215, 262, 268
temperature discrepancy, 268
thermal conduction, 235, 261, 263, 267, 269
trajectories, 18, 44, 133, 226, 318, 322, 323, 328, 346
transport, 6

tumbling bulk motion, 254
velocity, 36, 62, 86, 101, 117, 124, 127, 129, 175, 185, 192, 200, 204, 226, 246, 252, 253, 285, 290, 291, 294, 295, 297, 307, 309, 331, 348, 351, 357
components, 291
distribution, 290
void fraction, 86
volume, 80–82, 92
Peak values, 182–184, 200, 235, 238, 242, 249–251, 254, 306, 335, 344
Pebble flow, 11, 13
Pebble motion, 114
Pentagonal particles, 351
Pentagonal shaped particles, 280
Periodic
motion, 249
oscillation, 155–157, 294
pulsed motion, 227
tumbling bulk motion, 252
tumbling motion, 254
Polygonal particles, 347
Prevailing interparticle collision, 127
Probability density function (PDF), 130, 235, 291, 304
Pulsed fluidization, 177, 179, 181–183, 185

R

Radial distribution function (RDF), 91, 207, 208, 216, 221, 222, 224, 237
Rectangular hopper, 329
Rectangular particles, 348
Relative motion, 28
Relative motion particles, 171
Restitution coefficient, 33, 35, 36, 42, 50, 66, 67, 72, 73, 75–77, 100, 101
Reynolds number, 3, 8, 93, 96, 99, 114, 123, 126, 145, 150, 155, 156, 164
Reynolds number particles, 8, 86, 99, 100
Reynolds stress models (RSM), 15
Rocking, 330, 331, 337, 338, 340, 343, 344, 346
amplitude, 330–344, 346
beds, 331, 335, 344
frequency, 332, 334–336, 338–344, 346
hopper, 343
motion, 331, 338, 346
Rotation
drum, 290, 300
motion, 48
particles, 125
speeds, 217, 220–225, 237, 239, 240, 243, 256, 273, 290, 306, 309, 318
velocity, 227, 229–231, 241, 242
velocity augments particle, 229
Rotational
kinetic energy, 34, 73, 342
motion, 4, 33, 34, 50
Reynolds number, 100
velocities, 307

S

Sampling particles, 18
Secondary collision, 247, 254
Shaken hopper, 329
Shannon information entropy, 208, 222, 225, 302, 316
Silo orifice, 62, 335
Single particle drag force, 18
Single sampled particle, 231
Sliding frictions, 43
Sliding motion, 27, 43, 227
Smoothed voidage, 203
Soft sphere model (SSM), 9
Sparsely dispersed particles, 245
Spherical
granular particle, 225
particle, 21, 29, 50, 67, 256, 289, 290, 293, 294, 298, 304, 305, 307–309, 320, 329
particle collisions, 29
particle shapes, 292
spheres, 303
Spout, 191, 192, 195, 199–201, 204
Spout radius, 195
Spouted beds, 196
Spouted flow pattern, 190
Spouted flow pattern particles, 191
Spouting, 190
Spouting flow patterns, 192
Static base, 217, 218, 224, 226, 229, 232, 280, 284, 285
Static friction, 43, 44
Static friction coefficient, 44
Stationary hopper, 331
Steel particles, 80
Stochastic motion, 246, 322
Subatomic scaled particles, 6
Subgrid turbulence models, 19
Subparticle, 204
cell, 92
fine scales, 92
mesh scale, 89
scale, 91, 92, 190, 197, 203, 204
Subspheres, 54, 55, 57–60, 62, 354
collision, 53, 54
covering, 53, 54
motion, 54
Superparticle scale, 92
Suspended particles, 9, 132
Synchronous oscillation, 156

T

Tangential relative velocities, 25
Temperature discrepancy function (TDF), 214, 215, 235
Temperature probability density function (TPDF), 215
Tetrahedral particle, 36, 298, 303–305, 308
Tetrahedrons, 298, 300, 302–308
  filling level, 300
  flow regime, 300
  rotational motions, 305
  velocities, 307
Thermal conduction particles, 235, 261, 263, 267, 269
Thermal information entropy (TIE), 216, 217
Thermal radial distribution function (TRDF), 216
Tracer particles, 74, 75
Transitional particle Reynolds number, 99
Translational kinetic energy, 34, 72, 342
Translational motions, 306, 307
Triangular particle, 352
Tumbler
  boundary, 225, 243
  circular, 207, 246
  configuration, 244
  motion, 244
  revolutions, 232
  wavelike boundary, 247
  wavy, 245, 246
Tumbling
  bulk motion, 253, 254, 256
  flow pattern, 290, 292–295
  motion, 217, 252, 255, 256, 293
  motion particles, 221
Turbulence
  augmentation, 172
  flow, 9
  fluctuation, 166, 167, 172
  fluctuation energy, 166
  gas, 130, 131
  integral, 8
  intensity, 18, 126
  modification, 8
  modulation, 8, 9, 132, 158, 166, 168, 172
  spectrum, 7
  transport, 171
Turbulent kinetic energy (TKE), 167–169, 171
Turbulent motion, 156

U

Unsteady turbulence, 124
Upcasting motion, 273

V

Velocities, 22, 33, 60, 62, 67, 125, 150, 151, 157, 158, 177, 243, 284, 304, 333, 334, 348
  gas, 201, 203
  particles, 252, 309, 357
  rotation, 227, 229–231, 241, 242
  tetrahedrons, 307
Virtual mass distribution function (VMDF), 89–91
  discretized, 91
Virtual void fraction, 88, 89, 92, 190, 194, 203
Void fraction, 85, 87–89, 91, 92, 114, 173, 190, 198–200, 203
Void fraction oscillation, 199
Void fraction particles, 86
Voidage, 181, 200–204, 354, 355
Voidage distribution, 355
Voidage variations, 354
Vortex breakdown, 4, 5, 148, 150–152, 155, 156, 161

W

Wavelike boundary, 249, 251
Wavelike boundary tumbler, 247
Wavy
  drum, 243, 256, 257, 261–265, 268, 269
  tumbler, 245, 246
  wall, 243, 245–247, 261–263
  wall motion, 246
  wall tumblers, 256
Weighted temperature (WT), 214

Printed in the United States
By Bookmasters